DESIGN OF WATER-BASED FIRE PROTECTION SYSTEMS

DESIGN OF WATER-BASED FIRE PROTECTION SYSTEMS

Robert M. Gagnon, P.E., SET

DELMAR
CENGAGE Learning

Albany • Bonn • Boston • Cincinnati • Detroit • London • Madrid
Melbourne • Mexico City • New York • Pacific Grove • Paris • San Francisco
Singapore • Tokyo • Toronto • Washington

Design of Water-Based Fire Protection Systems

Robert M. Gagnon, P.E., SET

Publisher:
Robert D. Lynch

Acquisitions Editor:
Mark Huth

Developmental Editor:
Jeanne Mesick

Project Editor:
Thomas Smith

Production Coordinator:
Toni Bolognino

Art and Design Coordinator:
Michael Prinzo

Cover Photo Courtesy of:
Grinnell Corporation, Exeter, NH

© 1997 Delmar Cengage Learning

ALL RIGHTS RESERVED. No part of this work covered by the copyright herein may be reproduced, transmitted, stored or used in any form or by any means graphic, electronic, or mechanical, including but not limited to photocopying, recording, scanning, digitizing, taping, Web distribution, information networks, or information storage and retrieval systems, except as permitted under Section 107 or 108 of the 1976 United States Copyright Act, without the prior written permission of the publisher.

For product information and technology assistance, contact us at
Cengage Learning Customer & Sales Support, 1-800-354-9706

For permission to use material from this text or product,
submit all requests online at **cengage.com/permissions**
Further permissions questions can be emailed to
permissionrequest@cengage.com

Library of Congress Control Number: 96-46987

ISBN-13: 978-0-8273-7883-4

ISBN-10: 0-8273-7883-1

Delmar Cengage Learning
5 Maxwell Drive
Clifton Park, NY 12065-2919
USA

Cengage Learning products are represented in Canada by Nelson Education, Ltd.

For your lifelong learning solutions, visit **delmar.cengage.com**

Visit our corporate website at **www.cengage.com**

Notice to the Reader
Publisher does not warrant or guarantee any of the products described herein or perform any independent analysis in connection with any of the product information contained herein. Publisher does not assume, and expressly disclaims, any obligation to obtain and include information other than that provided to it by the manufacturer. The reader is expressly warned to consider and adopt all safety precautions that might be indicated by the activities described herein and to avoid all potential hazards. By following the instructions contained herein, the reader willingly assumes all risks in connection with such instructions. The publisher makes no representations or warranties of any kind, including but not limited to, the warranties of fitness for particular purpose or merchantability, nor are any such representations implied with respect to the material set forth herein, and the publisher takes no responsibility with respect to such material. The publisher shall not be liable for any special, consequential, or exemplary damages resulting, in whole or part, from the readers' use of, or reliance upon, this material.

Printed in the United States of America
12 13 14   11 10 09

Fire is a cruel thief that steals our very young, our elderly, and even the fittest among us. This book is dedicated to all professionals who work toward solving the problem of fire in our society.

Fire service professionals take significant risks to perform their duties, and they get the immediate satisfaction of knowing that the risks they take pay off in saving human lives. No reward could be greater.

Fire protection design professionals make a contribution that is no less significant. They are not only responsible for saving the lives of people whom they may never meet, they help to significantly lessen the risk to fire service professionals by designing fire protection systems that prevent a fire from getting out of control during rescue and extinguishment operations.

Fire science educators responsible for training fire protection professionals have a profound effect on the safety and welfare of our society. Their contribution to the safety of our world is cumulative and grows exponentially with each graduating class.

Fire protection design professionals, fire service professionals, fire science educators, and all others who are working actively to solve the fire problem in the world deserve to be proud of their positive influence on life safety and the protection of property. We are bound by a common goal. We are motivated by a common pride. We are inspired by a common vision.

I believe that there exists a special gate to heaven for people who dedicate their lives to the safety and welfare of others. May that gate be opened for each of you.

Robert M. Gagnon

Contents

Foreword	xiii
Preface	xvii
About the Author	xix
Acknowledgments	xxi
Introduction	xxiii

WHO SHOULD USE THIS BOOK/xxiii ■ HOW TO USE THIS BOOK/xxiii ■ BASIC MATHEMATICS REQUIRED FOR THIS BOOK/xxiii ■ Metric Conversions/xxiv ■ Metric Units/xxiv ■ SIGNIFICANT FIGURES AND ROUNDING/xxv ■ NFPA PUBLICATIONS/xxv ■ FIRE PROTECTION AS A PROFESSION/xxvi ■ FIRE PROTECTION TECHNICIANS AND ENGINEERS/xxvii ■ ORAL, WRITTEN, AND GRAPHIC COMMUNICATION/xxviii ■ GRAPHIC COMMUNICATION—DRAWING FIRE PROTECTION SYSTEMS/xxviii ■ The Blank Page/xxix ■ Drafting Equipment/xxx ■ Drawing Surface/xxx ■ Drawing Parallel Lines/xxxi ■ Scales/xxxi ■ Drawing Media/xxxii ■ Drawing Arrangement/xxxiii ■ Computer-Aided Design/xxxiv

Chapter 1 Contract Drawings   1

CONTRACT DRAWINGS COVER SHEET/2 ■ SITE DRAWINGS/3 ■ ARCHITECTURAL PLANS/8 ■ Building Sections/10 ■ Reflected Ceiling Plans/10 ■ Finish Schedule/12 ■ STRUCTURAL DRAWINGS/12 ■ Foundation Plans/12 ■ Framing Plans/12 ■ Branch Line Orientation/13 ■ Using Framing Plans to Dimension a Sprinkler System/13 ■ HVAC DRAWINGS/15 ■ PLUMBING DRAWINGS/17 ■ FIRE PROTECTION CONTRACT DRAWINGS/18 ■ ELECTRICAL DRAWINGS/19

Chapter 2 Contract Specifications   23

SPECIFICATION TYPES/24 ■ Computer Specification Database/24 ■ Performance Specifications and Detailed Specifications/24 ■ SPECIFICATION FORMAT/25 ■ Division 0: Instructions to Bidders/26 ■ Division 1: General Requirements/26 ■ Division 2: Site Work/26 ■ Divisions 3–6: Concrete, Masonry, Metals, and Wood and Plaster/26 ■ Division 7: Thermal and Moisture Protection/27 ■ Divisions 8 and 9: Doors and Windows, and Finishes/27 ■ Divisions 10–12, 14: Specialties, Equipment, Furnishings, and Conveying Systems/27 ■ Division 13: Special Construction/27 ■ Division 15: Mechanical/28 ■ Division 16: Electrical/29 ■ SAMPLE SPECIFICATION/29

Chapter 3 Fire Protection Survey Procedure   35

GET YOUR BEARINGS/36 ■ DETERMINE THE GENERAL BUILDING LAYOUT/36 ■ BRING PROPER SURVEY EQUIPMENT/37 ■ SURVEY BUILDING DETAILS/38 ■ DEVELOP A SYSTEM DESIGN STRATEGY/39 ■ Building Elevations/39 ■ Ceiling Measurements/39 ■ Water Supply Information/40 ■ DRAW THE BUILDING AND LAY OUT THE SYSTEM/40 ■ FIELD CHECK THE DRAWING/40 ■ INSPECT THE SYSTEM AFTER INSTALLATION/41

Chapter 4 Underground Piping Design Procedure   43

OBTAIN REFERENCE MATERIAL AND TRACE OR SCAN BACKGROUND/45 ■ DETERMINE PIPE ROUTING/46 ■ DRAW CUT-IN DETAIL/47 ■ ADD INVERT PIPING ELEVATIONS AND PLAN DIMENSIONS/47 ■ SPECIFY PIPING AND FITTINGS/50 ■ DETAIL THRUST BLOCKS AND PIPING RESTRAINTS/51 ■ DETAIL LOCATIONS OF FIRE HYDRANTS/52 ■ PERFORM A WATER FLOW TEST/53 ■ Flow Test Tools/56 ■ Flow Test Protocol/56 ■ Flow Test Procedure/57 ■ DETAIL WATER SUPPLY INFORMATION ON PLANS/65

Chapter 5 Automatic Sprinkler Systems   70

SPRINKLER SYSTEMS—MYTH VERSUS FACT/71 ■ STANDARDIZED AUTOMATIC SPRINKLER SYSTEM DESIGN/72 ■ SPRINKLER SYSTEM DEFINITIONS/73 ■ "Approved" versus "Listed"/73 ■ "Shall" versus "Should"/73 ■ "Fire Control" versus "Fire Suppression"/74 ■ SPRINKLER TYPES/74 ■ Old Style and Spray Sprinklers/74 ■ Upright, Pendent, and Sidewall Sprinklers/75 ■ Sprinkler Coverage/75 ■ Sprinkler Response/75

■ Sprinklers for Storage Applications/78 ■ SPRINKLER SYSTEM TYPES/78 ■ Wet Pipe Systems/78 ■ Dry Pipe Systems/78 ■ Preaction and Deluge Systems/79 ■ SPRINKLER SYSTEM COMPONENTS/81 ■ Sprinklers/81 ■ Piping and Tubing/82 ■ Fittings/82 ■ Hangers/84 ■ Valves/84 ■ DEVELOPING A SPRINKLER PLAN FOR APPROVAL/87 ■ Collect and Review Data/88 ■ Review Project Schedule/88 ■ Trace or Scan Building Outline/89 ■ Draw Necessary Building Detail/90 ■ Determine Branch Line Logic/91 ■ Determine System Type and Configuration/93 ■ Determine the Hazard Class of the Occupancy/95 ■ Determine the Area Protected by Each Sprinkler (A)/96 ■ Determine the Number of Branch Lines/98 ■ Determine the Distance between Branch Lines (L)/100 ■ Determine the Maximum Allowable Distance between Sprinklers (S)/100 ■ Determine the Minimum Number of Sprinklers on Each Branch Line/101 ■ Determine the Actual Distance between Sprinklers (S_{actual})/102 ■ Dimension Sprinklers and Piping on Your Drawing/102 ■ Verify Sprinkler Coverage/102 ■ Determine Crossmain Location/103 ■ Assign Elevations to Piping/104 ■ Locate System Risers/105 ■ Select Hangers and Determine Hanger Rod Lengths/105 ■ Determine Hanger Spacing/106 ■ Add Final Details/108 ■ Use the NFPA 13 Design Checklist/109 ■ Perform Precalculation Field Check/109 ■ Add Pipe Lengths and Verify Accuracy/109

Chapter 6　Hydraulic Calculation of Sprinkler Systems　114

THE FUNDAMENTALS OF HYDRAULICS/115 ■ The Advantages of Water/115 ■ The Properties of Water/116 ■ Hydrostatics/120 ■ Hydrokinetics/123 ■ Friction Loss/126 ■ Friction Loss Tables/128 ■ Reducing Friction Loss/129 ■ Friction Loss in Simple Pipe Loops/129 ■ Friction Loss in Simple Loops—Equivalent Length/132 ■ Analysis of Loop Calculation Results/136 ■ HYDRAULIC CALCULATION METHOD FOR SPRINKLER SYSTEMS/137 ■ Hand Calculations versus Computer Calculations/137 ■ Choosing the Hydraulically Most Demanding Area/138 ■ Hydraulic Calculations—Selection of Occupancy/138 ■ Selecting a Hydraulic Density/139 ■ Determining the Length of the Design Area/142 ■ Number of Sprinklers Flowing along the Length of the Design Area/143 ■ Configuration of Sprinklers in the Design Area/145 ■ Minimum Flow at the Hydraulically Most Demanding Sprinkler/146 ■ Determining the Minimum Pressure at the Hydraulically Most Demanding Sprinkler/146 ■ Determining Friction Loss/147 ■ An Example Calculation/148 ■

Fire Hose Streams/156 ■ Graphing Results/156 ■ Submitting Calculations for Approval/157 ■ Velocity Pressure Calculations/157 ■ Quick Response Sprinklers and the Design Area/158 ■ Extended Coverage Sprinklers and the Design Area/159 ■ COMPUTERIZED HYDRAULIC CALCULATION PROGRAMS/160 ■ Use of the Computerized Hydraulic Program Included in This Book/160

Chapter 7 Specialized Piping and Calculation Methods 163

PIPE SCHEDULE DESIGNS/164 ■ RESIDENTIAL SPRINKLER SYSTEMS/165 ■ One- and Two-Family Dwellings and Manufactured Homes/165 ■ Residential Occupancies up to and Including Four Stories/173 ■ Residential Occupancies Greater Than Four Stories/173 ■ Freeze Protection for Residential Sprinkler Systems/173 ■ SPRINKLER PROTECTION OF HIGH-PILED STORAGE OCCUPANCIES/176 ■ ESFR Sprinklers/179 ■ STANDPIPE AND HOSE SYSTEMS/179 ■ FIRE PUMPS/182 ■ Fire Pump Components/183 ■ Calculating Pressure Differentials/184 ■ When to Add a Fire Pump/186 ■ Graphing a Fire Pump Supply/186 ■ Combining a Pump and City Supply/187 ■ Sizing a Fire Pump/188 ■ GRAVITY TANKS/189 ■ Gravity Tank Calculation/189 ■ Graphing a Gravity Tank Supply/190

Chapter 8 Water Spray System Design for Transformers 193

WATER SPRAY SYSTEMS/194 ■ Water Spray Protection for a Power-Generating Plant/194 ■ The Rankine Cycle/196 ■ WATER SPRAY PROTECTION FOR TRANSFORMERS/197 ■ Transformer Hazards/197 ■ Personnel and Other Hazards/198 ■ Power Generation/198 ■ Oil Recovery and Reclamation/198 ■ Extinguishment/Control Mechanism/198 ■ TRANSFORMER DESIGN PROCEDURE/200 ■ Compile Reference Data/200 ■ Commence Layout/202 ■ Transformer Loop Design/204 ■ Calculation of Transformer Surface Area/206 ■ Water Spray for Nonabsorbing Ground Surfaces/208 ■ Recommendations for Water Spray System Layout/209 ■ Hydraulic Calculation Procedure for Transformers/213 ■ Detection Systems/217

Chapter 9 Water Spray System Design for Flammable Liquid Storage Tanks or Vessels 224

HAZARDS ASSOCIATED WITH TANKS AND VESSELS/225 ■ Piloted and Autoignition/225 ■ Pool Fires/225 ■ Pool Fire Containment/225 ■ Pressure Spray Fires/225 ■ Deflagration and

Rupture/226 ■ VESSEL PROTECTION MECHANISMS/226 ■ Exposure Protection/226 ■ Surface Cooling/227 ■ Rundown/227 ■ Prevention of Boiling Contents/227 ■ Supplementary Protection for Vessels/229 ■ VESSEL WATER SPRAY DESIGN PROCEDURE/230 ■ Surface Area Calculation/230 ■ Estimating the Minimum Water Supply Requirement/231 ■ Total Water Requirement/232 ■ Estimation of Water Supply Adequacy/232 ■ Pressure Estimation/233 ■ Loop and Nozzle Spacing Procedure/234

Chapter 10   Ultra High Speed Water Spray Systems   246

ULTRA HIGH SPEED WATER SPRAY SYSTEMS/247 ■ Definition/247 ■ Ultra High Speed Water Spray Systems and Explosion Suppression Systems/247 ■ Why Water?/248 ■ Deflagrations and Detonations/249 ■ Uses for Ultra High Speed Water Spray Systems/249 ■ Reaction Time/249 ■ Human Senses/249 ■ Detectors for Ultra High Speed Water Spray Systems/250 ■ Design Approaches/250 ■ Types of Ultra High Speed Water Spray Systems/250 ■ THE SQUIB-ACTUATED ULTRA HIGH SPEED WATER SPRAY SYSTEM/252 ■ Definition/252 ■ Preprimed Piping System/252 ■ The Squib-Actuated Valve/253 ■ System Response Time/254 ■ Squib-Actuated Ultra High Speed Water Spray System Design/255 ■ THE SOLENOID-OPERATED ULTRA HIGH SPEED WATER SPRAY SYSTEM/255 ■ System Arrangement/255 ■ The Discharge System/255 ■ The Pilot System/256 ■ Sequence of Operation for a Pilot-Actuated System/258 ■ Factors Influencing the Speed of a Pilot-Actuated System/260 ■ Response Time for Solenoid-Operated Systems/260

Chapter 11   Water Mist System Design   262

WATER MIST PERFORMANCE OBJECTIVES/263 ■ WATER MIST DROPLETS/263 ■ Droplet Size/263 ■ Droplet Size Performance Objectives/265 ■ Droplet Size and the Fire Service/265 ■ Measuring Droplet Size/266 ■ Water Mist Pressure and Droplet Size/266 ■ WATER MIST SYSTEM DESIGN/267 ■ Water Mist System Types/267 ■ Water Mist System Calculation Criteria/268 ■ WATER MIST APPLICATIONS/268 ■ Submarine Applications/269 ■ Surface Ship Applications/270 ■ Aircraft Applications/271 ■ Total Flooding Electronics and Telecommunications Applications/272 ■ Commercial and Residential Applications/273

Chapter 12	**Quality Assurance and Coordination**	279

QUALITY ASSURANCE PROCEDURES/280 ■ Quality Assurance Review/280 ■ COORDINATION/281 ■ Scheduling the Coordination Process/282 ■ Prefabrication of Fire Protection Systems/282 ■ The Coordination Meeting/283 ■ Sanctioning a Coordination Meeting/283 ■ Coordination Negotiations/284 ■ Putting the Coordination Agreements into Action/284 ■ Coordination Failures/285 ■ Coordination Drawings/285 ■ The Value of Coordination/286

Chapter 13	**Ethics and Professional Development**	288

GOOD VERSUS EVIL/289 ■ The Ethical Dilemma/289 ■ RESPONSIBILITIES OF A FIRE PROTECTION PROFESSIONAL/290 ■ Personal Values/290 ■ Company Values/290 ■ Societal Values/291 ■ Values of the Client/292 ■ Values of the Fire Protection Profession/293 ■ PROFESSIONAL STANDARDS OF CONDUCT/293 ■ Code of Responsibility for NICET-Certified Engineering Technicians and Technologists/293 ■ The Code of Ethics for Engineers/297 ■ The Profession of Engineering/298 ■ LEGAL/ETHICAL CONFLICTS/299 ■ RESOLVING ETHICAL DILEMMAS/299 ■ Rationalization/299 ■ Whistle-Blowing/300 ■ RESPONSIBILITIES OF A STUDENT/301 ■ RESPONSIBILITIES OF A PROFESSIONAL/301 ■ Professional Development/301

Appendix A	**Reference Tables**	307
Appendix B	**c-Factors**	324
Appendix C	**Pipe Diameters**	326
Appendix D	**Friction Loss Tables**	328
Appendix E	**Computerized Hydraulic Program Documentation**	415
Appendix F	**Selected NFPA Standards**	436
	Acronyms	439
	Glossary	441
	Additional Readings	449
	Index	454

Foreword

My interest in fire protection started over 65 years ago when I hung around the neighborhood fire house in New York City. I was fascinated by the bell alarm system by which companies were dispatched. The system was very detailed. Any unit could be called to any box location in the greater city. More than 30 companies could be called to a location with a single bell signal. The ordinary signal was the location of a fire alarm box. For instance, ✱✱✱✱✱✱✱ ✱✱✱✱✱✱✱✱✱✱ ✱✱✱✱ indicated fire alarm box 704 located at 7th Avenue and 33rd Street. An assignment card indicated the units that were to respond.

One day I counted off a strange signal, ✱✱✱ ✱✱✱✱✱✱✱ ✱✱✱✱✱✱✱✱✱✱ ✱✱✱✱ ✱ ✱✱✱, 3-704-13. My mentor explained, "That's a special building box. It's an alarm from a building near 704 box. It can originate from a private manual fire alarm box in the building, from an automatic fire alarm system, or from a water flow on a sprinkler system." He showed me the card. "See," he said, "it's a sprinkler alarm. Engine 1 and Ladder 24 will respond. It's most unlikely we'll hear any more from it. The sprinklers will have the fire out or at least controlled so two companies can handle it."

Similar experiences relative to the effectiveness of sprinklers are recounted in "Sprinkler Experience in High Rise Buildings (1979)," SFPE Technology Report 79-1, by W. Robert Powers, PE, then Superintendent of the New York Board of Fire Underwriters.

Insurance engineers had the same absolute confidence in sprinklers. During World War II, the availability of lift trucks made rack storage a reality. At Norfolk, Virginia, the U.S. Navy had huge sprinklered warehouses with stock stacked in racks 16 feet high. I set up prefire plans for heavy-caliber streams to confine the fire to the area of origin. Navy fire protection engineers with insurance company experience, but no fire experience, scoffed, "There is no need to plan for a major fire—the building is sprinklered." They never answered the question, "How will a half-inch spray of water work its way down through this stack to a fire in the bottom of the stack?"

Although we did not face the possibility of lawsuits for negligence, we did face an exhaustive investigation into the circumstances of any major fire. Our engine companies were sternly warned to hook up to domestic hydrants only because the high-pressure hydrants were on the mains feeding the sprinkler systems, and we would be charged with robbing the sprinklers of water, possibly causing a huge loss.

In varying degrees, the attitude that sprinklers will always extinguish fires still persists in the fire service. It is reinforced by another myth, "Insurance companies require that buildings stay up to code." [For a frank discussion of insurance—fire department relations, see pages 618–619 in Chapter 14 of *Building Construction for the Fire Service*, 3d edition, by

Francis L. Brannigan, published by NFPA, 1992.] A common viewpoint is that sprinklers are just a matter between the building owner and the insurance company. Answers usually given to the question "What action is taken by the fire department when notified that a sprinkler system is out of service?" indicate serious complacency. [See "Fire Department and Sprinklers" in Chapter 13 of *Building Construction for the Fire Service*, 3d edition, by Francis L. Brannigan, published by NFPA, 1992.]

This complacency should be countered by the following considerations:

- The record of huge losses in warehouses insured by highly protected risk insurers, and presumably protected by "state-of-the-art" sprinkler protection. [Refer to pages 605–612 in *Building Construction for the Fire Service*, 3d edition, by Francis L. Brannigan, published by NFPA, 1992.]
- Major fire losses are not a private matter. They can seriously impact the economic health of the community.
- The fact that some "sprinkler" installations have been completely fraudulent in that sprinkler heads were visible but there was no water supply.
- Articles such as Richard Shulte's (R. C. Shulte, "Quality Control in the Sprinkler Industry," *Plumbing Engineer*, March 1996, page 14) describe practices that may be common in the construction industry, but we somehow think that because human lives are at stake, the sprinkler contractor would be beyond reproach. Shulte tells of an early work experience. He discovered that a large duct would completely mask the sprinkler discharge. He was told "Forget it unless the fire department or insurer objects."
- Fire departments have taken over the function of plan review from the building department in a number of cities.
- Many significant reductions in static fire protection and life safety provisions are permitted if sprinklers are installed. The life safety code includes the concept that this writer has preached, and enforced where he had the authority, for 45 years, that where sprinklers are installed for life safety, very special precautions, up to and including closing the building, are indicated if the sprinkler protection is impaired.
- Professor Vincent Brannigan, JD, has pointed out that the substitution of "Performance Codes" for "Prescriptive Codes" means that the public authorities will have "cradle to the grave" responsibilities to see that the provisions of the performance code remain in operation.

All these considerations point to the necessity for developing fire officers with a more sophisticated understanding of sprinkler systems design and maintenance than they generally have.

One fire chief excellently expressed the function of a sprinkler system as "My First Due Engine Company," meaning that the sprinkler system activates before the first fire engine arrives at the scene. The municipality spends a sizable sum to keep an engine company in service. Therefore, the modest cost of educating some personnel so they can properly ensure that the sprinkler system will function adequately when needed is more than justified.

In *Design of Water-Based Fire Protection Systems*, Robert M. Gagnon, PE, has made available his more than 20 years of experience in the engineering of sprinkler systems and, more importantly, of teaching the subject in both fire protection engineering courses and fire science programs. The book is vital for those who design sprinkler systems, but it is certainly appropriate for fire officers who approve sprinkler plans or inspect installed systems.

The discussions on ethical situations are most pertinent. There is no question that building contracting is a money-driven operation in which margins are often very thin. The fire officer, not faced with such pressures, might otherwise be naive in what he or she expects from the contractor.

I recommend this book for unit or self-study in fire science programs where we would hope that the vision goes beyond simply manually "putting the wet stuff on the red stuff." It should certainly be in all fire libraries for the benefit of those future readers who will go beyond the basic required courses.

Francis L. Brannigan, SFPE
Professor (retired) and founder
Fire Science Curriculum
Montgomery College
Rockville, Maryland

Preface

This book was written for all fire protection professionals who wish to learn or polish their skills in fire protection system design. All fire protection professionals who work with fire protection systems in some capacity must learn the basic principles upon which fire protection systems are designed.

Fire service personnel and fire service administrators, fire protection technicians, authorities having jurisdiction, fire protection engineers, sprinkler system installing contractors, plumbing contractors, architectural firms, and engineering firms will benefit considerably from the design methodology described in this book.

There are very few prerequisites for understanding the material contained in this book, because the mathematical demands are quite modest and the presentation of the material assumes little or no previous exposure to fire protection system design.

For this reason, fire service training academy students, fire science students at community colleges, university-level fire protection or fire safety engineering students, and entry-level technicians, architects, and engineers will find the book of considerable assistance in developing fire protection design skills and understanding the basic principles of fire protection system design.

For those who have never performed a fire protection design, the Introduction provides a background on the tools and techniques to get started. Fire protection design is a valuable skill that will bolster your ability to acquire a wide variety of responsibilities as a fire protection professional.

Chapters 1, 2, and 3 provide a basic foundation for obtaining the minimum requisite design information from contract drawings, specifications, and surveys. Regardless of where your career in fire protection takes you, the material in these three chapters provides valuable information to guide you through even the most challenging assignments.

Chapters 4 through 11 are specific design chapters that outline a methodical approach for the design and calculation of the most commonly encountered types of fire protection systems. Knowledge of the basics of fire protection system design is a fundamental building block to a career as a fire protection professional.

Some designs look good on paper but fail in the field. Chapter 12 provides information on the methodology that can be used to successfully coordinate the installation of fire protection systems and ensure that the system serves as a vital component of a completed building.

Chapter 13 may be the most important chapter of all, with the groundwork it lays with respect to the ethical practice of fire protection system design and layout. It may be tempting to concentrate on the technical data contained within this book, but don't neglect chapter 13, because knowledge of ethical practice is as important as knowledge of technical information.

A significant amount of material has been included in the appendices and is intended to provide useful reference information for the reader long after this book has been read or completed as a text. Extensive friction loss tables and a computer disk of a hydraulic calculation program are included at the back of the book.

One feature of the book that may be very helpful for readers is its methodical approach to the solution of design problems and the numerous examples that should solidify understanding of the subject matter. The book is generously illustrated, which should serve to take the mystery out of fire protection system design.

The material in this book can be successfully completed in one semester at a community college, fire service academy, or university-level fire science curriculum. Performing the problems at the end of each chapter will help the reader to understand and retain the material.

Performing the drawing and calculation assignments at the end of the pertinent chapters is essential to a solid understanding of fire protection design. Although the book can be successfully completed without performing drawings and calculations, readers would be dealt a considerable disservice if an instructor did not require them. For entry-level designers and technicians using this book, the material can be applied to the development of design projects that have been assigned.

Instructors using this book should obtain the Instructor's Resource Guide, which contains solutions to review questions, suggestions for presentation of the subject matter, and examination materials.

To all instructors and users of the book, I welcome your comments and suggestions for improvement in the next edition. Please contact the author:

> Robert M. Gagnon, P.E., SET
> Gagnon Engineering
> 10110 LaBelle Court
> Ellicott City, Maryland 21042-6203

About the Author

Robert M. Gagnon, P.E., SET, is president of Gagnon Engineering and performs fire protection system design and calculation, detection system design, fire protection engineering, expert witness work, and code consulting. He teaches courses in water-based fire protection system design and special hazard and detection system design at the University of Maryland Department of Fire Protection Engineering, and teaches three fire protection and detection system design courses at Montgomery College. He worked for over 20 years as a fire protection systems designer for "Automatic" Sprinkler Corporation of America and the Fireguard Corporation before forming his consulting firm.

He holds a B.S. and an M.S. in fire protection engineering from the University of Maryland, a B.A. in mathematics from Western Maryland College, and NICET Level IV Certifications in Automatic Sprinkler Systems Layout and Special Hazards Systems Layout. He is a registered professional engineer in Maryland, Virginia, Pennsylvania, and the District of Columbia, and has served as president of the Howard County Chapter of the National Society of Professional Engineers, president of the Baltimore Alumni Chapter of the Tau Beta Pi Engineering Honor Society, vice president of the Chesapeake Chapter of the Society of Fire Protection Engineers, and as a member of the National Fire Sprinkler Association and the American Fire Sprinkler Association.

Bob is a principal member of the NFPA Committee on Water Spray Fixed Systems (NFPA 15), the NFPA Committee on Foam-Water Sprinkler Systems (NFPA 16 and NFPA 16A), the NFPA Committee on Water Tanks (NFPA 22), the NFPA Committee on Private Water Supply Piping Systems (NFPA 24 and NFPA 291), and the NFPA Committee on Water Cooling Towers (NFPA 214). He is the author of chapter 6–15 in the 18th edition of the NFPA *Fire Protection Handbook*, entitled Ultra High Speed Suppression Systems for Explosive Hazards, and is the author of numerous journal articles.

He is also the author of *Design of Special Hazard and Fire Alarm Systems*. (Delmar, 1997).

Acknowledgments

A persons' outlook on the world and the manner in which that outlook is expressed is to a large degree attributable to ones' education. For this reason, I want to cumulatively thank all teachers who helped me to develop and express my opinions.

Dr. Clyde Spicer and Dr. James Lightner in the mathematics department of Western Maryland College challenged and inspired me, and I will always be guided by the high standards that they set.

Ced Dr. John L. Bryan, Dr. James A. Milke, and Dr. Frederick W. Mowrer provided wisdom and knowledge during my undergraduate education at the University of Maryland Department of Fire Protection Engineering, and Dr. James G. Quintiere, Dr. Vincent M. Brannigan, and Dr. Steven M. Spivak provided the foundation for my graduate education at the Department of Fire Protection Engineering. These men are outstanding teachers and role models.

Dr. Richard H. McCuen of the Department of Civil Engineering at the University of Maryland was my honors professor in the engineering honors program at the University of Maryland and was the inspiration for Chapter 13. Dr. McCuen is an internationally renowned expert on engineering ethics and he was kind enough to edit and add his experience and wisdom to this very important chapter.

The professors and staff at the Department of Fire Protection Engineering at the University of Maryland continue to be of enormous support, assistance, and inspiration to me in my current duties as lecturer with the department. Those reading this book are strongly encouraged to further their education at this truly exceptional and unique school.

Dr. John L. Bryan deserves special recognition, not only for founding the Department of Fire Protection Engineering at the University of Maryland in 1956, but for helping it to grow and prosper. Dr. Steven M. Spivak has, since Dr. Bryan's retirement, ably guided the department into an exciting period of discovery and growth. On behalf of all graduates of the Department of Fire Protection Engineering, I wish to offer the highest accolades to these two distinguished individuals for their leadership and for their friendship.

So much of a person's expertise is obtained on the job. For a significant portion of my career, "Automatic" Sprinkler Corporation of America provided excellent teachers and an atmosphere that encouraged me to expand my experience and knowledge. The leaders at "Automatic" allowed me to arrange my work schedule to return to school and study fire protection engineering and authorized the reimbursement of my tuition. To everyone at "Automatic", especially Terry Victor and Art O'Neil, I owe my deepest gratitude, and to all who have helped me perform my duties with Gagnon Engineering, my undying thanks.

Terry Victor, now a manager of Grinnell Fire Protection Systems Company, and Roland Buitron, formerly with "Automatic" and now president of Superior Automatic Sprinkler Corporation, were instrumental in supplying numerous photos and illustrations that improved this book tremendously. Kevin Kimmel and Mark Lentocha of FPE Software provided a demonstration model of "THE" Sprinkler Demo Program, computer program documentation for the appendix, and a computer hydraulic solution of example 6.17. Also, many thanks go to Frank Brannigan for writing the Foreword, and for his belief in this book.

To the National Fire Protection Association, I owe my deep gratitude, not only for their enormous contributions to fire safety in the world, but for their cooperation in supplying figures and tables for this book. I am truly proud to be a member, and I heartily recommend that all readers of this book become active members of this vital and extremely important organization.

The excellent team of experts assembled by Delmar Publishers to review this book includes the foremost leaders of our esteemed profession, and they invariably offered valuable advice that improved the quality of this book. They include:

Dr. Richard H. McCuen, University of Maryland, College Park, MD.

Franklin Clay, Sinclair Community College, Dayton, OH.

Mike Pickett, San Antonio College, San Antonio, TX.

Dr. Larry Collins, Eastern Kentucky University, Richmond, KY.

Jack Fenner, University of Cincinnati, Cincinnati, OH.

Clinton Smoke, Northern Virginia Community College, Annandale, VA.

Dr. John L. Bryan, University of Maryland, College Park, MD.

Wendy Johnson, Bexar County Fire Marshal, San Antonio, TX.

Terence Fitzpatrick, University of Cincinnati, Cincinnati, OH.

One's family is the rock upon which all accomplishments are based, and the fountain from which all inspiration flows. To my late father Robert, my mother Martha, my wife Martha, and her parents, Susan and Bill, my daughter, Rebecca, her husband John, and my granddaughter, Alana, I bestow my most sincere thanks for your love and encouragement. My mother-in-law, Susan Mason, provided countless hours of word processing and data entry, and was especially instrumental to the completion of this book. Anyone who can accurately decipher my rough handwritten notes is deserving of sainthood.

Everyone at Delmar Publishers was always friendly, professional, helpful, and encouraging during the entire process of writing and publishing this book. From the beginning, Delmar fervently believed in the value of this book, and this knowledge inspired me to meet and exceed every deadline. They made the difficult experience of writing a book a most enjoyable one, and my thanks go to all of them.

Introduction

WHO SHOULD USE THIS BOOK

This book provides design advice for students at community colleges, fire academies, and universities and is recommended for use by sprinkler, engineering, and architectural firms for training entry-level design personnel. The book has therefore been written not only for fire service personnel who encounter fire protection systems in their daily lives, but for practicing or aspiring fire protection technicians and engineers and other fire protection professionals aspiring to a greater understanding of fire protection systems.

HOW TO USE THIS BOOK

Most chapters in this book that relate to fire protection system design are based on conformance to an accepted national design standard. The **National Fire Protection Association** (NFPA) is the nation's principal source for the creation, modification, and publication of codes and standards applicable to fire protection.

Each applicable chapter refers the reader to one or more NFPA standards. It is possible to obtain a good foundation for designing fire protection systems using this book as the sole reference, but it is recommended that the applicable NFPA standards be obtained and used in conjunction with this book for greater understanding of the subject matter.

Although it may be possible for some readers to gain an introductory understanding of fire protection system design simply by reading this book, the author has found that the best way to learn fire protection system design is to perform fire protection system design using the level of detail shown in this book. Therefore, for the greatest understanding, the reader should do, as a minimum, the design assignments and activities at the end of each chapter.

You may never have done a scaled drawing before reading this book, but transferring the ideas in your head onto a blank drawing medium is among the most important and valuable skills expected of a fire protection professional. You can make your ideas and intentions clear by presenting them in a comprehensible drawing. The price paid for a lack of clarity could be the lives of the people who depend on your skill.

BASIC MATHEMATICS REQUIRED FOR THIS BOOK

Most readers with a background in introductory high school algebra and geometry should not find the mathematical demands of this book to be a problem. To perform the relatively sim-

National Fire Protection Association (NFPA)
the principal source and publisher of fire protection standards in the United States

Metric Conversion Act of 1975
legislation that created a requirement for conversion to the metric system for all federal projects by 1992

ple calculations required, the reader should have a calculator with square root, exponent, and inverse keys, because the highest level of math required for this book uses these functions.

Metric Conversions

Metric conversions can be made using the appropriate tables in Appendix A. The U.S. government has made a commitment to convert to the metric system, and some agencies have already done so. The position of the U.S. government was established by the **Metric Conversion Act of 1975**, which created a requirement for conversion to the metric system for all federal projects by 1992. President Bush issued Executive Order 12770 on July 25, 1991, which provided impetus for the transition to the metric system. Numerous federal departments, such as the U.S. Army Corps of Engineers, converted to the metric system immediately for all construction projects. Although the original timetable for metric conversion has not been met, federal construction will be uniformly converted to the metric system in the foreseeable future. The smoothness of the transition by contractors performing federal construction has surprised those that predicted chaos.

The reason for this conversion is that the United States is one of the few countries in the world that does not base its system of weights and measures on the metric system, the international standard. The ability of the United States to compete on an international scale depends on our nation's ability to design and manufacture commodities that can be sold and used overseas, and to design federal projects that can be installed with foreign tools and labor.

Using the metric system can create problems when renovating existing facilities that were not designed using the metric system. For example, a building undergoing renovation may use "soft" metric conversions, which consist of direct conversion of the existing feet-and-inch units to a fractional metric unit. Working with these decimals can be cumbersome, and it may be simpler to use "hard" metric, which involves rounding to the nearest whole metric unit for convenience.

Metric Units

For fire protection system designs, the following metric units are used:

Quantity	*Unit*	*Symbol*
Length	Meter	m
Mass	Kilogram	kg
Time	Second	s
Temperature	Kelvin	K

When very large or very small quantities are encountered, these units are broken down for ease of use by using decimal prefixes. Prefixes most commonly used in construction are:

Prefix	Symbol	Multiplier	Example
Kilo	k	1000	km, kg
Milli	m	0.001	mm, mg

In an effort to provide uniformity in construction units, the *Metric Guide for Federal Construction* advises against the use of the centimeter in construction. The meter and millimeter are to be used for fire protection system design, and the meter and kilometer are to be used for fire protection underground piping design.

Common conversions for the evaluation of U.S. and metric scales are listed below:

Inch–Foot Scale	Ratio	Closest Metric Scale
1/8" = 1'-0"	1:96	1:100
1/4" = 1'-0"	1:48	1:50
1/2" = 1'-0"	1:24	1:25

Other metric conversions can be found in Appendix A.

SIGNIFICANT FIGURES AND ROUNDING

Most fire protection calculations are performed with no more than two significant figures to the right of the decimal point. A significant figure is a digit, other than a zero used to locate a decimal point. As an example, the number 20.205 has 5 significant figures, and the number 0.000514 has 3 significant figures. Where greater levels of precision are appropriate, the recommended number of significant figures will be displayed in examples and sample problems. Rounding is performed by rounding up if the value to be rounded is 5 or more, and by rounding down if the value to be rounded is 4 or less.

NFPA PUBLICATIONS

The NFPA publishes over 290 codes, standards, recommended practices, and guides that apply to fire safety and the design of fire protection systems. For each standard published by the NFPA, a committee of volunteers, appointed by the NFPA, creates, modifies, and revises the standard. The committee membership is balanced to include such categories as special expert, system user, insurance, research, installer, manufacturer, laborer consumer, and enforcer, and each committee is assigned a full-time employee of the NFPA to serve as a nonvoting liaison to the committee. A chairperson and secretary are selected for the committee, and voting members, or principal members are permitted to have alternate members serve in their place, provided that an application as an alternate member is submitted and accepted.

The NFPA creates **codes**, **standards**, **recommended practices**, and **guides** for use by fire protection professionals. A code includes mandatory requirements that are suitable for adoption into law. A standard includes mandatory require-

---

codes
mandatory requirements suitable for adoption into law

standards
mandatory NFPA requirements that may be used by an approving authority to approve a fire protection system

recommended practices
an NFPA document that provides nonmandatory advice

guides
informative but nonbinding NFPA documents

ments that may be used by an approving authority to approve a fire protection system. A recommended practice is a document that provides nonmandatory advice. A guide is informative, but nonbinding.

Each NFPA document follows a standardized format. The document begins with an overview of the publication history of the document, a list of committee members, and a table of contents. The body of the document includes statements of scope and purpose for the document, a list of definitions, and detailed technical requirements.

The NFPA system of committee selection and document publication is a marvel for its efficiency and its ability to produce consensus documents that are capable of being readily adopted by local governments and approving authorities. The reader is strongly encouraged to join the NFPA and become a part of the standards-making process.

FIRE PROTECTION AS A PROFESSION

Professionals who dedicate their lives to the specification, design, installation, approval, and maintenance of fire protection systems and to the operation of fire departments are unique. The dedication and commitment exhibited by fire protection professionals is almost uniformly exceptional, and this level of commitment distinguishes the fire protection profession from many other occupations.

The dedication we display in our work is a reflection of our belief that we are making a difference in improving the safety and welfare of the world in which we live. There can be no more rewarding profession than to dedicate one's life to the welfare of others. This dedication characterizes our lives. It sustains us. We save the lives of people whose names we may never know.

Along with the rewards of the profession come responsibilities. As fire protection professionals, we must realize that if we make a serious error or knowingly provide less than the minimum level of fire protection, people may die. This responsibility is enormous, and I ask that all readers of this book accept this responsibility freely and fully, because it is the acceptance of this responsibility, more than any other factor, that differentiates our profession from most other occupations.

A fire protection system is so much more than just a collection of assorted hardware. If we are truly committed to the spirit and intent of the goals of our profession, we can never accept being associated with the design and installation of fire protection systems that are anything less than compliant with the applicable codes and standards. The vast majority of this book is intended to help the reader achieve this goal.

The application of technical knowledge to our profession is fundamental, but it does not comprise our total responsibility as fire protection professionals. Chapter 13 covers a topic rarely covered in a technical book: Ethical behavior among fire protection professionals is as important as the application of technical knowledge. Knowingly failing to apply technical knowledge in an ethical manner is the most fundamental error that a fire protection professional can make. All readers of this book are advised to become familiar with the code of ethics that applies to the practice of engineering and technology in your jurisdiction, and to follow it to the letter.

This book assists fire protection professionals in becoming technically proficient with respect to several of the most commonly installed types of fire protection systems and helps to familiarize the reader with the fundamentals of ethical practice. These two essential tools will assist the reader in achieving a successful and rewarding career as a fire protection professional.

FIRE PROTECTION TECHNICIANS AND ENGINEERS

A **fire protection engineer** is a person who is qualified by experience and education to perform **fire protection system design**, based on principles of physics, chemistry, thermodynamics, statics, dynamics, fluid dynamics, and upper-level mathematics. Fire protection design may involve the testing of a particular fire scenario or research of prior testing data to establish an engineering basis for the specification and design of a fire protection system whose criteria may or may not coincide with criteria found in an accepted national standard. Engineers are expected to remain current with the latest fire protection innovations and specify and design fire protection systems based on fundamental engineering principles, as opposed to simple reference to an NFPA standard. Engineers perform work in accordance with the Code of Ethics for Engineers, published by the National Society of Professional Engineers, and are legally required to comply with the laws, statutes, and code of ethics for the states in which they practice design.

Fire protection technicians are persons with the knowledge, skill, and training to perform **fire protection system layout**, using accepted national standards. Fire protection system layout is the act of performing a drawing by following the requirements of a standard whose requirements were derived from tests and data based on engineering principles. A fire protection layout, performed by a fire protection technician, is often performed in accordance with a fire protection design, performed by a fire protection engineer. Technicians are urged and sometimes required to seek certification as certified engineering technicians, by taking examinations, obtaining professional recommendations, and accruing experience as a practicing technician under supervision and in responsible charge. Technicians are expected to perform their services in accordance with the Code of Responsibility for NICET-Certified Engineering Technicians and Technologists, published by the National Institute for Certification in Engineering Technologies (NICET).

The fundamental difference between engineers and technicians in fire protection is, therefore, that engineers perform design and technologists perform layout. In an effort to avoid repeatedly using the term "design and/or layout," this book primarily uses "design" for simplicity.

fire protection engineer
a person who is qualified by experience and education to perform fire protection system design

fire protection system design
design of a fire protection system based on engineering criteria that may not always coincide with criteria found in an accepted national standard

fire protection technicians
persons with the knowledge, skill, and training to perform fire protection layout, using accepted national standards

fire protection system layout
the act of following the requirements of a standard to execute a drawing

Although this book is based on reference to and concurrence with recognized national standards, engineers or engineering students may use this book as a basis for their engineering design career. It is expected that engineers base their design decisions on engineering principles as well as being familiar with the requirements of the applicable national standards as a point of reference in their design.

ORAL, WRITTEN, AND GRAPHIC COMMUNICATION

Students taking courses in a fire protection curricula may wonder why they may be asked or required to take courses in writing, speech making, and literature. Success as a fire protection professional depends to a large degree on the ability to clearly communicate complex ideas to a wide audience. Communications skills include written communication, oral communication, and graphic communication, with each category being equally vital and important. To illustrate this point, the following scenarios are presented:

- Imagine a fire chief speaking before a group of citizens who is unable to clearly describe why residential sprinkler systems are beneficial.
- Imagine a fire protection engineer writing a specification who is unable to clearly communicate requirements for a preaction system.
- Imagine a fire protection technician attempting to defend a layout in the presence of a fire marshal without being clearly understood.
- Imagine any fire protection professional trying to interpret and clearly explain the fundamental concepts depicted on a fire protection drawing without understanding what is shown.
- Imagine trying to explain intricate concepts related to fire protection systems without the ability to effectively graphically represent those ideas.

For a fire protection professional to achieve effective oral, written, and graphic communication, he or she must receive specific training. It is strongly recommended that all fire protection professionals seek specific instruction on technical writing and effective oral communication and that all professionals practice these skills at every opportunity. It is further recommended that those reading this book perform as many of the drawing exercises in this book as possible and seek out advanced instruction on fire protection system design.

GRAPHIC COMMUNICATION—DRAWING FIRE PROTECTION SYSTEMS

A **drawing** is a graphical representation of a designer's ideas. Drawings are needed because attempts to verbally explain a mental idea can become cumbersome, and for some of the more complex designs, impossible.

Some people reading this book may have little or no drawing experience,

drawing
graphical representation of a designer's ideas

but it must be understood that fire protection professionals who lack the ability to graphically represent their ideas may be in for a frustrating career with limited options. It is the intent of this book to help all fire protection professionals become more well rounded and effective in their profession so that they can represent ideas in a clearly understood graphical form and that they can clearly understand drawings performed by others.

Technical drawing is a unique and fascinating form of communication. It is partly an art form and partly a technical media. The skillful combination of these two seemingly opposite concepts is fundamental to the success of a technician, an engineer, or any fire protection professional.

For those who have mastered the art and science of fire protection design and layout, success is ensured, and enjoyment of this wonderful profession is greatly enhanced.

The Blank Page

For the novice designer, few concepts may be more intimidating than starting a drawing with a blank page. Your ideas are in your head, and you are faced with a large piece of blank drawing medium, an enormous book of specifications, and a huge roll of drawings.

Starting a drawing begins with a thorough review of all contract drawings and specifications. Chapters 1 and 2 will help you to extract the necessary information from these references. Chapter 3 will help you obtain needed information from a fire protection survey, and the technical chapters will give you a method for system design and calculation. If possible, seek out a mentor, an instructor, or a person experienced in fire protection system design who can help you organize your thoughts and your reference material.

It may be helpful to begin layout of a floor, a building, or an area that is less complex than other areas. Applying basic skills to a simple drawing will provide experience for more complex designs.

Do not be afraid of erasing or changing your drawing: You may discover an error or a better or more efficient way of presenting your ideas. Draw lightly at first, then darken your lines as your drawing becomes finalized.

Regardless of your experience, remember that the technical information shown on the drawing is of little value if it cannot be read. Be as neat and as organized as possible on your first drawings. Your presentation style will improve with subsequent drawings.

Starting your first design may be somewhat intimidating, but there is a genuine feeling of triumph and exhilaration upon completing a design. The challenge of filling a blank page with your ideas and representing them in a clear and usable fashion is one of the most rewarding feelings that a fire protection professional can experience. Completion of your first drawing will give you the confidence to perform more difficult designs, so hang in there and you will succeed!

Introduction

Drafting Equipment

It may surprise you to know that it costs very little to get started on your first design. Although drafting equipment can be very sophisticated, performing a design with only the essential items, shown in Figure 1, can produce very effective results.

Drawing Surface

A flat surface, perhaps your kitchen table or your desk, may provide an acceptable surface for your drawing. The only requirement is that the surface be as large as your drawing and that the edge be square to facilitate the use of a T-square.

The next higher level of drawing surface sophistication is the portable drawing board, which may be purchased for well under fifty dollars. Drafting tables can be bought for significantly less than one hundred dollars in office sup-

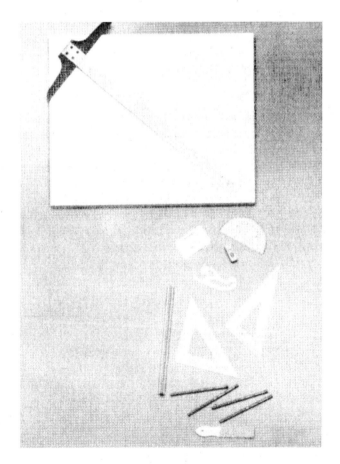

Figure 1 *A drawing board with T-square is at the top of the picture. Other essential equipment (shown from top to bottom) includes an erasing shield, protractor, eraser, french curve, drafting triangles, a scale, pencils, and a pencil sharpener. (Courtesy Teledyne Post)*

Introduction

ply and discount stores. A drafting table allows the drawing to be elevated, making it easier to reach the upper portions of the drawing.

Drafting lights are available for about twenty dollars or less, but any lighting that can be adjusted for clear view of your drawing will suffice.

Drawing Parallel Lines

The best way for beginning designers to ensure that horizontal lines are parallel would be to buy a T-square, shown on Figure 1. Cost for a T-square ranges from ten dollars up to fifty dollars, depending on length. It is recommended that a 42-inch T-square be purchased, because many drawings that you may do are this size.

Drafting triangles, shown in Figure 1, are very inexpensive, usually under five dollars, and provide reliable vertical parallel lines when aligned with the T-square.

Scales

Fire protection system drawings are drawn to scale. For this reason, at least one scale should be purchased. An architect's scale, shown in Figure 2, is needed for the design of fire protection systems, and an engineer's scale, shown in Figure 3, is needed for underground fire protection piping design, described in Chapter 4. These scales can be purchased for well under ten dollars each, with plastic scales being much less expensive than wooden scales.

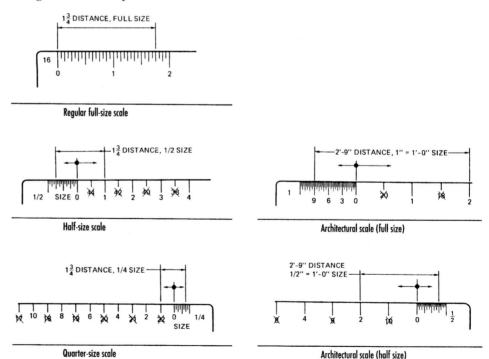

Figure 2 *An architect's scale is needed for fire protection system design. (Courtesy Delmar Publishers, Technical Drawing 3e)*

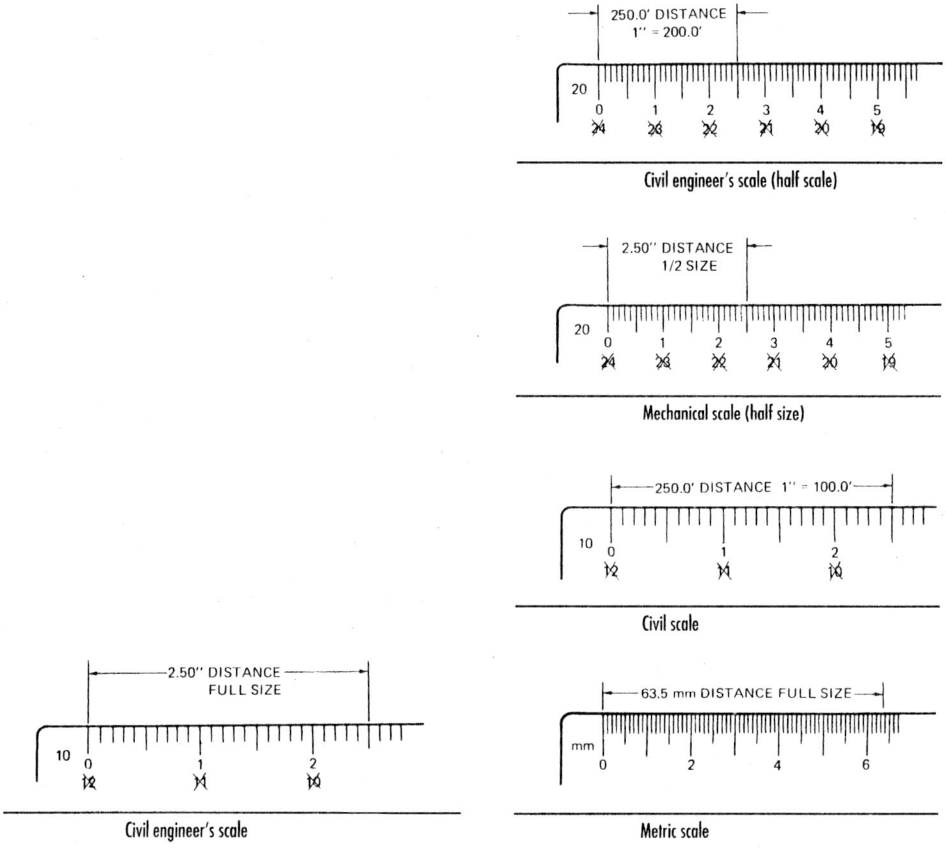

Figure 3 *A civil engineer's scale is required for underground piping drawings. (Courtesy Delmar Publishers, Technical Drawing 3e)*

Civil engineer's scale (half scale)

Mechanical scale (half size)

Civil scale

Civil engineer's scale

Metric scale

Drawing Media

The drawing medium is the reproducible surface upon which one draws. A large piece of plain white paper may suffice, but a blueprint of this medium can not be effectively made because it is not translucent. Some copying shops have the equipment to make large photocopies, but this process is usually four or more times as expensive as a blueprint.

Translucent media allow light to shine through them, which is the process used in making a blueprint. Translucent media include Mylar, vellum, and onion skin, with Mylar being the thickest and onion skin being the thinnest.

It is recommended that Mylar be used for your design projects, because erasing it is easier, you can bear down harder on the lead for a darker line, and sharp pencils are not likely to tear or damage it. Mylar is very forgiving of your mistakes, and the frustration of making mistakes is lessened. Mistakes requiring erasing occur to even the most experienced professional.

Pencils are offered in a wide, and possibly confusing, variety. For Mylar, H series lead (H, 2H, 3H, and 4H) has a tendency to smear easily. It is recommended that you obtain a P series lead (P1, P2, P3) for use with Mylar.

A mechanical pencil can be purchased for under five dollars and is usually offered in 0.5 mm- and 0.7 mm-lead sizes. You may find that 0.5 mm-lead, the thinner of the two, breaks easily, especially if you bear down as hard as I do. If you have a heavy hand, consider 0.7 mm P3 film lead.

Drawing Arrangement

Now that you have the equipment, let's fill that blank page. The size of the drafting medium that you select must be larger than the size of the drawing you are tracing, with room to spare for notes and a title block.

Orient the building or object to the upper left-hand corner of the drawing, allowing sufficient space for dimensions, which are drawn outside of the building walls. Additional floors or areas, should space permit, may be drawn below the floor already drawn, as shown in Figure 4. If the building is too big to fit on your drawing, break the building into sections or use a smaller scale. Drawings utilizing a scale smaller than 1/8" = 1'-0" should not be used for fire protection.

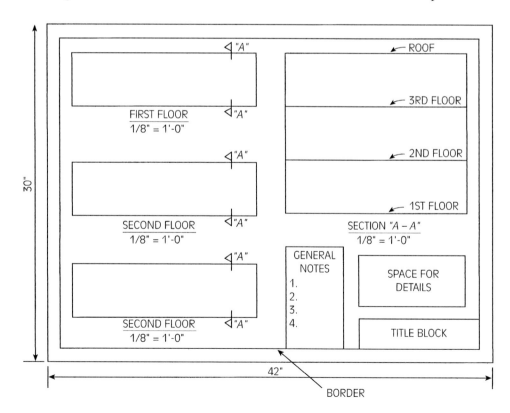

Figure 4 *General drawing arrangement.*

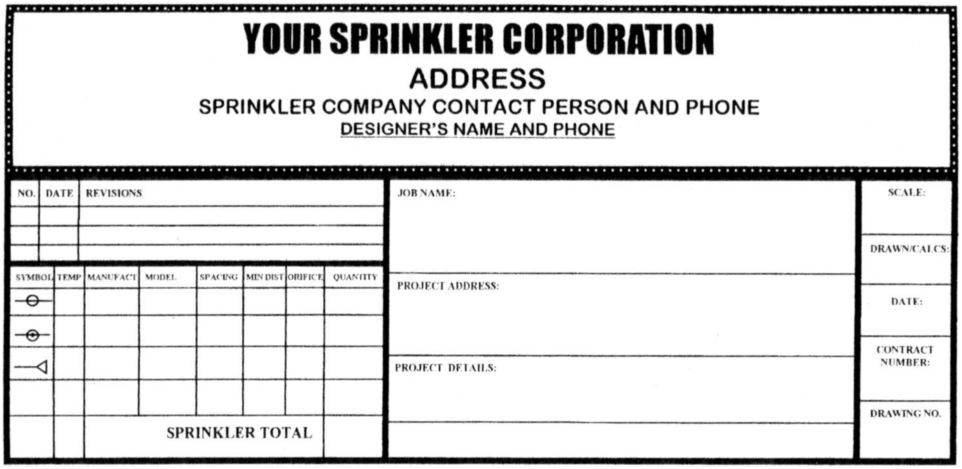

Figure 5 *Sample title block.*

A title block is added to the lower right-hand corner of the drawing. Using a format similar to Figure 5 should suffice for most design projects. Feel free to use your creativity to develop a title block with which you are comfortable.

Computer-Aided Design

Fire protection system design has been performed for many decades using the manual drafting method described previously and shown in Figure 6. Some design firms continue to use this method, and many readers of this book who are performing their first drawing may select this method.

The personal computer has revolutionized almost every facet of our personal lives, with no less effect on fire protection system design. Many sprinkler firms, architectural offices, and engineering offices have converted from manual drafting to **computer-aided design (CAD)**, a computerized method of preparing a drawing. Specifications are increasingly requiring CAD for fire protection system design.

It is hoped that all fire protection professionals become acquainted with CAD before graduation, or as early in their careers as possible. Most community colleges offer courses in CAD, and it is recommended that all fire protection professionals register for this instruction. Most CAD programs are available for use on the personal computer, shown in Figure 7.

The use of CAD requires specialized software and compatible hardware. The software may be selected from a variety of commercially available computer programs and is loaded onto the hard disk drive in your computer. Some CAD software is specially created for fire protection system design and contains the appropriate symbols and design tools needed, including hydraulic calculation software on some CAD programs.

computer-aided design (CAD)
a computerized method of preparing a drawing

Figure 6 *Manual drafting. (Courtesy Grinnell Fire Protection Systems Co.)*

Figure 7 *Computer-aided design (CAD) components. CAD equipment (shown from left to right) is: an instruction manual with software, a keyboard, computer and screen, and a digitizer tablet and puck. (Courtesy Autodesk Inc.)*

The hardware required includes a computer with sufficient speed, memory, and mathematics capability, a terminal screen, and a keyboard. Input of information may be facilitated with a digitizer and puck, which serves as a drawing tablet. Another handy tool for data input might be a scanner, which scans a print from a drawing into your computer. A drawing that has already been drawn on CAD can be loaded onto your hard drive to serve as a background for your fire protection system design.

Output of a completed design is performed with a plotter, which draws what you have designed into a reproducible medium. If you don't have a plotter, you can take your CAD disk to a copying shop for plotting.

CAD takes most people a while to learn, and must be used fairly often in order to become proficient. However, it has many advantages over manual drafting. Although some manual designs are quite attractive, CAD is neater than most manual designs, and because most CAD software adds plan dimensions for you, CAD is more accurate than manual drafting. Revisions and corrections are considerably easier on CAD, and disk storage is more efficient than the plan file storage rooms that once existed for manual designs.

People who hire fire protection professionals expect a solid background in fire protection system design, and with increasing regularity they are looking for people who have acquired experience with CAD.

Chapter 1

Contract Drawings

Objectives

Upon completion of this chapter, you should be able to:

- List the items that comprise a set of contract documents for the design of a fire protection system.
- Explain the role and responsibilities of a fire protection system designer.
- List the categories of drawings that comprise a contract drawing package.
- Explain the differences among the categories of contract drawings.
- Evaluate a set of contract drawings to determine the value of each drawing relative to the accurate development of a fire protection system design.
- Select those contract drawings that would be most beneficial to the design of an underground fire protection piping system.
- Explain the problems associated with the failure of a fire protection systems designer to reference the drawings relevant to fire protection systems in a contract drawing package.

Fire protection system designers are often given less information than they need to do a complete and competent job of designing a fire protection system. One of the purposes of Chapters 1 and 2 is to explain the value of each component of the contract documents, the primary source of reference for the design of a fire protection system. The inherent danger of failing to use these documents to full advantage is the possibility that critical information may be omitted from your fire protection design resulting in installation problems or failure to meet the applicable standards that could compromise or negatively impact the effectiveness of the fire protection system.

The contract document package, or the contract documents, consist of **contract drawings**, a set of plans that describes a project in pictorial form, discussed in this chapter, and **specifications**, a written description of project requirements, discussed in Chapter 2. Knowledge of the value of each component of the contract drawings and each section of the specifications is essential to the successful design of a fire protection underground piping plan or a fire protection system layout.

A fire protection system designer should never forget that the public expects and demands that a fire protection system installed in a building should meet its primary **performance objective** of properly activating in a timely manner, and should provide a minimum acceptable level of life and property protection when the system is called upon to function. A performance objective is an engineering basis for a predetermined end result of a given design.

Fire protection system design is a highly respected profession that carries enormous responsibility. Acceptance of this responsibility entails the complete investigation of all pertinent information that may affect the performance of the fire protection system.

CONTRACT DRAWINGS COVER SHEET

The cover sheet ordinarily contains a wealth of diverse useful information for the fire protection designer. The location map, a large-scale drawing showing roads and intersections, not only helps you get to the job site, but it can, if traced or scanned onto a sprinkler drawing, help speed your plans through the approval and permit process by assisting the **authority having jurisdiction** (AHJ) in referencing the building during the permit process. An authority having jurisdiction is the individual or agency placed in responsible charge of reviewing and approving drawings and completed installations.

The vicinity map, showing all buildings on a job site, helps to provide a visual identifier on your plans, to show the building or portion of building described by each plan. This feature is especially valuable for large buildings requiring more than one drawing to portray a single floor, or for contracts where more than one building is included in the fire protection design project. Sample location and vicinity maps are shown in Figure 1-1.

Other valuable facts that can be found on many cover sheets are the names and addresses of the owner and tenants. Many permit agencies require this infor-

---

contract drawings
a set of plans that describe a project in pictorial form

specifications
a written description of project requirements

performance objective
an engineering basis for a predetermined end result of a given design

authority having jurisdiction (AHJ)
the individual or agency placed in responsible charge of reviewing and approving drawings and completed installations

Chapter 1 Contract Drawings

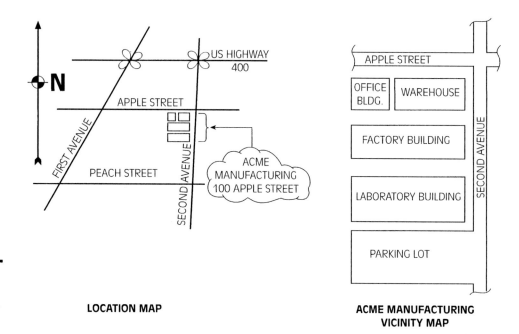

Figure 1-1 *Location map (left) and vicinity map (right).*

mation to be on your plans. Phone numbers of the architect, engineer, and key personnel may be listed on this sheet providing a source for assistance in the resolution of problems. The building permit number is often noted as well, and by referencing this number, your plans may be reviewed more expeditiously by the AHJ.

A drawing index, a list of all drawings in the contract drawing set as shown in Figure 1-2, usually appears on the cover sheet and can be used as a checklist against the drawings in your possession. If revision numbers are noted, these should be checked as well to ensure that the most recent edition of each drawing in the set is in your possession. If drawings are missing or if drawings have outdated revision numbers, new drawings must be requested immediately.

A symbol and abbreviation index serves as a reference in your interpretation and use of the drawings contained within the contract drawing set. **NFPA 170**, *Standard for Fire Safety Symbols*, should be used as the basis for symbols used in the design of fire protection systems.

NFPA 170

Standard for Fire Safety Symbols

civil drawings
scaled drawings that coordinate underground utilities entering and leaving a building or group of buildings

SITE DRAWINGS

Site plans, often called **civil drawings**, are scaled drawings used to lay out and coordinate underground utilities entering and leaving a building or group of buildings. Shown on this plan are existing utilities, usually shown dotted, and proposed new utilities, shown with solid lines.

COVER SHEET

CIVIL DRAWINGS:
- C-1.1 EXISTING CONDITIONS
- C-2.1 DEMOLITION PLAN
- C-3.1 LAYOUT PLAN
- C-4.1 SITE DEVELOPMENT PLAN
- C-4.2 SITE PLAN
- C-4.3 DRAINAGE AREA MAP
- C-5.1 SEWER LINE PROFILE
- C-6.1 WATER MAIN & STORM DRAIN PROFILES
- C-7.1 STORM DRAIN PROFILES
- C-8.1 EROSION & SEDIMENT CONTROL PLAN
- C-9.1 EROSION & SEDIMENT CONTROL DETAILS
- C-10.1 FUEL PUMP CANOPY

- L-1.1 LANDSCAPE PLAN

ARCHITECTURAL:
- A-1.1 ABBREVIATIONS & SYMBOLS
- A-1.2 SITE PLAN AND DETAILS
- A-1.3 DETAILS
- A-2.1 FIRST FLOOR PLAN-FIRE STATION
- A-2.2 FIRST FLOOR PLAN-POLICE STATION
- A-3.1 ELEVATIONS
- A-3.2 ELEVATIONS
- A-3.3 ELEVATIONS
- A-4.1 ROOF PLAN-FIRE STATION
- A-4.2 ROOF PLAN-POLICE STATION
- A-5.1 BUILDING SECTIONS
- A-5.2 BUILDING SECTIONS
- A-6.1 WALL SECTIONS
- A-6.2 WALL SECTIONS
- A-6.3 WALL SECTIONS
- A-7.1 DETAIL PLANS-FIRE STATION
- A-7.2 DETAIL PLANS-POLICE STATION
- A-7.3 DETAILS
- A-7.4 DETAILS
- A-7.5 DETAILS
- A-7.6 DETAILS
- A-8.1 REFLECTED CEILING PLAN-FIRE STATION
- A-8.2 REFLECTED CEILING PLAN-POLICE STATION
- A-9.1 DOOR SCHEDULE-DETAILS FIRE STATION
- A-9.2 DOOR SCHEDULE-DETAILS POLICE STATION PHASE 1
- A-9.3 DOOR SCHEDULE-DETAILS POLICE STATION PHASE 1&2

STRUCTURAL:
- S-1.1 PARTIAL FOUNDATION PLAN
- S-1.2 PARTIAL FOUNDATION PLAN
- S-2.1 PARTIAL ROOF FRAMING PLAN
- S-2.2 PARTIAL ROOF FRAMING PLAN
- S-3.1 GENERAL NOTES AND SCHEDULES
- S-3.2 SECTIONS AND DETAILS
- S-3.3 SECTIONS AND DETAILS
- S-3.4 SECTIONS AND DETAILS
- S-3.5 SECTIONS AND DETAILS
- S-3.6 SECTIONS AND DETAILS

MECHANICAL/PLUMBING
- M-1.1 MECHANICAL GEN.NOTES,SYMBOLS,ABBR.
- M-2.1 SITE PLAN - PHASE 1
- M-3.1 FIRST FLOOR PLAN-PIPING FIRE STATION
- M-3.2 FIRST FLOOR PLAN-PIPING POLICE STATION
- M-4.1 FIRST FLOOR PLAN-HVAC FIRE STATION
- M-4.2 FIRST FLOOR PLAN-HVAC POLICE STATION
- M-5.1 SECTIONS-FIRE STATION
- M-5.2 SECTIONS-POLICE STATION PHASE 1
- M-6.1 PART PLANS-PLUMBING FIRE STATION
- M-6.2 PART PLANS-PLUMBING POLICE STATION PHASE 1
- M-7.1 WATER RISER DIAGRAMS
- M-7.2 SANITARY RISER DIAGRAMS - FIRE AND POLICE STATION
- M-8.1 PART PLANS-MECHANICAL ROOM
- M-8.2 SECTIONS-MECHANICAL ROOM
- M-9.1 AUTOMATIC TEMPERATURE CONTROL DIAGRAM
- M-10.1 DETAILS
- M-10.2 DETAILS
- M-10.3 DETAILS

- M-11.1 SCHEDULES-PHASE 1
- FP-1.1 FIRST FLOOR PLAN-SPRINKLER FIRE STATION
- FP-1.2 FIRST FLOOR PLAN-SPRINKLER POLICE STATION PHASE 1

ELECTRICAL:
- E-0 ELECTRICAL-GEN.NOTES AND LEGEND
- E-1.1 SITE PLAN-REMOVAL WORK
- E-1.2 SITE PLAN-PHASE 1
- E-2.1 FIRST FLOOR PLAN-POWER FIRE STATION
- E-2.2 FIRST FLOOR PLAN-POWER POLICE STATION PHASE 1
- E-3.1 FIRST FLOOR PLAN-LIGHTING FIRE STATION
- E-3.2 FIRST FLOOR PLAN-LIGHTING POLICE STATION PHASE 1
- E-4.1 SPECIAL SYSTEMS-FIRE STATION
- E-4.2 SPECIAL SYSTEMS-POLICE STATION PHASE 1
- E-5.1 ROOF PLAN-LIGHTNING PROTECTION
- E-6.1 ONE LINE DIAGRAM-POWER
- E-6.2 ONE LINE FDAS, PAGING AND CCTV SYSTEMS
- E-7.1 PART PLANS
- E-7.2 MCC AND LIGHTING FIXTURES
- E-8.1 DETAILS
- E-8.2 DETAILS
- E-8.3 DETAILS
- E-9.1 PANEL BOARD SCHEDULES
- E-9.2 PANEL BOARD SCHEDULES

Figure 1-2 *A sample drawing index.*

Chapter 1 Contract Drawings

contour lines
lines indicating the elevation of the finished exterior grade

plan reference elevation
sea level, or an elevation specifically chosen for a building or group of buildings

Civil drawings are uniquely characterized by the presence of a series of lines, called **contour lines**, which indicate the elevation of the finished exterior grade with respect to a **plan reference elevation**.

The elevation of reference could be sea level, or a reference specifically chosen for a building or group of buildings (elevation 100′-0″ is commonly used as a reference elevation).

Contour lines are usually spaced in increments of one to ten feet, depending on the scale of the site drawing. Contour lines that are closely spaced indicate a hill or slope, whereas contour lines spaced far apart indicate a gently sloping grade. The severity of the slope of a hill can be determined by observing the number of feet between contour lines and the spacing between the lines. Refer to Figure 1-3 for a sample contour plan.

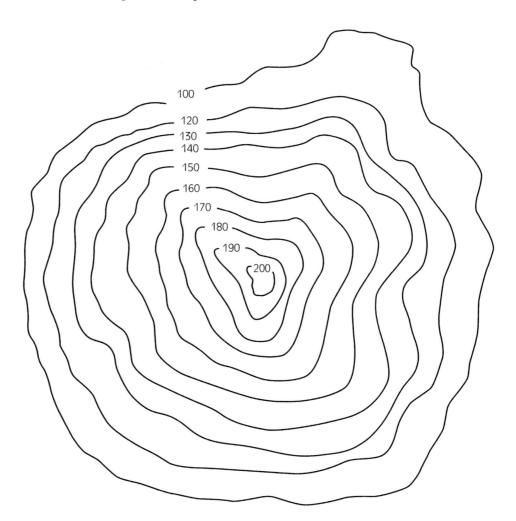

Figure 1-3 *Contour lines depicting a hill using elevation 100 as a reference elevation.*

reference grid
a system of parallel reference lines aligned either to magnetic north or plant north

Site plans can also be referenced to a **reference grid**. Reference grids, a sample of which is shown in Figure 1-4, provide a system of parallel reference lines, aligned either to magnetic north or plant north. Magnetic north is the direction a compass would point, whereas plant north is a reference direction most usually oriented toward the top of the page on most contract drawing sets. The use of reference grids is an issue of convenience for the architect or engineer, allowing for buildings to be oriented in a uniform manner throughout the contract drawing set, with plant north being the point of reference, and with all drawings oriented using plant north pointing toward the top of the page.

Reference grids make it easier to relate one building to another, using the site plan as an accurate road map. Some site plans use reference grids as the basis for precisely locating specific points on a site plan.

As an example of the use of a reference grid, a fire hydrant, identified as point N105'/E110' in Figure 1-4, would indicate that the hydrant is located 105 feet north of the north-south reference line and 110 feet east of the east-west reference line. The reference grid method can greatly simplify problems and is significantly more accurate and useful than scaling of ungridded site plans.

In another reference grid example using Figure 1-4, a 10-inch fire main runs east to west, parallel to the reference north-south line. Two fire hydrants on this main, referenced as points N105/E110 and N105/E210, are exactly 100 feet apart. Site plans referenced in this manner are of significant assistance to contractors responsible for estimating and installing underground utilities.

Utilities can also be assigned elevations. A site plan showing elevations for all proposed utilities can be an enormous benefit for all contractors installing utilities underground.

depth of cover
lineal distance from the top of an underground pipe to the finished grade

Site plans are used to graphically show the **depth of cover** for a pipe, or the depth of an underground pipe below finished grade. A plan that specifically shows reference elevations with respect to finished grade is called a **profile plan**. A profile plan of a fire protection underground pipe, shown in Figure 1-5, shows the entire pipe, from source to building entry, in a sectional view. A profile plan resembles a graph, with the Y axis representing elevation, in feet, and the X axis representing horizontal travel of pipe, in feet. An underground pipe elevation is identified as an **invert elevation**, an elevation referencing the bottom of the pipe with respect to the reference elevation.

profile plan
a plan that specifically shows reference elevations with respect to finished grade

invert elevation
an elevation referencing the bottom of a pipe with respect to the reference elevation

A good profile plan shows finished grade, the fire protection pipe, utilities running perpendicular to the fire protection pipe, and special features, such as roads, railroad tracks, and building foundations. These plans are the best way of coordinating the location of a fire protection pipe with respect to other underground trades and of identifying areas where special protection is needed.

boring table
compilation of data obtained by drilling cores of earth at several strategic locations

Another feature of site plans is a **boring table**. These very valuable tables display data obtained by drilling cores of earth at several strategic locations. A fire protection designer should refer to these tables to determine the composition of the earth through which the underground pipe must travel. A bore that yields solid

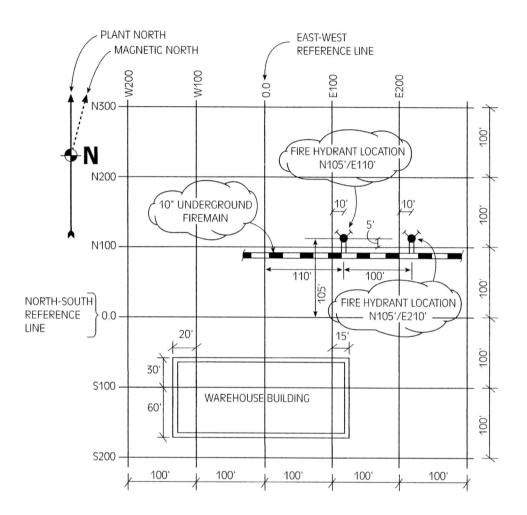

Figure 1-4 *Reference grid.*

rock indicates an area to be avoided with respect to placement of an underground fire protection pipe. A bore that yields very loose sand may indicate an area unsuitable for the placement of underground piping without special backfill. This table is of significant value when calculating thrust blocks for underground piping, as detailed in Chapter 4.

Other utilities that are likely to be represented on a site drawing are fire protection piping, storm water piping, sanitary piping, gas lines, telephone lines, and electrical lines. Aboveground features would include buildings, finished floor elevations, hydrants, streets, curbs, trees, light poles, manholes, and transformers. The locations of these utilities and features serve as points of reference and simplify the survey and design of a fire protection underground drawing.

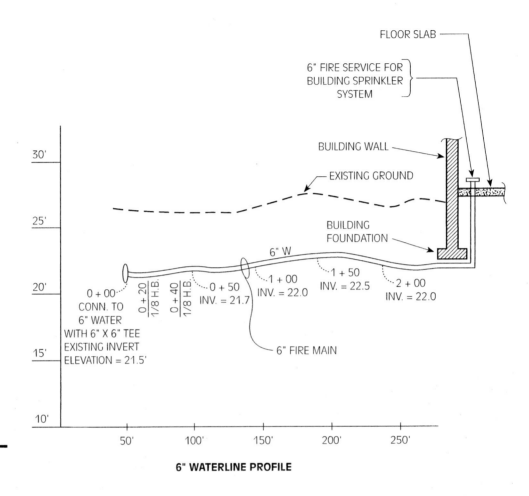

Figure 1-5 *Profile plan.*

ARCHITECTURAL PLANS

architectural drawings
drawings that show dimensions of walls, floors, ceilings, and other building features

plan job
design performed using new architectural plans as the basis for design

Architectural drawings, an example of which is shown as Figure 1-6, are identified by drawing numbers beginning with an A, such as A-1 and A-2, and are drawings that show dimensions of walls, floors, ceilings, and other building features. They can vary widely in the level of detail provided. Two basic types of architectural drawings are new-building architectural drawings and renovation architectural drawings.

A good set of new-building architectural drawings are ordinarily fully dimensioned plans that are quite usable for sprinkler design and involve little or no survey work. When a sprinkler designer has a complete set of well-dimensioned new-building architectural drawings, all design can be performed using the plans as reference. This procedure is commonly called a **plan job** in the sprinkler industry.

Chapter 1 Contract Drawings

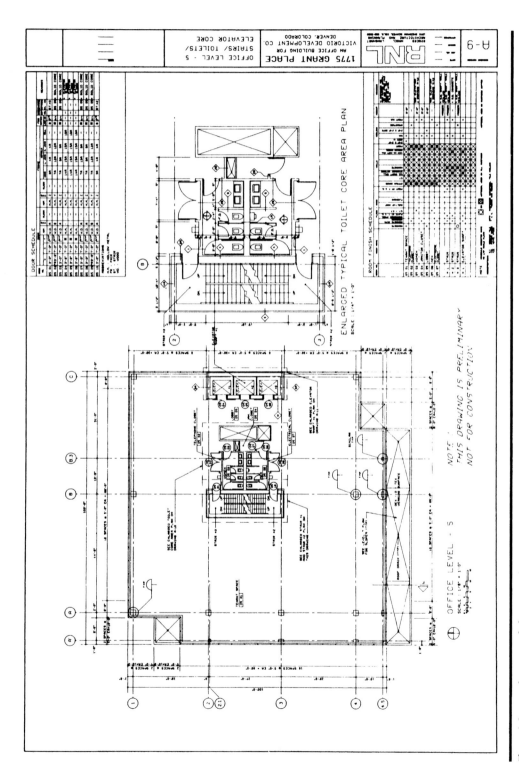

Figure 1-6 Architectural drawing. (Courtesy Sigma Design)

Renovation architectural drawings are plans that detail the renovation of an existing building and ordinarily concentrate on providing dimensions for only new rooms within the renovated building. They usually do not show dimensions for the portions of the building that remain unrenovated or are undergoing renovation without the addition of new walls. Renovation architectural drawings can be deceptively detailed with plenty of dimensions for the area being renovated, but with few dimensions in unrenovated areas and usually no complete line of dimensions from outside wall to outside wall.

A sprinkler designer, attempting to use renovation architectural drawings for the design of a prefabricated sprinkler system, may be considerably frustrated if the objective is to provide fire protection for renovated and unrenovated areas. The best tactic that can be used with renovation architectural drawings is to review the contract drawings well in advance of the commencement of design and request a set of the original architectural drawings of the building. Some foresight and persistence, applied judiciously, can save a significant amount of design time. If the original architectural drawings cannot be located, a thorough survey of the existing building is needed before layout of the fire protection system can be completed. If an existing fire protection system is present and modifications to the existing system are required, a survey of the existing fire protection system is required as well.

A project involving renovation architectural drawings is called a combination plan/survey job. A project involving an existing building for which plans cannot be obtained is a **survey job**.

survey job
a project involving an existing building for which plans cannot be obtained

Building Sections

Cutaway views through a building that are included within the architectural set and are designated with a drawing number beginning with an A are called **building sections**. Since NFPA 13 requires at least one sectional view on each fire protection shop drawing submittal, a building section, thoughtfully chosen to represent a typical area involving special handling, can most likely be traced or scanned directly from an architectural elevation detail. Fire protection piping can then be added to the section. A sectional view is required for each unusual condition related to the sprinkler piping, to assist the sprinkler plan reviewer in understanding the proposed design.

building sections
cutaway views through a building

Reflected Ceiling Plans

Plan views of a ceiling usually found in the architectural section of a contract drawing package and provided in the contract drawings when suspended ceilings are installed are called **reflected ceiling plans**. A reflected ceiling plan is likely to be one of the most valuable drawings in a contract drawing package for the design of a fire protection system in a building with suspended ceilings, because this plan provides the requisite data to be used as the basis for such a fire protection drawing.

reflected ceiling plan
a plan view of a suspended ceiling

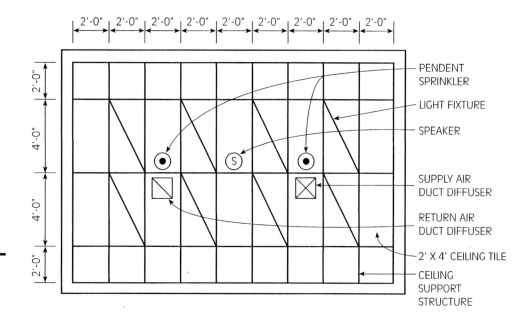

Figure 1-7 *A reflected ceiling plan for an office.*

Detailed on a reflected ceiling plan is the acoustical tile layout, lighting locations, air diffuser locations, skylights, and partitions that extend to the suspended ceiling. A sample reflected ceiling plan of an office is shown in Figure 1-7.

Sometimes sprinkler head locations are shown on a reflected ceiling plan. Be very careful about tracing or scanning these sprinkler head locations. Be certain to check these locations for compliance to NFPA requirements relative to the proper spacing of sprinklers, as detailed in Chapter 5. One cannot be successful in using a noncompliant sprinkler layout on a reflected ceiling plan as justification for failure to comply with NFPA spacing provisions. The best tactic is to use NFPA requirements as the basis for the spacing of sprinklers, as opposed to sprinkler locations shown on a reflected ceiling plan.

For most small rooms, the acoustic tile arrangement is usually symmetrical, making the spacing of sprinklers fairly straightforward. When the logic chosen for the arrangement of a reflected ceiling plan is not immediately evident, dimensions should be provided on the reflected ceiling plan for the location of the starting points for the ceiling grid system. Sprinkler spacing and lighting fixture spacing often coincide, requiring that sprinkler locations be carefully coordinated to avoid lighting fixtures, sometimes meaning that extra sprinklers are needed, and usually meaning that sprinklers are prevented from being installed in their ideal locations or in the exact centers of rooms.

For larger rooms, especially rooms with irregularly shaped walls, determining the exact location of acoustic tile runners or support structure can be a difficult exercise. Many such areas center the tile patterns with respect to building structural columns. Another ceiling alignment tactic is to arrange tile patterns to align to the centers of a point of reference, such as a skylight or corridor. If the starting point for tile installation is not immediately evident or not specifically noted on the reflected ceiling plan, a coordination meeting, as discussed in Chapter 12, is needed to determine the ceiling tile starting points.

Finish Schedule

A contract drawing, contained within the architectural drawing set, that lists all rooms in the building by room number, with details of several room features designated is called a **finish schedule**. Design details related to the floors, walls, and ceilings are included for each room.

Of particular interest to the fire protection designer are the architectural features related to the ceiling. The finish schedule indicates whether the room is exposed or whether the room has a suspended ceiling. The height of the ceiling is listed for each room. Ceiling construction, such as plaster, exposed, or acoustic tile, forms an important component of the schedule, and may affect the manner in which you approach the design of a particular area.

The finish schedule and the reflected ceiling plan are essential points of reference for buildings containing suspended ceilings.

STRUCTURAL DRAWINGS

For buildings without suspended ceilings, **structural drawings**, drawings that provide details related to the floors, roof, and structural elements, should be the central reference for fire sprinkler design. The structural set consists of foundation drawings and framing plans, and are identified with an "S" drawing number.

Foundation Plans

The **foundation plans** show finished floor structural details and sectional views with which one must coordinate when designing underground entries into the building. On multistory buildings, the foundation plans detail floor slab thicknesses and show building elevations that may contain some of the needed details not shown on the architectural elevations.

Framing Plans

An **I-beam** is a solid steel member that looks like an "I" from its end, and a **bar joist** is a webbed member supported by the I-beams, as shown in Figure 1-8. **Framing plans** show I-beam and bar joist size and elevation. Tops of I-beams are referenced to a baseline elevation, such as finished floor, or the reference elevations used for the reference grid and profile plans. Sloping of I-beams can be deduced by noting the change in the elevation of the tops of the beams, noted on the framing plans.

Since bar joists rest on top of the I-beams, a **bar joist bearing dimension**, or the depth of the top of a joist, must be ascertained in order for the elevation of the

---

finish schedule
a contract drawing that lists all rooms in a building and provides details of several room features

structural drawings
drawings that provide details related to the floors, roof, and structural elements of a building

foundation plans
plans that show floor and wall structural details and sectional views

I-beam
a solid steel member that looks like an "I" from its end

bar joist
a webbed member supported by I-beams

framing plans
plans that show beam and joist size and elevation

bar joist bearing dimension
the depth of the top of a joist

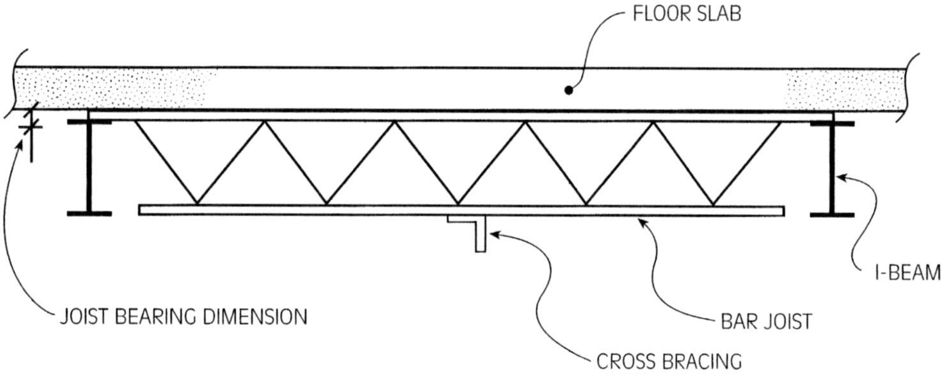

Figure 1-8 *Partial building sectional view showing structural components.*

[Diagram labels: FLOOR SLAB, I-BEAM, JOIST BEARING DIMENSION, BAR JOIST, CROSS BRACING]

top of the bar joist to be determined. This information can usually be found either in the structural general notes or on structural sectional details. The elevation of the top of the bar joist is of importance for the fire protection designer because it is from this point that sprinkler piping is supported.

If the joist bearing dimension is not shown, it can be obtained in a book of standard steel dimensions that is printed by most manufacturers of structural members. Should all else fail, the joist bearing dimension could be obtained in a field survey. Many steel bar joists have a joist bearing dimension of 2½", but this dimension could increase if the joists must support a heavy load, such as a roof heating, ventilating, and air conditioning (HVAC) unit or cooling tower, or for a roof in areas with unusually heavy snowfalls.

Other information that should be obtained from the structural drawings is the depth and location of **cross bracing**. This bracing is usually angle iron that provides structural rigidity between bar joists, usually to counteract wind load or earthquake seismic load. The location of this bracing could conflict with sprinkler pipes if the bracing is installed within the joists, and with sprinkler mains if the bracing is attached to the bottom of the bar joists. On the framing plans, a fire protection designer can find essential information needed for the design of the fire protection system, such as I-beam depth and bar joist depth.

cross bracing
supports to provide structural rigidity between bar joists

Branch Line Orientation

For projects that contain few suspended ceilings where the piping is to be installed primarily exposed, the sprinkler designer should trace the framing plan and use it as the basis for the fire protection layout. Sprinkler system **branch lines**, the pipes that have sprinklers installed directly on them, are oriented to run perpendicular to the bar joists for ease of hanging.

branch lines
sprinkler pipes that have sprinklers installed directly on them

Using Framing Plans to Dimension a Sprinkler System

The centers of the building columns should be the points of reference for all fire protection piping dimensions. This advice pertains to both exposed construction and concealed construction with suspended ceilings. A good practice for fire protection designers to follow is to have three lines of running dimensions in each direction, referenced to the structural members, as shown in Figure 1-9:

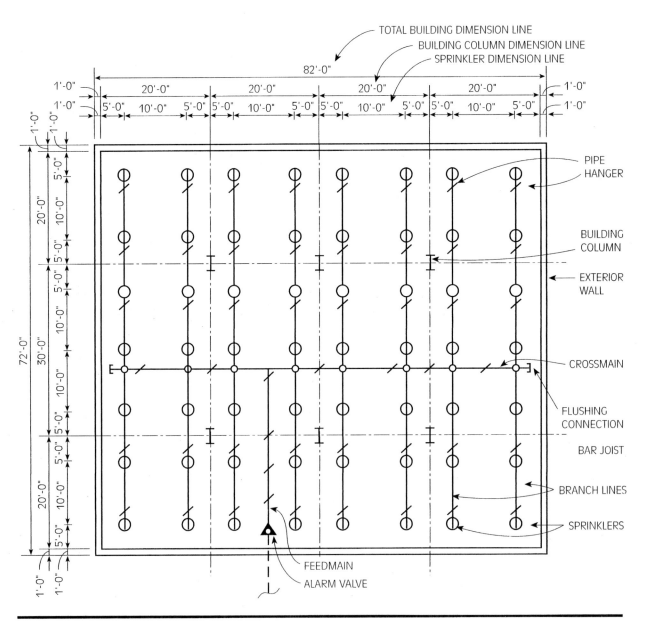

Figure 1-9 *Plan view showing dimensioning guidelines for a sprinkler system. Sprinkler branch lines always run perpendicular to bar joists.*

1. The total building dimension line is outside wall to outside wall. Of the three lines of dimensions, this line of dimensions is placed the furthest from the building wall.
2. A second line of dimensions is the building column dimension line, taken to the centers of each building column.
3. The third, or innermost line of dimensions, should be the sprinkler dimension line, which shows dimensions of all sprinklers and sprinkler pipes with respect to the building column centerlines.

Before submitting the drawing for approval and again before releasing the drawing for fabrication, the three lines of dimensions should be re-added and compared to dimensions shown for the piping system until all dimensions are in total agreement. Simple errors in addition of dimensions can be very costly but can be quickly isolated and corrected using this method. A feet-and-inch calculator is an essential time saver in performing this vital task.

HVAC DRAWINGS

There are two types of heating, ventilating, and air conditioning drawings, or **HVAC drawings**, identified with an "M" drawing number: (1) conceptual HVAC drawings, and (2) shop HVAC drawings.

HVAC drawings
heating, ventilating and air conditioning drawings

The most common type of HVAC plan, found in most contract drawing packages, is the conceptual HVAC drawing. These plans are ordinarily not dimensioned to show exact location of HVAC ductwork with respect to the steel beams and columns, but they do show the widths and depths of the ducts. The lack of dimensions and elevations of the HVAC duct can create conflicts with sprinkler piping unless the sprinkler designer initiates a coordination meeting with the HVAC contractor as discussed in Chapter 12.

A great deal of information can, however, be obtained from a conceptual HVAC drawing. A major difference among ductwork designs relates to the manner in which the ducts reduce in size. Ordinarily ducts are either flat on top, with ducts reducing in size on the bottom plane, or flat on bottom, with ducts reducing in size from the top plane, as shown in Figure 1-10. The HVAC plans or general notes should have a statement relative to the reduction strategy.

The other type of HVAC plan is the shop drawing or sheet metal drawing. In a manner similar to the creation of a sprinkler shop drawing, an HVAC designer uses a conceptual HVAC plan to derive a fully dimensioned and elevated duct drawing, in which all ducts are dimensioned to the building columns and all supply and return diffusers are dimensioned. These plans may appear somewhat complicated, because they contain additional information that the conceptual HVAC plans do not show.

Sprinkler designers use sheet metal drawings to derive a considerable amount of information with knowledge of the chosen reduction strategy. Sheet metal drawings are usually very well dimensioned and elevated. If the ducts are

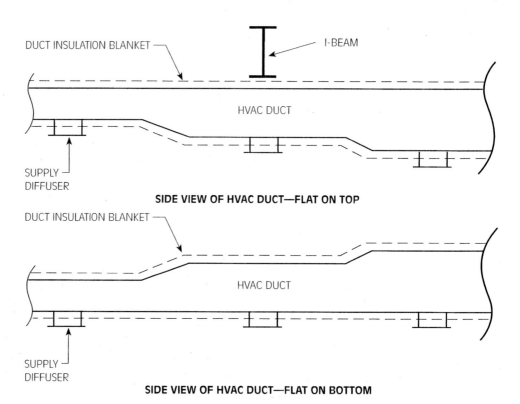

Figure 1-10 *Side views of an HVAC duct.*

flat on top and if a uniform elevation to top of duct is given, the bottom elevations can be determined. The flat-on-top arrangement usually provides more space for sprinkler pipes running below the ducts. In very tight spaces, sprinkler pipes may only be able to cross a duct at its smallest dimension, after the point where the duct reduces in size.

If no elevation to the top or bottom of the duct is given, a coordination meeting is necessary. Usually, ductwork that is flat on top is elevated no closer than 2 to 3 inches below the lowest steel beam that the duct must cross. The space between the duct and a beam is needed to allow for clearance required for the ductwork structural ribs and the ductwork insulation blanket.

If a duct is flat on bottom and space is at a premium, a considerable number of conflicts between HVAC and sprinkler piping may arise. The elevation of the bottom of the duct is a function of the lowest steel beam that the duct must cross and the largest duct dimension crossing that beam. A duct of considerable depth that must cross deep beams may take up most of the space above a suspended ceiling. The sprinkler piping must either be installed above the duct, or must rise over the duct in all areas of conflict. A sprinkler contractor may be able to persuade an HVAC contractor to modify duct locations or installation strategy in cases of severe conflict.

supply diffuser
ceiling element used to distribute fresh air to a room

return diffuser
ceiling element used to draw stale air from a room

plenum space
a space above a suspended ceiling that is kept under negative pressure for return air

Another important function of conceptual HVAC drawings is to show supply and return diffuser locations. A **supply diffuser** distributes fresh air to a room, and a **return diffuser** draws stale air from the room.

Supply diffusers are connected by a system of supply ductwork. For HVAC systems, the rate of air flow is a determined by the duct size, the duct friction coefficient, and the diffuser size.

Return diffusers may be connected to ductwork that draws air out of the rooms of a building and exhausts it outside. An alternate return air method consists of return air grilles mounted to the ceiling and a **plenum space** above the ceiling. The plenum space is the volume of air encountered above the ceiling. This air volume is kept under negative pressure by a fan, with air drawn through the return air grille into the plenum space and exhausted through the fan.

If a contract drawing package contains both an HVAC plan and a reflected ceiling plan, the supply and return diffusers may be shown on the reflected ceiling plan. Conflicts between the HVAC plan and reflected ceiling plan must be brought to the attention of the general contractor immediately. The locations shown on the reflected ceiling plan may take precedence over locations shown on a conceptual HVAC plan.

Working with duct shop drawings is, in the long run, much easier than working with conceptual plans, because much of the coordination has already been done. Your job is to identify locations where sprinkler piping crosses ductwork and to verify that sufficient clearance exists between HVAC ducts and sprinkler pipes. Areas of irreconcilable conflicts must be resolved in a coordination meeting.

PLUMBING DRAWINGS

A sprinkler designer should refer to the plumbing drawings for several reasons, but mainly because in some buildings, plumbing and sprinkler piping involve a common feed pipe, supplying both the domestic water system and the sprinkler system. The point of connection for fire protection water for such an arrangement may be shown on the plumbing plans.

Coordination between sprinkler piping and plumbing piping at this point of connection is an absolute necessity. In jurisdictions requiring water meters and/or backflow prevention devices, the method of connection between sprinkler and plumbing piping may require that the arrangement shown on the plumbing drawings be revised to comply with the requirements of the jurisdiction.

Even when plumbing and sprinkler piping are fed by separate piping runs into a building, fire protection riser details are often shown on the plumbing plans. A fire protection riser schematic, a fire department connection location, and the underground entry piping may be shown.

In some cases, a fire protection system piping layout or schematic may also be shown on the plumbing drawings, especially in the case of a high-rise building requiring a standpipe system. Water supply information might be found in the plumbing general notes.

Any fire protection features described by the plumbing plans or plumbing general notes should be closely studied. Conflicts between fire protection references and the specification, fire protection codes and standards, or the requirements of the authority having jurisdiction, must be brought to the attention of the general contractor.

Piping carrying plumbing water is also shown. Only in very rare cases will this piping be dimensioned with respect to the building columns. The sprinkler designer must recognize two varieties of plumbing piping shown on a plumbing drawing, buried and exposed pipe.

Buried plumbing pipe is installed in the floor slab. Such piping is most likely to be floor drains or toilet and sewer piping. **Rain leaders** and **domestic plumbing piping** are probably exposed below the ceiling slab or concealed between a suspended ceiling and the ceiling slab. Rain leaders are rainwater drainage piping, and domestic plumbing piping is a water supply for drinking water and toilets.

A rain leader is often given a slope that is specifically noted on the plans. Sometimes an elevation is given for the high point of all rain leaders, simplifying coordination. Undimensioned or unelevated rain leaders, usually 4- or 6-inch in diameter, should have their locations and elevations determined in a coordination meeting.

Domestic piping is ordinarily considerably smaller in diameter than rain leaders, usually 1/2-inch to 2-inch diameter. Coordination with exposed domestic piping is always a very good idea, since this piping is usually installed well before sprinkler piping arrives on the job site.

FIRE PROTECTION CONTRACT DRAWINGS

On larger projects, a set of contract drawings may include a set of preliminary fire protection plans. If such a drawing category is included, information that can be obtained might be the underground entry into the building, a riser schematic, a fire department connection location, and a floor plan containing some requirements for fire protection piping or sprinkler location.

If the building has a suspended ceiling, a reproduction of the reflected ceiling plan might be included, with sprinkler locations shown. In exposed structures, a piping layout may be shown. When estimating a sprinkler layout, check to see if the piping layout shown is perpendicular to the bar joists. Many such plans show sprinkler branch lines parallel to the bar joists, making hanging of the branch lines impossible or extremely impractical. The piping should be reoriented such that the branch lines run perpendicular to the bar joists. The feed main arrangement is also affected by such a discrepancy.

Some fire protection contract drawings contain only the minimal details necessary to begin a fire protection piping layout. It is not unusual to find a fire protection concept drawing with only the underground entry pipe shown, and per-

---

rain leaders
rainwater drainage piping

domestic plumbing piping
piping that supplies water fountains, sinks, and toilets

ELECTRICAL DRAWINGS

Information on electrical drawings of use to sprinkler designers centers upon lighting and device locations and power requirements.

Lighting fixture locations are shown in this section of the contract drawings. In exposed areas, the fixture location and type can be ascertained. The hanging method of the exposed light fixtures should be determined. **Flush mounted lights**, or lights that are mounted below the ceiling and flush to the ceiling, depending upon their depth, could interfere with the sprinkler water spray distribution below the ceiling. **Recessed fixtures**, or lights whose faces are flat to the ceiling, but are recessed to protrude into the ceiling space, could interfere with branch lines, if the branch lines are installed in the joist space, and with feedmains or crossmains, if mounted close to the ceiling. **Suspended fixtures**, or lights suspended on rods or chains, could be a source of piping or sprinkler water spray distribution conflict, depending on the elevation and dimension of the light. If the lighting dimensions, height, or mounting method are not shown, a coordination meeting is needed.

Fixture layouts for rooms with suspended ceilings are likely to correspond to the fixture locations shown on the reflected ceiling plans. Discrepancies must be brought to the attention of the general contractor.

Locations of other electrical devices can also be shown. Included might be some equipment associated with the sprinkler system, such as **water flow switches**, that give electrical indication of water movement in a sprinkler pipe, **tamper switches**, that give electrical indication of a sprinkler valve closure, **solenoids**, that are used to open and close electrically actuated fire protection valves, **water level switches**, that indicate the level of water in a fire protection water storage tank, and **pressure switches**, that give electrical indication of high or low pressure in a water tank or in a sprinkler system. Careful coordination with these locations is essential. Should it be necessary to install these devices in locations other than those specified by the electrical drawings, the changes must be confirmed in a letter to the general contractor and approved before installation of the fire protection system. Failure to do so could result in an additional charge by the electrical contractor.

Coordination between the power availabilities shown on the plans and the power requirements of electrical fire protection devices is extremely important. If a fire pump is required, it must be ordered in such a way that congruence of the power to the controller and motor is ensured. Other fire protection devices, such as control panels, air compressors, flow and tamper switches, and alarm devices require the same level of coordination to ensure applicability to a fire protection system in a particular building.

---

flush mounted lights
lights that are mounted below and flush to the ceiling

recessed fixtures
lights whose faces are flat to the ceiling and whose bodies protrude into the ceiling space

suspended fixtures
lights suspended on rods or chains

water flow switches
electronic devices that indicate water movement in a sprinkler pipe

tamper switches
electronic devices that indicate a sprinkler valve closure

solenoids
switches used to open and close electrically actuated fire protection valves

water level switches
devices that indicate the level of water in a fire protection water storage tank

pressure switches
electrical devices that indicate high or low pressure in a water tank or in a sprinkler system

Electrical riser diagrams may detail the locations of electrical devices associated with a fire protection system, especially in high-rise buildings. Careful study of the electrical drawings could avoid situations that may delay the job. A case in point could be a fire pump room with no power shown running to it, or an array of system subdivision control valves and flow switches required for each floor of a building with no power shown running to them. The sprinkler contractor should notify the general contractor in such cases.

Summary

The contract documents consist of the contract drawings and the specifications. The contract drawing package contains the cover sheet, site drawings, architectural plans, reflected ceiling plans, the finish schedule, structural drawings, HVAC drawings, plumbing drawings, fire protection contract drawings, and electrical drawings.

It is vitally important that all components of the contract drawing package be used to minimize design or installation conflicts or omissions that could negatively affect the performance of a fire protection system. A fire protection designer must assume the serious responsibility of obtaining all pertinent information and studying the effect of this information on the proper activation and operation of a fire protection system.

Review Questions

1. List the items contained within the contract documents.
2. List and explain the value of each of the categories of drawings that comprise the contract drawings.
3. Give examples of information that would be missed if only the plumbing drawings were made available to the fire protection system designer.
4. Describe what could happen if a fire protection system designer neglected to take full advantage of the wealth of information available in a complete set of contract drawings.
5. Explain the difference between a location map and a vicinity map, a reference elevation and a reference grid, a renovation architectural drawing and a new-building architectural drawing, a plan job and a survey job, a return diffuser and a supply diffuser, and a finish schedule and a reflected ceiling plan.
6. Identify the NFPA standard that should be used as reference for fire protection symbols.

Activities

1. Interview one or more experienced fire protection designers to examine their experience with contract drawings. What contract drawings do they ordinarily receive when they are assigned a project? What methods do they use to obtain a complete set of contract drawings? What difficulties have they had in obtaining a complete set of contract drawings? What experiences have they had when using an incomplete set of contract drawings?
2. Obtain or borrow a complete set of contract drawings from a sprinkler contractor, builder, architect, or engineer. Examine and compare the drawings contained within the set. Are any drawings discussed within this chapter not present? Would the absence of any of the

drawings discussed negatively affect the proper performance of a fire protection design?

3. Contact a sprinkler contractor and determine the reasons why a sprinkler contractor might be given an incomplete set of contract drawings by a builder or architect.

4. Obtain a reflected ceiling plan. Does the plan show dimensioned locations for the ceiling tiles? If not, is the location of the tiles centered in rooms or centered with respect to column lines?

Chapter 2

Contract Specifications

Objectives

Upon completion of this chapter, you should be able to:

- List the sixteen divisions of the contract specifications.
- Explain the relationship between a set of contract drawings and the contract specifications.
- Determine the divisions of the specifications that should be of most value to the development of a fire protection system design.
- Identify the problems that could occur if the contract specifications conflict with the contract drawings.
- Explain the limitations that could be placed on a fire protection designer if only the specification section directly applicable to a fire protection system were supplied.
- In cases of conflict between the contract drawings and the specifications, determine whether the specifications or the contract drawings usually take precedence.
- Write a specification specifically tailored for a project, using the sample specification in this chapter.

Contract specifications, along with the contract drawings, comprise the contract documents. The specifications are usually issued in book form, but for smaller projects they are sometimes printed on the contract drawings. Understanding the specifications and their relation to the contract drawings is essential to ensuring a successful fire protection design.

SPECIFICATION TYPES

Computer Specification Database

Architects and engineers develop specifications from either a standardized computer specification database or from a revision to the specifications developed for a previous project of similar scope. This revision process sometimes results in the inclusion of provisions in the specifications for a project that may conflict with the contract drawings.

A fire protection designer must carefully review the set of contract documents by comparing the contract drawings to the specifications. Any conflicts that may exist must be revealed and resolved before beginning the design of a fire protection system.

Contract specifications are a functional component of the contract documents, no less important than the contract drawings. In most cases, the specifications prevail when a conflict arises between contract drawings and the specifications, even when an extremely detailed set of contract drawings is compared to a broadly general set of specifications. The best approach is to discover and correct discrepancies before the system is designed and installed in order to prevent major changes to the sprinkler system in the field.

Many sprinkler contractors receive or request only the section of the specifications that directly relates to fire protection systems. This oversight can dramatically increase the probability of a conflict between the contract drawings and the specifications. The sprinkler contractor is legally responsible for compliance with all sections of the specifications and all drawings in the contract documents.

performance specification
the minimum information necessary to estimate, design, and install a fire protection system

detailed specification
in-depth requirements for the design of a fire protection system, such that very little latitude is available for interpretation or alternative design proposals

Performance Specifications and Detailed Specifications

A specification writer selects one of two methods to develop specifications for a fire protection system. A **performance specification** is a general specification that provides the minimum information necessary to estimate, design, and install a fire protection system. A performance specification contains broad requirements that offer few restrictions to a sprinkler contractor, allowing for the provision of economical or innovative design methods that comply with the applicable codes and standards.

A **detailed specification** provides in-depth requirements for the design of a fire protection system, such that very little latitude is available for interpretation or implementation of alternative design proposals. A specification writer who selects this method may see a heightened fire scenario or a specific application that may not be adequately addressed by a performance specification. A detailed specification may exclude or prohibit the use of newer or innovative piping or sprinkler technologies. Variations from the fire protection concept outlined in a detailed specification should be proposed by the sprinkler contractor in advance of the submission of a cost estimate for the system.

There is a growing consensus among architects and engineers that the performance-based specification is preferred for new buildings. Many building owners find that performance-based specifications provide a less expensive fire protection system when detailed requirements are removed and the latest technologies are employed.

SPECIFICATION FORMAT

Most specifications follow a format standardized by the Construction Specifications Institute (CSI), consisting of sixteen **specification divisions**, or broad categories of building component groupings. Each division consists of **specification sections** that outline the detailed requirements of each division. The sixteen divisions provide a road map for the building of a structure and its internal components to meet the needs of the owner and to comply with the applicable codes and standards.

> **specification divisions**
> broad categories of building component groupings
>
> **specification sections**
> detailed requirements for each division

Division 0—Instructions to Bidders

Division 1—General Requirements

Division 2—Site Work

Division 3—Concrete

Division 4—Masonry

Division 5—Metals

Division 6—Wood and Plaster

Division 7—Thermal and Moisture Protection

Division 8—Doors and Windows

Division 9—Finishes

Division 10—Specialties

Division 11—Equipment

Division 12—Furnishings

Division 13—Special Construction

Division 14—Conveying Systems

Division 15—Mechanical

Division 16—Electrical

Although all divisions may not apply and may not be included in a specification for a proposed building, most divisions of a specification contain information of interest to the sprinkler designer or contractor.

Division 0: Instructions to Bidders

Division 0 is primarily for use by a person responsible for the development of a cost estimate of a sprinkler system, but some information of interest to the designer can be found here. In many cases, addresses and phone numbers of key people are provided in this section. Requirements related to scheduling submittal of shop drawings and the project construction schedule may also be included.

Division 1: General Requirements

One of the most important divisions in the specifications, Division 1 contains information that applies to all contractors installing mechanical systems within a building, including the sprinkler contractor. The division is divided into several sections, many of which are vital to the successful installation of a fire protection system.

The summary of work, or **scope**, describes what is covered by the contract documents and what is not. If it is a job requirement that progress meetings or coordination between trades must take place, it should be stated here.

A requirement for **coordination meetings**, meetings where the exact locations of all mechanical equipment for a particular area are determined before installation, should make it easier for you to get cooperation from the other trades. A job progress schedule is found in this division, and if applicable, a requirement for the payment of **liquidated damages**, which are monetary fines for failure to perform completion of work within a specific time frame.

Division 2: Site Work

Division 2 applies only if underground piping is included in the contract. Specific requirements for excavating, trenching, backfilling, and the laying and support of underground piping are included. Any discrepancies between this division and the requirements of the authority having jurisdiction should be noted.

If a requirement for **cathodic protection**, an electronic method for combatting pipe corrosion, applies, then the procedures and minimum criteria are outlined in this division.

Any specific requirements for underground vaults for fire protection valves are included, and if a disinfection procedure for underground piping applies, it is also found in this division.

Divisions 3–6: Concrete, Masonry, Metals, and Wood and Plaster

These divisions apply to a fire protection contract only if an underground vault, valve house, or auxiliary building, was required for the protection of fire pumps or equipment. Should subcontractors be employed to perform such construction, the applicable divisions must be included as a contractual requirement of the subcon-

---

scope
part of a specification that describes what is and what is not covered by the contract documents

coordination meetings
meetings where the exact locations of mechanical equipment are determined before installation

liquidated damages
monetary fines for failure to perform completion of work within a specific time frame

cathodic protection
an electronic method for combating pipe corrosion

tract. If new or existing buildings require penetrations by fire protection pipe, these sections should be referenced for repair or patching of the floors, walls, or roof.

Division 7: Thermal and Moisture Protection

firestop systems
sealants for fire protection piping penetrations through walls or other building components

Firestop systems that may be required to seal fire protection piping penetrations through walls or other building components are outlined in this division. Moisture protection, usually involving standpipe penetrations through roofs for roof hose headers, have requirements delineated in Division 7.

Divisions 8 and 9: Doors and Windows, and Finishes

Any doors or windows required for a valve or pump enclosure must comply with these divisions. The finish division applies if floor coverings or special wall surfaces are required. Requirements for painting of sprinkler piping or equipment are referenced to Division 9.

Divisions 10–12, 14: Specialties, Equipment, Furnishings, and Conveying Systems

The types of equipment described by these divisions are unlikely to be supplied under a sprinkler contract. Reference may need to be made, however, if protection of these items is required. Examples of such items are: elevator and escalator protection (Division 14), toilet partition dimensions and heights (Division 10); saunas, cold storage rooms, or x-ray rooms (Division 13); compactors, food service, cooking, or washing (Division 11).

Division 13: Special Construction

The standardized specification format has been revised to accommodate fire protection system and fire alarm system specifications within Division 13, which represents a significant departure from previous practice. Before this change, fire suppression systems were specified in Division 15, the mechanical division, while fire alarm systems were detailed in Division 16, electrical.

By moving these two separate specification provisions into Division 13, section 13900, entitled Fire Suppression and Protection Systems, and section 13850, Fire Alarm and Detection Equipment, better coordination and reduced conflict will result by perhaps having one contractor perform or be responsible for both suppression and detection services, or by having two contractors being subcontracted by the same general contractor or mechanical contractor. Under the previous format, coordination of fire protection components was considerably more difficult when the sprinkler contractor reported to a mechanical contractor under Division 15, and the fire alarm contractor reported to the electrical contractor under Division 16. Provisions for fire protection piping, valves, specialties, and standpipes continue to be included in section 15300, and cross-references are made between the two divisions.

A complete list of sections that may be of interest to a fire protection designer under the new specifications format:

Section 13700—Security Access and Surveillance includes intrusion detection and security systems.

Section 13800—Building Automation and Control includes vital information relative to elevator control systems, that are used when a requirement exists for the shutdown of elevators by the fire detection system.

Section 13850—Detection and Alarm is essential to every fire protection systems designer, because it includes alarm and occupant warning systems, and fire, smoke, and gas detectors.

Section 13900—Fire Suppression and Protective Systems includes standpipe systems in sprinklered buildings. All fire protection systems designers should reference this section.

Section 13910—Sprinkler Systems includes wet, dry, preaction, water spray, and deluge fire protection systems.

Section 13950—Extinguishing Systems includes carbon dioxide systems, clean agent, dry chemical, foam, halon, and water mist systems.

Division 15: Mechanical

In the past, this division had the greatest effect on sprinkler system design. With the revision of placing suppression and alarm systems into Division 13, Division 15 only applies when coordinating with other mechanical systems, or when an older specification format is used. Engineers and architects not using the latest revision of the standardized specification format may continue to specify fire protection systems in this division. The following paragraphs apply in these cases.

A section that sadly is too often ignored by fire protection system designers is the **mechanical general requirements**, section 15100. This section includes requirements applying to all mechanical trades, such as submittal procedure, applicable documents, tests, and drawings. Section 15140 gives general information on hanging of piping that may not appear in the sprinkler section.

Section 15300 applies directly to fire protection systems under the old specification format:

Section 15310 outlines requirements for fire protection piping.

Section 15320 covers fire pumps.

Section 15325 details sprinkler system requirements.

Section 15365 outlines requirements for carbon dioxide or other special agent systems.

Section 15375 specifies requirements for standpipe and hose systems.

mechanical general requirements
requirements applying to all mechanical trades

The sectional numbering scheme listed above has been standardized by the Construction Standards Institute, but some specifications combine sections or employ an alternate numbering scheme for sections within Division 15.

Under the new specification format, section 15300 continues to be entitled Fire Protection Piping, and cross-references are made to sections 13850 and 13900.

It is unfortunate that quite often the only specification section that is distributed to fire protection system design professionals is the section that applies to sprinkler systems. So many other sections and divisions apply to fire protection work that an extremely limited, and perhaps misleading, viewpoint of fire protection requirements could be obtained.

Division 16: Electrical

Division 16, covering electrical requirements, begins with general electrical requirements, then covers specialized electrical installations and applications.

The new specification format has moved fire alarm and detection systems to section 13850, as previously described. Some engineers and architects may continue to use the older format, in which case fire protection wiring for pumps and switches would have applicable sections in Division 16 that include conduit, wiring, junction boxes, cabinets, grounding, lightning protection, and fire alarm and detection systems.

SAMPLE SPECIFICATION

A sample specification, using the new 13910 specification series, is shown as Figure 2-1. This sample specification provides fire protection professionals with an outline of what a specification may look like and may be used by engineers as a basis for a specification of a wet pipe sprinkler system.

Caution is urged in the modification of an existing specification, because serious problems could ensue if the specification is not properly and thoroughly edited to conform to the specific requirements of a particular project.

SECTION 13910
SPRINKLER SYSTEMS

PART 1—GENERAL

1.1 SCOPE

A. The fire protection system supplied under this specification is a new wet pipe sprinkler system designed and installed for all areas in the building covered by this contract, as shown on the contract drawings, and shall be hydraulically calculated, complete in all respects, and ready for operation.

1.2 RELATED WORK

A. All electrical work required for electrical devices included in this specification is included in the scope of work.

B. Sealing of all penetrations of walls is included in the scope of this contract.

C. All drawings, calculations, and submittals required by this specification are included in the scope of work.

1.3 QUALITY ASSURANCE

A. The fire protection contractor shall institute a system of quality assurance to assure that the design, installation, materials, inspection, and testing of all components are in accordance with the provisions of this specification and the applicable codes and standards listed in paragraph 1.6. The quality assurance program shall require that all drawings, calculations, and other submittals are checked by a supervisor possessing a minimum of NICET level III certification, and all submittals shall be signed by the supervisor who checked the work.

B. Hydraulic calculations are to be based upon the following criteria, and in accordance with NFPA 13:

  1. All areas classified as light hazard occupancy are to be hydraulically calculated to assure that 0.10 gpm per square foot is applied over the most remote 1500 square feet of area.

  2. All areas classified as ordinary hazard occupancy are to be hydraulically calculated to assure that 0.20 gpm per square foot is applied over the most remote 1500 square feet of area.

  3. All areas in the building are classified as light hazard occupancy, except for the following areas which are to be designed in accordance with ordinary hazard occupancy:

     a. mechanical rooms
     b. elevator machine room
     c. storage rooms
     d. garage
     e. serving pantry
     f. vending
     g. stage

C. Provide sprinklers in shafts, such as elevators and stairwells, in accordance with NFPA 13 and as shown on drawings FP-1, FP-2, and FP-3.

D. In addition to the sprinkler system demand, 250 gpm outside hose demand shall be added to the hydraulic calculations at the connection to the street water main.

E. A water flow test was performed on April 8th 1996, 10:00 A.M., and the results are as follows:

  a. Static pressure is 80 psi.
  b. Residual pressure is 70 psi.
  c. Flow at the residual pressure is 1030 gpm.
  d. Test hydrant elevation is 430.2'.
  e. Flow hydrant elevation is 435.1'.
  f. Finished floor elevation is 443.2'.
  g. Hydrant locations are shown on contract drawing FP-1.

F. Drawings FP-1, FP-2, and FP-3 show the positions of sprinklers that accommodate the proposed architectural renovation of the building. The new sprinkler system installed under this contract is to be designed to meet all codes, standards, and the performance objective and requirements of this specification. Drawings FP-1, FP-2, and FP-3 show piping and sprinkler locations that meet this performance criteria.

G. All devices used on this system shall be listed by Underwriters Laboratories in the UL Fire Protection Equipment Directory.

H. Cutout disks created by cutting holes in a pipe for welded connections or water flow switches, shall be recovered and wired to the pipe near the point of cutting. All

Figure 2-1 *Sample specification.*

disks shall be accounted for to assure that no blockage of the pipe would occur as the result of the displacement of these disks.

1.4 QUALIFICATIONS

A. Installation of the entire sprinkler system is to be performed by a sprinkler contractor with a minimum of five years of design and installation experience with sprinkler systems similar to the one specified under this contract. Evidence of such experience must be submitted with the bid for this project.

B. Layout to be performed by a person working under the direct supervision of a person with a minimum NICET III certification in Automatic Sprinkler System Layout. Evidence of such experience must be submitted with the bid for this project.

C. Quality Assurance checking of the completed design, performed in accordance with Paragraph 1.3A, is to be performed by a person other than the person who performed the layout, possessing a minimum NICET certification of level III in Automatic Sprinkler System Layout. Evidence of such experience must be submitted with the bid for this project.

D. Quality Assurance checking of the completed installation, performed in accordance with Paragraph 1.3A by the person who performed the Quality Assurance review of the completed layout, shall possess a minimum NICET certification of level III in Automatic Sprinkler System Layout. Evidence of such experience must be submitted with the bid for this project. The completed installation is to be compared to the completed design for completeness and correctness. Any deviations from the design are to be corrected, and the individual certifying the system shall sign the contractor's material and test certificate.

1.5 SUBMITTALS

A. All submittals required by this specification shall be submitted simultaneously as one package.

B. All submittals shall be reviewed and signed in accordance with paragraphs 1.3A and 1.4C.

C. Catalogue cut submittals shall be bound, and each component submitted shall reference the applicable paragraph number from part 2 of this specification, and shall represent all components proposed for the system, including:
  1. pipe and fittings
  2. valves
  3. ball drips
  4. all models and types of sprinklers on the project
  5. inspector's test connection detail
  6. sprinkler cabinets
  7. pressure and flow switches
  8. pipe hangers
  9. backflow preventer
  10. alarm valve and retard chamber
  11. fire department connection
  12. pipe penetration sealing system
  13. sleeves
  14. water motor gong
  15. tamper switches

D. Detailed shop drawings of the system are required, and shall include all information required by NFPA 13 paragraph 6.1, and shall be in complete compliance with all documents listed under paragraph 1.6 of this specification.

E. Hydraulic calculations of all remote areas of the system are required, and shall include calculations for at least one hydraulically most demanding light hazard area and at least one ordinary hazard area. Calculations shall include all information required by NFPA 13 paragraph 6.2 and shall be in complete compliance with all documents listed under paragraph 1.6 of this specification.

F. All required certificates of compliance shall be submitted, including:
  1. A certificate confirming compliance to paragraph 1.3, quality assurance.
  2. A certificate confirming compliance to paragraph 1.4A, experience qualifications.
  3. A copy of the contractor's material and test certificate.
  4. A copy of the NICET III certificate for the supervisor of the layout technician in accordance with paragraph 1.4B.

Figure 2-1 *(Continued)*

5. A copy of the NICET III certificate for the person checking the design work in accordance with paragraphs 1.3A and 1.4C.

6. A copy of the NICET III certificate for the person checking the completed installation in accordance with paragraphs 1.3A and 1.4D.

G. As-built drawings and as-built hydraulic calculations of the completed system are required.

H. Operation and maintenance manuals for the completed system are to be enclosed in a waterproof pouch and attached to the system riser at the alarm valve, and shall include:

1. As-built drawings and as-built hydraulic calculations of the completed system.

2. A set of catalogue cuts for all devices, including complete operating instructions and parts ordering information.

1.6 APPLICABLE DESIGN AND INSTALLATION DOCUMENTS

A. NFPA 13 *Standard on the Installation of Sprinkler Systems.*

B. The most recent edition of the county regulation on the fire safety code for fire protection systems.

C. NFPA 24 *Standard on the Installation of Private Fire Service Mains and Their Appurtenances.*

D. NFPA 72 *National Fire Alarm Code.*

E. *Underwriters Laboratories Fire Protection Equipment Directory.*

F. NFPA 70 *National Electric Code.*

G. Contract drawings FP-1, FP-2, and FP-3.

PART 2—PRODUCTS

2.1 PIPING

A. All wet pipe system sprinkler piping shall be in accordance with NFPA 13 and listed by the American Society for Testing of Materials.

2.2 VALVES

A. All control valves shall be UL listed for fire protection service, OS&Y type or butterfly type, with tamper switch.

B. Check valves shall be UL listed for fire protection service.

C. Drain valves shall be threaded bronze.

D. Alarm valve shall be UL listed for fire protection service, with retard chamber and connection for water motor gong.

E. Backflow preventer shall be UL listed.

2.3 SPRINKLERS

A. All sprinklers shall be UL listed for fire protection service, with quick response elements.

B. Sprinkler locations are to be in conformance with the performance objectives outlined in paragraph 1.3F.

C. Sprinkler temperature is to be appropriate for the room in which it is installed.

D. Each sprinkler shall be listed for the occupancy classification of the room in which the sprinkler is installed.

2.4 ELECTRICAL WIRING

A. The water flow switch and all tamper switches are to be wired to the main building fire alarm control panel, in accordance with NFPA 70 and 72.

Figure 2-1 *(Continued)*

Summary

Conflicts arising between the contract drawings and the specifications are ordinarily resolved with the specifications taking precedence over the contract drawings. Such conflicts are serious and should be discovered and resolved in advance of the design of a fire protection system. A performance specification provides a broadly general outline for the design and installation of a fire protection system that allows a sprinkler contractor great latitude. A detailed specification usually addresses a heightened hazard or unique situation, and allows a sprinkler contractor very little latitude. A specification is divided into sixteen divisions, each consisting of several sections that contain detailed requirements for each division. Specifications for fire suppression systems and fire alarm systems have been moved to sections 13900 and 13850, respectively, with provisions for fire protection piping remaining in section 15300.

Review Questions

1. Indicate the specification divisions that are most useful when designing a fire protection system.
2. Evaluate the differences between a performance specification and a detailed specification.
3. What problems could ensue if an incomplete set of specifications were used for the design of a fire protection system?
4. What section is referenced by a sprinkler designer to find information relative to fire protection systems under the new specification format? Why has the specification format changed?

Activities

1. Obtain a complete book of specifications from a sprinkler contractor, architect, or engineer, and review each division for its relevance to the design and installation of a fire protection system.
2. Locate a set of contract documents and compare the specification to the contract drawings to determine whether any inconsistencies exist. Discuss the problems that may ensue for a fire protection designer when presented with such conflicts.
3. Interview an architect or engineer to determine his or her criteria for the use of a performance specification versus a detailed specification. Determine whether a performance specification or a detailed specification supersedes the contract drawings in a case of conflict.
4. Review a specification that contains the revised provisions for fire protection systems in section 13900 and detection systems in section 13850. Compare the new format to

the older format that specified fire protection systems in Division 15, and detection systems in Division 16, and evaluate their similarities and differences.

5. Write a specification for a dry pipe sprinkler system in a cold storage warehouse, using the sample specification in Figure 2-1 as a basis. Be certain to carefully edit the specification to ensure that all provisions of the specification are in conformance with the project description.

Chapter 3

Fire Protection Survey Procedure

Objectives

Upon completion of this chapter, you should be able to:

- List the items that should be found in every designer's survey kit.
- Make arrangements for a survey of a fire protection system and develop a survey strategy.
- Perform a survey of a building for the design of a fire protection system.
- Determine a reference elevation in a building.
- Survey and accurately dimension a reflected ceiling plan.
- Survey proposed locations for fire protection piping with respect to existing mechanical systems installed above a suspended ceiling.
- Survey a building using the structural elements as the primary points of reference.
- Field check a drawing of a fire protection system.

survey
a thorough investigation of a building and its components, for the purpose of taking detailed measurements of the building to serve as reference for a fire protection drawing

In the course of our jobs as fire service officials, sprinkler layout technicians, and fire protection engineers, we are often called upon to perform a **survey** of an existing facility and to generate an original drawing that clearly depicts the building conditions, fire protection features, and fire protection systems. A survey is a thorough investigation of a building and its components for the purpose of taking detailed measurements of the building to serve as reference for a fire protection drawing.

A fire protection designer performing a survey for the first time may need a methodical survey procedure to ensure a successful effort. This chapter provides such a procedure, and takes some of the mystery out of an experience that can at first seem overwhelming. We begin by breaking the big picture into smaller, more manageable parts.

GET YOUR BEARINGS

Before taking measurements, it is usually wise to obtain some basic information that is required on your plans by the applicable NFPA standard. Start by obtaining the official name of the building, the juxtaposition of the building with respect to other buildings on the property and with other adjacent buildings, and the orientation of the building with respect to magnetic north. Careful compiling of this sort of basic information helps to speed your plans through the permit process. It is also necessary to obtain the name, mailing address, and phone number of the building owner, the building tenant or tenants, and the owners' representative or contact person.

DETERMINE THE GENERAL BUILDING LAYOUT

point of contact
a person who meets you at the job site, shows you the area to be surveyed, and remains available if any questions arise

Prior planning before your arrival on the job site saves many frustrating hours of confusion. Be certain to make an appointment and to arrange for a **point of contact**. Your point of contact is a person who meets you at the job site, shows you the area to be surveyed, and remains available if any questions arise. Having a point of contact is especially important if security procedures are involved. Your point of contact should ideally be a person familiar with the provisions of your contract and with the portion of the building affected by your contract.

Preparations should be made for a guided tour of areas that are not in the immediate area of work but that affect your system, such as mechanical rooms containing fire protection control equipment. Additional assistance must be prearranged to open locked doors, lift ceiling tiles, and obtain security clearances.

Time taken to get a general understanding of the layout of the building makes the detailed measurements to follow more meaningful. Walk every area, taking note of the occupancy of each room. Develop a solid understanding of the limits of your contract.

Although you quite possibly had been assured that no plans of the building exist, it never hurts to diplomatically request a complete set of plans for the building. Many buildings maintain files, records, and drawings. They could be there for the asking. Any drawings that you are able to obtain, no matter how outdated or incomplete, save you a considerable amount of survey time.

BRING PROPER SURVEY EQUIPMENT

As illustrated in Figure 3-1, equipment found in every technician's or engineer's survey kit should include as a minimum:

- Two 8-foot folding rules. Two rules can, with practice, be used to take quick elevations in the 8- to 12-foot range. Two rules are especially handy in the event that one breaks.
- A 50-foot tape measure with a hook to attach to columns or square-edged walls. A 100-foot tape is a highly recommended option. An infrared or ultrasonic measuring device, which sends signals to a specified target and determines the distance traveled by the sound waves, can be of assistance for quick approximations or in unusually tight quarters.

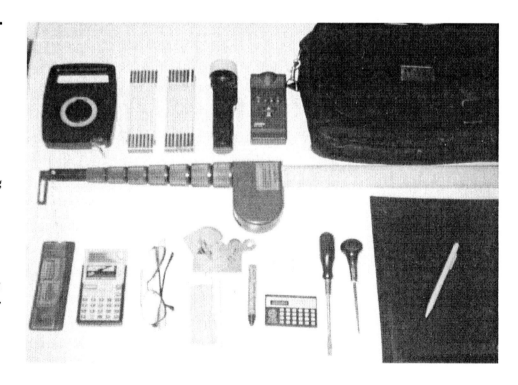

Figure 3-1 *Survey equipment. Top row left to right: 50-foot roll tape, two 8-foot folding rules, flashlight, sonic distance measuring device, survey bag. Middle row: 30-foot digital elevation measuring device. Bottom row: manual feet-and-inch adding device, electronic feet-and-inch adding device, safety glasses, pipe diameter measuring gauge, string, crayon, ear protection, calculator, screwdriver, awl, notebook*

● **CAUTION** When using a tape measure in a building, be very careful to avoid measuring near rotating or moving equipment in which the tape may become entangled. Avoid using metal tape measures, and avoid sources of electrical discharge that may cause a spark or transmit a shock. When measuring, keep a watchful eye for moving vehicles or machinery and other hazards, such as floor openings.

- Two or more clamps to attach an end of a tape rule to a column or other reliable point of measure. These can turn a two-person survey into a one-person survey.

- A **pipe diameter measurement gauge**, used to determine the diameter of an existing pipe into which you are connecting a new sprinkler main. This is an essential tool, especially since 3½-inch pipe and 5-inch pipe are usually extremely hard to identify by other measurement methods.

- A **telescoping elevation measurement pole**, capable of measuring elevations in the 30- to 50-foot range. Once you have such a device, it quickly becomes inconceivable that a survey could have been successfully completed without it.

- A feet-and-inch or metric manual adding device or electronic calculator. Such a device can reduce or eliminate measuring errors and could save you an extra survey trip.

- A flashlight for use in surveying areas above ceilings or other unlit areas.

- A string at least 50-feet long with a weight on one end to use as a vertical frame of reference when it is suspended from a key point at the ceiling.

- Survey clothing, including personal protective equipment such as a hard hat, raincoat, safety eyeglasses, goggles, and ear protection. Long sleeves are needed in most industrial plants, since chemicals can drip from pipes at the ceiling. Old or worn clothes are needed if the area surveyed is especially dirty. Special steel-toed boots, hard hats, and safety glasses are required in most industrial buildings or buildings under construction.

pipe diameter measurement gauge
a device used to determine the diameter of an existing pipe into which you are tapping a new sprinkler main

telescoping elevation measurement pole
a device used to measure elevations in the 30- to 50-foot range

SURVEY BUILDING DETAILS

An important element of this crucial step is a sound **survey strategy**. The approach that yields the most success is to begin your survey with the layout of the centerlines of structural elements, such as building columns and structural beams, establishing a building skeleton from which all other measurements can be taken.

Once a comprehensive structural plan has been obtained, walls can be measured using the column centerlines as a point of reference. Be certain to identify any concealed spaces or small inaccessible enclosures where sprinkler protection is not provided. It is of crucial importance to determine whether these spaces contain combustibles or items of combustible construction. Providing an access panel and sprinkler protection within the space is recommended if combustibles are found.

survey strategy
the approach by which you begin your survey with the layout of all structural elements, such as building columns and structural beams, then reference all measurements to these elements

Chapter 3 Fire Protection Survey Procedure

● **CAUTION**   It is vitally important to note the presence of any asbestos found. Asbestos, a fibrous, usually white substance which, if inhaled may cause cancer, is usually found in the form of insulation on old pipes or as a component of older floor tiles or shingles. Giving formal, written notification to the owner of the presence of asbestos is the first step that must be taken before any plans are made to deal with this problem. Do not attempt to handle this material yourself.

The completion of this phase includes recording the location of ceiling obstructions and potential conflicts with other existing mechanical facilities.

DEVELOP A SYSTEM DESIGN STRATEGY

Determine the most logical system arrangement (gridded, looped, or tree system, discussed in Chapter 5) and develop a basic understanding of probable sprinkler main locations, line arrangement, and direction and riser location. This arrangement may change once layout begins, but a frame of reference needs to be established before you return to your office, especially since field conditions may dictate the system arrangement chosen.

Building Elevations

survey reference elevation
the lowest, flattest, most reliable elevation that can be found

Your first action in the measurement of building elevations should be to determine a **survey reference elevation** from which all other elevations are measured. This is the lowest, flattest, most reliable elevation that can be found. Finding a reliable reference elevation may not be a simple exercise in a ramped parking garage or in an extremely old building where sagging and settling has taken a considerable toll.

If all else fails, it may be necessary to use a laser transit to establish usable reference elevations. Stairwells are usually a reliable way of measuring floor-to-floor elevations in multistory buildings. Be certain to note potential mechanical or structural conflicts during this phase. A good tactic to follow is to take the elevator to the highest floor and measure down to the ground floor, being careful to tie your dimensions into the reference elevation that you have chosen.

Ceiling Measurements

line-of-sight
a visual path above a ceiling created by removing ceiling tiles at regular intervals along a length or dimension of particular interest, such as the location of a sprinkler feedmain, and by using a flashlight and a telescoping elevation pole

Many regularly spaced ceiling tiles can usually be accurately surveyed by measuring the end pieces adjacent to the walls, and counting the tiles from wall to wall. It is also important to tie ceiling tile dimensions into the building skeleton that you drew in an earlier phase of your work to serve as a dimensional check in your survey. Carrying and using a feet-and-inch calculator during your survey may uncover problems or discrepancies before you return to the office and may make a resurvey unnecessary.

Surveying above a ceiling can be a frustrating experience. It is recommended that a **line-of-sight** be created by removing tiles at regular intervals along a dimension of particular interest, such as the location of a sprinkler feedmain, and using a flashlight and your telescoping elevation pole to shoot a visual path along your line of sight. Extra tiles may require removal where potential conflicts are found.

Careful, respectful handling and replacement of the tiles, especially in the vicinity of others working below you, and most especially in a room where food is served, is essential. Whenever possible, it is recommended that you arrange to have the owner supply a person to handle ceiling tile removal for you.

Water Supply Information

Accurate surveying of the point of tie-in on surveys where an existing fire protection pipe supplies your system is one of the more important phases of your survey. The measurements you take must be related to your building skeleton and your chosen reference elevation. Precise measurement of the pipe size into which you intend to tap is essential.

When tapping into an underground main, it is important to obtain a drawing that accurately shows the size and location of the existing underground main. A strategy for installing the new underground pipe in the building must be laid out, preferably in coordination with the owner and trenching contractor. Choice of a location for the entry into the building must be coordinated with the owner, and location of the fire department connection must be in accordance with NFPA standards and local codes.

A waterflow test, as detailed in Chapter 4, should be conducted if water supply information is in excess of a year old, or where new buildings have been built since the date of the last water flow test. Failure to plan and execute a flow test or failure to ensure accurate flow test data could result in drawings and calculations not being approved.

DRAW THE BUILDING AND LAY OUT THE SYSTEM

At this juncture, you are ready to return to your office and begin layout of the building and the fire protection system in accordance with Chapter 5. Shortly after you begin your layout, you will probably begin to compile a list of items that were missed in your survey. Some things that seemed clear at the job site may suddenly seem mysterious on the drawing board or CAD screen. This is to be expected, especially on your first survey or on large or complicated surveys. Areas of conflict or areas in question should be clearly noted so that they can be clarified on your return trip. Complete the drawing to the fullest extent possible during this phase.

FIELD CHECK THE DRAWING

Whatever you do, don't fail to **field check** your drawing before installing the fire protection system. A field check is a thorough survey of a proposed fire protection system using a completed fire protection system design as a basis. Your responsibility

field check
a thorough survey of a proposed fire protection system using a completed fire protection system design as a basis

at this juncture is to check each pipe on your layout and try to find errors. It is much better that you fail-test your design and discover conflicts and problems than for someone using your drawing to find them. An excellent idea is to perform your field check with the fire protection system installer or foreperson in attendance.

Begin your field check from the point of connection to each remote point by measuring main locations, using your telescoping pole to establish the proposed main elevation, then walk the full length of the main to identify conflicts. On complicated surveys involving exposed or concealed areas, each branch line may also need to be similarly fail-tested.

Upon completion of the field check, you are ready to return to your office to complete the layout and perform hydraulic calculations in accordance with Chapter 6. On large or complicated surveys, make an appointment for future visits before leaving.

INSPECT THE SYSTEM AFTER INSTALLATION

Visiting the jobsite during or after installation can be very valuable, especially for the first projects performed by a fire protection designer. In doing so one can truly learn from any mistakes that were made. What might have been difficult to visualize on the drawing board should come to life when the finished product is seen.

You also have a sense of accomplishment when seeing the finished product. During the final field check before installation, a fire protection designer surveys only imaginary pipes and sprinklers. During the inspection of the completed system, one's accomplishments can be fully visualized and appreciated. Seeing the finished product of one's work, armed with the knowledge that the building is considerably safer after installation of the system, is truly the most rewarding benefit of performing fire protection system design.

Summary

An organized and well-planned strategy for performing a survey is your best assurance for a successful survey and your best chance for establishing a good relationship with the client. Advance planning before you arrive at the job site saves time and makes a good impression. Establishing a point of contact is essential to a trouble-free survey.

Successful use of the appropriate tools makes your job considerably easier. Beginning with a general building overview and the establishment of a survey strategy using the building skeleton as your first survey steps creates an organized and logical basis for all survey dimensions to follow. Using the line-of-sight method for surveying above ceilings makes the difficult task of surveying above ceilings easier. Development of a system layout strategy and performing a careful measurement of the water supply location and water supply availability serves you well throughout your survey. Field checking and fail-testing your completed layout will best ensure the success of your efforts, and viewing the completed system will develop a more meaningful understanding of fire protection system design.

Review Questions

1. List the tools used on a fire protection survey and explain their use.
2. Describe the best way to arrange for a survey appointment.
3. Explain how a surveyor should develop a survey strategy for a building.
4. Discuss the procedure for surveying above an existing suspended ceiling.
5. What cautions would you give to a person performing a fire protection survey?

Activities

1. Contact an experienced fire protection surveyor and arrange to accompany that person on a fire protection survey of a building. Take note of survey equipment used, strategies employed, and use of the point of contact.
2. Accompany an experienced person on a field check of a drawing that has been developed from survey measurements. Employ the fail-test method for checking fire protection piping locations.
3. Survey a completed fire protection system, and compare the system to the drawing from which it was installed.

Chapter 4

Underground Piping Design Procedure

Objectives

Upon completion of this chapter, you should be able to:

- Design a fire protection underground piping system for one or more sprinklered buildings.
- Use invert elevations to coordinate an underground piping system that avoids conflicts with other underground utilities and provides protection from freezing.
- Determine the size of a thrust block for support of an underground fitting.
- Calculate the pressure available from a gravity tank.
- List the types of water supplies to underground piping systems.
- Identify components of an underground piping system design by their symbols.
- List the types of piping and fittings commonly used for underground fire protection piping.
- Perform a water flow test, recording and plotting results.

Standard for Private Service Mains and Their Appurtenances

city water supply
a network of underground pipes pressurized and maintained by a public water authority, usually supplied by a reservoir or aboveground water storage tank

pressure tank
an enclosed water storage vessel that contains water pressurized by compressed air

gravity tank
an elevated water tank that utilizes the force of gravity to provide pressure to a water supply

fire pump
a specially designed and listed pump that increases the pressure of the water serving a fire protection system

Aboveground water-based fire protection systems are fed by an underground network of pipes designed in accordance with **NFPA 24**, *Standard for Private Service Mains and Their Appurtenances*. A fire protection system is pressurized by at least one of four water supply methods:

1. A **city water supply**—a network of underground pipes pressurized and maintained by a public water authority, usually supplied by a reservoir or aboveground water storage tank.
2. A **pressure tank**—an enclosed water storage vessel that contains water pressurized by compressed air.
3. A **gravity tank**—an elevated water tank that utilizes the force of gravity to provide pressure to a water supply.
4. A **fire pump**—a specially designed and listed pump that increases the pressure of the water serving a fire protection system.

Water-based fire protection system design projects involving the design and installation of underground piping should be scheduled so that the underground piping is installed before the sprinkler piping. In those instances, the underground piping plan must be designed before the sprinkler plan. Projects with especially tight schedules may require that the sprinkler plans and underground plans be developed simultaneously, using two or more designers.

Depending on the length, complexity, and configuration of the underground piping, it may be possible to fit the underground piping plan onto the same drawing as the sprinkler plan. Complicated layouts, especially layouts involving underground entries into more than one building, should be displayed on separate sheets.

This chapter does not set forth every provision of NFPA 24. Those using this chapter to design an underground piping system for fire protection should refer to NFPA 24 as well as the material presented in this chapter.

A methodical approach to the design of underground piping lends structure and organization to your design effort. A designer should use the following method to design underground fire protection systems:

1. Obtain reference material and trace or scan background.
2. Determine pipe routing.
3. Draw cut-in detail.
4. Add invert elevations and plan dimensions.
5. Specify piping and fittings.
6. Detail thrust blocks and pipe restraints.
7. Detail locations of fire hydrants.
8. Obtain water supply information.
9. Perform a water flow test.

Chapter 4   Underground Piping Design Procedure

OBTAIN REFERENCE MATERIAL AND TRACE OR SCAN BACKGROUND

The drawings in the contract drawing package that have the most relevance to the performance of underground piping design and are of the greatest assistance to the fire protection designer are found in the **civil section** of the contract drawings. The civil section is comprised of site and underground utility drawings primarily originated by civil engineers and landscape architects. Select a drawing from this section that best illustrates the buildings served by the underground piping, the point of connection for the new underground piping, and the route between point of connection and points of building entry.

Many civil plans, scaled 1" = 20'-0" or greater, are usually of a suitable scale for most fire protection underground piping layouts, and may be traced or scanned into your computer-aided design program. Some contract drawings in the civil or plumbing section may show a proposed route for the new fire protection underground piping.

Complicated underground plans, where many changes of direction are anticipated, may need to be enlarged to a 1" = 10'-0" scale. Plans involving relatively straight pipes, with few changes of direction may be successfully done in scales as large as 1" = 100'-0". The choice of scale is based on the detail required.

The locations of potential sources of interference with the fire protection underground piping should be obtained from the civil drawings and traced or scanned directly onto the fire protection plan. Conflicts that may be found on an underground plan include other pipes, underground pits or vaults, aboveground objects with underground footings, electrical cables, or cable vaults.

Natural conflicts such as trees, streams, and ponds should also be traced or scanned onto the plan. Only those items in the vicinity of the proposed underground pipe that cross, intersect, or pose a potential conflict to the pipe need to be traced. Identifiable aboveground landmarks that may serve as a point of reference or as a reliable point of measure may be helpful additions to the plan.

The plumbing plans or fire protection plans in the contract drawing set may be needed to determine the point of underground entry into each of the buildings that are to be supplied with new fire protection systems. In addition, the mechanical, fire protection, and sitework sections of the specifications must be referenced to obtain requirements related to pipe type, depth of bury, trenching, and pipe restraint. Catalogues supplied by underground pipe and fitting suppliers may provide significant information relative to the design criteria applied to pipe connections and restraints for the models of pipes and fittings used on a particular design.

The structural section of the contract drawings must be used to obtain an elevation of the bottom of **building footings**. Footings are underground structural elements, as shown in Figure 4-1, that support building walls and floor slabs. Footings are also commonly referred to as foundations. An elevation should be clearly shown on a plan that shows the relative elevation of the footing with respect to the underground pipe passing below it.

---

civil section
portion of contract drawings that contains site and underground utility drawings, primarily originated by civil engineers and landscape architects

building footings
underground structural elements that support building walls and floor slabs

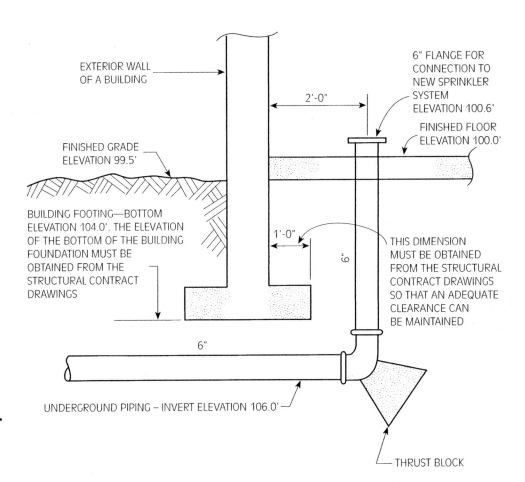

Figure 4-1
Coordination with building footings.

DETERMINE PIPE ROUTING

Lightly draw proposed fire protection underground piping locations, using the most direct route from the point of cut-in to the point of building entry. Draw potential sources of conflict, and revise fire protection piping locations where conflicts can be avoided. Sectional views or profile plans should be shown whenever a fire protection pipe changes direction to avoid a conflict, or when a pipe must traverse an area with numerous conflicts. The relative elevations of the conflicts should be noted lightly on the plan at this time.

Once a piping arrangement or strategy has been ascertained, the piping layout and details related to potential piping conflicts can be darkened, using piping symbols found in NFPA 170, *Standard for Fire Safety Symbols*. This standard also details symbols for fire protection valves, hydrants, elbows, and thrust restraints. Some of the symbols most commonly used for fire protection underground design are shown in Figure 4-2.

Chapter 4 Underground Piping Design Procedure

cut-in
procedure involving cutting into the existing underground main, removing a segment of existing pipe, installing a new underground tee, and reconnecting the completed assembly

tapping sleeve
a device that allows piping to be easily reconnected once a cut-in is complete

thrust block
a poured block of concrete that restrains an underground pipe at points where it changes direction, such as at elbows, tees, fire hydrants, and caps

DRAW CUT-IN DETAIL

A detail showing the point at which the new fire protection underground pipe is proposed to connect to an existing water supply main must be shown on the plan, preferably as an enlarged detail. A **cut-in** is performed by cutting into the existing underground main, removing a segment of existing pipe, installing a new underground tee, and reconnecting the completed assembly, as shown in Figure 4-3.

A **tapping sleeve** is used to facilitate cutting into the existing main, by allowing the piping to be easily reconnected once the cut-in is complete. A **thrust block** is a poured block of concrete that restrains an underground pipe at points where it changes direction, such as at elbows, tees, fire hydrants, and at caps.

ADD INVERT PIPING ELEVATIONS AND PLAN DIMENSIONS

Invert piping elevations for fire protection pipes and other pipes posing potential conflicts may now be added to the plan. An invert elevation refers to the elevation of the bottom of an underground pipe with respect to a reference elevation. Invert

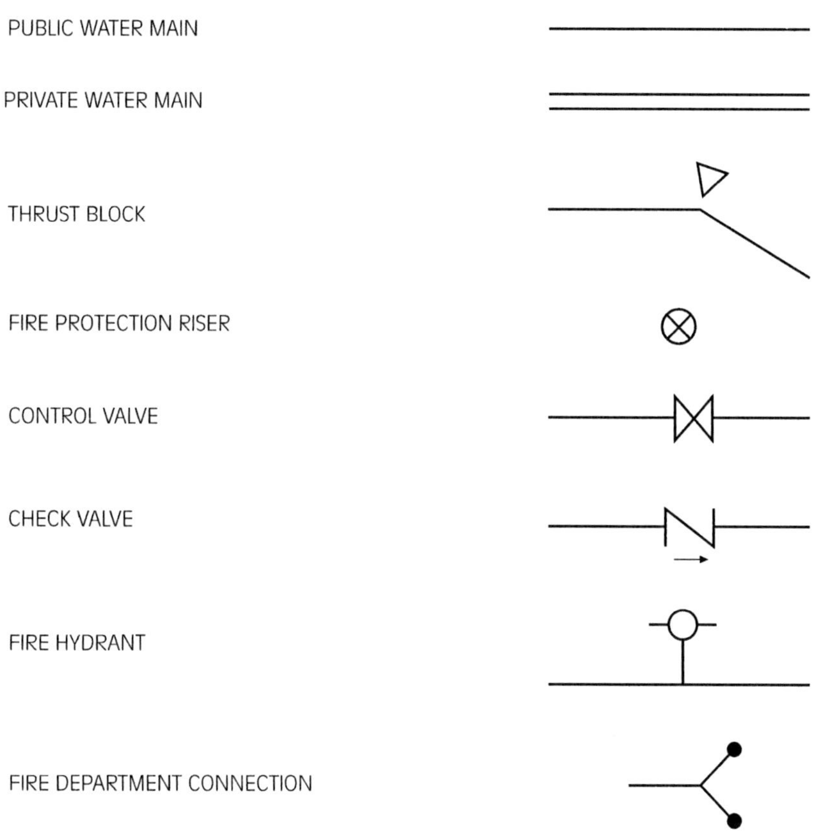

PUBLIC WATER MAIN

PRIVATE WATER MAIN

THRUST BLOCK

FIRE PROTECTION RISER

CONTROL VALVE

CHECK VALVE

FIRE HYDRANT

FIRE DEPARTMENT CONNECTION

Figure 4-2 *Common symbols used for underground design.*

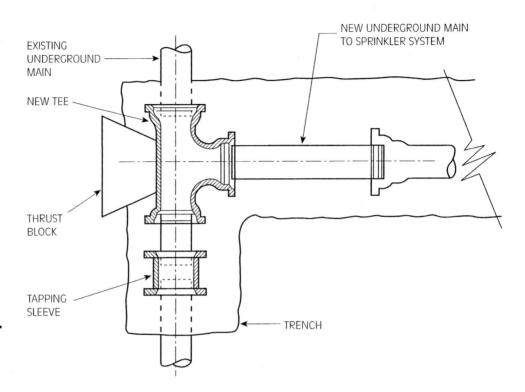

Figure 4-3 *Cutting into an existing underground main.*

elevations should be shown on the plan view and on any sectional views or profile plans that are shown. Invert elevations are used to coordinate the relative elevations of fire protection pipes, finished grade, and potential sources of conflict.

Fire protection underground designers must check the invert elevations of their piping to meet three sets of criteria:

1. All underground pipe must be positioned such that the depth of cover, or lineal distance from the top of the pipe to the finished grade, is sufficient to protect the pipe from freezing. An accepted depth of cover that meets this criteria is one foot below the frost line. Figure 4-4 shows a map of North America showing required minimum depth of cover to provide proper freeze protection for buried piping. The minimum depths of cover range from 8 feet in Canada to 2½ feet in Florida. Figure 4-5 illustrates the variables associated with depth of cover.

2. Underground piping must be positioned to avoid conflicts with other utilities.

3. Piping must be positioned to avoid physical damage. NFPA 24 prohibits routing underground piping under building slabs to avoid pipe damage from settling buildings. Underground piping entries to a building should be located adjacent to an outside wall. The elevation of the bottom of the footing and the width of the footing must be ascertained, and a safe clearance must be provided between

Chapter 4 Underground Piping Design Procedure

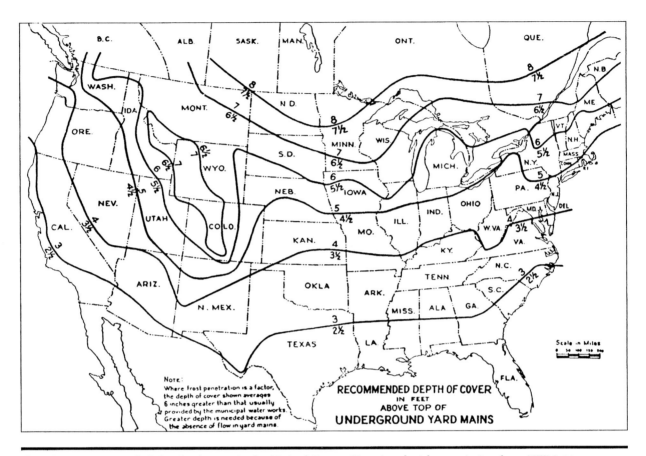

Figure 4-4 *Recommended depth of cover in the United States. (Reprinted with permission from NFPA 24, Installation of Private Fire Service Mains. Copyright © 1995, National Fire Protection Association, Quincy, MA 02269. This reprinted material is not the complete and official position of the National Fire Protection Association on the referenced subject, which is represented only by the standard in its entirety.)*

the pipe and the bottom and side of footing. Piping running under roads or railroad tracks requires extra protection, such as sleeves, structural devices, and extra depths of bury.

Dimensions from the center of fire protection pipes to identifiable points of reference should be added to the plan view and to all sectional views.

It is necessary to add a sectional view of each fire protection pipe entering a building, showing pipe size, invert elevations, foundation elevations, finished grade elevations, finished floor elevations, and entry flange size and elevations.

Any underground pits or vaults should be detailed carefully and drawn to a scale of 1/4" = 1'-0", for both plan views and sectional views.

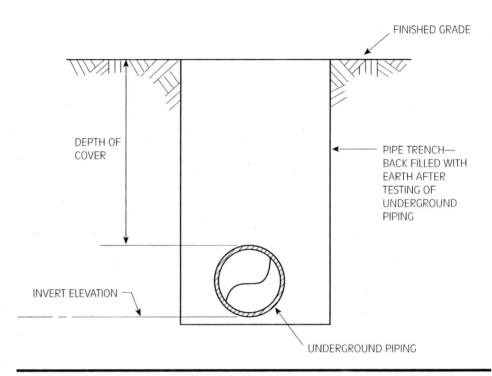

Figure 4-5 *Depth of cover. Depth of cover is the largest of the following dimensions:*
  1'-0" or greater below frost line. Frost line depth varies with geographic location.
  2'-6" or greater to prevent mechanical damage.
  3'-0" minimum below driveways.
  4'-0" below railroad tracks.
  No piping is to be run below building floor slabs.

SPECIFY PIPING AND FITTINGS

Details related to the type of piping and fittings should be added at this time. Common types of underground piping include:
- ductile iron underground piping, class 50,
- ductile iron underground piping, class 52,
- PVC (polyvinyl chloride) underground piping, class 150 plastic pipe,
- cement lined underground piping,
- cast iron underground piping,
- CPVC (chlorinated polyvinyl chloride) underground, plastic pipe.

Underground fittings that are often used for fire protection underground piping include:
- mechanical joint fittings,
- push-on joint fittings,
- PVC plastic fittings,
- CPVC plastic fittings.

The type of pipe and fittings can be handled in a general note if all pipes and fittings on the plan are of a uniform type. Mechanical joint pipe and fittings must be clamped and/or retained in accordance with NFPA 24 and the manufacturer's instructions.

It may be necessary to detail the trenching and bedding for the pipe. The bedding and **backfill** material should be shown in a sectional view. Backfill is the dirt or other material used to cover an underground pipe after installation. Ordinarily, special backfills, such as a gravel or compacted granular base, are needed only for very loose or wet soil, or for underground pipes larger than 14 inches in diameter.

backfill
the dirt or other material used to cover an underground pipe after installation

DETAIL THRUST BLOCKS AND PIPING RESTRAINTS

The composition of the soil, obtained from the boring table as discussed in Chapter 1, is needed when calculating thrust blocks for the underground main. The size of the block is dependent upon the soil content. Table 4-1 can be used to determine the sizes of thrust blocks for piping installed in relatively firm soil, such as a sand, gravel, and clay mixture.

Example 4.1 Using Table 4-1, one can quite simply determine that a 10-inch underground 90° elbow in firm soil requires a thrust block with 13 square feet of bearing area. **Bearing area** is the area of the vertical plane of the thrust block onto which the fitting must rest.

bearing area
the area of the vertical plane of the thrust block onto which the fitting must rest

For other soil composition factors, NFPA 24 requires that the result obtained in example 4.1 be multiplied by a factor of:

4 for soft clay,
2 for sand,
1.33 for sand and gravel,
0.4 for shale.

Example 4.2 If the soil composition is sand, the previous result must be multiplied by a factor of two. Thus, in sand, the 10-inch 90° ell in example 4.1 requires a bearing area of at least 26 square feet. Such a condition might be encountered near a body of water.

As an option to thrust blocks, underground piping can be restrained with tie rods and tie bolts. Tie rods and bolts clamp fittings together in a manner similar

Table 4-1 *Area of bearing face of concrete thrust blocks.*

Pipe Size In.	90° Bend		45° Bend		Tees, Plugs, Caps, and Hydrants	
	sq ft	m²	sq ft	m²	sq ft	m²
4	2	0.19	2	0.19	2	0.19
6	5	0.46	3	0.28	4	0.37
8	8	0.74	5	0.46	6	0.56
10	13	1.21	7	0.65	9	0.84
12	18	1.67	10	0.93	13	1.21
14	25	2.32	14	1.30	18	1
16	32	2.97	18	1.67	23	2

Source: Reprinted with permission from NFPA 24, *Installation of Private Fire Service Mains.* Copyright © 1995, National Fire Protection Association, Quincy, MA 02269. This reprinted material is not the complete and official position of the National Fire Protection Association on the referenced subject, which is represented only by the standard in its entirety.

Note: Areas in table have been derived using a water pressure of 225 pounds per square inch (15.5 bars) and a soil resistance of 2000 pounds per square foot (1.0 bars). The values given in this table include a design safety factor of 1.5.

to the illustration in Figure 4-6. The number, location, and diameter of the rods is prescribed in NFPA 24.

DETAIL LOCATIONS OF FIRE HYDRANTS

fire hydrant
a device that provides a water supply to fire department pumpers for use in combating structure fires

Fire hydrants provide a water supply to fire department pumpers for use in combatting structure fires. They are required to be adjacent to buildings and should be at least 40 feet away from buildings to provide fire service personnel with some degree of safety from potential building collapse or from falling debris. They must be supplied in sufficient number and at a spacing sufficient to deliver water to all structures in an area intended for protection. In addition, fire hydrants must be located to be immediately adjacent to roads, streets, alleys, or driveways accessible to fire department apparatus. Coordination of hydrant locations with the requirements of the authority having jurisdiction is essential.

Since the pressures created by fire department pumpers can be considerable, a thrust block is required at changes in direction to protect the hydrant and its associated piping and fittings from damage caused by pipe movement.

Chapter 4 Underground Piping Design Procedure

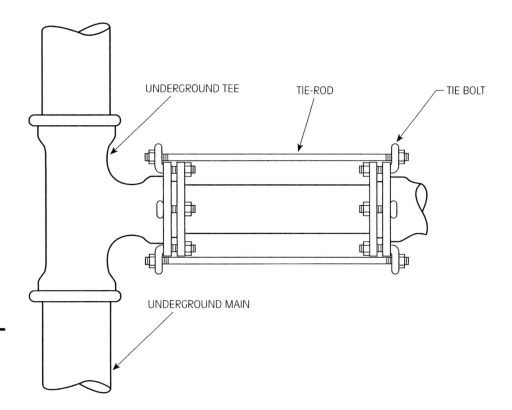

Figure 4-6
Underground piping joint restraint.

water hammer
excessive pressure caused by rapid valve or hydrant closure

water flow test
procedure to determine that a water supply of sufficient flow and pressure is available for a water-based fire protection system

Hydrants are equipped with an operating nut at the top of the hydrant, which opens and closes the hydrant when turned by a wrench. Closing a hydrant too quickly can create a phenomenon called **water hammer**, where excessive pressures caused by rapid valve or hydrant closure can cause major damage to the piping system.

In areas where water inside a hydrant is exposed to freezing, a hydrant seat controls the flow of water into the hydrant barrel. The hydrant seat is located at the base of the hydrant barrel and has a special drain to drain water from the barrel after use. Figure 4-7 illustrates a cross-sectional view of a hydrant assembly with a hydrant seat. In warm climates, hydrants may be supplied without a drain feature for the hydrant barrel.

PERFORM A WATER FLOW TEST

A **water flow test** must be performed for each source of water supply to ensure that a water supply of sufficient flow and pressure is available for a water-based fire protection system. It is essential that accurate water flow availability be obtained that represents the available water supply under the worst case scenario, where

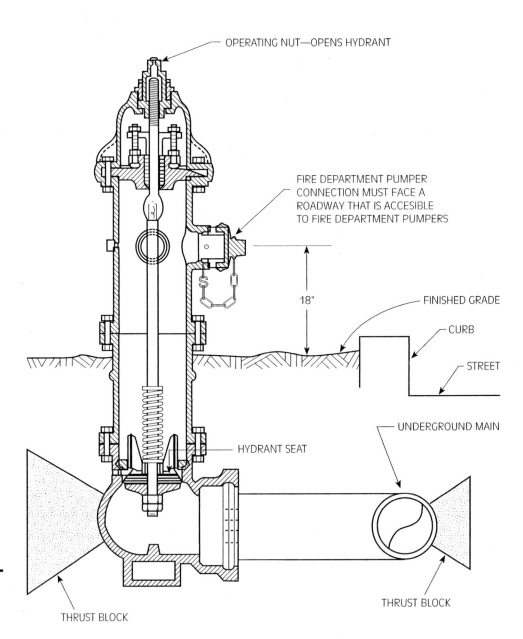

Figure 4-7 *Fire hydrant connection to underground.*

water demand is at its maximum. Failure to do so gives misleading results for a hydraulically calculated sprinkler system, as discussed in Chapter 6.

The purpose of a water flow test is to establish that an adequate water supply exists to supply the following requirements:

1. Water to support firefighting activities,

Chapter 4 Underground Piping Design Procedure

2. Water consumed during the peak domestic demand,
3. Calculated sprinkler system demand.

Supplies from gravity tanks, reservoirs, and pressure tanks are of a constant and reliable pressure. City supplies, however, vary as a function of the water consumption that is drawn from the piping network, as shown in Figure 4-8. Most city supplies have lowered pressures during times of high domestic water use, such as early in the morning and late in the evening, or at times when water-consuming industrial processes are in operation. In an area where the water consumption has increased over time, such as in an industrial park under development, the avail-

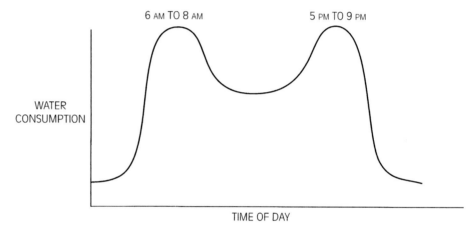

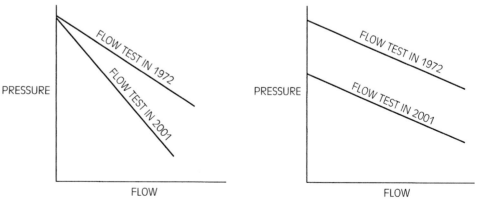

Figure 4-8 *Flow test changes.*

able water supply may decline with the additional water consumption created by the development of the industrial park.

Flow Test Tools

Water flow tests are ordinarily performed using fire hydrants. As a minimum, the equipment that must be available for a flow test is:

- A **pitot tube with bourdon gauge**, inserted into a water stream to measure the pressure of a water flow, shown on Figure 4-9,
- A hydrant wrench,
- A **hydrant cap with bourdon gauge**, mounted to a hydrant to measure the water pressure, shown on Figure 4-9,
- Calipers or other measuring device.

Flow Test Protocol

Before arranging a water flow test, it is necessary to follow the proper protocol to obtain the permission to run the test and to provide for maximum safety during the test. The protocol should include:

1. Check with the agency responsible for the water supply in the vicinity of the building in question. The authority may insist that flow tests be performed by

pitot tube with bourdon gauge
a device inserted into a water stream to measure the pressure of the water flow

hydrant cap with bourdon gauge
a device mounted to a fire hydrant to measure the water pressure

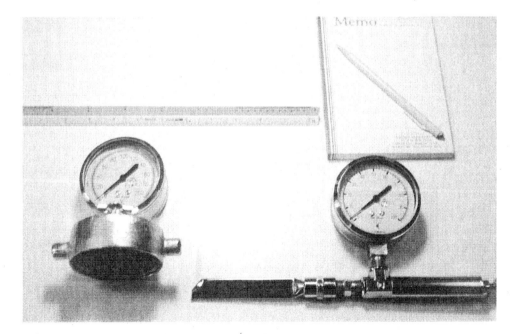

Figure 4-9 *Water flow test equipment. Clockwise from top left: ruler to measure hydrant opening, notebook to record flow test results, pitot tube and bourdon gauge, hydrant cap with bourdon gauge.*

the authority or under its supervision. A designer may wish to avoid liability for damage from a water flow test by arranging for others to perform the test, or by obtaining records of a recent flow test for a location near the building of interest.

2. Check with the building owner to schedule the test with minimal business interruption.

● **SAFETY**

3. Always perform a safety check. Among items to consider:
- **Are there any open street access hatches, sewer openings, or accessible underground utilities that could fill with water and endanger personnel?**
- **Could flow from a hydrant disrupt traffic?**
- **Could the force of the water stream damage automobiles or other property?**
- **Could low temperatures cause pooled water to freeze on roads or sidewalks?**
- **Will water drain to a safe location? Could electrical or mechanical problems ensue?**

4. Consider your liability. Designers should consider having others open and close the hydrants.

Flow Test Procedure

A water flow test should be conducted using the following methodology:

test hydrant
a hydrant selected for observing and recording water pressures available to the water supply

1. Select the **test hydrant**, also called the pressure hydrant. The test hydrant should be the one located closest to the building being considered for a water-based fire protection system. This hydrant is used to observe and record the water pressures available to the water supply.

flow hydrant
a hydrant selected to measure the water flow available from the water supply

2. Select the **flow hydrant**. This hydrant should be downstream from the test hydrant. It may be necessary to reference a set of underground piping plans to verify the direction of water flow in the underground piping system. The flow hydrant is used to measure the water flow available to the water supply. The choice of flow hydrant must consider possible damage that could occur when water flows from this hydrant. In areas with good water supplies, the force of water discharging from a flow hydrant can dislodge gravel or asphalt pavement or could damage automobiles or disrupt traffic.

3. Determine elevations of the test and flow hydrants with respect to the floor slab elevation of the building in question. Contract drawings may provide this information, or field measurements may be taken to determine these relative elevations.

4. Draw a sketch of the testing area. The sketch should show the test hydrant, flow hydrant, the existing underground main with pipe size and pipe type noted, and the dimensioned location of the building in question. Dimensional and elevational measurements should be noted for all items. A sample sketch is shown on Figure 4-10.

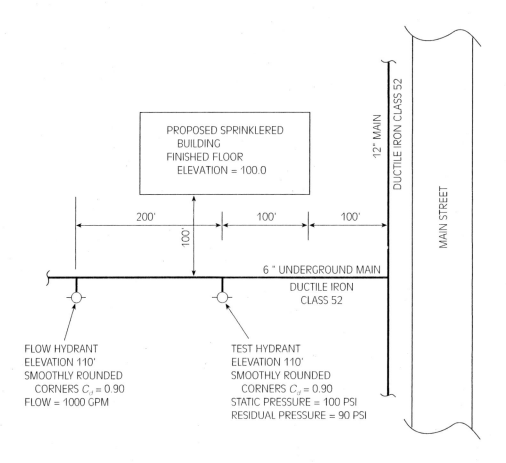

Figure 4-10 *Flow test arrangement.*

static pressure
the pressure available from a water supply with no water flowing

hydrant coefficient
a factor that describes the nature of the opening from which water flows from a fire hydrant

5. Remove a cap from the test hydrant, and install a hydrant cap with attached gage.

6. Slowly and completely open the test hydrant. Read and record the pressure on the gage, as shown in Figure 4-11. This is the **static pressure**, or the pressure available from the water supply with no water flowing.

7. Remove a hydrant cap from the flow hydrant and accurately measure and record the inside diameter of the opening to the nearest 1/16″, using calipers or a good ruler, as shown on Figure 4-12.

8. Using your fingers, determine the shape of the opening of the flow hydrant outlet, as shown on Figure 4-13, to determine the **hydrant coefficient** (C_d), a factor that describes the nature of the opening from which water flows from a hydrant, in accordance with Figure 4-14. Record this coefficient. Note that some AHJs have requirements for hydrant coefficients that differ from NFPA, and differing values are shown on Figure 4-14.

Chapter 4 Underground Piping Design Procedure

Figure 4-11 *Flow test procedure at test hydrant: Record the static pressure before opening flow hydrant then record the residual pressure after opening flow hydrant. (Courtesy Superior Automatic Sprinkler Corp.)*

Figure 4-12 *Flow test procedure at flow hydrant: Accurately measure the inside opening to the nearest 1/16". Record this value. (Courtesy Superior Automatic Sprinkler Corp.)*

Chapter 4 Underground Piping Design Procedure

Figure 4-13
Checking the hydrant coefficient. Flow test procedure at flow hydrant: Determine the hydrant coefficient (C_d) (rounded, square, or projected), by referring to Figure 4-14. Record this value. (Courtesy Superior Automatic Sprinkler Corp.)

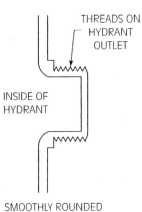

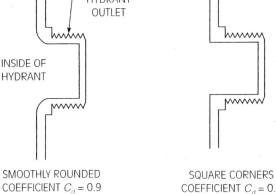

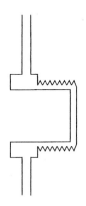

SMOOTHLY ROUNDED
COEFFICIENT $C_d = 0.9$

(FACTORY MUTUAL
$C_d = 0.8$)

SQUARE CORNERS
COEFFICIENT $C_d = 0.8$

(FACTORY MUTUAL
$C_d = 0.7$)

PROJECTIONS ON CORNERS
COEFFICIENT $C_d = 0.7$

(FACTORY MUTUAL
$C_d = 0.6$)

Figure 4-14
Coefficients of hydrant outlets.

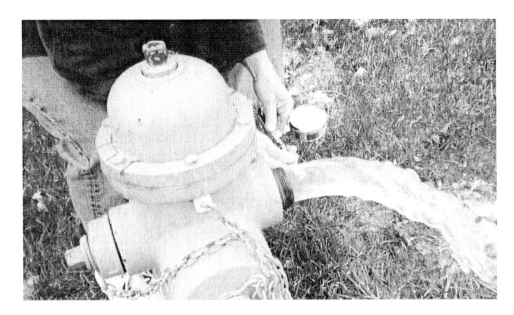

Figure 4-15 *The flow hydrant. Flow test procedure at flow hydrant: Position the pitot tube in the center of the stream per Figure 4-16. Record the pitot pressure reading displayed on the bourdon gauge. (Courtesy Superior Automatic Sprinkler Corp.)*

pitot pressure
the pressure reading (in psi) as shown on the bourdon gauge attached to the pitot tube

residual pressure
the pressure reading (in psi) at the test hydrant with water flowing from the flow hydrant

9. Open the flow hydrant slowly to its fully open position by turning the operating nut with the hydrant wrench. Count the number of turns that it takes to fully open the hydrant and record this number. Wait several minutes until the water runs clear before proceeding. Debris in the water flow stream could damage the pitot tube.

10. Insert the pitot tube into the water stream of the flow hydrant, as shown in Figure 4-15, using the method shown in Figure 4-16.

11. Record the pressure reading (in psi) as shown on the bourdon gauge attached to the pitot tube. This is known as the **pitot pressure**. If the pressure fluctuates, use the average reading. If the gauge reading is steadily dropping, wait until the gauge reading is steady.

12. Simultaneous to step 11, have another individual read and record the pressure at the test hydrant with the water flowing from the flow hydrant. This reading should be relatively steady, but if it fluctuates, record the range of readings and take an average of the readings. This is the **residual pressure**, and it is lower than the static pressure because of pressure loss due to friction within the pipe. The residual pressure relates to the specific rate of flow recorded in step 11 and calculated in step 15.

13. Close the flow hydrant slowly to avoid water hammer. Count the number of turns to close the hydrant, and compare to the number of turns recorded in step 9. If the numbers differ, the hydrant may require maintenance. Replace the hydrant cap. Make sure that water is draining from the barrel. This drainage process may be audible, or may be determined by feeling vibrations in the hydrant barrel. If the hydrant barrel fails to drain, call for repair immediately.

14. Close the test hydrant in a manner similar to step 13. Remove the cap and gauge.

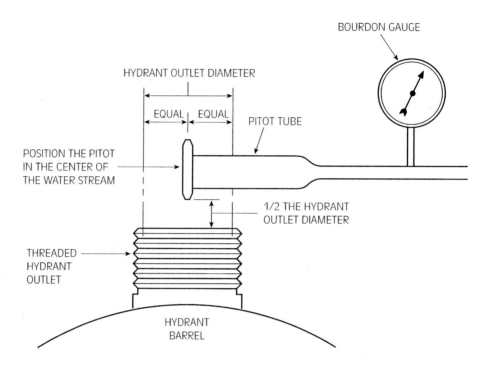

Figure 4-16
Positioning the pitot tube. (Note: Factory Mutual requires that the pitot be held tight to the edge of the threaded hydrant outlet. Make a note of the test method that is used.)

15. Calculate the flow based upon the information recorded in steps 7 through 11, using the formula:

$$Q = (29.83) \times (C_d) \times (D^2) \times (\sqrt{P})$$

where
Q = flow, in gallons per minute (gpm)
29.83 = a constant
C_d = the hydrant coefficient from step 8
D = the hydrant outlet diameter from step 7
P = the pitot pressure from step 11

Example 4.3 A fire hydrant outlet has a diameter of 2.51 inches and has smoothly rounded corners. The pitot pressure is 14.5 psi. Substituting into the flow equation above:

$$Q = (29.83) \times (C_d) \times (D^2) \times (\sqrt{P})$$
$$Q = (29.83) \times (0.9) \times (2.51^2) \times (\sqrt{14.5})$$
$$Q = 644 \text{ gpm}$$

Note that the result is rounded to the nearest gallon. The precision of a water flow test is not sufficient to warrant fractional gallonages.

Chapter 4 Underground Piping Design Procedure

logarithmic graph paper
lined paper used to show pressure in psi on the Y axis and flow in gpm on the X axis, with spacing between flow values increasing with flow

16. Plot the results of the flow test on **logarithmic graph paper**. Logarithmic graph paper, shown on Figure 4-17, also known as $N^{1.85}$ paper, shows pressure in psi on the Y axis, and flow in gpm on the X axis. Although the values of the pressure are evenly spaced on the graph paper, spacing between flow values increases with flow. This is a logarithmic function designed into the graph that permits the static and residual pressures to be connected with a straight line instead of a curve.

Example 4.4 The test used in example 4.3 showed a static pressure of 108 psi and a residual pressure of 80 psi. The flow, as calculated in example 4.3, was 644 gpm. These results must then be plotted on graph paper as shown on Figure 4.18.

17. Analyze the flow test graph. For the most usable and accurate flow tests, the pitot pressure should be at least 10 psi, and the residual pressure should drop

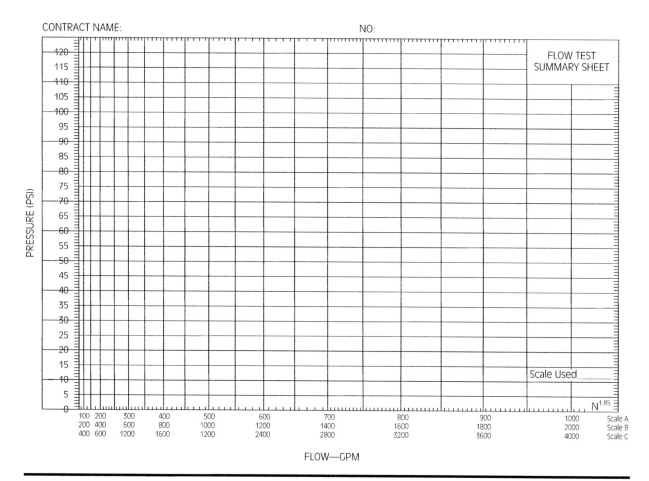

Figure 4-17 *Logarithmic graph paper for review question 1.*

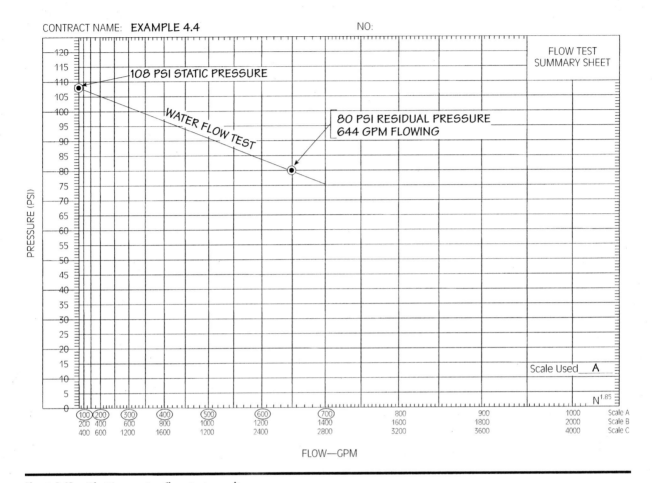

Figure 4-18 *Plotting water flow test results.*

25% or 15 psi below the static pressure. Pressure drops less than these values induce an element of inaccuracy to the test results, and a second flow test should be run with two or more outlets flowing. When more than one outlet is opened, the flow from each flowing outlet must be measured by the pitot gage, and each flow must be calculated per step 15. The total flow for the test is the sum of all flows flowing simultaneously during the test.

18. Prepare a formal water flow test report. The report must include a detailed sketch of the test area showing hydrant locations, main sizes, building locations, water flow directions in looped or dead-end mains, and dimensions of all hydrants and identifiable features. The report must also include calculations for pitot conversions, notes of test methods taken, and a plot of the flow test results.

DETAIL WATER SUPPLY INFORMATION ON PLANS

Once drawn, the underground plan must show at least one automatic water supply for the proposed fire protection system. The types of water supplies most commonly encountered are listed below:

• City water supply: Fire hydrants used to conduct a water flow test must be shown, with their elevations, pipe sizes, static pressure, residual pressure, pitot pressure, and gpm flowing. The distance from the fire department connection to the closest fire hydrant must be within 100′. Enough detail must be shown to document the relative locations and accessibility of the fire hydrants and fire department connection.

• Aboveground gravity water storage tank: Plans must show a sectional view, height of the bottom of the tank above grade, and size of underground piping from the tank to the fire protection system. The static pressure yielded by the storage tank in psi at a point of reference is computed by multiplying the height from the point of reference to the bottom of the tank, in feet, by 0.433.

$$\text{Pressure (psi)} = (\text{Height to bottom of tank}) \times (0.433)$$

The bottom of the tank elevation is used as a conservative reference for the worst case static pressure condition that would be expected from the elevated tank, as shown in Figure 4-19. NFPA 22, *Standard for Water Tanks for Private Fire Protection*, is to be used as the basis for this design.

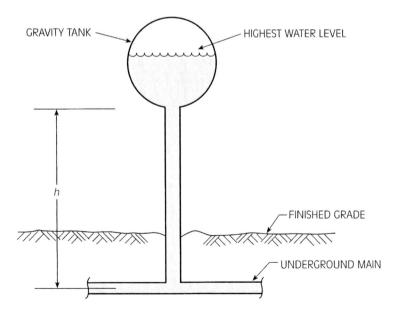

Figure 4-19 *Gravity tank. Pressure from a gravity tank is computed from the bottom of the gravity tank and is determined by the formula: Pressure (in psi) = (0.433) × (h, in feet).*

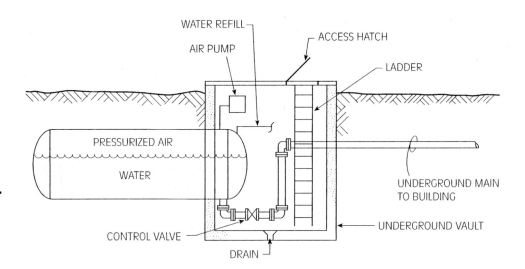

Figure 4-20 *Below ground pressure tank in an underground vault.*

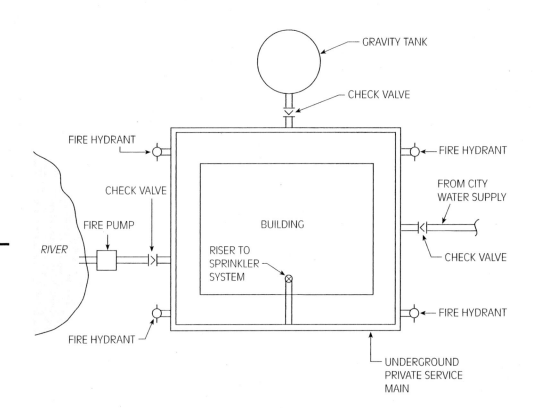

Figure 4-21 *Plan view of a building with three sources of water supply. Each source of water supply requires a check valve to prevent backward flow to any source.*

Chapter 4   Underground Piping Design Procedure

Standard for the Installation of Centrifugal Fire Pumps

- Pressure tank: Plans must show a sectional view of the tank end penetrating the basement of a building or an underground vault, as shown in Figure 4-20, with details of piping and valving associated with the tank. Pressure tanks may also be installed in basements. NFPA 22 is to be used for this design.
- A fire pump supplied by a reservoir, pond, or body of water: A plan view and section of the fire pump, strainer, and valve pit is required. Fire pump installations are to be designed in accordance with **NFPA 20**, *Standard for the Installation of Centrifugal Fire Pumps*. Figure 4-21 shows a fire pump drawing water from a pond, combined with two other sources of water supply.

Summary

A methodical approach to the design of underground piping for water-based fire protection systems, using the method covered in this chapter, ensures good results. Underground piping must be designed to avoid conflicts and provide physical protection and protection from freezing. Thrust blocks and pipe restraints are required to protect the pipe from excessive pipe movement. Understanding the types of water supplies available for the supply of an underground piping system is important information for each designer. Testing a water supply is essential to verify the adequacy of every water supply.

Review Questions

1. A water flow test has been conducted, and one hydrant outlet, 2.62 inches in diameter with square-edged corners, was flowed. The static pressure was recorded as 165 psi, the residual pressure was recorded as 87 psi, and the pitot pressure was recorded as 21.3 psi.
   a. Determine the flow (Q), in gpm.
   b. Plot the flow test results on logarithmic graph paper, using Figure 4-17. Label the static pressure, residual pressure, and flow, and draw the water supply graph.

2. List the types of water supplies used to supply a water-based fire protection system.

3. A flow test was conducted using four identical hydrant outlets flowing simultaneously. Each hydrant outlet is 2.50 inches in diameter and protrudes into the hydrant barrel. The pitot readings were as follows:
   a. outlet 1: pitot pressure = 14.7 psi
   b. outlet 2: pitot pressure = 15.5 psi
   c. outlet 3: pitot pressure = 16.1 psi
   d. outlet 4: pitot pressure = 16.6 psi
   Determine the total flow in gpm, and graph the results.

4. Calculate the minimum thrust bearing area required for the following situations:
   a. a 10-inch underground tee installed in sand and gravel
   b. an 8-inch underground cap installed in firm soil
   c. a 16-inch underground 45° ell installed in shale
   d. a 4-inch underground 90° ell installed in sand

5. A 16-inch underground fire protection water main is to be installed in firm soil. The finished grade elevation is 100.00′, the frost line elevation is 96.00′, and a 10-inch sewer main with an invert elevation of 95.00′ crosses the fire protection underground main. Determine the highest invert elevation at which the 16-inch fire protection underground pipe may be installed, assuming a minimum depth of cover of 4 feet.

Activities

1. Obtain a set of civil drawings and determine the invert elevations of all utilities on the plan. Take note of the invert elevation of the water main adjacent to a particular building of interest, and determine the finished floor elevation of the building. Select a routing for a proposed underground fire protection main to the building and determine the invert elevation of the main.

2. Obtain a set of structural drawings of a building and determine the elevation and width of the building footings from the structural drawings.

3. Contact a sprinkler company or an engineering or architectural firm and make arrangements to witness a water flow test. Take notes of all test data, calculate the flow, and plot flow test results.

4. Arrange to visit a jobsite during the installation of an underground fire protection main. Make notes relative to underground piping installation, trench preparation, thrust block construction, underground pipe testing, and backfill procedures.

Chapter 5

Automatic Sprinkler Systems

Objectives

Upon completion of this chapter, you should be able to:

- Properly space and dimension sprinklers for a given occupancy.
- Describe the major differences and similarities between wet pipe, dry pipe, preaction, and deluge sprinkler systems.
- Explain the following terms as defined by NFPA 13: shall, should, approved, listed, fire suppression, and fire control.
- Explain the differences between old style and standard spray sprinklers, upright and pendent sprinklers, quick response and standard response sprinklers, and ESFR and large drop sprinklers.
- Detail the uses and applications for cast iron and malleable iron fittings used on sprinkler systems.
- Describe the difference between Mylar, vellum, and onion skin drawing media.
- Explain why sprinkler branch lines are oriented perpendicular to the bar joists or purlins of a roof.

- Determine the differences between a gridded sprinkler system, a tree sprinkler system, and a looped sprinkler system.
- Discuss the meaning and importance of occupancy relative to fire protection design, and determine an occupancy classification for a given facility.

Standard for the Installation of Sprinkler Systems

Automatic sprinkler systems are designed in accordance with **NFPA 13**, *Standard for the Installation of Sprinkler Systems*. NFPA 13 provides the minimum requirements for the design of sprinkler systems protecting buildings of a wide variety of uses. In buildings used for storage, NFPA 13 pertains to buildings where storage height does not exceed 12 feet, or where high-hazard storage, such as plastics and rubber tires, does not exceed 5 feet. NFPA 13 mandates that sprinklers be installed in all areas of a building and does not sanction partial sprinkler protection for a structure, with the exception of some rooms in residential occupancies, as described in this chapter, where sprinklers are permitted to be omitted.

SPRINKLER SYSTEMS—MYTH VERSUS FACT

Before reviewing the details necessary for the proper spacing of sprinklers, all sprinkler system designers must know the facts about sprinkler systems and be able to explain the facts relative to several commonly held myths about sprinkler systems.

Myth When a sprinkler system actuates, all sprinklers on the system go off at the same time.

Fact Each sprinkler has a heat-sensitive element with a predetermined temperature and sensitivity that responds individually to heat from a fire. Only those sprinklers in the immediate vicinity of the fire actuate and discharge water. It is not uncommon for only one or two sprinklers to go off in a fire.

Myth Sprinkler systems cause excessive water damage.

Fact In most cases, water flowing from a sprinkler causes much less damage than the fire would have caused in the absence of the sprinkler. A fire has the potential to completely destroy a building, and sprinkler systems have a solid record of performance in saving lives and property, with minimal water damage. It is also important to note that the water damage from a fire hose operated by the fire service during fire fighting operations in an unsprinklered building could greatly exceed water damage from a sprinkler, because the rate of discharge from a fire hose is several times the rate of discharge from a sprinkler.

Myth Sprinkler systems don't work.

Fact Most studies of sprinkler system effectiveness show that sprinkler systems are between 98% and 99.8% effective in the control of fire. The majority of incidences of ineffectiveness are related primarily to the failure of building owners in keeping sprinkler control valves in the open position.

Myth Automatic fire detection systems are an acceptable substitute for sprinklers.

Fact Automatic fire detection systems do not control or suppress a fire and are not a substitute for an automatic sprinkler system. Detection systems have a good record of providing notification, but do not provide suppression.

Myth Accidental sprinkler discharge is common.

Fact Sprinklers have an impressive history of reliable service in an emergency, and discharge in the absence of a fire is very rare.

Myth Sprinkler systems cost too much.

Fact Sprinkler systems are not prohibitively expensive. They usually are only a small fraction of the total cost for a building. It is not uncommon for an owner of a commercial property to recover the cost of the sprinkler system in 5 to 10 years through insurance rate reductions. A residential sprinkler system connected to a public water supply usually costs between $1.00 and $1.50 a square foot, less than the cost of most carpets!

Myth Sprinklers are ugly.

Fact Sprinkler systems can be designed to maintain the beauty of a building. Pipes can be concealed above ceilings or behind soffits, and sprinklers can be selected from a wide range of aesthetic models. Ornamental and decorative sprinklers are available that allow concealment above the ceiling, with only a small plate showing below the ceiling, with a wide range of factory-applied colors. While sprinklers can be concealed, management of a building may elect to use the visible presence of sprinklers as a sales tool to emphasize life safety in a building, especially in hotels.

STANDARDIZED AUTOMATIC SPRINKLER SYSTEM DESIGN

Automatic sprinkler system design and installation standardization is necessary to:

- Establish a minimum criteria for life and property protection for a building and its occupants.
- Maintain an impressive level of sprinkler system performance.

Chapter 5 Automatic Sprinkler Systems

- Maintain a single baseline reference for the proper use of the latest technologies.
- Provide a minimum level of protection for fire service personnel who enter a sprinklered building and rely on the protection that sprinkler systems deliver.
- Maintain public trust in automatic sprinkler systems. The public has every right to expect that a sprinkler system works properly when necessary. When the public sees sprinklers in a building, it should feel safe within that building.
- Provide a reference for an AHJ to ensure that minimal requirements are met.

SPRINKLER SYSTEM DEFINITIONS

A sprinkler systems designer must understand the language used in an NFPA document. Each document published by the NFPA begins with a list of definitions that help a designer to interpret the provisions of that document.

"Approved" versus "Listed"

Sprinkler system designers are required to know the role of the National Fire Protection Association. The NFPA approves nothing, but develops consensus fire protection standards that may then be adopted in whole or in part to become code or law by an AHJ, such as a state, a city, a county, a municipality, or an insurance company. The AHJ is the individual or agency responsible for the review and testing of sprinkler designs and installations for compliance with the code or law. If a sprinkler system meets the minimum requirements, the system is **approved** by the AHJ.

NFPA 13 requires that sprinkler components be **listed** for use in sprinkler systems by an approved testing laboratory, such as Underwriters Laboratories (UL), and is listed in a document, such as the UL *Fire Protection Equipment Directory*. The directory is testimony that a component has been tested against an established set of criteria and performed in accordance with that minimum criteria.

"Shall" versus "Should"

Any paragraph in an NFPA standard that uses the word **shall** constitutes a mandatory requirement, and any paragraph using the word **should** is a recommendation or suggestion but not a mandatory requirement.

With regard to NFPA standards, paragraphs using the word "shall" appear in the **body of the standard**, that portion of the standard containing the mandatory requirements for sprinkler system design and installation. Paragraph numbers followed by an asterisk (i.e., paragraph 1-1.1*) have supplementary material in the

---

approved
meets the minimum requirements of the authority having jurisdiction

listed
included in a document published by an approved testing laboratory, such as Underwriters Laboratories

shall
identifies wording that constitutes a mandatory requirement

should
identifies a recommendation or suggestion, but not a mandatory requirement

body of the standard
that portion of the standard containing mandatory requirements for sprinkler system design and installation

appendix
portion of the standard containing explanatory material and generally using the word "should"

fire control
application of water to limit the size of a fire and allow safe occupant egress and structural protection until the fire service arrives

fire suppression
application of water such that the release of heat from a fire is sharply reduced and the regrowth of the fire is prevented

fire service response time
the time it takes for the fire department to arrive at the fire scene after notification

old style sprinkler
a sprinkler that sprayed 40 to 60% of its water upward, with the remainder sprayed below

spray sprinkler
a sprinkler designed to discharge all of its water downward, with a deflector that creates smaller water droplets than old style sprinklers

appendix of the standard. Each paragraph in the appendix is identified by a paragraph number preceded by a letter, (i.e., paragraph A-1-1.1 is located in Appendix A of the standard). Paragraphs in the appendix contain explanatory material and generally use the word "should."

"Fire Control" versus "Fire Suppression"

NFPA 13 design requirements are intended to provide a minimum application of water to a fire that should achieve **fire control**, or the application of water that limits the size of a fire and allows safe occupant egress and structural protection, until the fire service arrives. Sprinklers for nonresidential occupancies work by structural cooling and prewetting fuel not yet burning. Residential sprinklers provide the minimum densities necessary to allow time for occupant egress.

Fire suppression is the application of water such that the release of heat from a fire is sharply reduced and the regrowth of the fire is prevented. The early suppression fast response (ESFR) sprinkler system is an example of an NFPA 13 system designed for fire suppression.

Fire service response time is the time it takes for the fire department to arrive at the fire scene after notification, and varies according to the distance from the firehouse to the fire, the road conditions between the station and the fire, fire service equipment condition, fire service personnel training, and numerous other factors, but is usually between 5 and 10 minutes for many areas. An AHJ may consider response time as a factor when evaluating system design criteria for a specific jurisdiction.

SPRINKLER TYPES

The growth in the variety of sprinklers to protect various applications is one of the more noticeable changes in sprinkler system technology in the last 25 years. For a sprinkler designer to understand the staggering number of sprinklers, it is necessary to study the basic classifications of available sprinklers.

Old Style and Spray Sprinklers

Prior to World War II, buildings were predominately built of wood, and sprinklers manufactured at that time provided protection above the sprinkler for the a combustible roof and provided direct water application below the sprinkler for the protection of combustibles at the floor. The **old style sprinkler** was designed for the protection of this hazard, and sprayed 40 to 60% of its water upward, with the remainder sprayed below.

Water fog usage for fire fighting during World War II caused researchers to believe that putting finer sprays directly on a fire creates steam for cooling of the flame. Also, as steel construction became predominant, a sprinkler was needed that provided no spray above the sprinkler. The **spray sprinkler**, first approved in

Chapter 5 Automatic Sprinkler Systems

standard spray upright (SSU)
a sprinkler mounted upright above a branch line, usually in an a room with exposed structural elements

deflector
metal plate on a sprinkler that distributes water from the sprinkler

standard spray pendent (SSP)
a sprinkler mounted below the branch line, usually at or below the surface of a suspended ceiling

sidewall sprinkler
a sprinkler with a specially designed deflector that allows the sprinkler to discharge water from a wall-mounted position

extended coverage sprinkler
a sprinkler designed to discharge water over larger areas than standard sprinklers

quick response sprinkler
a sprinkler with a specially designed response element that allows the rapid transfer of heat to the heat responsive element

1952 by Factory Mutual Engineering Corporation, was designed to discharge all of its water downward, with a deflector that created smaller water droplets than the old style sprinkler. The spray sprinkler was recognized in the 1955 edition of NFPA 13 and remains the current standard sprinkler in use today. Old style sprinklers are still manufactured for use in special applications, such as piers and wharves, where the hazard is both above and below, and fur vaults, where ceilings are often made of cedar, and where storage contents are chemically treated garments wrapped in plastic. Illustrations of the basic sprinkler types are shown in Figure 5-1a and 5-1b.

Upright, Pendent, and Sidewall Sprinklers

Spray sprinklers are manufactured in three basic styles. A **standard spray upright** (SSU) sprinkler is mounted upright above a branch line, usually in a room with exposed structural elements, and has a **deflector**, a metal plate whose edge is distinctively bent to deflect water downward from the sprinkler. A **standard spray pendent** (SSP) sprinkler is mounted below the branch line, usually mounted at or below the surface of a suspended ceiling and is characterized by a flat deflector. SSU and SSP discharge patterns are designed to be the same. **Sidewall sprinklers** have a specially designed deflector that allows the sprinkler to discharge water from a wall-mounted position.

Variations on upright, pendent, and sidewall sprinklers are the dry upright, dry pendent, and dry sidewall sprinklers. These special sprinklers are manufactured with a seal at the inlet that prevents water from entering the nipple until the sprinkler actuates. These sprinklers can be extended from a wet pipe system into an unheated area, such as a walk-in freezer, an outside loading dock, or an unheated attic space, or can be used in the pendent position on a dry pipe system to prevent condensation from entering the sprinkler and freezing.

Sprinkler Coverage

Another recent development in sprinkler manufacturing is the **extended coverage sprinkler**. The deflectors on these sprinklers are designed to discharge water over larger areas than standard sprinklers. While this means that each sprinkler must have a higher flow, it results in increased allowable distances between sprinklers, with the possibility that fewer sprinklers may be required in a compartment of given size. The 1996 edition of NFPA 13 added information that details spacing requirements for extended coverage sprinklers.

Sprinkler Response

A major development in sprinkler technology was the **quick response sprinkler**, a sprinkler with a specially designed response element that allows the rapid transfer of heat to the heat-responsive element and provides a more expeditious deliv-

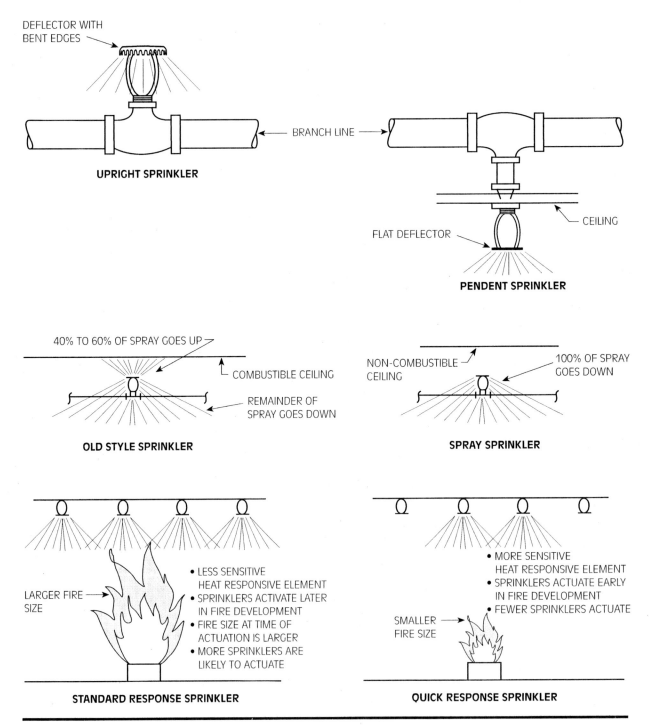

Figure 5-1a *Sprinkler differences.*

Chapter 5 Automatic Sprinkler Systems

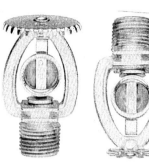

STANDARD SPRAY UPRIGHT (SSU) STANDARD RESPONSE EUTECTIC ELEMENT

STANDARD SPRAY PENDENT (SSP) STANDARD RESPONSE EUTECTIC ELEMENT

STANDARD SPRAY UPRIGHT (SSU) QUICK RESPONSE

STANDARD SPRAY PENDENT (SSP) QUICK RESPONSE

OLD STYLE SPRINKLER

DRY PENDENT SPRINKLER

SIDEWALL SPRINKLER STANDARD RESPONSE EUTECTIC ELEMENT

SSU-BULB SPRINKLER STANDARD RESPONSE GLASS BULB ELEMENT

SSP-BULB SPRINKLER STANDARD RESPONSE GLASS BULB ELEMENT

IN-RACK SPRINKLER

ESFR SPRINKLER

RESIDENTIAL SPRINKLER

FLUSH CONCEALED SPRINKLER

Figure 5-1b *Types of sprinklers. (Courtesy Grinnell Fire Protection Systems Co.)*

ery of water to the fire. At the time of sprinkler actuation, the fire should be smaller with a quick response sprinkler than with a standard response sprinkler, resulting in fewer sprinklers likely to actuate. Sprinkler elements are either glass bulbs or eutectic (soldered) elements.

The response of quick response and standard response sprinklers can be modeled by designers to estimate the time for a sprinkler to actuate, given ceiling height, fire size, and response time index (RTI), using the DETACT computer program, developed by the Building and Fire Research Laboratory at the National Institute of Standards and Technology (NIST). Response time index is a measure of the sensitivity of a sprinkler heat responsive element, with low RTI values being very sensitive to actuation by heat, and high RTI values being less sensitive.

Sprinklers for Storage Applications

ESFR and **large drop** sprinklers are for specialized high challenge fire hazards. An ESFR sprinkler has a quick response element and an extra large orifice that allows for greater amounts of water to be applied early in the fire development, usually on a high challenge fire. Large drop sprinklers have large deflectors with widely spaced teeth to facilitate the creation of large water drops intended to penetrate a high velocity fire plume without evaporating, resulting in some water reaching the burning materials. The fire plume is the column of smoke and gas traveling upward from a fire.

SPRINKLER SYSTEM TYPES

Sprinkler systems are designed to successfully operate in a variety of temperature conditions and for a variety of system functions.

Wet Pipe Systems

A **wet pipe system** has water in the pipes in the ambient or normal condition and has heat responsive elements on all sprinklers. Thus, water is instantaneously discharged from a sprinkler when it actuates.

Dry Pipe Systems

In areas where low temperatures could cause a wet pipe system to freeze, a **dry pipe system** is intended for use. Dry pipe systems are pressurized with air in the ambient condition and experience an inherent delay in the discharge of water to allow the pressurized air in the system to escape. When a sprinkler actuates, air is released through the sprinkler, allowing water to flow into the piping system through the dry pipe valve. NFPA 13 mandates that the time for the water to reach the most remote sprinkler be no longer than 60 seconds. This time delay allows the fire to grow larger than it would with a wet pipe system of similar design, and the larger fire size results in more sprinklers in the fire area actuating.

ESFR (early suppression fast response) sprinkler a sprinkler with a quick response element and an extra large orifice that allows for larger amounts of water to be applied early in the fire's development

large drop sprinklers sprinklers with deflectors that create large water drops intended to penetrate a fire plume without evaporating

wet pipe sprinkler systems systems with sprinklers having individual heat-responsive actuating devices and with piping filled with water in the ambient or normal condition

dry pipe sprinkler systems systems with air-filled piping in the ambient condition and sprinklers with individual heat-responsive actuating devices

Chapter 5 Automatic Sprinkler Systems

quick opening device
a device installed to rapidly remove air from the system and speed the operation of the dry pipe valve

preaction sprinkler systems
systems having sprinklers with individual heat-responsive actuating devices, and piping filled with pressurized air

deluge sprinkler systems
systems with open sprinklers that discharge water simultaneously from all sprinklers on the system

Figure 5-2a *Wet pipe system schematic. (Courtesy Grinnell Fire Protection Systems Co.)*

Figure 5-2b *Dry pipe system schematic. (Courtesy Grinnell Fire Protection Systems Co.)*

To limit the size of dry pipe system, a volumetric limitation with a maximum capacity of 750 gallons is placed on dry pipe systems. A **quick opening device**, such as an accelerator or exhauster, is installed to rapidly remove air from the system and speed the operation of the dry pipe valve and is required when the system volumetric capacity exceeds 500 gallons.

Figure 5-2a, b, c, and d illustrates the four basic types of sprinkler systems.

Preaction and Deluge Systems

Preaction systems and **deluge systems** require fire detectors for the actuation of the system. A deluge system uses open sprinklers or nozzles, so that all flow water is discharged when the deluge valve actuates. Deluge systems can be used for occupancies where the hazard is considered severe, such as with flammable liquid hazards where the fire could spread over a large floor area.

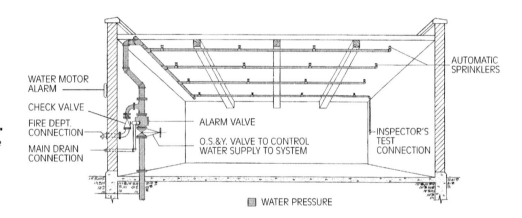

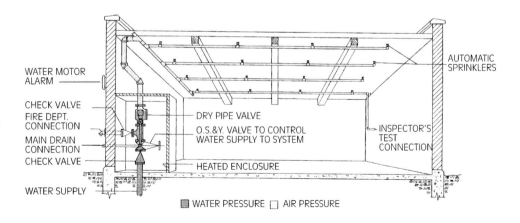

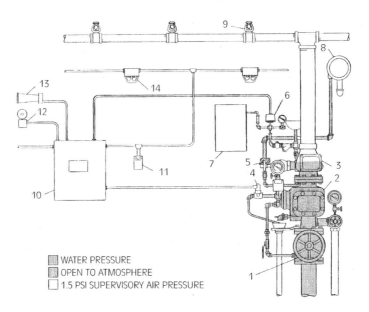

Figure 5-2c *Preaction system schematic. 1, OS&Y valve; 2, deluge valve with basic trim; 3, check valve; 4, solenoid valve and electric actuation trim; 5, water pressure alarm switch; 6, 1.5 psi low air pressure alarm switch; 7, 1.5 psi supervisory air pressure control; 8, water motor alarm; 9, automatic sprinklers; 10, deluge releasing panel; 11, electric manual control stations; 12, fire alarm bell; 13, trouble horn; 14, heat detectors. (Courtesy Grinnell Fire Protection Systems Co.)*

Preaction systems have closed heads and pipes filled with pressurized air that supervise a piping system, and can be considered for the protection of valuable or irreplaceable property. The detection system for a preaction system can be designed to prevent water discharge in cases of a false alarm from the detection system, or in case of a sprinkler whose element has encountered mechanical damage.

The detection system on a preaction system can be designed with a preaction logic capable of meeting one of the following objectives:

- Actuation of a fire detector trips a deluge valve to admit water into the sprinkler piping to await the actuation of a sprinkler.
- Actuation of a fire detector or actuation of a heat-responsive element on a sprinkler trips a deluge valve to admit water into the sprinkler piping.
- Actuation of a fire detector and actuation of a heat-responsive element on a sprinkler trips a deluge valve to admit water into the sprinkler piping.

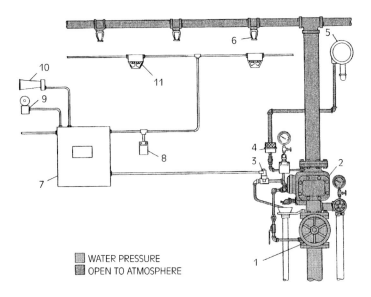

Figure 5-2d *Deluge system schematic. 1, OS&Y valve; 2, deluge valve with basic trim; 3, solenoid valve and electric actuation trim; 4, pressure alarm switch; 5, water motor alarm; 6, spray nozzles or open sprinklers; 7, deluge releasing panel; 8, electric manual control stations; 9, fire alarm bell; 10, trouble horn; 11, heat detectors. (Courtesy Grinnell Fire Protection Systems Co.)*

SPRINKLER SYSTEM COMPONENTS

Components used on an automatic sprinkler system must be listed for fire protection service and rated to withstand the maximum pressures anticipated for the system. The minimum pressure rating required by NFPA 13 for all system components is 175 psi.

Sprinklers

Sprinklers installed in a building must be new, must be of a temperature rating commensurate with its anticipated environment, must have factory-applied coatings where corrosive atmospheres exist, and must have permanent markings that indicate year of manufacture, temperature rating, listing mark of Underwriters Laboratories or Factory Mutual, sprinkler model number, and manufacturer. Once a sprinkler actuates, it may not be reused. Spare sprinklers are to be provided in a cabinet near the riser, for replacement of actuated sprinklers in a timely manner.

Piping and Tubing

Steel piping and copper tubing and their associated fittings must meet the standards established by the American Society for the Testing of Materials (ASTM) and the American National Standards Institute (ANSI), who list numerous items, including pipe and fittings for fire protection service. When a pipe diameter is given, a standard dimension exists for the exterior diameter of the pipe so that fittings can fit snugly and reliably on sprinkler pipes. The interior diameter of a steel pipe varies in accordance with its pipe schedule number, with schedule 5 having the greatest internal diameter and the thinnest pipe wall thickness, and schedule 80 having a considerably smaller internal diameter. A low pipe schedule number allows for more water-holding capacity and enhanced hydraulic characteristics, as illustrated in Figure 5-3.

Most sprinkler systems with 175 psi maximum pressures use schedule 5, 10, or 40 pipe. Pipe encountering pressures in excess of 175 psi may require schedule 80 pipe.

Fittings

Fittings are manufactured in many varieties. Cast iron fittings and malleable iron fittings are commonly used threaded fittings for fire protection service. A cast iron threaded fitting could break if hit with a hammer or thrown against a masonry wall, whereas a malleable iron threaded fitting would not. Malleable iron fittings are more expensive and manufactured in a limited variety of fitting sizes, but are preferred in cases where damage to the sprinkler system is probable, such as in storage racks, where forklift trucks frequently hit the rack uprights or the sprinkler pipe when placing loads in the racks. Cast iron fittings, when used in this situation, may crack upon direct impact or upon deflection of the pipe when the rack is hit.

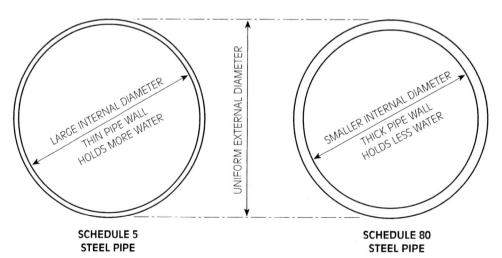

Figure 5-3
Illustration of pipe thickness.

Joining methods for sprinkler fittings include:
- threaded, cast iron or malleable, shown in Figure 5-4,
- welded, with shop-welded outlets, shown in Figure 5-4,
- flanged, where fittings are bolted together, shown in Figure 5-4,
- crimped, using a listed pipe crimping tool,
- drilled hole, with rubber-gasketed fittings clamped over the hole,
- cut or roll-grooved, with rubber-gasketed fittings, per Figure 5-4,
- brazed or soldered, for copper tube, shown in Figure 5-4,
- pipe bending, per Figure 5-5,
- glued, for plastic pipe.

THREADED ENDS

THREADED ENDS ARE TAPPED WITH ANSI STANDARD FEMALE TAPER PIPE THREADS

WELDED ENDS

WELDED END FITTINGS ARE RECOMMENDED WHERE HIGH PRESSURES ARE ENCOUNTERED AND TIGHT, LEAKPROOF CONNECTIONS MUST BE MAINTAINED OVER A LONG PERIOD OF TIME

FLANGED ENDS

FLANGED ENDS MAKE A STRONG, TIGHT JOINT AND ARE GENERALLY USED FOR PIPE SIZES ABOVE 3" THAT ARE FREQUENTLY DISASSEMBLED AND ASSEMBLED

SOLDERED

SOLDERED ENDS ARE USED WITH TYPES K, L, AND M COPPER TUBING

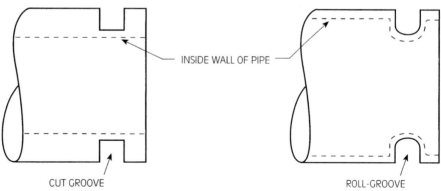

CUT-GROOVED PIPE
- PIPE MATERIAL IS REMOVED AT GROOVE
- PIPE THICKNESS IS DECREASED AT GROOVE

ROLL-GROOVED PIPE
- PIPE IS COMPRESSED INTO GROOVE
- NO PIPE MATERIAL REMOVED
- PIPE THICKNESS REMAINS APPROXIMATELY UNIFORM

Figure 5-4 *Piping connections.*

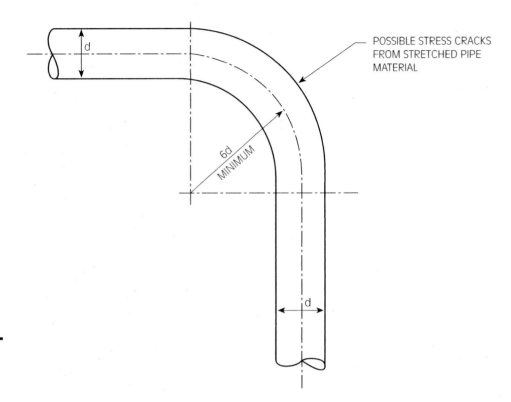

Figure 5-5 *Pipe bending per NFPA 13.*

Hangers

Hangers supporting sprinkler pipes must be capable of holding five times the weight of the sprinkler pipe, fittings, and water supported by the hanger, plus 250 pounds. Figure 5-19 on page 107 illustrates a wide variety of hangers commonly used with sprinkler piping. Sprinkler piping must be oriented perpendicular to the bar joists, as shown in Figure 5-8 on page 92, to facilitate the hanging of the pipe. Crossmains are ordinarily oriented parallel to the bar joists. A method for determining the hanging of crossmains follows later in this chapter.

Valves

Sprinkler systems have been shown to be 98 or 99% effective when called upon to respond in a fire situation. Of the 1 or 2% of instances where sprinkler systems fail to control a fire, closed valves are the most prominent source of failure. In order to try to ensure that valves remain open, valves controlling sprinkler systems must be listed and of a type that can be visually identified as open or closed. All listed valves controlling fire protection systems are designed to close in no less than 5 seconds to minimize a phenomenon known as water hammer, which is characterized by a rapid increase in pressure within the pipe. Forces created by

Chapter 5 Automatic Sprinkler Systems

OS&Y valve
a listed indicating control valve that is determined to be in the open or closed position by observing the position of the stem with respect to the valve wheel

rapid valve closure can cause pipe movement that could result in the breaking of fittings and hangers and the possibility of the entire system falling to the floor.

Outside screw and yoke valves, or **OS&Y valves**, can be determined to be in the open or closed position by observing the position of the stem with respect to the valve wheel, as shown in Figure 5-6a.

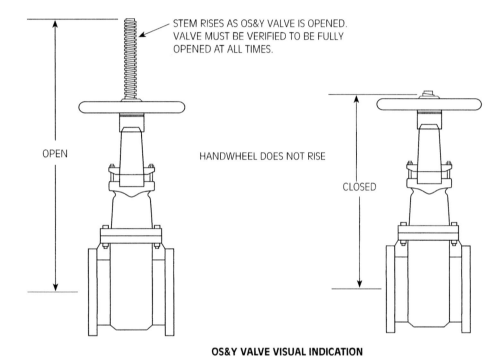

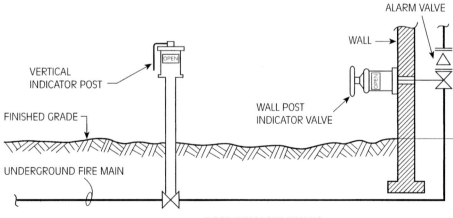

Figure 5-6a *OS&Y valves and post indicator valves.*

Figure 5-6b   *Fire protection valves. (Courtesy Grinnell Fire Protection Systems Co.)*

butterfly valve
a listed indicating control valve that has a dial showing the position of the valve

post indicating valve
a listed indicating valve with a shaft extending through a wall or from an underground fire protection main

A **butterfly valve**, shown in Figure 5-6b, is a type of listed indicating valve permitted for fire service that has a dial that visually indicates the position of the valve.

A **post indicating valve**, shown in Figure 5-6b, is a valve with a window that reads "open" or "shut" and is manufactured in two varieties:

1. A wall post indicator valve, with a shaft that extends through a wall for the operation of a sprinkler control valve from outside a building.
2. A post indicator valve, with a shaft that extends above grade from an underground fire protection main, for the operation of an underground water supply.

OS&Y, butterfly, and post indicator valves can also be locked open with padlocks and chains, or monitored electrically in the open position with tamper switches that send a signal when the valve is not fully open.

DEVELOPING A SPRINKLER PLAN FOR APPROVAL

To prepare a sprinkler plan for approval, it is necessary to:
- Collect and review data.
- Review project schedule.
- Trace or scan building outline.
- Draw necessary building details.
- Determine branch line logic.
- Determine system type and configuration.
- Determine the hazard class of the occupancy.
- Determine the area protected by each sprinkler.
- Determine the number of branch lines.
- Determine the distance between branch lines.
- Determine the maximum allowable distance between sprinklers.
- Determine the minimum number of sprinklers on each branch line.
- Determine the actual distance between sprinklers.
- Dimension sprinklers and piping on your drawing.
- Verify sprinkler coverage.
- Determine crossmain location.
- Assign elevations to piping.
- Locate system risers.
- Select hangers and determine hanger rod lengths.
- Determine hanger spacing.
- Add final details.
- Use the NFPA 13 design checklist.
- Perform precalculation field check.
- Add pipe lengths and verify accuracy.

Collect and Review Data

Chapters 1 and 2 detailed the wealth of information contained within the contract documents. Attempting to design a fire protection system with an incomplete set of contract documents is somewhat akin to driving with blinders.

A fire protection system is an integral part of a building and must therefore be designed with consideration for all other building components. Insisting upon anything less than a complete set of the most recent contract documents could jeopardize the ability of the fire protection system to function properly. By studying the contract drawings in advance of performing system layout, conformance to the documents should be enhanced.

Forming an adjunct to the contract documents is the contract between the fire protection design firm and the owner. The contract consists of a fire protection design and installation contract, fire protection system estimate sheets, a fire protection system preliminary plan, and preliminary hydraulic calculations. A building owner will insist that conflicts between the sprinkler contract and the contract documents be resolved in favor of the contract documents, therefore all conflicts should be resolved before the commencement of the design of a fire protection system.

The fire protection system designer must determine whether shop drawings, such as HVAC shop drawings, are going to be produced. While these drawings are not likely to be completed before the commencement of the fire protection system design, knowledge of their existence determines the method of coordination that is performed, as discussed in Chapter 12. It is imperative that an understanding or commitment be made with respect to the manner in which the mechanical trades are coordinated before installation.

The applicable codes and standards are likely to be referenced by the contract drawings or specifications and serve as your guide throughout the design process. These codes and standards are amended by the requirements of the contract documents, the AHJ, and the insurance company. Balancing and resolving these sometimes conflicting requirements is an integral part of the design of a fire protection system.

The best way to ensure that all relevant information is available and in hand is for the fire protection system designer to arrange or attend a project commencement meeting with representatives from all mechanical trades in attendance. At this meeting, a review of all relevant data can be performed, responsibilities of each contractor can be assigned, and agreements can be made. Each attendee should be prepared to take action once agreements are finalized.

Review Project Schedule

Sadly, too often the sprinkler design and installation contract are among the last contracts to be issued by a general contractor or building owner. The result is that the sprinkler design and installation schedule is almost always tight. Architects and engineers would be well advised to plan ahead and schedule the fire protection aspects of a building concurrently with its design. By reviewing the schedule in detail, it can be determined how many designers are needed on a particular project and how much time each designer has available to perform the work.

Project schedules are usually developed in reverse starting at the date of delivery of fire protection materials to the job site. The amount of time to perform design and installation tasks is determined by apportioning the time available between the receipt of the contract and the scheduled date of completion.

Special order items, those requiring several weeks to arrive, should be ordered as soon as the contract is signed. Fire pumps, air compressors, tanks, strainers, and electrical items with unusual power requirements would probably fall into this category.

To make your scheduling of the work more accurate, an initial building progress visit at this time is a good idea to determine whether the building progress matches the project schedule.

Trace or Scan Building Outline

Choose an appropriate size reproducible medium for your fire protection system design. Reproducible media come in several varieties. Mylar is a durable clear plastic that preserves the drawing for years. Other translucent or semitransparent media, such as onion skin or vellum, are thinner and more prone to pencil damage, wrinkling, or being crushed or folded in the plan storage drawer. Mylar is more expensive than vellum and onion skin, but it provides a much more stable background for making changes and has a longer life span. Single matte Mylar is shiny on one side and dull on the other. The dull side is used for drawing. Double matte Mylar is dull on both sides.

The size of your drawing is determined by several factors. It must be at least as big as the drawing of the building or portion of building that you are tracing. It is recommended that a scale no smaller than 1/8″ = 1′-0″ be used for sprinkler design work. Enough space must be left for general notes, title blocks, sections, and hydraulic results. Should space permit, perhaps more than one floor or area may fit on the drawing with a key plan that shows a diagram of the entire area protected by a fire protection system, and with the area detailed by a particular drawing indicated. The introduction of this book provides additional information for laying out a drawing.

> **It is recommended that a scale no smaller than 1/8″ = 1′-0″ be used for sprinkler design work.**

Another consideration in choice of drawing size is plan storage arrangement. A small drawing in a large plan storage drawer may gravitate to a position within the drawer such that the drawing might become prone to damage or difficult to find. Using a uniform drawing size makes plan storage easier.

The building outline chosen for tracing depends on whether the fire protection system is exposed or concealed. If piping is to be exposed, the roof framing plans, from the structural section of the contract drawings, should be traced or scanned. If the system is to be concealed above a suspended ceiling, the reflected ceiling plan should be traced or scanned.

Draw Necessary Building Detail

There is an abundance of information in the contract drawings that applies to your fire protection layout. Sorting through this information and knowing what minimum information is needed is key to the success of your design.

For a drawing of a building with a suspended ceiling, your drawing should show the building outline, the roof structure, and the reflected ceiling plan. The reflected ceiling plan should show all of the walls that extend to the ceiling, the ceiling tile arrangement, and lighting locations.

For a drawing of a building with no ceiling, the drawing should show all walls that extend to the roof deck, and any **draft curtains** that suspend from the roof of the structure to limit the spread of smoke. These should be drawn as light solid lines. Walls that do not extend to the ceiling or up to the structure may be ignored if they do not extend to within 18 inches of the ceiling. No other features need to be drawn unless they pose a potential conflict.

Duct supply and return diffusers should be shown on your reflected ceiling plan. In exposed areas such as warehouses, **unit heaters**, modular heating units that project heat, must be drawn. Sprinklers located in zones shown in Figure 5-7 must reflect the appropriate temperature ratings shown. It is a good idea to make a scaled template of this figure in clear plastic or cardboard, in various scales, so that these temperature patterns can be quickly and efficiently traced onto the drawing.

Sources of conflict with the installation of sprinkler piping or sprinkler discharge should be drawn. Examples of conflicts are lights and ducts in exposed or concealed areas. Drawing the duct outline lightly onto the fire protection system drawing could save time in the coordination process. If the potential sources of conflict are drawn lightly, they will not appear on a print if the print is run through the reproduction machine at a fast speed. By running the print slowly through the machine, the ducts will appear on the print for reference at coordination meetings and field checks. By showing the ducts, you are making the process of revising, updating, and recoordinating the drawings much simpler.

It may help to use a color coding scheme for your drawing, such as light pencil for the building outline and structure, thin red lines for the ceiling tiles, and thin green lines for the ducts and diffusers. The sprinkler piping and heads can then be drawn in bold, dark pencil or erasable pen.

If computer-aided design is used, showing the necessary building detail is considerably simpler. Each source of potential conflict can be shown on the drawing for coordination purposes, then removed when making a print of the drawing.

---

draft curtains
barriers suspended from the roof of the structure to limit the spread of smoke

unit heaters
modular units that project heat in exposed areas

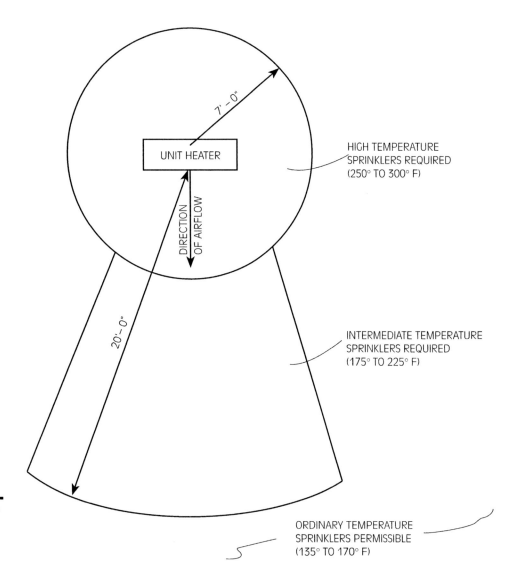

Figure 5-7 Sprinkler temperature profiles. Source: NFPA 13

Determine Branch Line Logic

Branch lines are pipes to which sprinklers are directly attached and are always installed perpendicular to the bar joists or purlins. They are fed by crossmains and feedmains, as shown in Figure 5-8. NFPA 13 requires that branch line hangers be spaced per Figure 5-8. By running perpendicular to the joists, spacing of branch lines is more flexible and NFPA hanging requirements are enhanced. Bar joists are not always installed straight and true, and a joist in one bay may not line up with

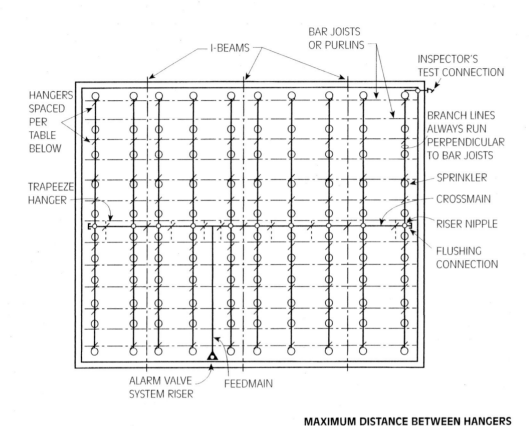

Figure 5-8 *Branch line logic for a sprinkler system.*

PIPE DIAMETER		MAXIMUM DISTANCE BETWEEN HANGERS PER NFPA 13
STEEL PIPE	1"–1 1/4"	12'-0"
	1 1/2" – 8"	15'-0"
THREADED LIGHTWALL	1"–3"	12'-0"
COPPER TUBE	3/4"–1"	8'-0"
	1 1/4"–1 1/2"	10'-0"
	2"–3"	12'-0"

a joist in an adjacent bay. For this reason, installing branch lines parallel to a bar joist is not recommended.

Branch lines attached to concrete ceilings capable of supporting sprinkler pipes may be run in either direction, and branch line logic in this case would therefore be primarily a function of system configuration, discussed in the next section.

Determine System Type and Configuration

The major system configurations commonly used for sprinkler systems are the **tree**, the **grid**, and the **loop**. Configurations are selected by considering building geometry and the potential for hydraulic advantage.

The tree is a system of dead-end branch lines that is centrally fed by a crossmain. The crossmain is the pipe that supplies water to the branch lines. Long branch lines should, if possible, be fed so that the crossmain is equidistant from the ends of the branch lines. A drawing showing a center-fed wet pipe tree system is shown in Figure 5-9.

An end-fed tree may be chosen when feeding branch lines that are installed on a flat roof or on a peaked roof where branch lines slope up to the peak of a building, and with the mains at the low point. Figure 5-10 shows an end-fed dry pipe tree system.

tree system
system of dead-end branch lines centrally fed by a crossmain

grid system
system of branch lines interconnected by crossmains on both ends of the branch lines

looped system
a system with crossmains connected at two or more locations

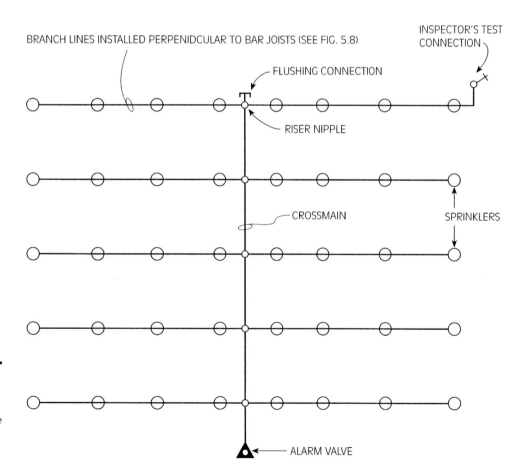

Figure 5-9 *Center-fed wet pipe tree system. Dry pipe systems may also be designed as tree systems.*

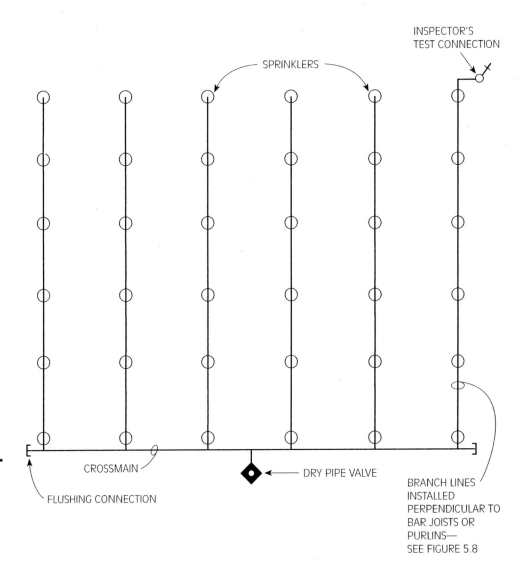

Figure 5-10 *End-fed dry pipe tree system. Wet pipe systems can also be designed as tree systems.*

A grid is a system of branch lines interconnected by crossmains on both ends of the branch lines, as illustrated by Figure 5-11. A grid can be used only for wet pipe systems to provide a hydraulic advantage for systems with numerous branch lines by providing water flow to sprinklers from more than one direction. Dry pipe and preaction grid systems are prohibited because the gridded arrangement makes it harder for the water to push air out of the system. Usually, systems with less than eight branch lines and less than ten sprinklers per branch line are more amenable to a tree system arrangement than to a grid arrangement.

Chapter 5 Automatic Sprinkler Systems

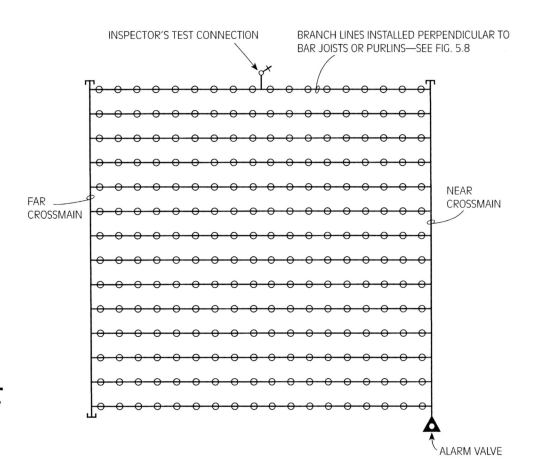

Figure 5-11 *Wet pipe grid system. Dry pipe grids are prohibited.*

occupancy
a function of the expected level of severity of a fire in a particular building or room

fire load density
a function of the quantity, arrangement, combustibility, and rate of heat release of the combustibles in a space

A looped system connects crossmains at two or more locations, as illustrated by Figure 5-12. This arrangement can provide a hydraulic advantage over the traditional tree system by providing water flow to sprinklers from more than one direction.

Determine the Hazard Class of the Occupancy

The most fundamental concept related to sprinkler spacing is the concept of **occupancy**. Occupancy is a function of the expected level of severity of a fire in a particular building or room, given the fire load densities that are associated with a building of a particular use. The **fire load density** is related to the combustible items contained within a building or room and is a function of the quantity, arrangement, combustibility, and rate of heat release of the material. Figure 5-13 illustrates the concept of relative occupancy.

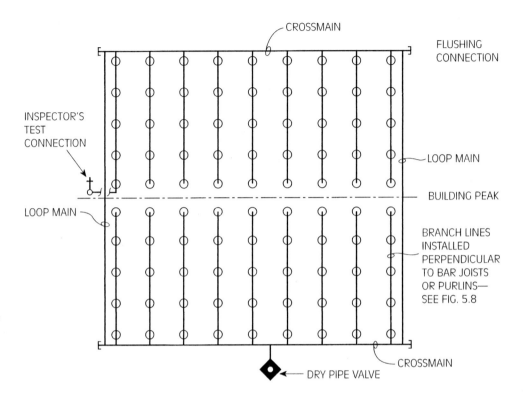

Figure 5-12 *Looped dry pipe system. Wet pipe systems may also be looped.*

The occupancy within a building may differ from room to room, and it is therefore necessary to carefully classify the occupancy of each room or space within a building.

A school classroom, for instance, would contain several small desks spaced a few feet apart. The desks may be made of wood or metal, or a combination thereof. Although each desk may be combustible in whole or in part, the energy required to heat a desk to ignition may require a high ignition temperature or extra sources of combustibles, such as a considerable amount of stacked or crumpled papers, or a liquid fuel source. Since most classrooms do not contain these items, a classroom is customarily considered a light hazard occupancy.

A classroom that has been converted to a storage room, containing stacks of loose papers, boxes of books or papers, lawn mowers, motor oil, gasoline cans, tall shelves filled with plastic or paper items, no longer can be classified as light hazard occupancy. A careful building survey is essential for the accurate assignment of occupancy classifications to a room or building.

Determine the Area Protected by Each Sprinkler (*A*)

Each occupancy classification for a fire protection system has a maximum floor area limitation. A limitation of 52,000 square feet is imposed by NFPA 13 for light

Chapter 5 Automatic Sprinkler Systems

COMPARISON OF:
- FIRE LOAD DENSITY
- COMBUSTIBILITY OF CONTENTS
- RATE OF HEAT RELEASE

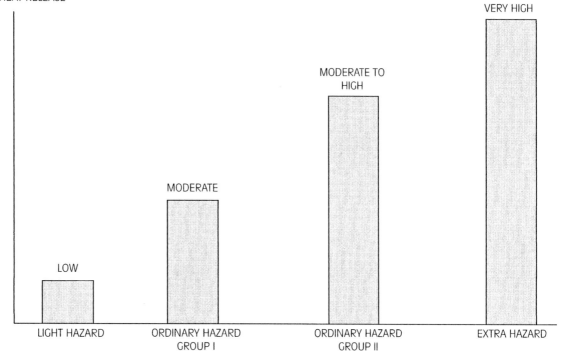

LIGHT HAZARD	ORDINARY HAZARD GROUP I	ORDINARY HAZARD GROUP II	EXTRA HAZARD
EXAMPLES:	**EXAMPLES:**	**EXAMPLES:**	**GROUP I:**
CHURCHES	PARKING GARAGES	CEREAL MILLS	FLAMMABLE LIQUIDS
CLUBS	CAR DEALERS	CHEMICAL PLANTS	FLAMMABLE METALS
EDUCATIONAL	BAKERIES	CONFECTIONARY	WOOD DUSTS
HOSPITALS	BEVERAGE MANUFACTURING	DISTILLERIES	PRINTING INK
PRISONS	CANNERIES	DRY CLEANERS	(LOW FLASH POINTS)
LIBRARIES	DAIRIES	FEED MILLS	PLASTIC FORMS
MUSEUMS	ELECTRONIC PLANTS	HORSE STABLES	
NURSING HOMES	GLASS MANUFACTURING	LEATHER GOODS	**GROUP II**
OFFICES	LAUNDRIES	LIBRARIES-LARGE STACKS	ASPHALT SATURATING
RESIDENTIAL	RESTAURANT SERVICE AREAS	MACHINE SHOPS	FLAMMABLE LIQUIDS
RESTAURANT SEATING		METAL WORKING	SPRAYING
THEATERS		RETAIL AREAS	SOLVENT CLEANING
ATTICS		PAPER MILLS	VARNISH & PAINT
		PIERS & WHARVES	DIPPING
		POST OFFICES	
		PRINTING & PUBLISHING	
		REPAIR GARAGES	
		STAGES	
		TEXTILE MANUFACTURE	
		TOBACCO PRODUCTS	
		WOOD SHOPS	

Figure 5-13 *Occupancy classifications. Source: NFPA 13.*

hazard and ordinary hazard occupancies, while a limitation of 40,000 square feet is imposed for extra hazard occupancies.

Spacing between sprinklers follows the formula:

$$A = S \times L$$

where A = area covered by the sprinkler, in square feet,
S = distance between sprinklers on the branch lines, in feet, and
L = distance between branch lines, in feet.

Maximum square footage limitation per sprinkler (A_{max}), in accordance with NFPA 13, is as follows:

225 square feet for a light hazard calculated system, per Chapter 6,

200 square feet for a light hazard pipe schedule system, per Chapter 7,

168 square feet for a light hazard system installed in a building of combustible construction,

130 square feet for an ordinary hazard system,

100 square feet for an extra hazard system.

The maximum allowable distance between branch lines (L_{max}) and the maximum allowable distance between sprinklers on the lines (S_{max}) cannot exceed 15 feet for light or ordinary hazard, and 12 feet for extra hazard, in accordance with NFPA 13.

Determine the Number of Branch Lines

In exposed construction, the distance between the beams on the column lines is used to determine the L dimension. The space between the beams along the column lines in exposed construction is called a **bay**. In a room with smooth ceilings, the dimension between walls may be considered in a manner similar to a bay, provided no interferences, such as lights or duct diffusers, exist. When these interferences exist, the location of each sprinkler must be coordinated with the obstructions shown in a reflected ceiling plan.

The number of branch lines per bay is determined by dividing the width of the bay, or the distance between column lines in feet, by the maximum allowable distance between sprinklers, (L_{max}). If the result is not a whole number, round up to the nearest whole number. The formula below can be used for this calculation:

$$\text{Number of branch lines} = \frac{\text{(Width of bay)}}{(L_{max})}$$

bay
space between the beams along the column lines in exposed construction

Example 5.1 An ordinary hazard system installed in a bay that is 35'-0" wide would require three branch lines, since NFPA 13 mandates a maximum of 15'-0" between branch lines:

$$\frac{35'}{15'} = 2.3, \text{ rounded to 3 branch lines (see Figure 5-14)}$$

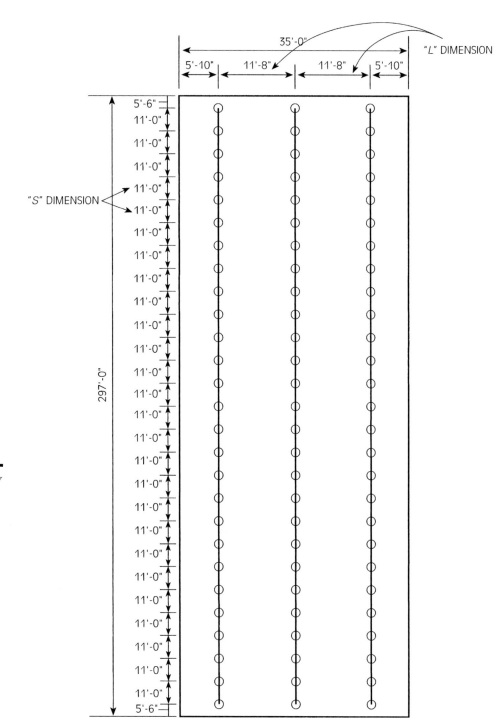

Figure 5-14 *Ordinary hazard spacing example in one bay of a building. Always check to be certain that sprinkler spacing is acceptable:*
- $A = S \times L$
- $11'\text{-}0'' \times 11'\text{-}8'' = 128.3$ *sq. ft.*
- A_{max} *for ordinary hazard is 130 sq. ft.*
- *Layout is acceptable*

Determine the Distance between Branch Lines (L)

The equidistant spacing between the lines in the bay is obtained by dividing the width of the bay by the number of branch lines, using the formula:

$$L = \frac{\text{(Total width of bay)}}{\text{(Number of branch lines)}}$$

Example 5.2 Continuing with the building in example 5.1, to determine the L dimension:

$$L = \frac{(35')}{3 \text{ branch lines}}$$

$$L = 11.667 \text{ feet, or } 11'\text{-}8''. \text{ See Figure 5-14.}$$

This distance, $11'\text{-}8''$, is the L dimension. The distance from the end lines to the walls is:

$$1/2(L) \text{ or } 5'\text{-}10''. \text{ See Figure 5-14}$$

When dealing with foot-and-inch calculations, it is recommended that a foot-and-inch calculator be used for the conversion of decimals to fractions, and vice versa. A conversion table, Figure 5-15, is included for manually converting these units.

Determine the Maximum Allowable Distance between Sprinklers (S)

Although NFPA 13 allows S_{max} to be 15 feet for light hazard and ordinary hazard and 12 feet for extra hazard, S is also constrained by the formula:

$$A = S \times L$$

Example 5.3 For the building used in examples 5.1 and 5.2, we know $A = 130$ square feet maximum for ordinary hazard and $L = 11'\text{-}8''$ as computed above, so we can solve for S, the maximum distance allowed between sprinklers on a branch line. In this case,

$$S = \frac{A}{L}$$

or, for the building in example 5.1:

Chapter 5 Automatic Sprinkler Systems

Figure 5-15 *Decimals of a foot for each quarter of an inch.*

Inches	Decimal of a Foot	Inches	Decimal of a Foot	Inches	Decimal of a Foot
0	0	4	.3333	8	.6667
0-1/4	.0208	4-1/4	.3542	8-1/4	.6875
0-1/2	.0417	4-1/2	.3750	8-1/2	.7083
0-3/4	.0625	4-3/4	.3958	8-3/4	.7292
1	.0833	5	.4167	9	.7500
1-1/4	.1042	5-1/4	.4375	9-1/4	.7708
1-1/2	.1250	5-1/2	.4583	9-1/2	.7917
1-3/4	.1458	5-3/4	.4792	9-3/4	.8125
2	.1667	6	.5000	10	.8333
2-1/4	.1875	6-1/4	.5208	10-1/4	.8542
2-1/2	.2083	6-1/2	.5417	10-1/2	.8750
2-3/4	.2292	6-3/4	.5625	10-3/4	.8958
3	.2500	7	.5833	11	.9167
3-1/4	.2708	7-1/4	.6042	11-1/4	.9375
3-1/2	.2917	7-1/2	.6250	11-1/2	.9583
3-3/4	.3125	7-3/4	.6458	11-3/4	.9792

$$S = \frac{(130 \text{ sq. ft.})}{(11'\text{-}8'')}$$

$S = 11.14$ feet, or $11'\text{-}1^{11}/_{16}''$.

Given A_{max} of 130 square feet for ordinary hazard and a computed L dimension of $11'\text{-}8''$, the maximum distance between sprinklers on a branch line for this example is $11'\text{-}1^{11}/_{16}''$, as computed above.

Determine the Minimum Number of Sprinklers on Each Branch Line

The number of sprinklers on a line in a bay is obtained by dividing the length of the bay by the maximum distance allowed between sprinklers on the branch lines (S):

$$\text{Sprinklers on branch line} = \frac{(\text{Total length of bay})}{(\text{Max. distance between sprinklers})}$$

Example 5.4 Continuing with the building used in examples 5.1, 5.2, and 5.3:

$$\frac{(297'\text{-}0'')}{(11.14')} = 26.7 \text{ sprinklers.}$$

Since it is impossible to have a fractional sprinkler, the total above is always rounded up to the nearest whole sprinkler. For this example we would have 27 sprinklers on each line.

Determine the Actual Distance between Sprinklers (S_{actual})

The actual spacing of sprinklers on each line is determined by dividing the length of the bay by the number of sprinklers on the line:

$$S_{actual} = \frac{\text{(Total length of bay)}}{\text{(Number of sprinklers on line)}}$$

Example 5.5 Continuing with the building used in examples 5.1, 5.2, 5.3, and 5.4:

$$S_{actual} = \frac{297'\text{-}0''}{27} = 11'\text{-}0''. \text{ See Figure 5-14.}$$

The distance from the end sprinkler to the wall is:

$$1/2(S) = 5'\text{-}6''. \text{ See Figure 5-14.}$$

The total number of sprinklers in the bay is:

$$(27 \text{ sprinklers per line}) \times (3 \text{ lines}) = 81 \text{ sprinklers.}$$

Dimension Sprinklers and Piping on Your Drawing

The branch lines and sprinklers may now be dimensioned on the drawing with respect to the bays using light pencil guide lines showing the dimension between each line and sprinkler. In order to provide the most efficient system layout, the distance from a sprinkler to a wall or column should not exceed $1/2(L_{actual})$ or $1/2(S_{actual})$, or one-half the actual distance between sprinklers.

Verify Sprinkler Coverage

Verify at this time that the actual square foot limitation is in accordance with NFPA 13.

Example 5.6 Continuing with the building used for examples 5.1, 5.2, 5.3, 5.4, and 5.5:

$$S = 11'\text{-}0'' \text{ from example 5.5}$$
$$L = 11'\text{-}8'', \text{ from example 5.3}$$
$$A = S \times L$$
$$= (11'\text{-}0'') \times (11'\text{-}8'')$$
$$= 128.3 \text{ square feet.}$$
$$A_{max} = 130 \text{ square feet for ordinary hazard, from example 5.1.}$$

Since A is less than A_{max} in this example, the layout complies with NFPA 13.

This actual sprinkler coverage area, as shown in Figure 5-14, is less than the maximum permitted by NFPA 13 for the occupancy used in this example and is therefore an acceptable layout.

The maximum square footage is a function of the greatest spacing encountered.

Example 5.7 In order to demonstrate the importance of regular, equidistant spacing of branch lines and sprinklers within a bay, this example details a situation where the sprinkler spacing is not uniform. If the sprinkler closest to the wall in Figure 5-14 shifts 6 inches, increasing the distance from sprinkler to wall from 5'-6" to 6 feet, with all other dimensions in the bay remaining the same, the shifted sprinkler is protecting a larger square footage than the other sprinklers on the branch line. S_{actual} would have doubled to become 12 feet, and the actual area for the shifted sprinklers would be:

$$S \times L = A$$
$$12'\text{-}0'' \times 11'\text{-}8'' = 140 \text{ square feet.}$$

Because 140 square feet exceeds the maximum area per sprinkler for ordinary hazard, it is not compliant with NFPA 13.

Any obstructions encountered by the lines or sprinklers requiring shifting of the sprinklers mandate that the areas protected by the shifted sprinklers be reevaluated, and may result in the addition of extra sprinklers to conform with the maximum protection area allowed per sprinkler (A_{max}).

Determine Crossmain Location

The crossmains should be located at the low points of a sloped roof to allow the system to drain properly. If a tree system is being designed, the crossmain should be centrally located between the ends of the branch lines. A grid system would have crossmains at both ends of the branch lines.

If it is possible to align a crossmain parallel to a beam, hanging of the crossmain is facilitated. It is not recommended that hangers be aligned parallel to and hung to the side of a bar joist or purlin. Hanging the crossmain in this manner may twist these relatively light structural members and negatively affect their load-bearing capability, especially when large pipe sizes are being hung. It may be feasible and advantageous to arrange crossmains so that the number of branch lines penetrating brick or masonry walls is minimized.

Assign Elevations to Piping

It may not be possible to locate crossmains at the same elevation as the branch lines because of interference with beams. If this is the case, the crossmain is installed at an elevation below the beams, and the branch lines are elevated above the crossmain to ensure proper deflector distances. The crossmain would then feed each branch line with a **riser nipple**, a short vertical pipe that connects the crossmain with each branch line, shown in Figure 5-16.

The branch lines must be elevated with respect to the minimum and maximum **deflector distances** of 1 inch to 12 inches for unobstructed construction, and 1 inch to 6 inches below beams in obstructed construction, up to 22 inches maximum below the ceiling. The deflector distance is the dimension from face of ceiling to the top of the sprinkler deflector. Designing within these distances expedites proper sprinkler actuation.

Interferences may affect branch line elevation. Should the lines be required to be elevated below the range of deflector distances listed above, the sprinklers must be installed on **sprigs**, or vertical pipes rising from the branch line to sprinklers extended above the branch line. In addition, the sprinklers need to be coordinated with horizontal and vertical obstructions as detailed in NFPA 13. In a building with a suspended ceiling, vertical pipes must drop down from the branch lines to sprinklers mounted to the ceiling. These pipes, shown in Figure 5-17, are called **drops**.

riser nipple
short vertical pipe that connects the crossmain with a branch line

deflector distance
the dimension from the face of a ceiling to the sprinkler deflector

sprigs
vertical pipes rising from the branch line to sprinklers extended above the branch line

drop
vertical pipes that drop from the branch lines to sprinklers mounted to the ceiling

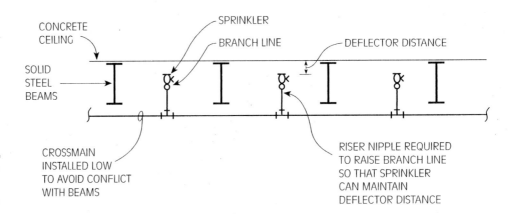

Figure 5-16 *Deflector distance and riser nipple arrangement.*

Chapter 5 Automatic Sprinkler Systems

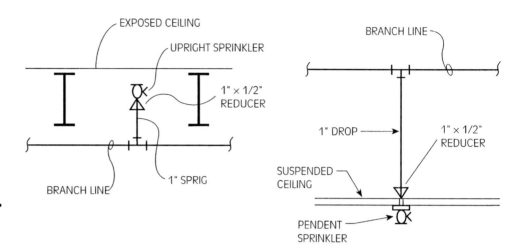

Figure 5-17 *Sprigs and drops.*

If the system being drawn is a dry pipe system, notes must be added to the drawing that show the branch lines being sloped 1/2 inch every 10 feet, and crossmains and feedmains being sloped 1/4 inch every 10 feet, to facilitate draining of the system. Ideally, to minimize maintenance problems after installation, all system piping should be designed to drain back to the main riser. Piping that does not drain to the riser must be supplied with auxiliary drains.

Locate System Risers

The riser locations may already have been designated by the underground entry locations specified by the site plans, plumbing plans, or fire protection plans in the contract drawing set. The risers and control valves would rise from these designated locations. Feedmains can then be drawn from the risers, connecting the crossmains in the most expeditious and practical way.

A scaled and dimensioned sectional view of the riser can then be drawn on your plan, using a minimum scale of 1/4″ = 1′-0″. An example of a riser section is shown in Figure 5-18. The building section may be selected such that the riser section is shown on the building section. A fire department connection must be connected to the system riser for systems having more than 20 sprinklers.

Select Hangers and Determine Hanger Rod Lengths

A selection of commonly used hangers for a variety of hanging applications is shown in Figure 5-19 on page 107, and many additional specialized hanger configurations are available from hanger manufacturers. NFPA 13 stipulates that the hangers must be listed and that the application and arrangement be in accordance with that section.

A very handy set of tables is included in Figure 5-20 on page 108 for the proper sizing of a **trapeze hanger**, a piece of angle iron or pipe used to span two struc-

trapeze hanger
a piece of angle iron or pipe that spans two structural members to facilitate the hanging of crossmains and feedmains

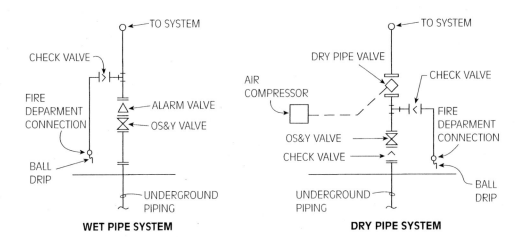

Figure 5-18 *Example riser sections.*

tural members to facilitate the hanging of crossmains and feedmains. These tables summarize the results of structural load calculations that have been performed for varying spans and pipe sizes.

It is important that fire protection system designers know how to use the trapeze hanger selection procedure properly to ensure proper support for fire protection feedmains. A method for using the trapeze tables is to:

1. Determine the span between the structural members (joists, purlins, or beams).
2. Determine the pipe size being hung.
3. Determine the pipe schedule of the pipe being hung (schedule 10 or 40).
4. Select a section modulus value for a pipe being supported by a trapeze bridging a given span from the table in Figure 5-20 entitled Section Modulus Required for Trapeze Hangers. A section modulus is a number that represents the ability of a structural member to resist bending. Low section moduli would indicate that a trapeze hanger would undergo only a slight deflection if a pipe would be hung from it. Very large section moduli indicate that the trapeze hanger would be more prone to bending or deflection, requiring a more substantial trapeze.
5. Using the section modulus obtained, find a section modulus on the table in Figure 5-20, entitled Available Section Moduli of Common Trapeze Hangers, that is equal to or greater than this value corresponding to a specific size of an angle, iron trapeze hanger, or a schedule 10 or 40 pipe trapeze hanger.
6. The trapeze hanger must be of the minimum size found in step 5.

Determine Hanger Spacing

By performing a quick review of the hanger requirements of NFPA 13, the reader can confirm that running the branch lines perpendicular to the joists is the correct

Chapter 5 Automatic Sprinkler Systems

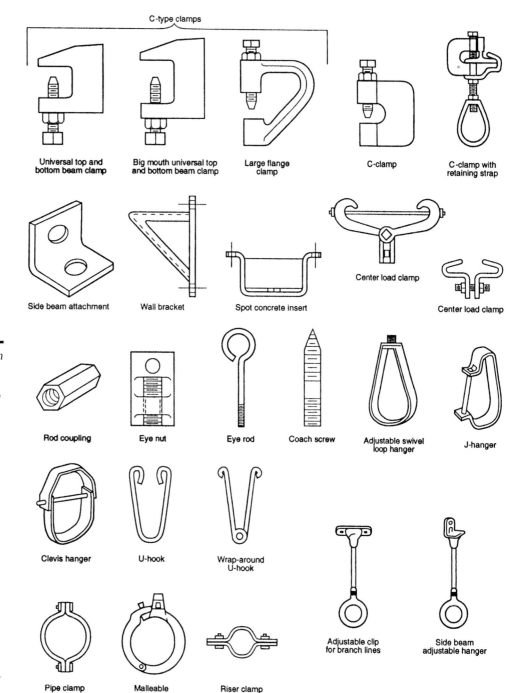

Figure 5-19 *Common types of acceptable sprinkler pipe hangers. (Reprinted with permission from NFPA 13, Installation of Sprinkler Systems. Copyright © 1996, National Fire Protection Association, Quincy, MA 02269. This reprinted material is not the complete and official position of the National Fire Protection Association on the referenced subject, which is represented only by the standard in its entirety.)*

Table 2-6.1.5(a) Section Modulus Required for Trapeze Members (in.³)

Span of Trapeze	1 in.	1¼ in.	1½ in.	2 in.	2½ in.	3 in.	3½ in.	4 in.	5 in.	6 in.	8 in.	10 in.
1 ft 6 in.	.08	.09	.09	.09	.10	.11	.12	.13	.15	.18	.24	.32
	.08	.09	.09	.10	.11	.12	.13	.15	.18	.22	.30	.41
2 ft 0 in.	.11	.12	.12	.13	.13	.15	.16	.17	.20	.24	.32	.43
	.11	.12	.12	.13	.15	.16	.18	.20	.24	.29	.40	.55
2 ft 6 in.	.14	.14	.15	.16	.17	.18	.20	.21	.25	.30	.40	.54
	.14	.15	.15	.16	.18	.21	.22	.25	.30	.36	.50	.68
3 ft 0 in.	.17	.17	.18	.19	.20	.22	.24	.26	.31	.36	.48	.65
	.17	.18	.18	.20	.22	.25	.27	.30	.36	.43	.60	.82
4 ft 0 in.	.22	.23	.24	.25	.27	.29	.32	.34	.41	.48	.64	.87
	.22	.24	.24	.26	.29	.33	.36	.40	.48	.58	.80	1.09
5 ft 0 in.	.28	.29	.30	.31	.34	.37	.40	.43	.51	.59	.80	1.08
	.28	.29	.30	.33	.37	.41	.45	.49	.60	.72	1.00	1.37
6 ft 0 in.	.33	.35	.36	.38	.41	.44	.48	.51	.61	.71	.97	1.30
	.34	.35	.36	.39	.44	.49	.54	.59	.72	.87	1.20	1.64
7 ft 0 in.	.39	.40	.41	.44	.47	.52	.55	.60	.71	.83	1.13	1.52
	.39	.41	.43	.46	.51	.58	.63	.69	.84	1.01	1.41	1.92
8 ft 0 in.	.44	.46	.47	.50	.54	.59	.63	.68	.81	.95	1.29	1.73
	.45	.47	.49	.52	.59	.66	.72	.79	.96	1.16	1.61	2.19
9 ft 0 in.	.50	.52	.53	.56	.61	.66	.71	.77	.92	1.07	1.45	1.95
	.50	.53	.55	.59	.66	.74	.81	.89	1.08	1.30	1.81	2.46
10 ft 0 in.	.56	.58	.59	.63	.68	.74	.79	.85	1.02	1.19	1.61	2.17
	.56	.59	.61	.65	.74	.82	.90	.99	1.20	1.44	2.01	2.74

For SI Units: 1 in. = 25.4 mm; 1 ft = 0.3048 m.
Top values are for Schedule 10 pipe, bottom values are for Schedule 40 pipe.
NOTE: The table is based on a maximum allowable bending stress of 15 KSI and a midspan concentrated load from 15 ft (4.6 m) of water-filled pipe, plus 250 lb (114 kg).

Table 2-6.1.5(b) Available Section Moduli of Common Trapeze Hangers

Pipe	Modulus	Angles					Modulus
Schedule 10							
1 in.	.12	1½	×	1½	×	³⁄₁₆	.10
1¼ in.	.19	2	×	2	×	⅛	.13
1½ in.	.26	2	×	1½	×	³⁄₁₆	.18
2 in.	.42	2	×	2	×	³⁄₁₆	.19
2½ in.	.69	2	×	2	×	¼	.25
3 in.	1.04	2½	×	1½	×	³⁄₁₆	.28
3½ in.	1.38	2½	×	2	×	³⁄₁₆	.29
4 in.	1.76	2	×	2	×	⁵⁄₁₆	.30
5 in.	3.03	2½	×	2½	×	³⁄₁₆	.30
6 in.	4.35	2	×	2	×	⅜	.35
		2½	×	2½	×	¼	.39
		3	×	2	×	³⁄₁₆	.41
Schedule 40		3	×	2½	×	³⁄₁₆	.43
1 in.	.13	3	×	3	×	³⁄₁₆	.44
1¼ in.	.23	2½	×	2½	×	⁵⁄₁₆	.48
1½ in.	.33	3	×	2	×	¼	.54
2 in.	.56	2½	×	2	×	⅜	.55
2½ in.	1.06	2½	×	2½	×	⅜	.57
3 in.	1.72	3	×	3	×	¼	.58
3½ in.	2.39	3	×	3	×	⁵⁄₁₆	.71
4 in.	3.21	2½	×	2½	×	½	.72
5 in.	5.45	3½	×	2½	×	¼	.75
6 in.	8.50	3	×	2½	×	⅜	.81
		3	×	3	×	⅜	.83
		3½	×	2½	×	⁵⁄₁₆	.93
		3	×	3	×	⁷⁄₁₆	.95
		4	×	4	×	¼	1.05
		3	×	3	×	½	1.07
		4	×	3	×	⁵⁄₁₆	1.23
		4	×	4	×	⁵⁄₁₆	1.29
		4	×	3	×	⅜	1.46
		4	×	4	×	⅜	1.52
		5	×	3½	×	⁵⁄₁₆	1.94
		4	×	4	×	½	1.97
		4	×	4	×	⅝	2.40
		4	×	4	×	¾	2.81
		6	×	4	×	⅜	3.32
		6	×	4	×	½	4.33
		6	×	4	×	¾	6.25
		6	×	6	×	1	8.57

For SI Units: 1 in. = 25.4 mm; 1 ft = 0.3048 m.

Figure 5-20 *Tables used for sizing trapeze hangers. (Reprinted with permission from NFPA 13,* Installation of Sprinkler Systems. *Copyright © 1996, National Fire Protection Association, Quincy, MA 02269. This reprinted material is not the complete and official position of the National Fire Protection Association on the referenced subject, which is represented only by the standard in its entirety.)*

approach to branch line orientation. Maximum distances between hangers vary with pipe size and pipe type, but steel pipe feedmain pieces (usually 20 to 25 feet long) usually require no fewer than two hangers.

Each section of pipe must have at least one hanger. Although NFPA 13 allows the elimination of a hanger on crossmains under certain conditions, eliminating this hanger may make the crossmain more difficult to install.

Add Final Details

inspector's test connection
test valve used to simulate the flow of the hydraulically most demanding sprinkler

Every system must have an **inspector's test connection**, a test valve used to simulate the flow of the hydraulically most demanding sprinkler, located at the end of the hydraulically most demanding branch line on the system, as determined in Chapter 6. Every portion of a system monitored by a water flow switch must have an inspector's test connection in order to test the switch.

Systems with more than 20 sprinklers must have either an electric bell or a mechanical water motor gong and your drawing must indicate a method of valve supervision, using one of three methods:

- Valves locked in the open position.
- Valves electronically monitored with a tamper switch.
- Valves located in a constantly inspected locked fenced area.

Such measures are intended to insure the reliability of the system by ensuring that fire protection valves remain open at all times.

Use the NFPA 13 Design Checklist

NFPA 13, Chapter 6, provides a convenient listing of items that must be on all sprinkler plans submitted for approval. This list is used by many authorities having jurisdiction to review your plans, and it is suggested that this list be used as a checklist before submitting a plan for approval. Any special requirements of AHJs would be added to or would supersede items on this list.

Perform Precalculation Field Check

Should the building progress permit, a visit to the jobsite before performing hydraulic calculations is a good idea. A method for hydraulically calculating a sprinkler system is included in Chapter 6.

In your jobsite visit you may find that the underground water supply inlet has been installed and can be measured. Note the diameter of the inlet and verify that it is the same size shown on the site plans in the contract drawing set. Any installed trades can be surveyed, and other building features that might have been unclear on the contract drawings, or that might have changed, can be investigated.

Add Pipe Lengths and Verify Accuracy

Before performing hydraulic calculations, it is a good idea to add pipe lengths and check conformance to the building dimensions. Errors in adding of dimensions that are discovered will make your hydraulic calculations more accurate. Your system is now ready to be hydraulically calculated, as detailed in Chapter 6.

Summary

When designing sprinkler systems, its is best to follow a design method to ensure accuracy and conformance to the applicable standards. It is necessary to know the key terms used in sprinkler system design and to be able to counteract the commonly held myths about sprinkler systems with factual data.

Review Questions

1. A building is divided into six bays, with each bay having dimensions of 44'-0" wide by 196'-0" long, as shown in Figure 5-21. If branch lines run parallel to the 196'-0" direction, answer the following questions, assuming that the occupancy of the building is light hazard and that standard coverage sprinklers are used:
   a. Determine the number of branch lines in each bay.
   b. Determine the spacing of the branch lines in the bays.
   c. Determine the number of sprinklers on each branch line.
   d. Determine the spacing between sprinklers on the branch lines, and the distance from the end sprinklers to the wall.
   e. Determine the total number of sprinklers in each bay, and in the entire building.
   f. Dimension the sprinklers and piping in a manner similar to Figure 5-14.
   g. Check your spacing to ensure that the spacing is in accordance with the requirements of NFPA 13 for light hazard occupancy.

2. Discuss the differences between wet pipe sprinkler systems and dry pipe sprinkler systems. List the restrictions that are placed on dry pipe sprinkler systems, and explain why such limitations are necessary.

3. For the reflected ceiling plan shown in Figure 5-22:
   a. Space sprinklers in the room, avoiding conflicts with all lights, diffusers, and speakers assuming ordinary hazard occupancy and standard coverage sprinklers.
   b. Use the minimum number of sprinklers allowable per NFPA 13, and do not exceed A_{max}, L_{max}, or S_{max}, while maintaining a proper distance from all walls.
   c. Dimension the final sprinkler locations.
   d. Verify the spacing by proving that the greatest spacing between sprinklers, or the greatest spacing from sprinklers to walls is in accordance with ordinary hazard occupancy.

4. Write a paragraph contrasting and comparing each set of terms:
   a. Approved and listed
   b. Control and suppression
   c. Shall and should
   d. Old style sprinkler and spray sprinkler
   e. Upright sprinkler and pendent sprinkler
   f. Quick response sprinkler and standard response sprinkler

Chapter 5 Automatic Sprinkler Systems

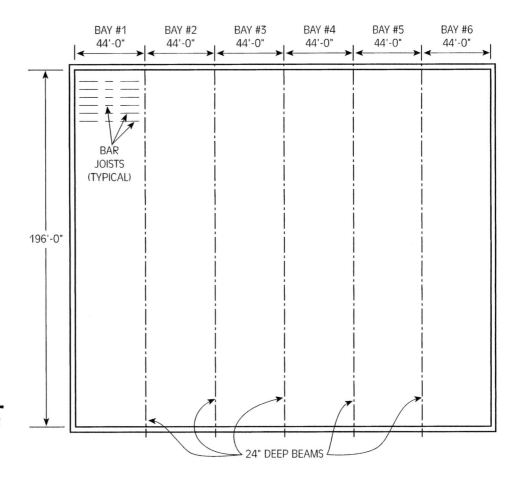

Figure 5-21 *Building for review question 1.*

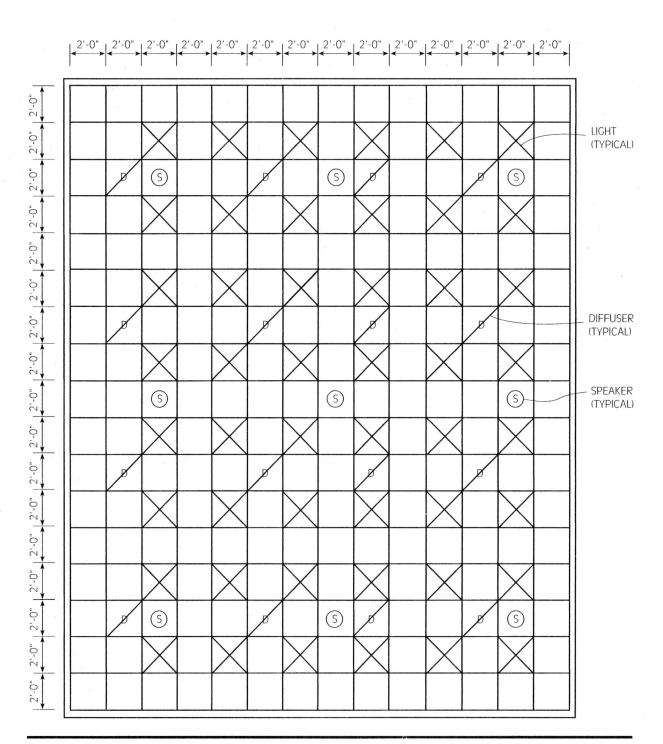

Figure 5-22 *Reflected ceiling plan for review question 3.*

g. ESFR sprinkler and large drop sprinkler
h. Mylar, vellum, and onion skin drawing media
i. Cut grooved and roll-grooved
j. OS&Y valve, post indicator valve, and butterfly valve
k. Cast iron fittings and malleable iron fittings
l. Sprig, drop, and riser nipple
m. Tree system, grid system, and looped system

5. Select the occupancy of the following buildings:
   a. Church
   b. Shoe store
   c. Insurance company
   d. Movie theater
   e. Whiskey manufacturing
   f. Coin-operated laundry
   g. Newspaper press room
   h. College classroom
   i. Paint spray booth

6. Compare a 2" schedule 5 pipe to a 2" schedule 80 pipe. In each case, provide an explanation for your answer.
   a. Which pipe has the largest internal diameter?
   b. Which pipe has the largest external diameter?
   c. Which pipe holds the most water per foot?
   d. Which would provide better water flow and hydraulic characteristics?
   e. Which pipe would withstand the higher system pressure?
   f. Which pipe would not be suitable for cut grooves?

Activities

1. Interview a representative of a fire department and determine the response time for that fire department. Evaluate this response time with respect to the concepts of fire control and fire suppression. Is the fire service response time adequate for the design of sprinkler systems for fire control, or should system design assuming fire suppression be considered?

2. Obtain a copy of the DETACT computer program from NIST or the NFPA and determine the time for a sprinkler to actuate if a quick response sprinkler with a response time index (RTI) of 50 is used, the ceiling height is 12 feet, the sprinkler temperature is 165°, sprinkler spacing is $S = 15'-0"$ and $L = 15'-0"$, and a fire associated with a light hazard occupancy is anticipated.

3. Obtain a reflected ceiling plan for a portion of a building, and space sprinklers in that area.

4. Obtain a structural plan for a building that has no ceiling, orient branch lines perpendicular to the bar joists, and space sprinklers in each bay as appropriate for the spacing requirements of the occupancy.

Chapter 6

Hydraulic Calculation of Sprinkler Systems

Objectives

Upon completion of this chapter, you should be able to:

- Perform calculations using the physical properties of water.
- Calculate the pressure lost to elevation in a vertical pipe.
- Convert feet of head to psi, and vice versa.
- Calculate friction loss using the Hazen-Williams formula and by using friction loss tables.
- Calculate the split flow and friction loss in a pipe loop.
- Calculate a sprinkler system and determine the adequacy of a water supply.
- Use the computerized program in Appendix E to perform a hydraulic calculation of a sprinkler system.

Chapter 6 Hydraulic Calculation of Sprinkler Systems

Performing hydraulic calculations is as much an art form as a science. It is among the most elegant of disciplines, based on the most beautiful laws of nature and the most fundamental laws of physics. Although the performance of hydraulic calculations is relatively straightforward, the implication of those calculations is powerful and persuasive. Without physically flowing a single drop of water, a fire protection system designer performing hydraulic calculations can accurately predict the performance of a water-based fire protection system.

THE FUNDAMENTALS OF HYDRAULICS

A drop of water holds incredible mystery and power, yet we are so familiar with it that we tend to take it for granted. If a scientist wished to invent a brand new fire protection fluid, it would not be possible to duplicate the impressive properties of water. Water is the fluid most predominately used for fire protection for numerous reasons.

The Advantages of Water

fire triangle
a concept identifying the interdependent entities of oxygen, heat, and fuel as they relate to sustaining combustion

fire tetrahedron
a concept that considers the inhibition of chain reactions in addition to the three points of the fire triangle

Btu
British thermal unit, the unit of heat energy needed to raise the temperature of 1 pound of water 1°F at constant pressure of 1 atmosphere

1. Water is an astonishingly effective extinguishing media. The physical properties listed under item 4 below make water effective for all three elements of the **fire triangle**, consisting of the interdependent entities of oxygen, heat, and fuel, as they relate to the sustaining of the combustion phenomena, illustrated in Figure 6-1. The concept of the fire triangle is extended to the **fire tetrahedron** when agents that are not water based, such as halon and dry chemical, are considered. The fourth point of the fire tetrahedron consists of the inhibition of chain reactions and intermolecular collisions that sustain combustion, by using an agent such as dry chemical.
2. Water is an economical choice, with its price per unit significantly less than fire suppression agents that are not water based.
3. Water is readily available. Public water supplies are installed adjacent to most commercial buildings and are available for many residential buildings.
4. Water has very reliable and predictable physical qualities.
   a. It freezes at 32°F.
   b. It boils at 212°F.
   c. It weighs 62.4 pounds per cubic foot.
   d. When water boils at 212°F, it turns to steam, at a ratio of 1:1600 (1,600 units of steam produced for each unit of water). One gallon of water becomes 223 cubic feet of steam. Water absorbs heat from a fire at a rate of 9,330 **Btu** per pound of water. A Btu, or British thermal unit, is a unit of heat energy needed to raise the temperature of 1 pound of water 1°F at constant pressure of 1 atmosphere.

Figure 6-1 *The classical fire triangle. Each leg represents an element which, if removed, will extinguish a fire.*

e. Water occupies a predictable space: 7.48 gallons of water fills a container that has a volume of 1 cubic foot. This means that 1 gallon of water occupies 0.1337 cubic feet of space.

The Properties of Water

Example 6.1 Using the predictable properties of water, determine the weight of water contained in a reservoir, shown in Figure 6-2, that is 40 feet wide, 100 feet long, and 15 feet deep.

Solution Volume of reservoir:

$$V = (\text{width}) \times (\text{length}) \times (\text{height})$$
$$= (40') \times (100') \times (15') = 60,000 \text{ cubic feet}$$

Using the relationship in item 4c above:

$$\text{Unit weight of water} = 62.4 \text{ pounds/cubic foot}$$

Chapter 6 Hydraulic Calculation of Sprinkler Systems

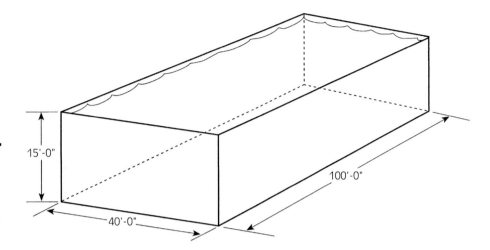

Figure 6-2
Illustration for Example 6.1. A reservoir filled with water. Calculate the weight of the water.

Weight of water in reservoir:

$$W = (62.4 \text{ pounds/cubic foot}) \times (60{,}000 \text{ cubic feet}) = 3{,}744{,}000 \text{ pounds}$$

Clearly, with a weight of 3.7 million pounds, this is not a portable reservoir! The reservoir must withstand the force of the weight of the water within and must therefore be constructed of materials of significant strength.

We may now examine the force of the water on the bottom of the reservoir in further detail. To do this, select at random an area at the bottom of the reservoir 1 inch wide by 1 inch long. The force of the water on this 1 square inch area is the weight of 15 feet of water above. We are purposefully selecting this area for study because the standard unit of pressure measurement in hydraulic calculation is the psi, or pound per square inch.

Pressure is expressed as force divided by area.

$$P = \frac{F}{A}$$

where P = pressure, in psi
F = force of water, in pounds
A = area, in square inches

Example 6.2 Determine the force of the water in the reservoir described in example 6.1 and shown in Figure 6-3, on an area of 1 square inch at the bottom of the reservoir.

Solution First, let's convert the floor of the reservoir into square inches.

40 feet wide = (40 feet) × (12 inches/1 foot) = 480 inches

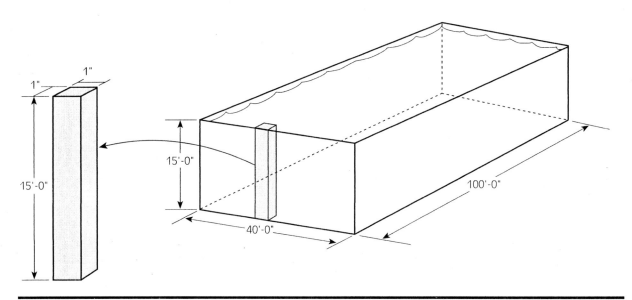

Figure 6-3 *Illustration for Example 6.2. Select a random area at the bottom of the reservoir 1 inch long by 1 inch wide and determine the force of the water on this area.*

100 feet long = (100 feet) × (12 inches/1 foot) = 1200 inches

Area of the floor of the reservoir in square inches:

$$A = (480 \text{ inches}) \times (1200 \text{ inches})$$
$$= 576{,}000 \text{ square inches}$$

Substituting in

$$P = \frac{F}{A}$$

we have $F = 3{,}744{,}000$ pounds from example 6.1 (the force is the total weight of water on the floor of the reservoir) and $A = 576{,}000$ square inches, as previously calculated. Thus,

$$P = \frac{(3{,}744{,}000 \text{ pounds})}{(576{,}000 \text{ square inches})}$$
$$= 6.5 \text{ psi, or pounds per square inch.}$$

The pressure of 15 feet of water on 1 square inch at the bottom of the reservoir is 6.5 psi. We will discover in this chapter that knowing the weight of 1 foot of water on 1 square inch of area is a valuable tool for calculating the pressure in a vertical pipe.

Chapter 6 Hydraulic Calculation of Sprinkler Systems

Example 6.3 Determine the force of a column of water 1 foot high on an area of 1 square inch.

Using the result in example 6.2,

$$P = \frac{(6.5 \text{ psi})}{(15 \text{ feet})}$$

$$= 0.433 \text{ psi/ft}$$

This is an extremely powerful finding that allows us to calculate the pressure of water in psi for any height, using the formula

$$P = (h) \times (0.433 \text{ psi/ft})$$

where P is the pressure of the water on 1 square inch of area in psi and h is the height of the water column above the 1 square inch in feet.

Example 6.4 A new 100-foot tall building, shown in Figure 6-4, requires a new sprinkler system. The top floor of the building is fed by a 4-inch sprinkler riser

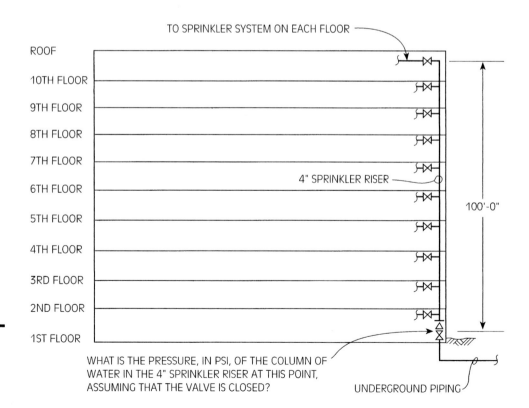

Figure 6-4 *A cross-sectional view of a new sprinklered high-rise building.*

filled with water. Assuming no pressure contribution from the water supply, calculate the pressure in psi of the water column at the bottom of the 4-inch riser.

$$P = (h) \times (0.433 \text{ psi/ft})$$
$$= (100 \text{ feet}) \times (0.433 \text{ psi/ft})$$
$$= 43.3 \text{ psi}$$

This means that the water supply at the lowest floor level of the building must be capable of providing at least 43.3 psi to overcome the weight of water in the riser feeding the uppermost floor. This chapter discusses additional pressure considerations that must be added to this minimum water supply requirement.

Hydrostatics

hydrostatics
the study of water (hydro) at rest (static)

Hydrostatics is the study of water (hydro) at rest (static). Our discussion of the reservoir and the high-rise building in the previous section dealt with the weight of water in a static, or nonflowing state. Examples 6.1, 6.2, and 6.3 concerning the reservoir illustrate an important cornerstone in the study of hydrostatics:

The pressure exerted by the weight of water is directly proportional to the height of the water.

As we learned, a foot of water height exerts a pressure of 0.433 psi on a 1-square-inch area.

An important corollary of this principle is that as long as the depth is constant, the pressure exerted by the water is constant in any direction at a given depth, as shown in Figure 6-5.

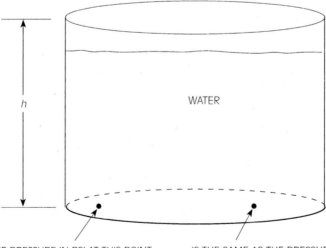

Figure 6-5 *Pressure is constant in any direction for a given water depth.*

Chapter 6 Hydraulic Calculation of Sprinkler Systems

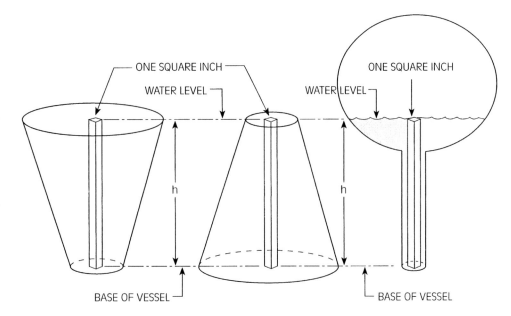

Figure 6-6 *Shape of a vessel. The shape of the vessel does not change the pressure in psi exerted by the water on the base of the vessel because the 1 square inch water column of depth h is identical for all three vessels shown.*

Another principle of hydrostatics is one that some students find difficult to understand:

Given a constant depth of water, the pressure created by the weight of water is not affected by the shape of the vessel containing the water.

Figure 6-6 illustrates that even though the vessels appear markedly different, the pressure in psi at the base of each vessel is identical, because the height of the water column in each vessel is the same.

It is sometimes useful to speak of water pressure in terms of **head**, rather than in psi. Head is the pressure of water described in terms of feet of water column, instead of psi. Fire pump pressure ratings are often expressed in feet of head.

Recall that for pressure in psi, we used the formula:

head

the pressure of water described in terms of feet of water column instead of psi

$$P = (h) \times (0.433 \text{ psi/ft})$$

where P is the pressure of the water on 1 square inch of area, in psi, and h is the depth of the water, or the height of the water column above the 1 square inch, in feet.

Rearranging this formula:

$$h = \frac{(P)}{(0.433 \text{ psi/ft})}$$

The depth of the water, h is in units of feet, and when expressed in this way, h is the pressure of the water in feet of head. This formula can also be expressed as:

$$h = (1 / 0.433 \text{ psi/ft}) \times (P)$$

Dividing 1 by 0.433, we obtain

$$h = (2.31 \text{ ft/psi}) \times (P)$$

Readers should memorize these two important formulas:

$$P = (h) \times (0.433 \text{ psi/ft})$$
$$h = (2.31 \text{ ft/psi}) \times (P)$$

Example 6.5 A fire pump has a pressure rating of 150 feet of head. Determine the fire pump rating in psi.

$$\begin{aligned} P &= (h) \times (0.433 \text{ psi/ft}) \\ &= (150 \text{ ft}) \times (0.433 \text{ psi/ft}) \\ &= 65 \text{ psi} \end{aligned}$$

The fire pump pressure rating is usually rounded up to the nearest pressure rating available from a fire pump manufacturer, usually in 5-psi increments.

Example 6.6 A fire pump has a pressure rating of 125 psi. Determine the fire pump pressure rating in feet of head.

$$\begin{aligned} h &= (2.31 \text{ ft/psi}) \times (P) \\ &= (2.31 \text{ ft/psi}) \times (125 \text{ psi}) \\ &= 289 \text{ ft of head} \end{aligned}$$

Example 6.7 A gauge on a pipe reads 10 psi, as shown in Figure 6-7. Assuming that water is not flowing in the pipe, what would the pressure reading be at an elevation 14 feet below this gauge?

$$\begin{aligned} P &= (h) \times (0.433 \text{ psi/ft}) \\ &= (14 \text{ ft}) \times (0.433 \text{ psi/ft}) \\ &= 6.1 \text{ psi} \end{aligned}$$

This static pressure must be added to the gauge pressure on the top gage.

$$\begin{aligned} P &= 10 \text{ psi} + 6.1 \text{ psi} \\ &= 16.1 \text{ psi} \end{aligned}$$

Chapter 6 Hydraulic Calculation of Sprinkler Systems

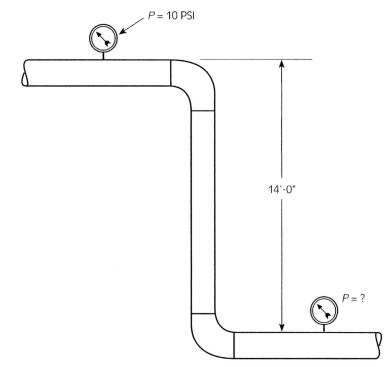

Figure 6-7
Illustration for Example 6.7. Pressure readings at different elevations.

Hydrokinetics

hydrokinetics
the study of water (hydro) in motion (kinetics)

Up to this point, we have discussed water at rest, where the velocity of the water is zero. **Hydrokinetics** is the study of water (hydro) in motion (kinetics). Water moves in a pipe at a given speed or velocity. The flow rate in a pipe is a function of the velocity of the water in the pipe and the internal cross-sectional area of the pipe, as illustrated by Figure 6-8.

$$Q = (A) \times (v)$$

where Q = water flow in gallons per minute (gpm), and v = the velocity of the water in feet per second. To convert cubic feet to gallons, use the relationship:

$$1 \text{ cu. ft} = 7.48 \text{ gallons}$$

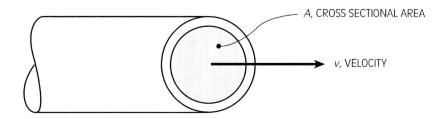

Figure 6-8
Illustration of $Q = Av$

The internal cross-sectional area of the pipe, in square feet is represented by A. To calculate A, use the formula:

$$A = (3.1416) \times (r^2)$$

where r = the internal radius of the pipe, in feet.

Example 6.8 Water is moving at a velocity of 10 feet per second in a pipe whose internal diameter is 6.065 inches. Determine the water flow in gpm.

$$Q = (A) \times (v)$$
$$= (3.1416) \times (r^2) \times (v)$$

The diameter of the pipe is 6.065 inches, so the radius is 1/2 the diameter, or 3.0325 inches

$$r = (3.0325 \text{ inches}) \times (1 \text{ foot}/12 \text{ inches})$$
$$= 0.2527 \text{ feet}$$
$$Q = (3.1416) \times (0.2527 \text{ ft})^2 \times (10 \text{ ft/sec})$$
$$= 2.006 \text{ ft}^3/\text{sec}$$

This result is expressed in cubic feet per second (ft³/sec), but most fire protection designers prefer to use values in gpm in their calculations. To convert cubic feet per second to gpm:

$$Q = (2.006 \text{ ft}^3/\text{sec}) \times (7.48 \text{ gallons/ft}^3) \times (60 \text{ seconds/minute})$$
$$= 900.3 \text{ gpm}$$

Note that cubic feet and seconds cancel in the above equation, yielding gallons per minute, or gpm.

The relationship between velocity and flow is a cornerstone in the understanding of the principles of hydrokinetics. Using what we've just learned, we can begin to solve more complex problems.

Let's rewrite the formula, solving for the velocity:

$$Q = Av,$$

or

$$Q = (3.1416) \times (r^2) \times (v),$$

or

$$v = \frac{Q}{(3.1416) \times (r^2)}$$

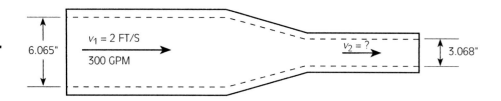

Figure 6-9
Illustration for Example 6-8.

By examining this formula, we can see that a given flow, in gpm, travels more rapidly in a small diameter pipe than in a larger diameter pipe. This principle is illustrated in example 6.9.

Example 6.9 A horizontal pipe reduces from an internal diameter of 6.065″ to 3.068″, as shown in Figure 6-9. If 300 gpm is moving in the larger pipe at a velocity of 2 ft/sec, determine the velocity of the water in the smaller pipe.

Solution If there are no openings along the length of the pipe, the same gallonage that flows into the 6.065″ pipe must flow through the 3.068″ pipe. Therefore:

$$Q_1 = Q_2$$

Since $Q = Av$ and $Q_1 = Q_2$,

$$A_1 v_1 = A_2 v_2$$

In this problem, we must solve for v_2.

$$v_2 = \frac{(A_1 v_1)}{(A_2)}$$

$$= \frac{(3.1416 \times (r_1^2) \times (v_1)}{(3.1416) \times (r_2^2)}$$

$$= \frac{(r_1^2) \times (v_1)}{(r_2^2)}$$

where v_1 = 2 ft/sec (given)
r_1 = 6.065/2 = 3.0325 inches
r_1 = (3.0325 inches) × (1 ft/12 inches) = 0.2527 ft
r_2 = 3.068/2 = 1.534 inches
r_2 = (1.534 inches) × (1 ft/12 inches) = 0.1278 ft

$$V_2 = \frac{(0.2527 \text{ ft})^2 \times (2 \text{ ft/sec})}{(0.1278 \text{ ft})}$$

$$= \frac{(0.1277)}{(0.01633)}$$

$$= 7.82 \text{ ft/sec}$$

As we expected, the velocity in the smaller pipe is greater than the velocity in the larger pipe. This makes sense when you consider that for the reduced diameter pipe to maintain a constant rate of water flow of 300 gpm, the velocity of the water must increase to maintain this rate of water delivery.

Friction Loss

friction loss
the frictional resistance of the interior walls of a pipe to the flow of the water within

In addition to the pressure loss associated with the weight of water, another significant source of pressure loss in hydraulic calculations is **friction loss**, or the frictional resistance of the interior walls of a pipe to the flow of water within. Numerous pipe flow experiments have been conducted to develop an equation that allows us to determine the friction loss in a pipe segment.

Hazen-Williams formula
standard formula for calculating friction loss in pipes

One of the most popular friction loss formulas is the **Hazen-Williams formula**, recognized by NFPA 13 and considered a standard formula for the calculation of friction loss. Derived from flow tests performed within underground water mains around 1905 by Allen Hazen and Gardner S. Williams, the Hazen-Williams friction loss formula was built on the established work of Antoine de Chezy (1775) and is as follows:

$$P_f = \frac{(4.52) \times (Q^{1.85})}{(C^{1.85}) \times (D^{4.87})}$$

where P_f = friction loss of one foot of pipe, in psi per foot
Q = water flow, in gpm
C = Hazen-Williams roughness coefficient, obtained by referencing Appendix B.
D = internal diameter of the pipe, in inches, obtained from Appendix C.

c-factor
the Hazen-Williams roughness coefficient

The Hazen-Williams roughness coefficient, called the **c-factor**, is a dimensionless measure of the relative roughness or smoothness of the internal surface of a pipe, as compared to other pipes. This factor was developed by Hazen and Williams to compare the frictional resistance of pipe walls of various materials, manufacture, and condition. A high c-factor represents a smooth pipe wall, whereas a low c-factor represents a rougher pipe wall. The number 4.52 in the formula is a numerical constant derived by Hazen and Williams during their hydraulic experimentation.

Example 6.10 Determine the friction loss factor, in psi per foot, for 200 gpm flowing through a new 6-inch schedule 40 pipe serving a wet pipe sprinkler system.
  Solution

$$P_f = \frac{(4.52) \times (Q^{1.85})}{(C^{1.85}) \times (D^{4.87})}$$

From Table B-1 in Appendix B, the c-factor of a new black steel pipe in a wet pipe sprinkler system is C = 120.

From Appendix C, Table 1, the exact internal diameter for a 6-inch schedule 40 pipe is D = 6.065″.

$$Q = 200 \text{ gpm (given)}$$

$$P_f = \frac{(4.52) \times (200^{1.85})}{(120^{1.85}) \times (6.065^{4.87})}$$

$$= \frac{(4.52) \times (18067.8)}{(7022.4) \times (6492.11)}$$

$$= 0.0018 \text{ psi/ft}$$

To determine the total friction loss in a pipe segment, multiply P_f (in psi/ft) by the number of feet in the pipe segment, L, to obtain the total friction between two points, P_t:

$$P_t = \frac{(4.52) \times (Q^{1.85}) \times (L)}{(C^{1.85}) \times (D^{4.87})}$$

$$L = \text{the length of the pipe segment, in feet}$$

Example 6.11 For the pipe in Example 6.10, determine the total friction loss, in psi, if the pipe is 780 feet long.

Solution

$$P_t = \frac{(4.52) \times (Q^{1.85}) \times (L)}{(C^{1.85}) \times (D^{4.87})}$$

$$= \frac{(4.52) \times (200^{1.85}) \times (780)}{(120^{1.85}) \times (6.065^{4.87})}$$

$$= (0.0018 \text{ psi/ft}) \times (780 \text{ ft})$$

$$= 1.4 \text{ psi}$$

P_t represents the total pressure lost to friction in the pipe segment described. The pipe diameter, D, is raised to the 4.87 power in the Hazen-Williams equation and is therefore a very powerful function of friction loss. As can be seen by examining the Hazen-Williams formula, as D increases, friction loss decreases. Conversely, as D decreases, friction loss increases.

In essence, when we hydraulically calculate a sprinkler system, we are adjusting sprinkler pipe sizes in such a way that the frictional losses are modified by changing pipe sizes, until the pressure and flow available from the water supply matches or exceeds the pressure and flow requirements of the sprinkler system.

Friction Loss Tables

friction loss tables
compilations of data representing the results of numerous friction loss calculations using the Hazen-Williams friction loss formula

The **friction loss tables** in Appendix D are given in psi per foot for various pipe diameters, c-factors, and flows. The reader is encouraged to select one or more values from Appendix D and demonstrate that the values actually represent the results of friction loss calculations using the Hazen-Williams friction loss formula.

Once a friction loss value is obtained from a table in Appendix D, it is multiplied by the length of a given pipe in feet to obtain the total friction loss in a pipe segment, in a manner similar to example 6.11. This relationship is represented by the formula:

$$P_t = (P_f \text{ value selected from Appendix D, in psi/ft}) \times (L)$$

Example 6.12 Using the friction loss tables in Appendix D, determine the friction loss in a 6-inch schedule 10 steel pipe, 500 feet long, supplying a dry pipe sprinkler system, with 1,150 gpm flowing.

Solution From Appendix B, Table B-1, dry pipe sprinkler systems have a roughness coefficient of c = 100. Dry pipe systems are more likely to be conducive to internal pipe corrosion over a period of several years, resulting in a rougher internal pipe wall surface and a lower c-factor than for wet pipe systems.

From Appendix D, page 345, for $Q = 1{,}150$ gpm:

$$P_f = 0.051 \text{ psi/ft for } c = 100.$$
$$P_t = (P_f \text{ value selected from Appendix D, in psi/ft}) \times (L)$$
$$= (0.051 \text{ psi/ft}) \times (500 \text{ ft})$$
$$= 25.5 \text{ psi}$$

As can be seen, compared to solving the Hazen-Williams formula, using the tables in Appendix D is a much simpler way of calculating friction loss, especially for a piping system in which numerous friction loss calculations must be performed.

Reducing Friction Loss

Before we move on, let's prove that we can reduce friction loss in a sprinkler system by increasing pipe sizes.

Example 6.13 Determine the friction loss if an 8-inch, schedule 10 pipe is substituted for the 6-inch, schedule 10 pipe in Example 6.12.

Solution From Appendix D, page 346, $P_f = 0.014$ psi/ft for $c = 100$ at $Q = 1{,}150$ gpm:

$$P_t = (P_f \text{ value selected from Appendix D, in psi/ft}) \times (L)$$
$$= (0.014 \text{ psi/ft}) \times (500 \text{ ft})$$
$$= 7.0 \text{ psi}$$

Remembering that the friction loss in the 6-inch pipe was 25.5 psi, increasing the diameter of this pipe from 6 inches to 8 inches has saved us 18.5 psi in friction loss, a value that could be significant in designing the system to match the flow and pressure available from a given water supply.

Friction Loss in Simple Pipe Loops

Underground pipes, especially in urban areas, are designed in a looped or gridded arrangement to provide a more reliable water supply. Loops provide a significant hydraulic advantage over a dead-end feed by supplying water to any point in the loop from two directions.

Designers of fire protection systems often use loops or grids to provide more reliable and efficient delivery of water to a sprinkler system. This section deals with **simple loops**, or loops with one inflow, one outflow, and two paths for water flow, as shown in Figure 6-10.

Analysis of a simple loop is an outgrowth and extension of our previous study of friction loss in a single pipe segment. For the two pipe segments shown

simple loops
loops with one inflow, one outflow, and two paths for water flow

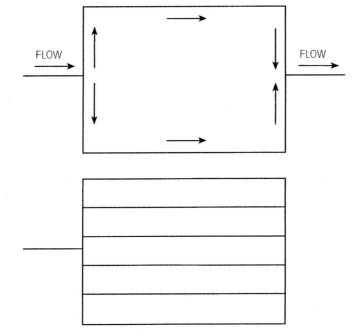

Figure 6-10 *Piping loops. Top: Simple pipe loop with one inflow, one outflow, and two paths for water flow. Bottom: Complex pipe loop (grid) with many paths and one or more inflows.*

in Figure 6-11, the friction loss for each pipe segment can be described by using the Hazen-Williams formula.

$$P_{f1} = \frac{(4.52) \times (Q_1^{1.85}) \times (L_1)}{(C_1^{1.85}) \times (D_1^{4.87})}$$

$$P_{f2} = \frac{(4.52) \times (Q_2^{1.85}) \times (L_2)}{(C_2^{1.85}) \times (D_2^{4.87})}$$

P_{f1} is the friction loss in pipe segment 1

P_{f2} is the friction loss in pipe segment 2

It must be strongly emphasized that the flow splits at point A and divides its flow to pipe segments 1 and 2 so that the pressure for pipe segment 1 at B is identical to the pressure for pipe segment 2 at B. This must be true, because the laws of nature demand that two pressures cannot exist at point B at the same time. This statement is profound yet deceptively simple. Nature and the laws of physics require that pressures be equalized at a given point in space. It must therefore be true that $P_{f1} = P_{f2}$.

By dividing these two equations, assuming that the c-factors and diameters are identical for the two pipe segments of the simple loop, and manipulating

Chapter 6 Hydraulic Calculation of Sprinkler Systems

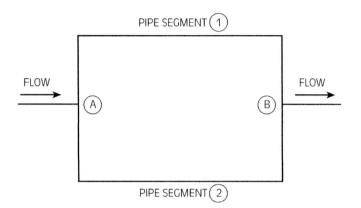

Figure 6-11 *Analysis of a simple loop.*

terms, we obtain a very useful formula for the calculation of simple loops, called the *split flow formula*:

$$Q_1 = \frac{Q_t}{1 + (L_1/L_2)^{0.54}}$$

where Q_1 = the flow in pipe segment 1, in gpm
Q_t = the total flow in the loop, in gpm, $Q_1 + Q_2 = Q_t$
L_1 = the length of the pipe in pipe segment 1
L_2 = the length of the pipe in pipe segment 2

The split flow formula can be used directly to determine how flow splits in the two pipe segments when the total flow, Q_t, is known, and when pipe diameters of the two pipe segments are the same. Once Q_1 is obtained, Q_2 is determined by the conservation of mass formula:

$$Q_1 + Q_2 = Q_t$$

Example 6.14 As shown in Figure 6-12, 500 gpm enters point A and splits into pipe segment 1 and pipe segment 2. The diameter of both pipe segments is 6-inch, schedule 40 (6.065 inches). The c-factor of both pipe segments is $C = 120$. Pipe segment 1 is 500 feet long and pipe segment 2 is 300 feet long. Determine the flow in each pipe segment in gpm.

$$Q_1 = \frac{Q_t}{1 + (L_1/L_2)^{0.54}}$$

where Q_t = 500 gpm (given)
L_1 = 500 feet (given)
L_2 = 300 feet (given)

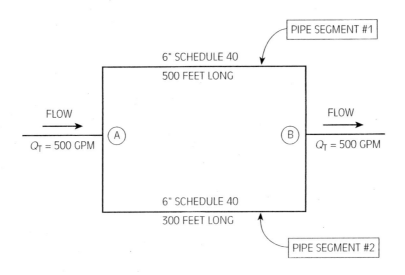

Figure 6-12
Illustration for Example 6.13.

$$Q_1 = \frac{500}{1 + (500/300)^{0.54}}$$

$$= \frac{500}{1 + (1.31764)}$$

$$= 215.7 \text{ gpm}$$

$$Q_1 + Q_2 = Q_t$$
$$215.7 \text{ gpm} + Q_2 = 500 \text{ gpm}$$
$$Q_2 = 284.3 \text{ gpm}$$

The formula $Q_1 + Q_2 = Q_t$ must be true if no flow is removed from either pipe segment. This principle, known in physics as *conservation of mass*, recognizes that mass is neither created nor destroyed within either pipe segment, but is merely transported. In a simple loop, the rate of flow that comes in goes out.

Friction Loss in Simple Loops—Equivalent Length

L_1 and L_2 in the formula below assume that the diameters of the two pipe segments are the same, and the formula therefore works only under this assumption.

$$Q_1 = \frac{Q_t}{1 + (L_1/L_2)^{0.54}}$$

However, a situation could exist in which the diameters of the two pipe segments in a simple loop are not the same. In this case, we must introduce the con-

Chapter 6 Hydraulic Calculation of Sprinkler Systems

equivalent length
a length of pipe of a given diameter whose friction loss is equivalent to the friction loss of a pipe of differing diameter

cept of the **equivalent length**. An equivalent length is a length of pipe of a given diameter whose friction loss is equivalent to the friction loss of a pipe of differing diameter. The equivalent length formula describes this relationship:

$$L_e = (L_1) \times (D_e/D_1)^{4.87} \times (C_e/C_1)^{1.85}$$

where L_e = the equivalent length, in feet
L_1 = the original length, in feet
D_1 = the original diameter, in inches
D_e = the diameter of the new equivalent length, in inches
C_1 = the original c-factor
C_e = the c-factor of the new equivalent length

The best way to illustrate the equivalent length concept is to give an example.

Example 6.15 A 4-inch, schedule 40 pipe (4.026 inches) is 1,000 feet long, and has a c-factor of 120. What equivalent length of 3-inch, schedule 40 pipe (3.068 inches) with a c-factor of 120 would give the same friction loss as the 4-inch pipe?
Solution

$$L_e = (L_1) \times (D_e/D_1)^{4.87} \times (C_e/C_1)^{1.85}$$

where L_1 = 1000'
D_1 = 4.026"
D_e = 3.068"
C_1 = 120
C_e = 120

$$L_e = (1000') \times (3.068"/4.026")^{4.87} \times (120/120)^{1.85}$$
$$= 1000' \times (.2662) \times (1)$$
$$= 266.2 \text{ feet}$$

Let us analyze the results of example 6.15. We found that 266.2 feet of 3-inch pipe has the same friction loss as 1,000 feet of 4-inch pipe. When we think about this important finding, it makes sense. The reader is encouraged to use the Hazen-Williams formula to prove that the friction loss of the 3-inch pipe is greater than the friction loss of the 4-inch pipe when all other factors are the same. Since the friction loss of the 3-inch pipe is greater than the friction loss of the 4-inch pipe, the length of 3-inch pipe that would give the same friction loss as the 4-inch pipe (the equivalent length) must be shorter than the 4-inch pipe. What we have done in example 6.15 is to convert 1,000 feet of 4-inch pipe into an equivalent length of 266.2 feet of 3-inch pipe.

We can now use the knowledge we have gained so far in this section, using the following four formulas, to analyze a simple pipe loop where the pipe diameters and/or c-factors of the two loop segments are not equal.

$$L_e = (L_1) \times (D_e/D_1)^{4.87} \times (C_e/C_1)^{1.85}$$

The equivalent length formula

$$Q_1 = \frac{Q_t}{1 + (L_1/L_2)^{0.54}}$$

The split flow formula

$$Q_1 + Q_2 = Q_t$$

The conservation of mass formula

$$P_f = \frac{(4.52) \times (Q^{1.85}) \times (L)}{(C^{1.85}) \times (D^{4.87})}$$

The Hazen-Williams formula

Example 6.16 A simple loop, shown in Figure 6-13, has 1,000 gpm flowing into it at point A, and splits flow into two pipe segments. Pipe segment 1 is 4-inch schedule 40 pipe, 1,000 feet long, with a c-factor of 120. Pipe segment 2 is 3-inch schedule 40 pipe, 1,000 feet long, with a c-factor of 120. Determine the flow in pipe segment 1, the flow in pipe segment 2, the friction loss in pipe segment 1, and the friction loss in pipe segment 2.

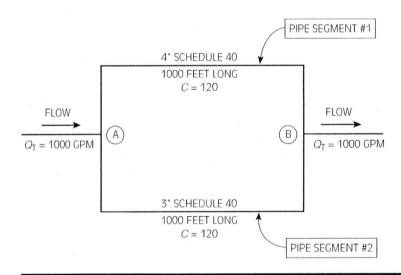

Figure 6-13 *Illustration for Example 6.16. Method of solution:*
- *Convert the 4" pipe on pipe segment 1 into an equivalent length of 3" pipe, using the equivalent length formula.*
- *Solve for flow Q_1, for a 3" simple loop, using the split flow formula.*
- *Solve for flow Q_2, using the conservation of mass formula.*
- *Determine loop friction loss, using the Hazen-Williams formula.*

Solution Because the two diameters differ, we must use the equivalent length concept to solve the problem. The 4-inch schedule 40 pipe is converted into an equivalent length of 3-inch schedule 40 pipe, and the loop is solved as a 3-inch simple loop.

We will make use of the calculation we performed in example 6.15, where we determined that 1,000 feet of 4-inch, schedule 40 pipe is equivalent to 266.2 feet of 3-inch, schedule 40 pipe. By using the equivalent length of 266.2 feet as length L_1, we have converted L_1 from 4-inch to an equivalent length of 3-inch pipe. Since L_2 was given in the problem as 3-inch pipe, we have created a 3-inch simple loop. Remember that we are creating a simple loop of uniform diameter solely for the purpose of solving the split flow formula below, because this equation only works when the diameters are the same.

We can now substitute directly:

$$Q_1 = \frac{Q_t}{1 + (L_1/L_2)^{0.54}}$$

where Q_1 = the flow in pipe segment 1 (unknown)
  Q_t = the total flow, 1000 gpm (given)
  L_1 = the equivalent length calculated in example 6.15, 266.2 feet
  L_2 = 1000 feet, given

Substituting in the equation we have:

$$Q_1 = \frac{1000}{1 + (266.2/1000)^{0.54}}$$

$$= \frac{1000}{1 + (0.4893)}$$

$$= 671.44 \text{ gpm}$$

Solving for Q_2, using the conservation of mass formula:

$$Q_1 + Q_2 = Q_t$$

$$Q_2 = Q_t - Q_1$$
$$= 1000 \text{ gpm} - 671.44 \text{ gpm}$$
$$= 328.56 \text{ gpm}$$

Now that we have determined how the flow has split in the simple loop, we now must use the actual diameter, 4.026", and the actual length of 4-inch pipe, 1,000', to solve for the friction loss in pipe segment 1:

$$P_{f1} = \frac{(4.52) \times (Q^{1.85}) \times L}{(C^{1.85}) \times (D^{4.87})}$$

$$= \frac{(4.52) \times (671.44^{1.85}) \times (1000)}{(120^{1.85}) \times (4.026^{4.87})}$$

$$= \frac{(4.52) \times (169810) \times (1000)}{(7022.4) \times (882.452)}$$

$$= 123 \text{ psi}$$

Solving for the friction loss in pipe segment 2:

$$P_{f2} = \frac{(4.52) \times (Q^{1.85}) \times L}{(C^{1.85}) \times (D^{4.87})}$$

$$= \frac{(4.52) \times (328.56^{1.85}) \times (1000)}{(120^{1.85}) \times (3.068^{4.87})}$$

$$= \frac{(4.52) \times (45119.7) \times (1000)}{(7027.4) \times (234.955)}$$

$$= 123 \text{ psi}$$

Analysis of Loop Calculation Results

Analysis of these results illustrates your understanding of loop calculation. For each loop calculation problem solved, the reader must ask, Does it make sense? This section helps the reader to place the results of a loop calculation into logical perspective. Time taken at this juncture to analyze the results helps to eliminate obvious errors.

> For each loop calculation problem solved, the reader must ask, "Does it make sense?"

First, let's look at the split flows. Does it make sense that 1,000 feet of 4-inch pipe is carrying more water (671.44 gpm) than 1,000 feet of the 3-inch pipe (328.56 gpm)? Of course it does. The 4-inch pipe has more area available to hold and transport water, as illustrated by the formula:

$$Q = Av$$

Does it make sense that the friction loss in pipe segment 1 is equal to the friction loss in pipe segment 2? The answer must be yes, since the pressure for pipe segment 1 at point B must be equal to the pressure for pipe segment 2 at point B. It is impossible for two pressures to coexist at the same point. Note that because of rounding of significant digits (digits to the right of the decimal point) in your calculations, the calculated friction losses in the two pipe segments of the simple loop may differ by one or two tenths of a psi. For loop calculations, carrying at least two significant digits to the right of the decimal point makes your calculation reasonably accurate. In nature, the friction losses in each pipe segment are exactly identical.

HYDRAULIC CALCULATION METHOD FOR SPRINKLER SYSTEMS

We are now prepared to proceed to a methodical approach for the hydraulic calculation of a sprinkler system. The procedure consists of the following steps:

- Selecting occupancy,
- Selecting a hydraulic density,
- Determining the length of the hydraulically most demanding area,
- Determining the number of sprinklers flowing along the length of the design area,
- Determining the configuration of sprinklers in the hydraulically most demanding area,
- Determining the minimum flow at the hydraulically most demanding sprinkler,
- Determining the minimum pressure at the hydraulically most demanding sprinkler,
- Determining friction loss in each pipe segment.

Hand Calculations versus Computer Calculations

At this juncture you may ask a very relevant question: "Why should I learn to hydraulically calculate a sprinkler system by hand, using the method listed above, when using a computer hydraulics program looks so much simpler?" The answer may not necessarily be comforting, but it drives to the heart of the moral and ethical responsibilities of a fire protection system designer.

Those who learn to manipulate a computerized hydraulics program without understanding the underlying principles of hydraulics upon which the computer programs are founded find themselves unable to intelligently interpret the results of a computerized hydraulic calculation. They may find themselves unable to make judicious modifications or alterations, and may find themselves trapped in their attempts to troubleshoot hydraulic problems.

Fire protection systems designers almost universally use computerized hydraulic programs in the performance of their work, but designers who understand the beauty, the power, and the elegance of the principles of hydraulics, become more reliable servants of the public trust. Don't ever forget that the lives of people you may never meet depend upon the depth of your expertise.

Choosing the Hydraulically Most Demanding Area

geographically most remote area
the area whose linear distance from the sprinkler system riser is the longest

The hydraulically most demanding area may be the **geographically most remote area**, which is the area whose linear distance from the sprinkler system riser is the longest, in feet.

Sometimes, however, the hydraulically most demanding area is not the geographically most remote area. Consider, for instance, the extreme case of a water-based suppression system that protects an office area 5,000 feet from the system riser, but also protects a nitroglycerine manufacturing facility directly adjacent to the system riser. It may well be that, even though the office is the geographically most remote area, the nitroglycerine manufacturing facility, being a significantly more hazardous occupancy, may demand considerably more water for fire suppression than the office area, making it the **hydraulically most demanding area**. The only way to know for sure whether the office area is more or less hydraulically demanding than the nitroglycerine manufacturing facility is to perform a hydraulic calculation on both areas to determine which is the hydraulically most demanding area.

hydraulically most demanding area
a design area protected by a sprinkler system that demands more water for fire suppression than any other design area on the system

At least one hydraulic calculation must be performed for the hydraulically most demanding area of each occupancy classification in a given building.

The motto here is, "When in doubt, calc it out," which means that if you are unsure which is the hydraulically most demanding area, perform calculations for other potential hydraulically most demanding areas until all uncertainty is removed.

Hydraulic Calculations—Selection of Occupancy

The correct selection of the occupancy classification of a building or a portion of a building is the foundation for meaningful and reliable hydraulic calculations. Careful selection of the occupancy is the most critical decision that a fire protection systems designer makes during the hydraulic calculation process.

A list of occupancies can be obtained by referencing Figure 5-13. Remember that occupancy classifications appear in the appendix of NFPA 13, as opposed to being placed in the body of the standard, because the occupancy conditions may vary dramatically between buildings of apparently similar occupancy. It must also be emphasized that NFPA 13 does not address such special occupancies as aircraft hangars and high-piled storage. Such occupancies have their own NFPA standard.

To clarify the concept of occupancy, Figure 5-13 lists a church as a light hazard occupancy, but all churches are not exactly alike. One church could have interior walls made of solid marble and another church could have interior walls lined with combustible plastic, designed to resemble marble. The difference in the relative severity of a fire originating from a trash can located against a wall in these two buildings could be dramatic. Careful analysis of the occupancy is the most crucial and important function of the fire protection systems designer, and coordination with the authority having jurisdiction is essential to the assurance that the occupancy has been correctly evaluated and identified.

Selecting a Hydraulic Density

Once the occupancy has been carefully evaluated and chosen, the area density curves shown in Figure 6-14 may be used as the basis for your calculation.

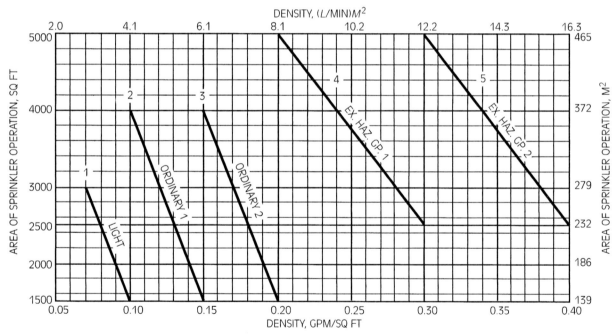

FOR SI UNITS: 1 SQ FT = 0.0929 M^2; 1 GPM/SQ FT = 40.746 (L/MIN)/M^2

Figure 6-14 *Area/density curves. (Reprinted with permission from NFPA 13,* Installation of Sprinkler Systems. *Copyright © 1996, National Fire Protection Association, Quincy, MA 02269. This reprinted material is not the complete and official position of the National Fire Protection Association on the referenced subject, which is represented only by the standard in its entirety.*

design density
the quantity of water per square foot of design area that has been found by experience to be effective in controlling a fire

design area
an area whose size is related to the occupancy, wherein all sprinklers in the area are expected to actuate

empirical evidence
the use of field experience or history as the basis for a design standard or criteria

These curves are a function of a **design density** and the total **design area** of sprinkler operation. The design density is the quantity of water per square foot of remote area that has been found by experience to be effective in controlling a fire of given occupancy. The design area is an area whose size is related to the occupancy, wherein all sprinklers in the area are expected to actuate.

When a fire occurs, only the sprinklers in the immediate area of the fire actuate. The area/density curves were based on years of recorded sprinkler actuation experience for differing occupancies. By examining the area/density curves, one can see that for each area of sprinkler operation, a density is associated with that area for light hazard, ordinary hazard group 1, ordinary hazard group 2, extra hazard group 1, and extra hazard group 2. The use of field experience or history as the basis for a design standard or criteria is referred to as using **empirical evidence** or data.

Example 6.17 Given an occupancy of ordinary hazard group 1, determine the density associated with a design area of 1,500 square feet.

Solution By referencing Figure 6-14 and observing the area/density curve for ordinary hazard group 1, we can see that the minimum density for operation of 1,500 square feet of area is 0.15 gpm per square foot.

The density is the minimum gpm per square foot that must discharge from every sprinkler in the design area. The minimum required flow, Q, that must be applied to the entire design area, is determined by the formula:

$$Q = (\text{area of operation}) \times (\text{density})$$

Example 6.18 Determine the minimum flow that must be applied to a design area of 1,500 square feet at a minimum density of 0.15 gpm/sq. ft.

$$\begin{aligned} Q &= (\text{area of operation}) \times (\text{density}) \\ &= (1500 \text{ sq. ft.}) \times (0.15 \text{ gpm/sq. ft.}) \\ &= 225 \text{ gpm} \end{aligned}$$

Note that the area and density we chose is for an ordinary hazard group 1 occupancy, 1,500 square feet and 0.15 gpm per square foot, at the bottom of the area/density curve. Let's determine what happens if we choose an area and density at the top of the curve.

Example 6.19 Determine the minimum density that must be applied over a hydraulically most demanding 4,000 square feet for ordinary hazard group 1, and calculate the minimum flow that must be applied over the entire area.

Solution Figure 6-14 shows that for ordinary hazard group 1 a density of 0.10 gpm per square foot is required for an area of 4,000 square feet. The minimum flow that must be applied to this area is:

$$Q = \text{(area of operation)} \times \text{(density)}$$
$$= (4000 \text{ sq. ft.}) \times (0.10 \text{ gpm/sq. ft.})$$
$$= 400 \text{ gpm}$$

We can conclude that by choosing an area/density point at the top of an area/density curve, we are requiring a higher minimum flow for the design area. Should the density or design area not be specified by the contract documents or mandated by the authority having jurisdiction, selecting a value at the bottom of each curve would yield the lowest minimum flow. The reader is encouraged to prove this for each occupancy curve.

NFPA 13 permits a designer to select any point along the area/density curve to use as the basis of a hydraulic calculation. As we have found, a designer may select a more demanding area/density point, should occupancy conditions warrant a certain level of conservatism. Most designers select a point at the bottom of the area/density curve, because it is the least demanding area/density point that would lead to a successful hydraulic result. A specification in a contract package may, however, mandate a required area/density point that eliminates a designer's ability to select an area/density point. For instance, some authorities having jurisdiction require a minimum design area of 3,000 square feet.

NFPA 13 stipulates that no less than 1,500 square feet of design area be calculated for light hazard and ordinary hazard. Extra hazard requires a minimum design area of 2,500 square feet.

room design method
a design area where only the sprinklers in a particular room need to be calculated

It should be noted that NFPA 13 also has provisions for a **room design method**, a design area where only the sprinklers in a particular room need to be calculated. Caution is advised, however, since the room design method depends heavily upon wall fire ratings, door fire ratings, and the presence of automatic or self-closing doors. A fire protection designer may not have control or knowledge of wall or door ratings. Wall ratings are often compromised by holes or openings in the walls, and door ratings can be violated by door stops or other methods of defeating self-closing doors. An AHJ responsible for approving a sprinkler system designed using the room design method may disapprove a system in the field if the walls have excessive holes, if the doors or door hardware are not available for inspection, if the doors are propped open, or if the doors or walls do not match the plans.

The design areas shown in Figure 6-14 are intended to be used for wet pipe systems only. Dry pipe systems require a design area 30% larger than wet systems, because they take longer to deliver water to the sprinklers, resulting in more sprinklers actuating over a larger area of operation. The 30% area increase for dry pipe systems affects the area only. The density is not modified.

Example 6.20 For an area of 1,500 square feet, chosen from Figure 6-14, determine the design area for a dry pipe system.

Solution

$$A_{dry} = (A) \times 1.30$$
$$= (1500 \text{ sq. ft.}) \times (1.30)$$
$$= 1950 \text{ sq. ft.}$$

The design density is not modified for this increased area.

Determining the Length of the Design Area

It has been determined, using empirical evidence, that a rectangular remote area, oriented parallel to the branch lines, is hydraulically more demanding than a perfectly square remote area. NFPA 13 provides a formula with which the hydraulically most demanding area is sized:

The minimum length of the rectangle is 1.2 times the square root of the design area:

$$L = (1.2) \times \sqrt{A}$$

Example 6.21 A designer has selected a design area of 1,500 square feet from the area/density curves in Figure 6-14. Determine the minimum length of the design area, L, shown in Figure 6-15.

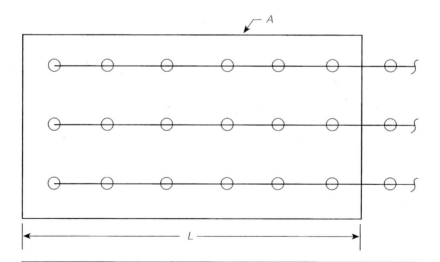

Figure 6-15 *Determining length of the design area. Select A from area/density curves (Figure 6-14).*

$$L = (1.2) \times \sqrt{A}$$

Note that L is always oriented parallel to the branch lines as shown, and is always rounded up to fall at the midpoint between two sprinklers.

$$L_{min} = (1.2) \times \sqrt{A}$$
$$= (1.2) \times \sqrt{1500}$$
$$= 46.5 \text{ feet}$$

This length is a minimum, because the length of the design area must be increased so that the edge of the design area falls at the midpoint between two sprinklers. We will make this adjustment in the next step.

Number of Sprinklers Flowing along the Length of the Design Area

The number of sprinklers flowing along the length of the design area is determined by dividing L_{min} by the spacing between sprinklers, S:

$$N_s = \frac{(L_{min})}{(S)}$$

Example 6.22 The 1,500 square foot remote area calculated in Example 6.21 had a minimum length, L_{min}, of 46.5 feet. If the sprinklers in that design area were spaced uniformly along the branch lines at a spacing (S) of 12′ between sprinklers, calculate the number of sprinklers that would flow along the length of the rectangular design area.

$$N_s = \frac{L_{min}}{S}$$
$$= \frac{(46.5 \text{ feet})}{(12 \text{ feet/sprinkler})}$$
$$= 3.88 \text{ sprinklers}$$

Partial sprinklers do not exist; we must round up to the nearest whole sprinkler. Therefore four sprinklers would flow along the length of the rectangle, making $N_s = 4$ sprinklers.

The actual length of the design area, L_{act}, is determined by multiplying the number of sprinklers, N_s, by the spacing between sprinklers, S:

$$L_{act} = (N_s) \times (S)$$

Substituting values in this example:

$$L_{act} = (4) \times (12')$$
$$= 48 \text{ feet}$$

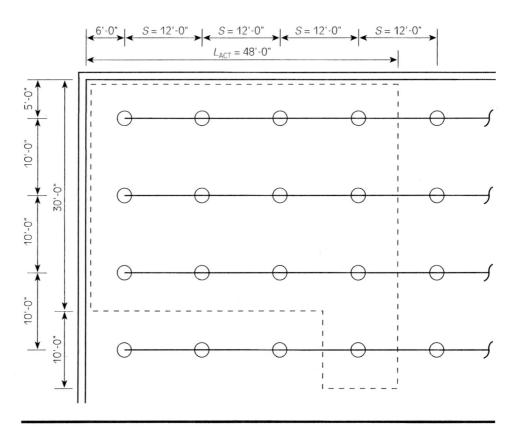

Figure 6-16 *Illustration for Examples 6.17, 6.21, 6.22, and 6.23.*

Given: A wet pipe sprinkler system

Occupancy: Ordinary hazard group I (Example 6.17)

Area of operation = 1,500 sq. ft. (Example 6.17)

Density = 0.15 gpm per square foot (Example 6.17)

L_{min} = 46.5 feet (Example 6.21)

N_s = 4 sprinklers (Example 6.22)

L_{act} = 48 feet (Example 6.22)

W = 31.25 feet (Example 6.23)

Width covered by 3 branchlines is 30'

48' × 30' = 1,440 sq. ft. (12 sprinklers)

1440 square feet is less than 1500 sq. ft.

Must add thirteenth sprinkler as shown to complete the design area.

This actual length is depicted in Figure 6-16. Note that while L_{min} was the minimum length of the rectangular design area, the actual length, L_{act}, represents an area that exactly divides the midpoint between two sprinklers, and therefore more accurately represents the area covered by the hydraulically most remote four sprinklers on the remote branch line.

Configuration of Sprinklers in the Design Area

Once we have determined the actual length of the design area, we must determine its width. Since the area of a rectangle is the product of the length and its width, the width (W) equals:

$$W = \frac{A}{L}$$

Example 6.23 Continuing with example 6.22, for a design area of 1,500 square feet, with an actual length of 48′:

$$W = \frac{A}{L}$$

$$= \frac{1500 \text{ sq. ft.}}{48 \text{ ft.}}$$

$$= 31.25 \text{ ft.}$$

The distance between branch lines is given as 10 feet. Let's look at an area 48 feet long (4 sprinklers long, as previously calculated) by 30 feet wide (three branch lines wide):

$$A = 48' \times 30' = 1440 \text{ square feet, containing 12 sprinklers.}$$

This falls slightly short of the required 1,500 square foot design area. Since each sprinkler covers a 12′ × 10′ area, or 120 square feet, we know that adding one sprinkler to 1,440 sq ft results in an area of 1,440 sq ft plus 120 sq ft, or 1560 sq ft, which meets the desired minimum criteria.

The question is, where is this thirteenth sprinkler added? This final sprinkler is always added closest to the crossmain to assure the hydraulically most demanding area. A complete calculation of the sprinkler system described by this example is included in this chapter, and it will be shown that the location of this 13th sprinkler shown in Figure 6-16 is placed in the hydraulically most demanding location.

Methods of selecting a hydraulically most demanding area are illustrated in NFPA 13, which provides several examples of hydraulically most demanding area shaping and configuration.

Minimum Flow at the Hydraulically Most Demanding Sprinkler

The flow from the hydraulically most demanding sprinkler and the pressure associated with this flow is a function of the design density, the area of sprinkler coverage, and the sprinkler discharge characteristics. To determine the minimum flow from the hydraulically most demanding sprinkler, multiply the design density by the area covered by the sprinkler. To ensure that the density is being correctly applied, the largest area covered by any sprinkler in the design area should be used in this calculation when the area covered by each sprinkler is not uniform for all sprinklers in the design area.

$$Q = (d) \times (A_s)$$

Example 6.24 Continuing with the example shown in Figure 6-16, a design density for an ordinary hazard group I occupancy obtained from Figure 6-14 was selected to be 0.15 gpm per square foot over the hydraulically most demanding 1,500 square feet. If each sprinkler in the design area is spaced uniformly at 12 feet by 10 feet, as shown in Figure 6-16, the minimum flow at the hydraulically most demanding sprinkler would be:

$$\begin{aligned} Q &= (d) \times (A_s) \\ &= (0.15 \text{ gpm/sq.ft.}) \times (120 \text{ sq.ft.}) \\ &= 18.0 \text{ gpm} \end{aligned}$$

This is the minimum required flow at the hydraulically most demanding sprinkler.

Determining the Minimum Pressure at the Hydraulically Most Demanding Sprinkler

The flow at a sprinkler head is determined by the formula:

$$Q = K \times \sqrt{P}$$

> **K-factor**
> a unique orifice coefficient that is calculated for a specific sprinkler

The flow at a sprinkler (Q) is equal to the sprinkler discharge coefficient (K) times the square root of the pressure (P). K is commonly referred to as the **K-factor**. Each sprinkler that is tested and listed for use on a sprinkler system has a unique K-factor, or orifice coefficient, that is calculated for that sprinkler. It is therefore necessary to select a make and model of a specific sprinkler before the pressure can be obtained. Sprinklers having nominal orifices of 1/2" generally have K-factors ranging from 5.3 to 5.8.

Example 6.25 A fire protection designer has selected a specific make and model of sprinkler with a K-factor of 5.6. From Example 6.24, we calculated the minimum flow at the hydraulically most demanding sprinkler to be 18.0 gpm. Determine the pressure required to deliver this minimum flow for the K-factor selected.

Solution If

$$Q = K \times \sqrt{P}$$

then

$$\sqrt{P} = \frac{Q}{K}$$

and

$$P = \left(\frac{Q}{K}\right)^2$$

Substituting the known values, we get

$$P = \left(\frac{Q}{K}\right)^2$$
$$= \left(\frac{18}{5.6}\right)^2$$
$$= (3.21)^2$$
$$= 10.33 \text{ psi}$$

Therefore, at the most remote sprinkler, a flow of 18 gpm at a pressure of 10.33 psi must be available.

NFPA 13 mandates that the minimum pressure at any sprinkler shall be 7.0 psi. If our resulting pressure had been less than 7.0 psi, we would have been required to recalculate the flow using a value of 7.0 psi:

$$Q = (5.6) \times \sqrt{7.0 \text{ psi}}$$
$$= 14.82 \text{ gpm}$$

Determining Friction Loss

Friction loss through the piping system is then calculated using the Hazen-Williams pipe friction formula, introduced previously in this chapter:

$$P_f = \frac{(4.52) \times (Q^{1.85})}{(C^{1.85}) \times (D^{4.87})}$$

node
a hydraulic reference point that correlates points of interest on a plan to points in a hydraulic calculation

The c-factor can be determined by Appendix B, and the exact internal diameter can be determined by Appendix C. Q is the minimum gpm as calculated in example 6.24.

The friction loss of each pipe segment is calculated. Figures 6-17 and 6-18 show a layout of a simple sprinkler system. Note that the design area in Figure 6-17 is the same area selected in Figure 6-16. Each of the thirteen sprinklers in the design area are assigned a number, called a **node**. A node is a hydraulic reference point that correlates points of interest on a plan to points in a hydraulic calculation. A hydraulic calculation determines friction loss between each flowing node, adding the flow of each flowing sprinkler in the remote area. The hydraulic calculation method is similar to an accounting table where pressure losses are totaled in one column and gallonage demands are totaled in another column.

As stated in the previous section with respect to the calculation of simple loops, the fire protection designer may calculate each pipe segment using the Hazen-Williams formula or may use the friction loss tables in Appendix D. Most design firms and sprinkler companies use the tables for hand calculations or computer programs that use the Hazen-Williams formula as their basis.

An Example Calculation

An example of a sprinkler system calculated by hand can be found in Figure 6-19a on page 151.

Note that on the hydraulic calculation work sheet shown in Figure 6-19, the minimum flow at the most remote sprinkler that we calculated in Example 6.24, 18.0 gpm, appears at the top of the column entitled Flow in gpm, and the minimum pressure at the most remote sprinkler that we calculated in Example 6.25, 10.33 psi, appears at the top of the column entitled Pressure Summary.

The pipe sizes shown in Figure 6-17 are pipe sizes whose diameters have been estimated by the hydraulic calculator. With experience, one can select pipe sizes that result in a solution that renders the available supply in excess of the system demand. If the system demand is greater than the system supply, pipe sizes must be increased until the supply exceeds the system demand.

Figure 6-17 *(Opposite) Layout of calculated sprinkler system.*
  Wet pipe system—ordinary hazard group I.
  0.15 gpm calculated over the most remote 1,500 sq. ft.
  13 sprinklers calculated in most remote area.
  Branch lines are schedule 40 black.
  Crossmain and feedmain is schedule 10 black.
  Underground is ductile iron class 52.

Chapter 6 Hydraulic Calculation of Sprinkler Systems

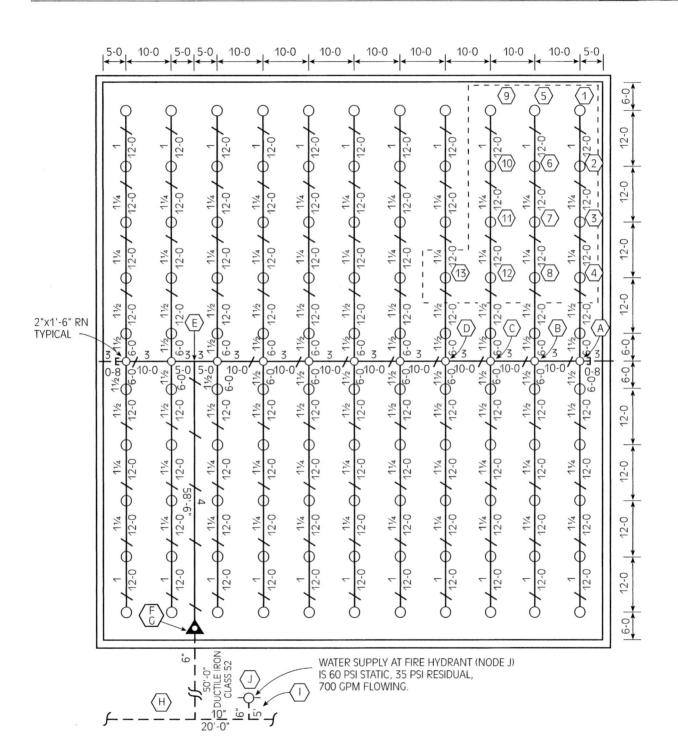

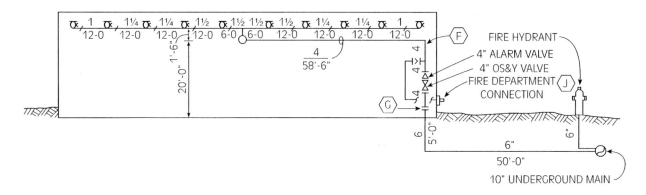

Figure 6-18 *Sectional view of building shown in Figure 6-17.*

Following the example calculation in Figure 6-19 should be fairly straightforward. Step 1 calculates the friction loss between the sprinkler at node 1 and the sprinkler at node 2. The branch line piping is given as schedule 40, with a c-factor of 120 for a wet pipe system. The designer has selected 1-inch pipe between these two nodes. Using the friction loss tables, (Appendix D, page 389), we find that for a flow of 18 gpm between these two nodes, $P_f = 0.107$ psi/ft. This value is then multiplied by the length of the pipe, 12 feet, resulting in 1.28 psi of friction loss. This loss is added to the minimum pressure required at the hydraulically most demanding sprinkler, 10.33 psi, to result in 11.61 psi being required at sprinkler 2.

Note that the column in Figure 6-19 entitled Pressure Summary on the hydraulic calculation form simply adds the friction loss in each pipe segment to determine the total pressure losses in the sprinkler system.

Since the pressure required at sprinkler 2 is 11.61 psi, the flow at sprinkler 2, which has a K-factor of 5.6, must be calculated using the formula:

$$Q = (K) \times \sqrt{P}$$
$$= (5.6) \times \sqrt{11.61}$$
$$= 19.08 \text{ gpm}$$

This flow for sprinkler 2 must be added to the flow of sprinkler 1, totaling 37.08 gpm.

Note that in the column in Figure 6-19 entitled Flow in GPM on the hydraulic calculation form that the flow of each flowing sprinkler is simply added to determine the total flow requirement for the sprinkler system.

Chapter 6 Hydraulic Calculation of Sprinkler Systems

SPRINKLER NODE NUMBER	FLOW IN GPM	PIPE SIZE	PIPE FITTINGS AND DEVICES	EQUIV. PIPE LENGTH	FRICTION LOSS PSI/FOOT	PRESSURE SUMMARY	NORMAL PRESSURE	SUPPLEMENTARY CALCULATIONS	END NODE
① MOST REMOTE SPRINKLER 0.15X120= Q=18.0 GPM	q — Q 18.0	1	SCH. 40 LINE	L 12.0 F T 12.0	0.107	Pt 10.33 Pe — Pf 1.28	Pt Pv Pn	$Q = K\sqrt{P}$ 18 GPM = 5.6 $\sqrt{P}$ P = 10.33 PSI	②
②	q 19.08 Q 37.08	1¼	SCH. 40 LINE	L 12.0 F T 12.0	0.112	Pt 11.61 Pe — Pf 1.34	Pt Pv Pn	$Q = K\sqrt{P}$ $Q = 5.6\sqrt{11.61}$ Q = 19.08 GPM	③
③	q 20.16 Q 57.24	1¼	SCH. 40 LINE	L 12.0 F T 12.0	0.245	Pt 12.95 Pe — Pf 2.94	Pt Pv Pn	$Q = K\sqrt{P}$ $Q = 5.6\sqrt{12.95}$ Q = 20.16 GPM	④
④	q 22.32 Q 79.56	1½	SCH. 40 LINE TEE	L 18.0 F 8.0 T 26.0	0.210	Pt 15.89 Pe Pf 5.46	Pt Pv Pn	$Q = K\sqrt{P}$ $Q = 5.6\sqrt{15.89}$ Q = 22.32 GPM	Aᴛ
Aᴛ AT= TOP OF RISER NIPPLE	q — Q 79.56	2	SCH. 40 LINE TEE	L 1.5 F 10 T 11.5	0.064	Pt 21.35 Pe 0.65 Pf 0.74	Pt Pv Pn	(1.5) X (.433) = Pe = 0.65 PSI	A_B
A_B AB= BOTTOM OF RISER NIPPLE	q — Q 79.56	3	SCHEDULE 10 CROSSMAIN	L 10.0 F — T 10.0	0.007	Pt 22.74 Pt — Pf 0.07	Pt Pv Pn	LINE K (HEADS 5-8) $Q = K\sqrt{P}$ $79.56 = K\sqrt{22.74}$ K = 16.68	B_B
B_B	q 79.66 Q 159.22	3	SCHEDULE 10 CROSSMAIN	L 10.0 F — T 10.0	0.024	Pt 22.81 Pe Pf 0.24	Pt Pv Pn	$Q = K\sqrt{P}$ $Q = 16.68\sqrt{22.81}$ Q = 79.66 GPM	C_B
C_B	q 80.08 Q 239.30	3	SCHEDULE 10 CROSSMAIN	L 10.0 F T 10.0	0.052	Pt 23.05 Pe — Pf 0.52	Pt Pv Pn	$Q = K\sqrt{P}$ $Q = 16.68\sqrt{23.05}$ Q = 80.08	D_B
D_B	q 25.62 Q 264.93	3	SCHEDULE 10 CROSSMAIN TEE	L 45.0 F 15.0 T 60.0	0.062	Pt 23.57 Pe Pf 3.72	Pt Pv Pn	SEE SHEET 3 FOR FLOW OF SPRINKLER ⑬	E
E	q — Q 264.93	4	SCHEDULE 10 CROSSMAIN 90° ELL	L 58.5 F 10 T 68.5	0.018	Pt 27.29 Pe — Pf 1.23	Pt Pv Pn		F
F	q — Q 264.93	4	SCHEDULE 10 CROSSMAIN ALARM VALVE OS & Y	L 20.0 F 20.0 2.0 T 42.0	0.018	Pt 28.52 Pe 8.66 Pf 0.76	Pt Pv Pn	20 X .433 = Pe = 8.66 PSI	G
						Pt 37.94			

CONTRACT NAME _____ CALCULATION OF SYSTEM SHOWN ON FIGURE 6.17 _____ SHEET 1 OF 3

MANUAL HYDRAULIC CALCULATION FORM

Figure 6-19a
Calculation of system shown in Figure 6-17.

CONTRACT NAME			CALCULATION OF SYSTEM SHOWN ON FIGURE 6.17					SHEET 2 OF 3	
SPRINKLER NODE NUMBER	FLOW IN GPM	PIPE SIZE	PIPE FITTINGS AND DEVICES	EQUIV. PIPE LENGTH	FRICTION LOSS PSI/FOOT	PRESSURE SUMMARY	NORMAL PRESSURE	SUPPLEMENTARY CALCULATIONS	END NODE
⬡ G	q — Q 264.93	6	CL. 52 D.I. UNDERGROUND ELL, TEE	L 55.0 F 18.62 39.90 T 113.52	0.0045	Pt 37.94 Pe 2.17 Pf 0.51	Pt Pv Pn	5' X .4331 Pe = 2.17	⬡ H
⬡ H	q — Q 264.93	10	CL. 52 D.I. UNDERGROUND TEE	L 20.0 F 66.50 T 86.50	0.0015	Pt 40.62 Pe — Pf 0.13	Pt Pv Pn		⬡ I
⬡ I	q — Q 264.93	6	CL. 52 D.I. UNDERGROUND ELL	L 10.0 F 18.62 T 28.62	.0045	Pt 40.75 Pe -2.17 Pf 0.13	Pt Pv Pn	5' X .4331 Pe = -2.17	⬡ J
⬡ J	ADD HOSE q 250.00 Q 514.93			L F T		Pt 38.70 Pe Pf	Pt Pv Pn		
						Pt	Pt		

CALCULATION SUMMARY:
514.93 GPM @ 38.70 PSI

	Q			T		Pf	Pn		
	q Q			L F T		Pt Pe Pf	Pt Pv Pn		
	q Q			L F T		Pt Pe Pf	Pt Pv Pn		
	q Q			L F T		Pt Pe Pf	Pt Pv Pn		
	q Q			L F T		Pt Pe Pf	Pt Pv Pn		
MANUAL HYDRAULIC CALCULATION FORM						Pt			

Figure 6-19b
Calculation of system shown in Figure 6-17.

CONTRACT NAME			CALCULATION OF SYSTEM SHOWN ON FIGURE 6.17					SHEET 2 OF 3	
NOZZLE IDENT. AND LOCATION	FLOW IN GPM	PIPE SIZE	PIPE FITTINGS AND DEVICES	EQUIV. PIPE LENGTH	FRICTION LOSS PSI/FOOT	PRESSURE SUMMARY	NORMAL PRESSURE	SUPPLEMENTARY CALCULATIONS	END NODE
⟨13⟩	q —	1½	SCH. 40 LINE TEE	L 18.0 F 8.0 T 26.0	0.013	Pt 10.33 Pe — Pf 0.34	Pt Pv Pn		⟨DT⟩
	Q 18.0								
⟨DT⟩	q —	2	SCH. 40 RN TEE	L 1.5 F 8.0 T 9.5	0.005	Pt 10.67 Pe 0.65 Pf 0.05	Pt Pv Pn	1.5 × .4331 Pe = 0.65 PSI	⟨DB⟩
	Q 18.0								
⟨DB⟩	q —			L F T		Pt 11.37 Pe Pf	Pt Pv Pn	23.05 PSI EXISTS AT DB – MUST ADJUST FLOW	
	Q 18.0								

ADJUST FLOW FOR SPRINKLER ⟨13⟩ AT NODE ⟨DB⟩

$$Q_{ADJ} = Q_L \sqrt{\frac{PH}{PL}}$$

$$Q_{ADJ} = 18.0 \sqrt{\frac{23.05}{11.37}}$$

$$Q_{ADJ} = 25.62 \text{ GPM}$$

MANUAL HYDRAULIC CALCULATION FORM

Figure 6-19c
Calculation of system shown in Figure 6-17.

interpolating
selecting a friction loss value between two values that appear on a friction loss table

equivalent fitting length
a length of straight pipe that has the same friction loss as a fitting where the water changes direction

The friction loss is similarly calculated for the pipe segments between sprinklers 2 and 3, 3 and 4, and 4 and the top of the riser nipple, node A_t. The designer has chosen to round the flow up to the nearest higher value that appears on the friction loss table. This practice provides a more conservative result than **interpolating** (selecting a friction loss value from between two values that appear on the table). Note that for the tee at the top of the riser nipple, a length of 8 feet was added to the pipe length. This length is called the **equivalent fitting length**, and represents a length of straight pipe that has the same friction loss as the fitting where the flow changes direction. A list of equivalent fitting lengths is found at the top of each page in Appendix D. Note that no equivalent fitting length is included for the tee into which sprinklers 1, 2, 3, and 4 are installed because the K-factor of a sprinkler includes the fitting into which the sprinkler is installed.

The vertical riser nipple is calculated between nodes A_t and A_b in step 5. A_t represents the top of the riser nipple, and A_b is the bottom of the riser nipple. As discussed previously in this chapter, we must account for the pressure loss resulting from this elevation change. The riser nipple is 1.5 feet, therefore the pressure loss is:

$$Pe = (h) \times (0.4331 \text{ psi/ft})$$
$$= (1.5 \text{ ft}) \times (0.433 \text{ psi/ft})$$
$$= 0.65 \text{ psi}$$

This elevation pressure loss is added to the friction loss for the riser nipple in the Pressure Summary column for step 5 between Nodes A_t and A_B.

When the calculation arrives at node B_b, we must add the flow of sprinklers 5, 6, 7, and 8. Note that the pipe sizes for these four sprinklers are the same as the pipe sizes for sprinklers 1, 2, 3, and 4. The friction losses and hydraulic results for sprinklers 4 through 5 must therefore be the same as the friction loss and results for sprinklers 1 through 4.

The difference is that we have lost 0.07 psi in friction loss between node A_b and node B_b. We can account for this difference by calculating a **branch line K-factor**, a K-factor that represents the flow of all flowing sprinklers on a branch line in the design area, using the equation:

branch line K-factor
a K-factor that represents the flow of all flowing sprinklers on a branch line in the design area

$$Q = (K) \times \sqrt{P}$$

Q is the total flow for sprinklers 5 through 8, which is the same as the gallonage we calculated for sprinklers 1 through 4 at node A_b, which was 79.56 gpm. P is the pressure at the bottom of riser nipple B_b, which is the same as the pressure we calculated for sprinklers 1 through 4 at A_b, which was 22.74 psi. We can now solve for K, which is the branch line K-factor:

$$Q = (K) \times \sqrt{P}$$
$$79.66 = (K) \times \sqrt{22.74}$$
$$K = 16.68$$

This K-factor represents the total flow of sprinklers 5 through 8 at node B_b, and can be visualized as a huge sprinkler installed at node B_b, with an orifice capable of discharging the total flow from four sprinklers at nodes 5 through 8.

We now must use this line K-factor to adjust the flow of sprinklers 5 through 8 to the higher pressure of 22.81 psi, shown on step 7. The 22.81 psi includes the friction loss for the crossmain between A_b and B_b. Using the formula:

$$Q = (K) \times \sqrt{P}$$
$$= (16.68) \times \sqrt{22.81}$$
$$= 79.66 \text{ gpm}$$

Note that 79.66 gpm, which represents the flow of sprinklers 5 through 8, is slightly more than the flow of sprinklers 1 through 4, which was 79.56 gpm. The difference between the flow of the two branch lines is created by the friction loss in the crossmain between nodes A_b and B_b.

A similar procedure occurs at node C_b. The K-factor of sprinklers 9 through 12 is the same as the K-factor for sprinklers 5 through 8, since the pipe sizes are the same. We now must use this line K-factor to adjust the flow of sprinklers 9 through 12 to the higher pressure of 23.05 psi, shown on step 8. The 23.05 psi includes the friction loss for the crossmain between B_b and C_b. Using the formula:

$$Q = (K) \times \sqrt{P}$$
$$Q = (16.68) \times \sqrt{23.05}$$
$$= 80.08 \text{ gpm}$$

At point D_b, we must consider the flow of sprinkler 13. The friction loss from sprinkler 13 to node D_b is shown in Figure 6-19c (sheet 3 of 3). To adjust for the pressure differential between the demand of sprinkler 13 at D_b, and sprinklers 1 through 12 at D_b, we use the balancing formula:

$$Q_{adjust} = (Q_{low}) \times \frac{\sqrt{P_{high}}}{\sqrt{P_{low}}}$$

Q_{low} and P_{low} are the gpm and psi associated with the lower demand, which in this case is sprinkler 13 at node D_b. P_{high} is the higher pressure associated with sprinklers 1 through 12 at node D_b. By accounting for the pressure difference between P_{high} and P_{low}, we are increasing the flow of sprinkler 13 from Q_{low} to a higher flow, Q_{adjust}. It is important to note that at node 13, we use the same minimum flow that we calculated for the most remote sprinkler at node 1. It is also essential that we use the same pipe sizes for the branch line on which sprinkler 13 appears as we have established for the other branch lines in the remote area.

$$Q_{adjust} = (Q_{low}) \times \frac{\sqrt{P_{high}}}{\sqrt{P_{low}}}$$

$$= (18.0) \times \frac{\sqrt{23.05}}{\sqrt{11.37}}$$

$$= 25.62 \text{ gpm}$$

This flow is added to the column entitled Flow in GPM in step 9, and because no additional flowing sprinklers are encountered in the calculation, we will continue to calculate friction loss back to the fire hydrant that was used to conduct the flow test. By doing this, we will be able to compare the calculated system demand to the water supply availability at the test hydrant.

Fire Hose Streams

Fire hose stream demands must be included in the hydraulic calculation. NFPA 13 requires a minimum hose stream demand of 100 gpm for light hazard occupancies, 250 gpm for ordinary hazard occupancies, and 500 gpm for extra hazard occupancies. For systems having hose connections inside the building, it is required that 50 gpm be added for each hose valve at the point of connection of the valve to the system, up to 100 gpm for inside hose flow. Any remaining hose flow requirement is added to the calculation at the nearest fire hydrant to the building. By including these flows, we will be simulating the expected use of inside hose valves and outside fire hydrants concurrent to the discharge of the sprinkler system.

Because the sprinkler system in Figure 6-17 was specified as ordinary hazard, 250 gpm is added at step 15 of the hydraulic calculation. The hose flow in this case represents the flow that could be drawn from a fire hydrant during a fire. If hose valves had been installed within the building, 50 gpm would have been added to the calculation for each of the most remote two hose valves with the remaining 150 gpm added at point J.

Graphing Results

The hydraulic results must be plotted with respect to the available water supply, as shown in Figure 6-20.

In this case, Figure 6-17 specifies a water supply of 60 psi static pressure, 35 psi residual pressure, and 700 gpm flowing at the residual pressure. This test data is plotted on the graph along with the sprinkler system demand to compare the system demand to the available water supply. Note that for this case, the system demand point falls below the available water supply indicating that the water supply can meet the system demand. If the system demand is greater than the avail-

Chapter 6 Hydraulic Calculation of Sprinkler Systems

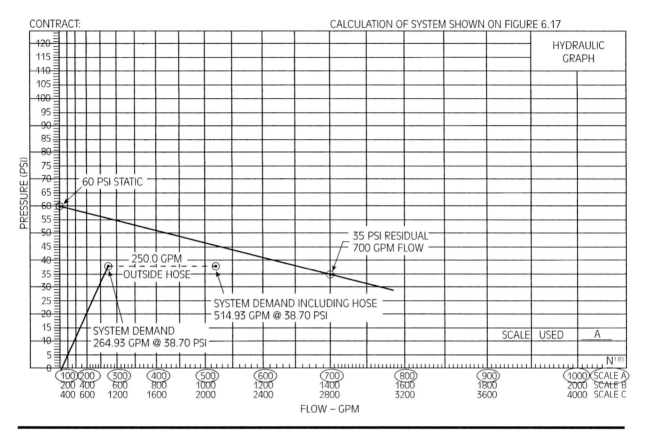

Figure 6-20 *Graphing calculated results.*

able water supply, we must increase the diameter of one or more pipes and recalculate the system until the demand is less than the available water supply. If the system demand falls significantly below the water supply curve, the system could be recalculated using smaller pipe sizes for a less expensive system.

Submitting Calculations for Approval

The completed calculations must meet the criteria listed in NFPA 13 before being submitted for approval by the authority having jurisdiction. NFPA 13 identifies the items that must be present on the hydraulic calculation submittal.

Velocity Pressure Calculations

NFPA 13 does not require the calculation of velocity pressure in hydraulic calculations submitted to an authority having jurisdiction. Velocity pressure is a measure of the kinetic energy of the flowing water, which may be added to the pressures calculated previously in Figure 6-19 at the designer's option.

Calculated velocity pressures for most sprinkler systems are relatively small, and because velocity pressures are subtracted from the available pressure at a node, the system pressure would be greater in most cases if velocity pressures are not considered. Therefore, by not calculating velocity pressures, we are adding a level of safety or a level of conservatism to the hydraulic calculations.

Velocity pressure is calculated by using the formula:

$$P_v = \frac{(0.001123) \times (Q)^2}{(D)^4}$$

where P_v = the velocity pressure, in psi
Q = the sprinkler flow, in gpm
D = the diameter of the pipe being calculated, in inches

In order to calculate the velocity pressure, one must determine the flow that travels from one node to another. In pipe segments where flow is being added, such as a segment between two flowing sprinklers, the flow must be guessed or rounded off, in order for the Q to be substituted in the equation.

Some authorities having jurisdiction may place a maximum velocity limitation in feet per second for water flow in a pipe.

The velocity may be calculated using the formula:

$$V = \frac{(0.4084) \times (Q)}{D^2}$$

NFPA at one time mandated a maximum velocity limitation of 32 feet per second, but deleted this reference. When this requirement was in force, a designer was required to increase the size of a pipe whose velocity exceeded 32 feet per second. NFPA 13 removed the requirement because the phenomena of velocity interference as a restriction to the flow of water in a pipe could not be justified with empirical evidence.

Some insurance companies, such as Factory Mutual, maintain a maximum velocity, but requirements for a maximum velocity are disappearing from many specifications and insurance requirements. Some insurance companies impose velocity maximums in an effort to provide a cushion in the calculations to account for pipe corrosion or possible inaccuracies in the calculations or in the calculation method.

Quick Response Sprinklers and the Design Area

The 1996 edition of NFPA 13 has modified the design area concept by giving the designer credit for the use of quick response sprinklers. Quick response sprinklers have heat-sensitive elements that allow the sprinklers to actuate faster than standard response sprinklers. These sprinklers would therefore actuate earlier in the fire development, and the total number of sprinklers that actuate would be fewer. NFPA 13 recognizes the potential for this smaller number of actuating sprinklers and permits the designer to reduce the size of the design area accordingly, without revising the density:

- For light hazard and ordinary hazard sprinkler systems with 10-foot-high ceilings, the design area may be reduced by 40%.
- For light hazard and ordinary hazard design areas with 20-foot-high ceilings, the design area may be reduced by 20%.
- For light hazard and ordinary hazard design areas with 30-foot-ceilings, the design area may be reduced by 10%.
- No reductions are allowed for ceiling heights greater than 30 feet.

Example 6.26 A designer has selected a design area of 1,500 square feet, with a corresponding density of 0.10 gpm per square foot, from the area/density curves in Figure 6-14. For a light hazard occupancy, determine the design area and design density if quick response sprinklers are used and if the ceiling height is 20 feet.

Solution The design area may be reduced by 20%. By multiplying the originally chosen area of 1,500 square feet by 0.80, or 80% of the original area, we obtain a new design area that is 20% smaller. This results in a design area of 1,200 square feet. The density is not modified, and remains 0.10 gpm per square foot. The use of quick response sprinklers in the design allows us to reduce the design area by 300 square feet.

Extended Coverage Sprinklers and the Design Area

A dramatic development in sprinkler technology has been the development of the extended coverage sprinkler. Material was added to the 1996 edition of NFPA 13 that provides advice for the design of sprinkler systems using extended coverage sprinklers.

Extended coverage sprinklers have larger orifices, higher K-factors, and differently designed deflectors than standard coverage sprinklers:

- NFPA 13 allows an extended coverage sprinkler protecting light hazard and ordinary hazard occupancies to cover an area up to 400 square feet per sprinkler.
- An extended coverage sprinkler protecting extra hazard and high piled storage occupancies may cover an area up to 196 square feet per sprinkler.

When performing a hydraulic calculation of a sprinkler system using extended coverage sprinklers, we can see that the number of sprinklers that we calculate in the design area is significantly less than for standard coverage sprinklers. However, since the density required for this design area remains the same, the minimum flow from each sprinkler is significantly greater than with standard sprinklers.

Example 6.27 Determine the minimum flow per sprinkler and the minimum number of sprinklers in the design area for an ordinary hazard sprinkler system with a hydraulically most demanding area of 1,500 square feet, using extended coverage sprinklers listed for the protection of 400 square feet per sprinkler.
Solution

$$\frac{1500 \text{ sq. ft.}}{400 \text{ sq. ft./sprinkler}} = 4 \text{ sprinklers}$$

This is a considerable reduction from the 13 standard coverage sprinklers we calculated for the 1,500 square foot design area in Figure 6-17. The minimum flow per sprinkler is (0.15 gpm/sq.ft.) × (400 sq.ft.) = 60 gpm minimum per sprinkler, a significant increase from the 18.0 gpm we calculated for the 13 flowing sprinklers for Figure 6-17.

COMPUTERIZED HYDRAULIC CALCULATION PROGRAMS

Numerous hydraulics programs are commercially available for use by a fire protection designer. Most programs are capable of being used on personal computers and are fairly easy to use. The programs require the fire protection designer to input the minimum information necessary for the program to solve the frictional and elevational losses in the system, using the estimated pipe sizes that were input by the designer. The computerized results show either a pressure cushion, indicating a demand pressure that is less than the available supply, or a pressure deficit, indicating a system demand that is greater than the available water supply. In a manner similar to hand hydraulic calculations, pipe diameters are then revised until the system demand is less than the available supply.

Use of the Computerized Hydraulic Program Included in This Book

A computer disc containing a demonstration model of a computerized hydraulic program, called "THE" Sprinkler Demo Program, is attached to the back of this book. Appendix E includes an instruction manual for the program. Readers should find this program relatively easy to use and interpret.

Appendix E also provides step-by-step instruction for the input and computer calculation of Figure 6-17, previously solved using manual calculation methods. Extensive explanation of the program and hardware requirements is also contained in Appendix E.

Summary

Water-based fire protection systems use the most amazing fire suppressant available—water. The properties of water make it perfect for most fire protection applications. Hydrokinetics is the study of water in motion, and hydrostatics is the study of water at rest. Friction loss is the phenomena related to the loss of water pressure associated with the frictional resistance of the interior pipe wall. Water flow in simple pipe loops can be solved using the split flow formula, and friction loss in either leg of a simple pipe loop can be demonstrated to be equal. Sprinkler systems may be hydraulically calculated using the method covered in this chapter. NFPA 13 has been modified for the reduction of the most remote area for systems using quick response sprinklers, and material has been added to NFPA 13 for extended coverage sprinklers. Computerized hydraulics programs can make hydraulic calculations easier, but the user must thoroughly understand the principles of hydraulics before using them.

Review Questions

1. A basement with dimensions 68 feet long by 42 feet wide by 10 feet deep is flooded to the ceiling. Calculate:
   a. The number of gallons of water in the basement.
   b. The weight of the water in the basement.
   c. The force exerted on 1 square inch of basement floor by the water.
   d. The force exerted on 1 square inch of basement floor by the water if the water level drops to 1 foot in depth.

2. A gauge at the top of a 76 foot tall vertical 4" schedule 10 pipe reads 65 psi. Determine the reading of a gauge at the bottom of the pipe assuming that water is not flowing in the pipe.

3. Determine the rating, in psi, of a fire pump with a rating of 214 feet of head.

4. Determine the rating, in feet of head, of a fire pump with a rating of 175 psi.

5. A horizontal 8" schedule 30 pipe reduces to a 4" schedule 40 pipe. In the 8" pipe, 640 gpm is moving at a velocity of 6 feet per second. Determine the velocity of the water flowing in the 4" pipe.

6. Determine the friction loss per foot in a 1" schedule 40 pipe, with 48 gpm flowing in a dry pipe system.

7. For the pipe in problem 6, determine the pressure differential, in psi/ft, if a 1¼" pipe is substituted for the 1" pipe.

8. A simple loop has 850 gpm flowing into the loop. Pipe segment 1 is 6" schedule 10 pipe, 420 feet long, c = 120. Pipe segment 2 is 6" schedule 10, 650 feet long, c = 120.
   a. Determine the flow in pipe segment 1.
   b. Determine the flow in pipe segment 2.
   c. Determine the friction loss in pipe segment #1.

d. Determine the friction loss in pipe segment #2.
9. Water enters a simple pipe loop at the rate of 1020 gpm. Pipe segment 1 is 8″ schedule 30 pipe, 730 feet long, c = 120. Pipe segment 2 is 10″ schedule 10, 450 feet long, c = 120.
   a. Determine the flow in pipe segment 1.
   b. Determine the flow in pipe segment 2.
   c. Determine the friction loss in pipe segment 1.
   d. Determine the friction loss in pipe segment 2.
10. Recalculate the sprinkler system shown in Figure 6-17 as a light hazard wet pipe sprinkler system.
11. Recalculate the sprinkler system shown in Figure 6-17 as an extra hazard group I dry pipe sprinkler system. Remember that a dry pipe system requires a most remote area 30% larger than for a wet pipe system, and that c = 100 is used in the calculations.
12. Recalculate the sprinkler system shown in Figure 6-17 as an ordinary hazard group I wet pipe sprinkler system, using quick response sprinklers and assuming a 10-foot ceiling height.
13. Calculate the velocity pressure, in psi, for 40 gpm flowing in a 1″ schedule 40 pipe.
14. Calculate the velocity, in feet per second, for a 6″ schedule 10 pipe with 930 gpm flowing.

Activities

1. One gallon of fresh water and one gallon of salt water are accurately weighed. Which do you think would weigh more? Why?
2. Measure the height of your house from ground level to the highest ceiling. What water supply must be available to supply 35 psi for a sprinkler at the highest ceiling?
3. Survey the sprinkler system in your school, or in a building convenient to your home. Obtain the water supply for the building and perform a hydraulic calculation of the system.

Chapter 7

Specialized Piping and Calculation Methods

Objectives

Upon completion of this chapter, you should be able to:

- Understand the limitations of pipe schedule sprinkler systems.
- Calculate a standpipe system or a combined sprinkler/standpipe system.
- Determine a combined fire pump and city supply curve.
- Determine a combined gravity tank and city supply curve.
- Design and calculate a residential sprinkler system.
- Understand the criteria required to design a sprinkler system for a high-piled storage occupancy.

Based on what we learned in Chapter 6, we may now expand our knowledge of hydraulic calculation to some specialized situations. In rare cases or for very small projects of limited scope, a pipe schedule system may be considered. Most fires involving death occur in the home, and residential sprinkler systems are likely to become more commonplace as AHJs make progress in enacting local requirements intended to reduce fire loss. Residential systems offer an interesting contrast to the systems we discussed in Chapter 6, which involved a hydraulically most demanding area. High piled storage occupancies are among the more challenging hydraulic calculations that a fire protection designer may perform, and for a wide variety of hazards, pumps and tanks may be considered in cases where the existing water supply requires enhancement.

PIPE SCHEDULE DESIGNS

Sprinkler piping is sized in accordance with NFPA 13 by one of two methods. By far the most common method is by hydraulic calculation, as explained in Chapter 6. **Pipe schedule** designs are also permitted under certain specific circumstances.

A pipe schedule is a chart in NFPA 13 that relates pipe size to a specific maximum number of sprinklers that can be supplied by that pipe. NFPA 13 limits the use of pipe schedules to new light hazard or ordinary hazard systems that do not exceed 5,000 square feet in coverage area and for extensions or additions to existing pipe schedule systems. Any new extra hazard system must be hydraulically calculated because NFPA 13 does not permit the use of pipe schedule design for extra hazard systems.

Pipe schedule systems must also meet the minimum pressure, flow, and duration requirements of NFPA 13. The minimum requirements for light hazard systems are:

- 15 psi at the highest sprinkler,
- A total flow between 500 and 750 gpm,
- A duration between 30 and 60 minutes.

The minimum requirements for ordinary hazard systems are:

- 20 psi at the highest sprinkler,
- A total between 850 to 1,500 gpm,
- A duration between 60 and 90 minutes.

The range of total flow and duration depends on whether the system is supervised by remote station or central station alarm services. These alarm services monitor water flow signals and help to alert fire department personnel more quickly, allowing for use of the minimum system flow and duration requirements listed above. Total flow and duration are concerns in determining the gallonage capacity of water storage tanks.

However, even with verification of the minimum water supplies required by NFPA 13, exact conformance to the water supply cannot be ensured in cases

pipe schedule
a chart in NFPA 13 that specifies pipe size to a specific maximum number of sprinklers that can be supplied by the pipe

involving unusual pipe configurations, in cases in which piping systems have corroded, or in systems with unusually high numbers of elbows and direction changes. Verification of performance based on available water supply can only be achieved with the use of hydraulic calculations.

Although pipe schedules were the norm for decades, their use is being increasingly restricted by NFPA 13 and will eventually become obsolete. A fire protection designer must become vigilant in assuring the adequacy of a water supply by hydraulically calculating a fire protection system.

Standard on the Installation of Sprinkler Systems in One and Two Family Dwellings and Manufactured Homes

Standard on the Installation of Sprinkler Systems in Residential Occupancies up to Including Four Stories in Height

RESIDENTIAL SPRINKLER SYSTEMS

The design of residential sprinkler systems falls into three categories, using three NFPA standards as reference:

NFPA 13D *Standard on the Installation of Sprinkler Systems in One and Two Family Dwellings and Manufactured Homes,*

NFPA 13R *Standard on the Installation of Sprinkler Systems in Residential Occupancies up to and Including Four Stories in Height,*

NFPA 13 *Standard on the Installation of Sprinkler Systems.* This standard applies to residential occupancies exceeding four stories in height.

One- and Two-Family Dwellings and Manufactured Homes

Most of the fires and fire deaths in this country occur in the home. AHJs are making a concerted effort to encourage the installation of residential sprinkler systems, especially in the lower-density dwellings covered by NFPA 13D, and some jurisdictions are passing laws that require the installation of residential sprinkler systems.

The concept of the hydraulically most demanding area is a cornerstone of the hydraulic calculation of sprinkler systems. For residential systems, the hydraulic design is based on the hydraulically most demanding 1, 2, or 4 sprinklers, as opposed to the performance of a hydraulically most demanding rectangular area. In effect, with residential systems we are dealing with a considerably smaller design area with lower flow requirements than the sprinkler systems covered in Chapter 6. The primary purpose of a residential sprinkler system is to control fire and prevent flashover in order to allow adequate time for the egress of the occupants of a residence. While life safety is the primary objective, residential systems have established an impressive record of fire control and property protection as well.

Residential sprinklers have quick response elements that are suitable for smoldering fires of relatively low heat release found in residential occupancies. Residential occupancies do not ordinarily have the high water pressures and gpm flows that are commonly available for many commercial buildings. A residential sprinkler is therefore designed to operate at relatively low pressures with relatively low flows. The combination of lighter fuel loads, along with the faster response provided by quick response elements, establishes a basis for design areas that are significantly smaller and more water efficient than for commercial and industrial occupancies.

NFPA 13D requires that two separate calculations be performed. The first calculation is for the two hydraulically most remote sprinklers to flow at least 13 gpm each, or at the minimum multiple sprinkler flow listed for the sprinkler, whichever is greater.

The second calculation is for the hydraulically most demanding sprinkler flowing a minimum of 18 gpm, or at the minimum single sprinkler flow listed for the sprinkler, whichever is greater.

Since the listed K-factors, minimum flows, spacing requirements, and areas of coverage vary widely among residential sprinklers, a specific residential sprinkler must be selected for use, and the design of the system must strictly follow the listing requirements for the sprinkler selected.

Example 7.1 Perform a hydraulic calculation for the residential sprinkler system for the single family residence shown on Figures 7-1a, b, c, and d. The available static pressure is 65 psi, CPVC piping is used inside the residence, type L copper tubing is used for the underground feedmain, and the sprinkler selected is a listed residential horizontal sidewall sprinkler having the following listing requirements:

Maximum coverage area: 15' width × 15' length (or throw),

Minimum single sprinkler flow listing: 27 gpm at 23.2 psi,

Minimum multiple sprinkler flow listing: 21 gpm at 14.1 psi.

Solution   The calculation on Figure 7-2 on page 171, shows the single sprinkler calculation, and the calculation on Figure 7-3 on page 172, shows the required two sprinkler calculation per NFPA 13D. The results of the calculation show both sprinkler demands to be less than the available supply of 65 psi. For NFPA 13D systems, the static pressure is usually accepted by AJHs for calculation, since residential sprinkler gallonage demands are usually very low compared to commercial and industrial sprinkler demands.

Three concepts that are very important to an understanding of residential sprinkler system design should be noted from the drawings of the residential system shown on Figures 7-1a, b, c, and d:

1. It is essential to note that the upper level of the residence shown on Figure 7-1a, is protected with sidewall sprinklers installed within interior wall spaces to eliminate the need to install wet pipes in the unheated attic above the second floor. NFPA 13D and NFPA 13R require that all wet pipes be installed in areas not subject to freezing, and attics in some locales have been found to develop freezing temperatures even in cases where insulation had been installed. The design shown provides reliable fire protection during even the coldest winter.

2. Carefully study the arrangement of the riser and valve assembly for this residence, as shown on Figure 7-1d. NFPA 13D requires that both the **domestic water supply** (the supply to toilets and sinks) and the sprinkler water supply be

domestic water supply
the water supply to toilets and sinks

Chapter 7 Specialized Piping and Calculation Methods

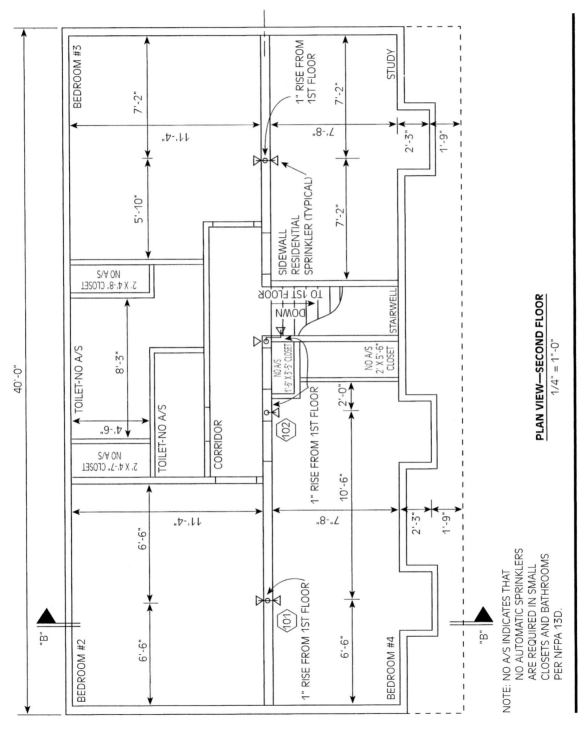

Figure 7-1a *The most remote area of a single family residence.*

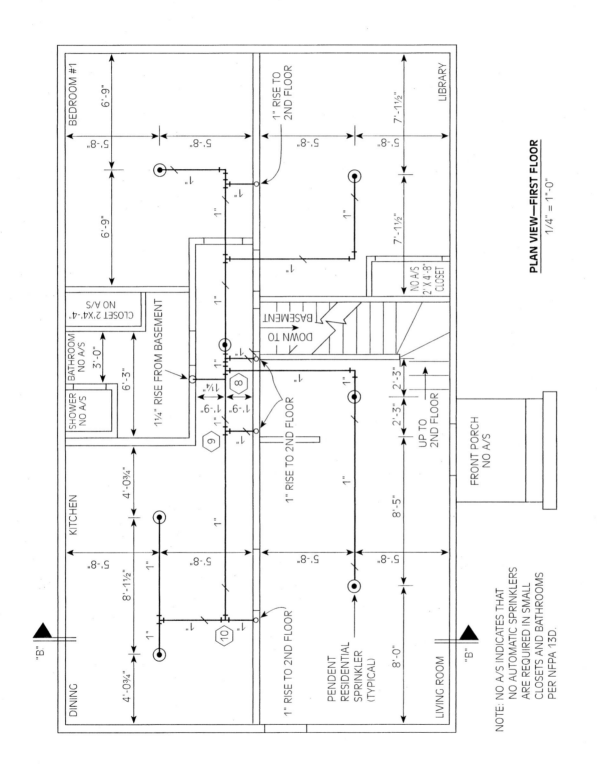

Figure 7-1b *First floor of a single family residence.*

Chapter 7 Specialized Piping and Calculation Methods

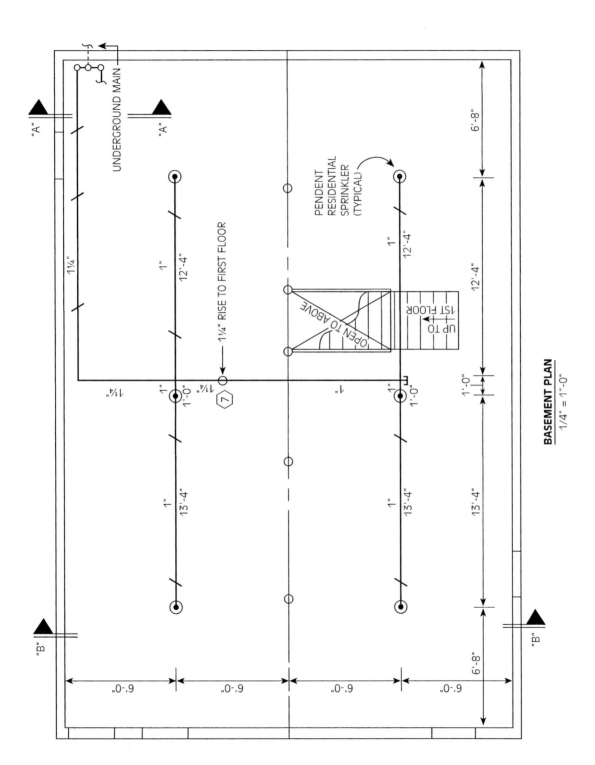

Figure 7-1c *Basement of a single family residence.*

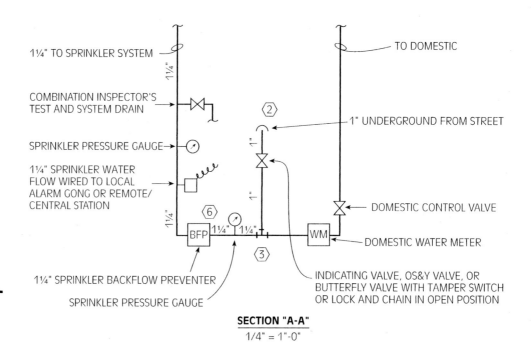

Figure 7-1d *Riser section for a single family residence.*

limited area dwellings
dwellings that do not exceed 2,000 square feet in total area

controlled by a single valve that shuts off both systems. A problem requiring the shutdown of the sprinkler system would necessitate that the domestic supply be turned off simultaneously, prompting residents to have the sprinkler protection restored immediately so that domestic water is restored.

3. Note that no sprinklers are shown in toilets and closets on Figure 7-1a and 7-1b. Since NFPA 13D systems are intended primarily for life safety of occupants, it permits the omission of sprinklers in normally unoccupied areas, such as small closets and bathrooms, and inaccessible concealed spaces. NFPA 13D also permits the elimination of sprinklers from normally unoccupied areas with the potential for freezing, such as attics, crawl spaces, garages, porches, and carports. The residential sprinkler system is intended to provide fire control in occupied areas and in routes of egress, allowing the occupants to safely exit the home during a fire, and is not intended to provide property protection in all areas of the dwelling.

A chapter was added to NFPA 13D in 1994 that details requirements for **limited area dwellings** that do not exceed 2,000 square feet in total area. Limited area dwellings are predominately manufactured homes, formerly referred to by the standard as being mobile homes. Specially designed residential sprinklers specifically listed for limited area dwellings must be used, since the system flow and pressure demands are even less than for the other residential occupancies previously covered. For limited area dwellings, NFPA 13D requires a minimum flow of 10 gpm at 25 psi for single sprinkler flow, or the minimum listed single sprinkler

Chapter 7 Specialized Piping and Calculation Methods

CONTRACT _SINGLE FAMILY RESIDENCE – SINGLE SPRINKLER CALCULATION_   SHEET ___ OF ___

SPRINKLER NODE NUMBER	FLOW IN G.P.M.	PIPE SIZE	PIPE FITTINGS AND DEVICES	EQUIV. PIPE LENGTH	FRICTION LOSS P.S.I./FOOT	PRESSURE SUMMARY	NORMAL PRESSURE	NOTES	END NODE
⟨101⟩	q — Q 27.0	1 CPVC	CPVC ELL, TEE	L 14.0 F 7.0 T 21.0	0.1144	Pt 23.2 Pe 5.20 Pf 2.40	Pt Pv Pn	12' X .433 = 5.20 PSI	⟨10⟩
⟨10⟩	q — Q 27.0	1 CPVC		L 11.0 F — T 11.0	0.1144	Pt 30.8 Pe — Pf 1.26	Pt Pv Pn		⟨9⟩
⟨9⟩	q — Q 27.0	1 CPVC	TEE	L 3.0 F 5.0 T 8.0	0.1144	Pt 32.06 Pe — Pf 0.92	Pt Pv Pn		⟨8⟩
⟨8⟩	q — Q 27.0	1¼ CPVC	ELL, TEE	L 10.0 F 9.0 T 19.0	0.0394	Pt 32.98 Pe 4.33 Pf 0.75	Pt Pv Pn	10' X .433 = 4.33 PSI	⟨7⟩
⟨7⟩	q — Q 27.0	1¼ CPVC	3 ELL	L 36.0 F 9.0 T 45.0	0.0394	Pt 38.06 Pe 3.46 Pf 1.77	Pt Pv Pn	8' X .433 = 3.46 PSI	⟨6⟩
⟨6⟩	q — Q 27.0	1¼ CPVC	TEE	L 1.0 F 6.0 T 7.0	0.0394	Pt 43.29 ✗ 5.0 Pf 0.28	Pt Pv Pn	ADD 5.0 PSI LOSS FOR BACKFLOW PREVENTER	⟨5⟩
⟨5⟩	q — Q 27.0	1 COPPER	ELL, TEE	L 38.0 F 7.0 T 45.0	.1796	Pt 48.57 Pe −1.73 Pf 8.08	Pt Pv Pn	4' X .433 = 1.73	⟨2⟩
	q Q			L F T		Pt 54.92 Pe Pf	Pt Pv Pn		
	q Q			L F T		Pt Pe Pf	Pt Pv Pn		
	q Q			L F T		Pt Pe Pf	Pt Pv Pn		
	q Q			L F T		Pt Pe Pf	Pt Pv Pn		

MANUAL HYDRAULIC CALCULATION FORM   Pt

Figure 7-2 *Single family residence—single sprinkler calculation.*

CONTRACT: SINGLE FAMILY RESIDENCE – TWO SPRINKLER CALCULATION   SHEET ___ OF ___

SPRINKLER NODE NUMBER	FLOW IN G.P.M.	PIPE SIZE	PIPE FITTINGS AND DEVICES	EQUIV. PIPE LENGTH	FRICTION LOSS P.S.I./FOOT	PRESSURE SUMMARY		NORMAL PRESSURE		NOTES	END NODE
⟨101⟩ q — Q 21.0		1 CPVC	ELL, TEE	L 14.0 F 7.0 T 21.0	.0719	Pt 14.1 Pe 5.20 Pf 1.51		Pt Pv Pn		12' X .433 = 5.20 PSI	⟨10⟩
⟨10⟩ q — Q 21.0		1 CPVC		L 11.0 F — T 11.0	.0719	Pt 20.81 Pe — Pf .79		Pt Pv Pn		$Q_{ADJ} = 21\sqrt{\frac{21.6}{20.81}}$ $Q_{ADJ} = 21.39$	⟨9⟩
⟨9⟩ q 21.39 Q 42.39		1 CPVC	TEE	L 3.0 F 5.0 T 8.0	.2707	Pt 21.60 Pe — Pf 2.17		Pt Pv Pn			⟨8⟩
⟨8⟩ q — Q 42.39		1¼ CPVC	ELL, TEE	L 10.0 F 9.0 T 19.0	.0908	Pt 23.77 Pe 4.33 Pf 1.72		Pt Pv Pn		10' X .433 = 4.33 PSI	⟨7⟩
⟨7⟩ q — Q 42.39		1¼ CPVC	3 ELLS	L 36.0 F 9.0 T 45.0	.0908	Pt 29.83 Pe 3.46 Pf 4.09		Pt Pv Pn		8' X .433 = 3.46 PSI	⟨6⟩
⟨6⟩ q — Q 42.39		1¼ CPVC	TEE	L 1.0 F 6.0 T 7.0	.0908	Pt 37.38 P̶e̶ 5.0 Pf .64		Pt Pv Pn		ADD 5.0 PSI LOSS FOR BACKFLOW PREVENTER	⟨5⟩
⟨5⟩ q — Q 42.39		1 COPPER	ELL, TEE	L 38.0 F 7.0 T 45.0	.4145	Pt 43.02 Pe −1.73 Pf 18.65		Pt Pv Pn		4' X .433 = 1.73	⟨2⟩
q Q				L F T		Pt 59.94 Pe Pf		Pt Pv Pn			
q Q				L F T		Pt Pe Pf		Pt Pv Pn			
q Q				L F T		Pt Pe Pf		Pt Pv Pn			
⟨102⟩ q — Q 21.0		1 CPVC	ELL, TEE	L 14.0 F 7.0 T 21.0	.0719	Pt 14.1 Pe 5.20 Pf 1.51		Pt Pv Pn		ADJUST FLOW AT NODE 9 ABOVE	⟨9⟩

MANUAL HYDRAULIC CALCULATION FORM   Pt 20.81

Figure 7-3 Single family residence—two sprinkler calculation.

flow, whichever is greater. A second calculation must also be performed using 6.5 gpm at 11.5 psi each for the flow of the two hydraulically most demanding sprinklers, or the minimum listed multiple sprinkler flow, whichever is greater.

Residential Occupancies up to and Including Four Stories

NFPA 13R is used to design residential sprinkler systems for residential occupancies such as low-rise hotels, dormitories, and apartment buildings using residential sprinklers.

The requirements of NFPA 13R are virtually the same as the requirements of NFPA 13D, with the primary differences being the design of the hydraulically most demanding area. Just as with NFPA 13D, NFPA 13R requires that two calculations be performed, with the single sprinkler minimum flow of 18 gpm being used. NFPA 13R requires that the system be designed for a multiple-sprinkler flow of four sprinklers flowing 13 gpm each, or the minimum listed multiple-sprinkler flow, whichever is greater.

Residential Occupancies Greater Than Four Stories

NFPA 13 is to be used for the design of sprinkler systems in buildings exceeding four stories. Tall buildings, especially those exceeding 75 feet, are vexing problems for fire protection professionals because of the difficulties associated with manual fire suppression in these buildings. Delivering personnel and fire-fighting materials to the fire scene is an extremely difficult job in high-rise fires, making the need for the installation of sprinkler systems in high-rise buildings even more compelling than for low-rise buildings.

Calculations for the hydraulically most demanding area in buildings exceeding four stories requires:

- The calculation of the four hydraulically most demanding residential sprinklers,
- The calculation of the hydraulically most demanding sprinkler, performed at the minimum listed flows for the sprinkler,
- A minimum hose stream demand of 100 gpm must be added to the calculations since most high-rise buildings are required to be fitted with standpipe systems that have hose valves installed on them.

Whereas NFPA 13D and 13R permit the omission of sprinklers in areas such as attics, NFPA 13 requires complete sprinkler protection in residential occupancies. NFPA 13, NFPA 13D, and NFPA 13R allow the omission of sprinklers in closets smaller than 24 square feet and bathrooms smaller than 55 square feet.

Freeze Protection for Residential Sprinkler Systems

NFPA 13D and NFPA 13R recognize four methods for protecting wet pipe systems from freezing:

- antifreeze systems,
- dry pipe systems,
- preaction systems,
- listed dry sprinklers (upright, pendent, or sidewalls), extended from piping located in heated areas.

antifreeze system
a sprinkler system where an additive is used in the water to prevent freezing

backflow preventer
a device designed to prevent sprinkler system water from being drawn backward into a public water supply

In an **antifreeze system**, an additive is used in the sprinkler system water to prevent freezing. These additives are antifreeze solutions similar to those you put in your car. Antifreeze systems are coming under increasing scrutiny because of the potential for contamination of water supplies. A backflow situation could occur during a fire if a fire truck pumps water at a faster rate than the water supply is capable of delivering, drawing water backward from piping systems in the adjacent area. If those systems contain antifreeze, the antifreeze could be drawn into the public drinking water supply.

Backflow preventers, consisting of two check valves, are designed to prevent sprinkler system water from being drawn backward into a public water supply. They are recommended for antifreeze systems, but many AHJs are still wary of approving an antifreeze connection to a public water supply, even with a backflow preventer, because the rubber seals on the check valves may become brittle after several years of service and fail under very high pressure. The AHJ and the water authority must be consulted before designing an antifreeze system. A designer should also be aware that some backflow preventers have very high pressure losses associated with them, and the pressure loss may be too great for a poor water supply or for a very tall building.

Dry pipe systems and preaction systems are permitted to be used for residential sprinkler systems, but at present, no residential sprinkler is approved for a dry pipe or preaction system, and CPVC piping is not listed for dry pipe or preaction system use. Wet pipe systems are therefore predominately used for residential sprinkler systems.

Dry sidewall sprinklers, dry pendent sprinklers, and dry upright sprinklers are not listed as being residential sprinklers, but they are recognized for the protection of areas where freezing may be a problem. NFPA 13D and NFPA 13R require that the pipe feeding these sprinklers be in a heated area, and it is strongly recommended that the length of the shaft of the dry sprinkler be a minimum of 8 inches, to minimize the possibility of freezing commonly encountered with very short sidewall sprinkler shaft lengths.

NFPA 13D and NFPA 13R allow the elimination of sprinklers from most of the areas in a residential occupancy that pose a potential for freezing, such as attics, garages, porches, crawl spaces, and exterior stairs. Even though sprinklers are not required in attics, some designers choose to run pipes in the attic to feed sprinklers on the top floor of a residence, with insulation draped over the pipe as a freeze protection scheme.

Placing this method into question in some cases is the large ventilation openings provided in attics by architects. An attic must be ventilated in order to prevent expanding air from causing ceiling damage during the summer. The failure of insulation as a freeze protection scheme can occur when the rate of ventilation into the attic is great, when a large distance from the pipe to the ceiling below allows for the dissipation of conducted or convected heat, when the insulation rating is insufficient to retain heat, when the insulation is installed in such a way as to allow heat to escape, or for other reasons that can make the choice of insulation as a freeze protection method risky. The problem involves the inability of many sprinkler system designers to accurately predict air flows, heat dissipation, insulation ratings, or insulation installation methodology.

Figure 7-1a (page 167) shows an alternate method used to feed the upper level without installing pipes in the attic and subjecting them to freezing. Another alternate method involves installing the pipes below the ceiling where the wall meets the ceiling. The pipe can be concealed with a soffit, as shown on Figure 7-4, and a sidewall sprinkler can be mounted to the side of the soffit.

It is essential that a fire protection designer recognize that all portions of a wet pipe system must be installed in an area that is not subject to freezing, in order for the system to provide reliable fire protection during the winter. Although frozen and bursting pipes are a property damage concern, the designer should be even more concerned about pipes that may be frozen or partially frozen without bursting, because these pipes may be incapable of delivering water to a fire when

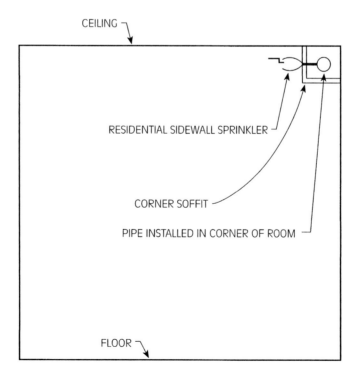

Figure 7-4 *Sectional view of a residential room with pipe concealed by a soffit.*

NFPA 231

Standard on General Storage

NFPA 231C

Standard on Rack Storage of Materials

NFPA 231D

Standard for Storage of Rubber Tires

NFPA 231F

Standard for the Storage of Rolled Paper

commodity
the item being stored plus its packaging and the pallet upon which it rests

pallet
a wooden, plastic, or metal platform used for transporting a commodity to its storage location

called upon to do so. Always remember that a sprinkler system is more than a just pile of parts. A well-designed sprinkler system can mean the difference between life and death for people who depend on the judgment of the fire protection system designer.

SPRINKLER PROTECTION OF HIGH-PILED STORAGE OCCUPANCIES

Residential sprinkler systems, being primarily life safety systems, require significantly less flow than commercial occupancies. By way of comparison, the protection of large quantities of stored combustibles is among the most challenging of fire protection problems that a designer must face, requiring flows significantly in excess of most systems covered by NFPA 13. NFPA 13 provides guidance for storage of ordinary combustibles up to 12 feet in height and for hazardous materials, such as aerosols, rolled paper, and rubber tires, up to 5 feet in height. When storage of materials exceeds these criteria, **NFPA 231**, *Standard on General Storage*, and **NFPA 231C**, *Standard on Rack Storage of Materials*, **NFPA 231D**, *Standard for Storage of Rubber Tires*, or **NFPA 231F**, *Standard for the Storage of Rolled Paper* may apply.

NFPA 231 establishes four classifications of commodities. A storage **commodity** is the item being stored plus the packaging and the **pallet** upon which it rests. A pallet is a wooden, plastic, or metal platform used for transporting the commodity to its storage location. Commodities are grouped into four classifications:

- Class I storage is storage of noncombustible items, such as glass and metals, on combustible pallets.
- Class II storage is Class I commodities in combustible packaging or in wooden crates.
- Class III storage includes combustible commodities such as leather, paper, textiles, limited amounts of plastics, and wooden products, without pallets.
- Class IV storage is Class I, Class II, or Class III commodities containing an appreciable amount of plastics, in cardboard cartons or on wooden pallets.

NFPA 231C has two sets of area/density curves for each class of storage commodity, with one set based on the use of 165°F sprinklers and the other set based on the use of 286°F sprinklers. The use of 286°F sprinklers provides the bonus of allowing lower densities, since fewer sprinklers are likely to actuate.

Systems using 286°F sprinklers must be selected from an area/density curve to deliver a density of no less than 2,000 square feet. Some examples of associated densities are:

- 0.15 gpm/sq. ft. over the hydraulically most demanding 2,000 square feet for a Class I commodity,
- 0.17 gpm/sq. ft. over the hydraulically most demanding 2,000 square feet for a Class II commodity,

- 0.21 gpm/sq. ft. over the hydraulically most demanding 2,000 square feet for a Class III commodity,
- 0.295 gpm/sq. ft. over the hydraulically most demanding 2,000 square feet for a Class IV commodity.

Systems designed with 165°F sprinklers must be selected from an area/density curve to deliver a density of no less than 2,000 square feet. Some examples of associated densities are:

- 0.205 gpm/sq. ft. over the hydraulically most demanding 2,000 square feet for a Class I commodity,
- 0.23 gpm/sq. ft. over the hydraulically most demanding 2,000 square feet for a Class II commodity,
- 0.285 gpm/sq. ft. over the hydraulically most demanding 2,000 square feet for a Class III commodity,
- 0.3855 gpm/sq. ft. over the hydraulically most demanding 2,000 square feet for a Class IV commodity.

The design area must be increased by 30% for dry pipe systems, without adjusting the density. The densities shown on the area/density curves apply only to storage heights of 20 feet, and a graph in NFPA 231C provides adjustment factors for the ceiling density when the actual storage height is greater than or less than 20 feet.

NFPA 231C provides requirements for the storage of materials on racks capable of stacking stored materials vertically. It includes design criteria for both ceiling sprinkler design and for the design of **in-rack sprinklers**, that have water shields to prevent the heat responsive elements of these sprinklers from being wetted by the roof sprinklers, as shown in Figure 7-5.

Ceiling sprinkler demands are selected from one of several area/density curves based on the storage classification, the width of the aisles separating the racks, the temperature rating of the sprinklers, the storage height, the presence of in-rack sprinklers, the dimension from the ceiling level sprinklers to the top of the storage, and the presence of **encapsulation** of the commodity. Encapsulation is the wrapping of a commodity in plastic in order for the commodity to remain rigid on its pallet while being lifted and transported by forklift trucks. The criteria for encapsulation is that the sides and top must be covered by the plastic. Commodities whose sides, but not top, are wrapped, are not considered encapsulated. Encapsulated commodities not only prevent the water from sprinklers to directly wet the commodity, but may provide a flammable surface for flaming brands or cinders to propagate flame. As a result, the required ceiling sprinkler design density is significantly increased for encapsulated commodities.

In-rack sprinklers can be spaced no further than 12 feet on center for Class I and II commodities and for Class III commodities with 4-foot-wide aisles, and 8 feet on center for Class III commodities with 4-foot-wide aisles and for Class IV commodities. NFPA 231C helps the designer determine whether in-rack sprinkler

in-rack sprinklers
sprinklers with water shields to prevent the heat responsive elements from being wetted by the roof sprinklers

encapsulation
the wrapping of a commodity in plastic on the sides and top in order for the commodity to remain rigid while being transported by forklift trucks

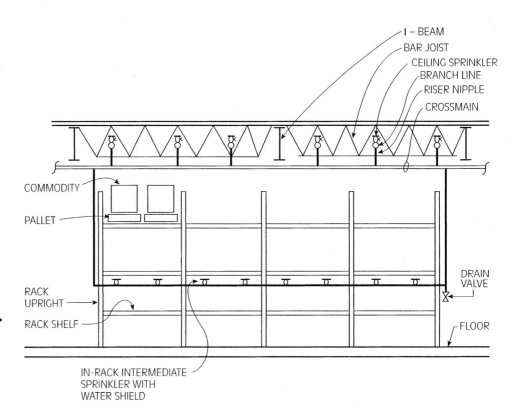

Figure 7-5 Sectional view of a building with ceiling sprinklers and in-rack sprinklers installed.

protection is mandatory and provides for significant ceiling sprinkler density reduction for the installation of one or more levels of in-rack sprinkler protection. Following is NFPA 231C criteria for in-rack sprinkler protection:

- The most remote six in-rack sprinklers are calculated with a minimum 15 psi starting pressure for Class I, II, or III commodities with one in-rack level installed.
- The most remote ten in-rack sprinklers, five on the top level and five on the next highest level, are calculated with a minimum 15 psi starting pressure for Class I, II, or III commodities with two or more in-rack levels installed.
- Eight in-rack sprinklers are calculated with a minimum 15 psi starting pressure for Class IV commodities with one level of in-rack sprinklers.
- Fourteen in-rack sprinklers are calculated, seven on the top level and seven on the next highest level for Class IV commodities with two or more levels of in-rack sprinklers.

Inside hose stations must be installed such that all portions of the storage facility may be serviced by water flow from the hose stations for manual fire fighting. A total hose stream demand of 500 gpm for inside and outside hose streams

Figure 7-6 *ESFR testing for storage applications. (Courtesy Grinnell Fire Protection Systems Co.)*

must be added to the sprinkler system calculations. Inside hose demand is calculated at 50 gpm for each of the two most remote inside hose stations for a total inside hose requirement of 100 gpm. The remaining 400 gpm hose stream requirement is for outside hose streams and is added to the hydraulic calculation at the fire hydrant closest to the building.

ESFR Sprinklers

ESFR sprinklers were specifically created for wet pipe sprinkler systems protecting Class I, II, III, and IV high-piled storage commodity applications up to 25 feet in height, installed in buildings not exceeding 30 feet in height with restrictions on the maximum allowable roof slope and the ceiling construction. The limitations on the building height and roof construction mean that many existing buildings are not amenable to the installation of ESFR sprinklers. ESFR sprinkler systems for high piled storage must be calculated at a minimum pressure of 50 psi for the most remote 12 ESFR sprinklers, with three branch lines flowing four sprinklers each. Figure 7-6 shows an ESFR sprinkler undergoing testing for high-challenge storage applications.

STANDPIPE AND HOSE SYSTEMS

NFPA 14, *Standard on the Installation of Standpipe and Hose Systems*, supplies the minimum criteria for the calculation of standpipes. A **standpipe** is a system of

---

NFPA 14

Standard on the Installation of Standpipe and Hose Systems

standpipe
a vertical piping riser that serves to supply water to hose connections for fire department or occupant use

pipes that supplies water to hose connections for fire department or occupant use. Standpipes are also permitted to be sized for combination standpipe and sprinkler system use.

Standpipes are divided into three classifications:

- Class I standpipes serve 2½-inch fire hose valves for fire department use. These valves are installed on a standpipe with no hose provided.
- Class II standpipes serve 1½-inch valves in fire hose cabinets for use by the building occupants. These cabinets are to be spaced throughout the building, fitted with enough hose to reach all areas of the building, and often contain fire extinguishers.
- Class III standpipes have both 2½-inch fire hose valves for fire department use, and 1½-inch fire hose cabinets for use by the building occupants.

Class I standpipes must serve 2½-inch hose valves at each intermediate floor landing between floor levels in each stairwell and in exit corridors when the stairwell does not lead occupants directly to the outside. The AHJ may permit standpipes to be installed at the main floor landing. See Figure 7-7 for the locations of these two landings. The travel distance between hose valves may not exceed 200 feet in a sprinklered building nor 150 feet in a nonsprinklered building.

The minimum diameter for a Class I or Class III standpipe is 4-inches. If the standpipe serves both hose and sprinklers, the standpipe diameter must be at least 6-inches if a pipe schedule is used, or 4-inches if hydraulically calculated. Although pipe schedules are included in NFPA 14 for sizing of standpipes, standpipes should be hydraulically calculated to assure adequacy of the water supply. A method for the calculation of a standpipe system is outlined below:

1. The most remote fire hose valve on the most remote standpipe in stairwell 1 (hydraulic node 1 on Figure 7-7) must provide 500 gpm at 100 psi at that valve.

2. If the building is completely sprinklered, it is not necessary to add the flow of the sprinkler system to the standpipe calculation. If the building is only partially sprinklered, the sprinkler system demand must be added to the standpipe demand at the point of connection of the most remote sprinkler system in the building.

Figure 7-7 *(Opposite) Sectional view of a high-rise sprinklered building with one Class I standpipe installed in each of three stairwells. NFPA 14 standpipe calculation procedure:*
- *500 gpm at 100 psi flows from most remote fire hose valve (Node 1).*
- *Friction and elevation losses between Node 1 and Node B are calculated.*
- *250 gpm is added for standpipe in stairwell 2 at Node B.*
- *Friction and elevation losses between node B and C is calculated.*
- *250 gpm is added for standpipe in stairwell 3 at Node C.*
- *Friction and elevation losses between Node C and Node D are calculated.*

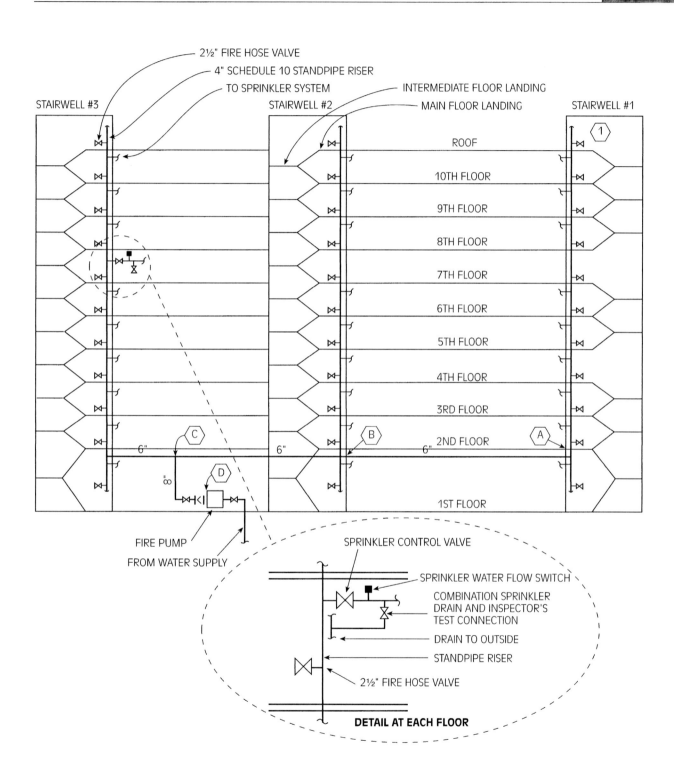

3. If Class II fire hose valves are attached to a standpipe serving Class I outlets, the Class II outlets are to be calculated at a minimum demand of 100 gpm at 65 psi for the most remote Class II hose valve.
4. Using the Hazen-Williams friction loss formula,

$$P_t = \frac{(4.52) \times (Q^{1.85}) \times (L)}{(C^{1.85}) \times (D^{4.87})}$$

the friction loss is calculated from the most remote hose valve (node 1) to the connection to the next standpipe (node B on Figure 7-7). Elevation static pressure,

$$P_{elev} = (h) \times (0.433 \text{ psi/ft})$$

is also calculated between these two nodes and added to the friction loss between the nodes.
5. To account for the additional standpipe flow from the standpipe in stairwell 2, 250 gpm is added at node B.
6. The friction loss of the combined flow of 750 gpm is calculated between node B and node C.
7. At node C, the flow of one standpipe at 250 gpm is added if the building is completely sprinklered. The flow of two standpipes at 250 gpm each is added at Node C if the building is unsprinklered or is only partially sprinklered and if four or more standpipes exist in the building. Flow from the standpipes must represent the actual number of standpipes in the building, not to exceed 1,000 gpm for a sprinklered building, or 1,250 gpm for a partially sprinklered or nonsprinklered building.
8. The friction loss for the combined standpipe flow is calculated back to the source of water supply, shown on Figure 7-7 as the pipe segment between node C and node D. The standpipe demand is compared to the available water supply by graphing the water supply curve and comparing it to the standpipe system demand curve, similar to the method we used for sprinkler system water supply analysis.
9. In cases where the standpipe system is supplied by a water tank, the tank must be sized for the calculated standpipe flow for a duration of 30 minutes.
10. If the standpipe demand exceeds the available water supply, pipe sizes must be increased until the standpipe system demand is satisfactorily supplied by the available water supply, or a method of boosting the water supply, such as a pump or tank, may be required.

FIRE PUMPS

NFPA 20, *Standard for the Installation of Centrifugal Fire Pumps*, applies to the design and installation of pumps for fire protection. **Centrifugal fire pumps** develop pressure by forcing water to rotate around a shaft, using centrifugal force to create pressure that boosts the water supply.

centrifugal fire pump
a pump that develops pressure by forcing water to rotate around a shaft, creating centrifugal pressure that boosts the water supply

Fire Pump Components

The suction side of a fire pump is the side connected to a water supply. The discharge side of a fire pump is the side to which the sprinkler or standpipe system is connected to receive pressurized water from the pump. Piping and components on the suction side must meet certain criteria:

- Piping on the suction side of the pump must be galvanized or specially coated to minimize corrosion or flakes of piping material from being drawn into the pump.
- Control valves on the suction side must be OS&Y valves.
- Also on the suction side, all piping, especially the reducer directly attached to the pump, must be designed to eliminate air pockets that could cause **cavitation**, or destructive vibrations caused by the churning of a fire pump when it is drawing air into the pump instead of water.

The discharge side of a fire pump must be rated for the pressures created by the pump. Since most sprinklers and valves are rated at 175 psi, a **pump relief valve** may be necessary to prevent overpressurization and damage to sprinkler components when pump discharge pressures exceed 175 psi. Pressure regulating valves are also permitted to be installed downstream of the fire pump to protect sprinkler components downstream of the pressure regulating valve. Such components may be necessary for the protection of sprinklers and valves in high rise buildings where the pressure required at the base of the riser to supply sprinklers at the top of the riser could exceed the rating of the components on the lower levels.

A fire pump test header with test valves is required on the discharge side of the pump to simulate the rated flow and pressure of the fire pump and compare the test pressures to the listed pressures of the pump. The number and size of test valves is commensurate to the gpm rating of the fire pump. Figure 7-8 shows a schematic layout of a fire pump piping system.

Fire pumps are designed to start when pressure drops in a sprinkler system and are not intended to run constantly. When a sprinkler opens, water flows from the sprinkler, and the pressure in the piping decreases. A switch on the pump notes the drop in pressure and starts the pump. When the pump is not running, the static pressure from the water supply pressurizes the system. A **jockey pump** is a very small pump that boosts the static pressure within the piping system when the fire pump is not running. A jockey pump is needed if the static pressure available from the water supply is insufficient to provide pressure at the most remote sprinkler, and minimizes operation of the fire pump by keeping pressures on the discharge side of the pump constant.

Fire pumps are needed when the city water supply is insufficient to supply the worst demand of the most remote sprinkler or standpipe system, or when a water supply of insufficient pressure exists, such as a water supply drawn from a pond or a reservoir.

cavitation
destructive vibrations caused by the churning of a fire pump when it is pumping air instead of water

pump relief valve
a valve installed to prevent overpressurization and damage to sprinkler components

jockey pump
a very small pump that boosts the static pressure within the piping system when the fire pump is not running

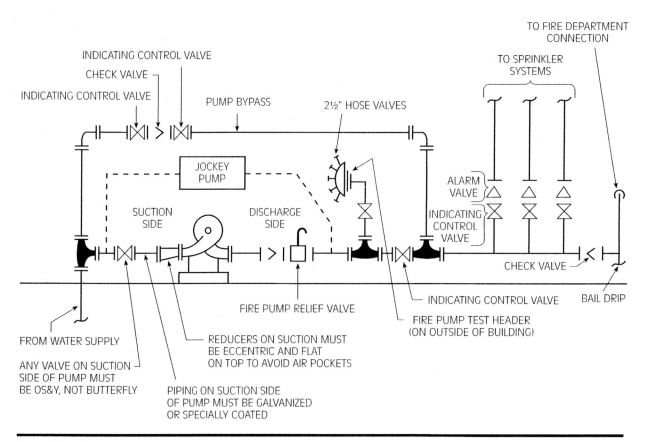

Figure 7-8 *Schematic of a fire pump layout.*

Calculating Pressure Differentials

A sprinkler demand is compared to a city water supply by graphing the two curves, in a manner similar to Figure 7-9.

As can be seen graphically, the system demand exceeds the available supply by about 7 psi. To determine the exact pressure differential, calculate the pressure available at the gpm flow of the sprinkler system, using the formula:

$$P_a = P_s - (P_s - P_r) \times (Q_a/Q_t)^{1.85}$$

where P_a = the pressure available from the water supply at the gpm flow of the sprinkler system, in psi
P_s = the static pressure, in psi
P_r = the residual pressure, in psi
Q_a = the calculated sprinkler demand, in gpm
Q_t = the test flow associated with the residual pressure

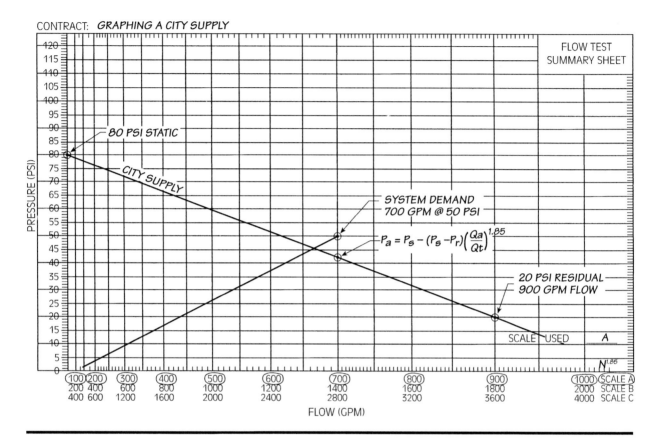

Figure 7-9 *Graphing a city supply.*

Example 7.2 Calculate the pressure differential between the sprinkler demand and the available supply for Figure 7-9.
Solution. Using the formula:

$$P_a = P_s - ((P_s - P_r) \times (Q_a/Q_t)^{1.85})$$

we can calculate the pressure available from the water supply.

$$P_a = 80 - ((80 - 20) \times (700/900)^{1.85})$$
$$= 80 - ((60) \times (.628))$$
$$= 80 - 37.7$$
$$= 42.3 \text{ psi}$$

The pressure differential, or safety factor, is the difference between P_a, the available pressure, and P_r, the sprinkler system pressure demand:

$$\text{Safety factor} = (P_a) - (P_r)$$
$$= 42.3 \text{ psi} - 50.0 \text{ psi}$$
$$= -7.7 \text{ psi}$$

A safety factor is required to be a positive number, such that the water supply exceeds the system demand. In this case, it is negative, indicating that the system demand exceeds the available supply.

When to Add a Fire Pump

Now that we know the demand pressure exceeds the supply pressure in Example 7.2, we have two options to solve this pressure differential:

1. Increase pipe sizes and recalculate until the system demand pressure is less than the available supply pressure.
2. Add a fire pump.

By this juncture, the reader should recognize that a pressure deficit of only 7.7 psi can be easily solved by increasing pipe sizes. Adding a fire pump to solve this minor pressure imbalance would be inappropriate.

A fire pump should be added only when the fire protection system designer has determined that the system is incapable of meeting the requirements of the authority having jurisdiction with pipe size increases alone, or if the cost of the large pipe sizes equal or exceed the cost of a fire pump. The pump should only be added as a last resort when all available hydraulic design efforts fail, and when the economic necessity for a fire pump is clearly demonstrated.

Graphing a Fire Pump Supply

A fire pump curve is based upon its rated point, x gpm at y psi, as shown on Figure 7-10. The three points of the fire pump curve shown on this figure are:

(x) gpm at (y) psi	pump rated capacity
$(1.4) \times (y)$ psi	shutoff (pressure at zero flow)
$(0.65) \times (y)$ psi, at $(1.5) \times (x)$ gpm	overload, or 150% point (maximum pump capacity)

Example 7.3 For a pump rated at 500 gpm at 60 psi, determine the fire pump curve.
Solution 500 gpm at 60 psi, pump rated capacity.
The pump curve is shown on Figure 7-10. The point of the pump curve intersecting the vertical graph axis is $(1.4) \times (60 \text{ psi}) = 84$ psi shutoff (pressure at zero flow).

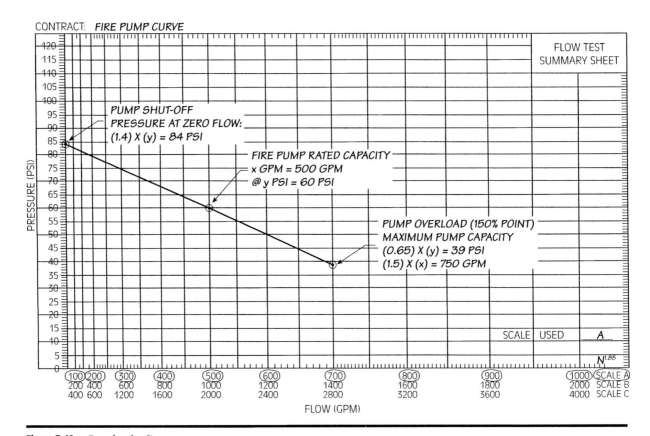

Figure 7-10 *Graph of a fire pump curve.*

The point to the right of the pump rated point is (0.65) × (60 psi) = 39 psi, overload, or 150% point (maximum pump capacity) and (1.5) × (500) = 750 gpm.

Note that the fire pump curve does not extend beyond the point on the far right of the curve, 750 gpm at 39 psi. This is the pump overload point, or maximum capacity of the fire pump, a point beyond which the pump is not designed to function. Designing sprinkler system demands beyond this overload point is not possible. If the sprinkler system demands more gpm than the pump overload point can provide, a larger capacity pump needs to be specified.

Combining a Pump and City Supply

Now that we know how to graph the performance curve of a fire pump, how do we determine the combined curve of a fire pump and a city water supply? We simply add them!

Example 7.4 A fire pump is rated at 500 gpm at 60 psi. The city supply has been tested to be 30 psi static, 15 psi residual, with 900 gpm flowing. The sprinkler system demand is 550 gpm at 78 psi. Determine the combined pump/city supply curve, and determine the pressure differential between the combined supply and the sprinkler system demand.

Solution Note that the sprinkler demand significantly exceeds the city supply. In the case of this problem, assume that a cost analysis has shown that a fire pump is economically justified. In order to add the pump curve and the city supply, we must know the pressure available from the city supply at the three fire pump rating points. At 500 gpm, the city supply can provide:

$$P_a = P_s - ((P_s - P_r) \times (Q_a/Q_t)^{1.85})$$
$$= 30 - ((30 - 15) \times (500/900)^{1.85})$$
$$= 30 - ((15) \times (0.33709))$$
$$= 24.9 \text{ psi}$$

At 750 gpm, the city supply can provide:

$$P_a = P_s - ((P_s - P_r) \times (Q_a/Q_t)^{1.85})$$
$$= 30 - ((30 - 15) \times (750/900)^{1.85})$$
$$= 30 - ((15) \times (0.713698))$$
$$= 19.3 \text{ psi}$$

The pump supply may now be added to the city supply by adding the two supplies at 0 gpm, 500 gpm, and 750 gpm:

At 0 gpm: Combined P is 84 psi + 30 psi = 114 psi.

At 500 gpm: Combined P is 60 psi + 24.9 psi = 84.9 psi.

At 750 gpm: Combined P is 39 psi + 19.3 psi = 58.3 psi.

The combined supply may now be plotted as shown on Figure 7-11. The sprinkler demand, as graphed, is less than the combined supply of the fire pump and city supply.

Sizing a Fire Pump

Note that the sprinkler demand falls to the right of the pump-rated gallons per minute of 500 gpm. This is the best design for fire pump supplies. If the sprinkler demand falls to the left of the rated capacity of the pump, a lower capacity pump could possibly be used. Also remember that the system demand may not exceed the overload, or 150% gpm of the fire pump. In the case of the pump in Figure 7-11, the sprinkler demand should be between 500 and 750 gpm for the most efficient use of the fire pump.

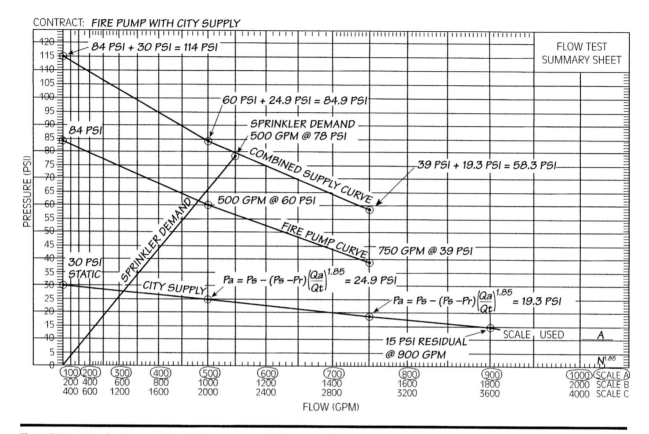

Figure 7-11 *Graph of a combined fire pump and city supply.*

GRAVITY TANKS

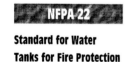

Standard for Water Tanks for Fire Protection

Gravity tanks are a valuable and popular method of boosting a city supply and are designed in accordance with **NFPA 22**, *Standard for Water Tanks for Fire Protection*. Gravity tanks can be quite costly compared to a fire pump and ground storage tank when considered for one building, but may be cost efficient for a large industrial plant or an entire community or industrial park.

Gravity Tank Calculation

The basic principles that apply to the calculation of gravity tanks combined with other sources of supply are:

1. There is no contribution from the gravity tank if another source of supply is providing a higher pressure than the gravity tank. The source with the higher pressure shuts the check valve of the gravity tank until the pressure of that source drops below the static pressure of the gravity tank. Conversely, the gravity tank shuts off other lower pressure sources of supply as long as the tank pressure exceeds the static pressure of the other supplies.
2. The pressure created by the elevated water in the gravity tank is computed from the bottom of the water storage portion of the tank, not from the average water storage level.
3. The contribution of gravity tanks may be determined by graphing the tank contribution and adding it to the other source of water supply, using the method outlined below.

Graphing a Gravity Tank Supply

The method for the graphical analysis of the combined contribution of a gravity tank and a city supply are shown on Figure 7-12 and example 7.4.

Example 7.4 A combined gravity tank and city supply provides water to a sprinkler system. The bottom of the gravity tank is 150 feet high and a calculation shows a 25 psi friction loss when the gravity tank is flowing 500 psi. The city supply is 45 psi static, 30 psi residual, at 750 gpm flow. The sprinkler system requires 1,200 gpm at 30 psi. Determine the combined tank/city supply using graphical analysis.
   Solution   The static pressure contribution of the gravity tank is:

$$P_{tank} = (150 \text{ ft}) \times (0.4331 \text{ psi/ft})$$
$$= 65 \text{ psi}$$

1. The tank static pressure is 65 psi. The tank encounters 25 psi friction loss at 500 gpm, so the residual pressure is:

$$65 \text{ psi} - 25 \text{ psi} = 40 \text{ psi at 500 gpm.}$$

   The gravity tank supply may now be graphed, as shown on Figure 7-12.
2. The city supply may now be graphed, as shown on Figure 7-12. Note that the two curves intersect at approximately 550 gpm at 36 psi.
3. The sprinkler demand is now graphed. Note that the sprinkler demand is significantly greater than both individual supplies.
4. Extend a dotted line from the point of intersection of the tank curve and the city supply curve at 550 gpm to a point that is twice this amount, or 1,100 gpm at 36 psi.
5. The combined city/tank curve is drawn through the point 1,100 gpm at 36 psi. Note that the combined city/tank supply now exceeds the sprinkler demand.

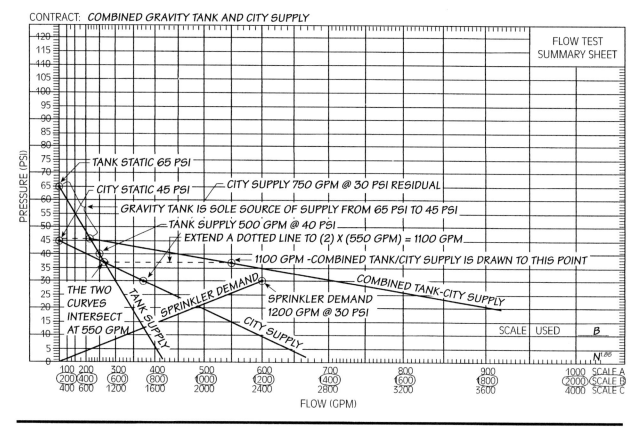

Figure 7-12 *Graph of a gravity tank with city supply.*

Summary

Although pipe schedules exist in NFPA 13, their use for new construction is severely restricted. Residential sprinkler systems fall into three categories and are designed in accordance with three NFPA standards. Residential sprinkler systems for single family homes and manufactured homes are designed in accordance with NFPA 13D, multifamily homes up to and including four stories are designed in accordance with NFPA 13R, and high-rise residential occupancies are designed in accordance with NFPA 13. Sprinkler systems for high-piled storage occupancies are designed for storage above 12 feet or high-hazard storage above 5 feet. High-rise buildings require standpipes in each stairwell, calculated in accordance with NFPA 14. Fire pumps and gravity tanks are methods of boosting the pressure of a water supply.

Review Questions

1. Calculate the demand of the most remote area of the first floor of the single family residence shown in Figure 7-1b in accordance with the correct NFPA standard.

2. Calculate the standpipe in Figure 7-7, assuming the building is fully sprinklered. The pipe between nodes 1 and A is 4-inch schedule 10, 100 feet long; the pipe between nodes A and B is 6-inch schedule 10, 150 feet long; the pipe between nodes B and C is 6-inch schedule 10, 75 feet long, and the pipe between nodes C and D is 8-inch schedule 30, 20 feet long.

3. A fire pump is rated 1000 gpm at 100 psi. Graph the fire pump curve.

4. A city supply of 45 psi static, 35 psi residual, and 1,200 gpm flowing, is connected to the fire pump supply described in review question 3. Graph the combined demand.

5. A gravity tank measures 225 feet to the bottom of the tank and 260 to the highest water line of the tank. Graph the gravity tank curve given a 40 psi friction loss with 350 gpm flowing.

6. A city supply has been tested to be 32 psi static, 20 psi residual, at 2,000 gpm flowing. Determine the combined water supply if this city supply is connected to the gravity tank in review question 5.

Activities

1. Contact the water authority that serves your house. Determine the available static pressure for your house. Is there enough pressure to supply pressure to the highest sprinkler?

2. Survey your home and design a residential sprinkler system for it. Perform a hydraulic calculation that solves the system demand with respect to the available water supply.

Chapter 8

Water Spray System Design for Transformers

Learning Objectives

Upon completion of this chapter, you should be able to:

- Describe the fire hazard associated with a transformer.
- List the methods by which water can affect extinguishment or control of a transformer fire.
- Use a transformer outline drawing to obtain information for a water spray system layout.
- Determine electrical clearances, locate piping loops, and space nozzles for a transformer water spray design.
- Calculate the surface area of a transformer and apply the correct density to the transformer.
- Determine the correct protection for any nonabsorbing ground surface that may surround a transformer.
- Complete a transformer water spray design in accordance with the provisions of NFPA 15.
- Calculate a transformer water spray system.

Sprinkler systems, discussed in Chapter 5, are the most common type of water-based fire suppression system. This chapter offers an extension of what we have learned about sprinkler systems by exploring a system that provides water delivery to numerous specialized hazards.

WATER SPRAY SYSTEMS

water spray system
a fixed pipe system permanently connected to a water supply providing directional water spray protection over a protected area

Standard for Water Spray Fixed Systems for Fire Protection

A **water spray system** is a fixed pipe system permanently connected to a water supply providing directional water spray protection over a protected area. Hazards protected by water spray include flammable liquids, flammable solids, dusts, and commodities capable of producing rapidly spreading fires. Water spray systems are designed in accordance with **NFPA 15**, *Standard for Water Spray Fixed Systems for Fire Protection*. NFPA 15 is not a detailed design document, but is primarily a performance standard that lists requirements on what is to be accomplished, without specifying how to design a system to accomplish the design objectives. The purpose of this chapter is to help the reader meet the design objectives for water spray protection by following a step-by-step method.

Water Spray Protection for a Power-Generating Plant

Power-generating plants provide a wide assortment of hazards that lend themselves to protection by water spray systems. Figure 8-1 gives a schematic diagram of a power-generating plant with some of the most commonly encountered water spray applications.

Figure 8-1 shows a coal car, pulled by a train, dumping coal onto a coal conveyor, which transports coal to provide fuel for a steam boiler. The coal conveyor requires water spray protection for the coal that is transported on the top belt. Below the top belt is the return belt, which also requires water spray protection. These belts can ignite if a belt roller stops rotating, creating friction with the belt, and creating a possible ignition scenario when the conveyor stops and the belt is heated by the defective belt roller. A fire moving on a conveyor belt creates a hazard that requires water spray. This hazard can be reduced if the conveyor can be stopped once the water spray system commences operation.

The boiler turns water into steam for the steam turbine. The boiler requires water spray protection on the burner front, the area where coal dust and lube oil can create a significant fire on the vertical burner fronts that are usually several stories tall.

The steam enters the steam turbine, which looks a lot like an enormous aircraft jet engine. The turbine rotates very rapidly and lubrication for the turbine bearings is supplied by oil pressurized by a lube oil pump. Water spray protection is required for both the turbine bearings and the lube oil pump. The rotation of the turbine forces the electric generator to rotate, creating electric power, which is sent to a step-up transformer.

Chapter 8 Water Spray System Design for Transformers

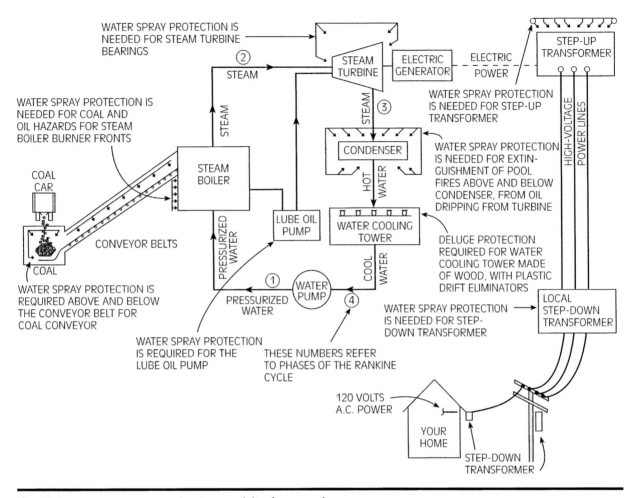

Figure 8-1 *Water spray protection in a coal-fired power plant.*

The steam leaving the turbine is sent to a condenser, in which the steam is condensed into a liquid. Water spray or sprinkler protection is required above and below the condenser to extinguish pools of flaming lube oil that may drip or spray from the turbine bearings above if a turbine bearing seal fails.

A water-cooling tower, shown in Figure 8-2, receives hot water from the condenser and sprays this water from pipes at the top of the tower. Mounted at the roof of the tower is a series of fans, resembling helicopter blades, that draw cool air into the tower as the hot water droplets fall to the ground. The water drips through plastic drift eliminators, intended to keep the downward path of the water as vertical as possible. The water then traverses through several layers of plastic fill, where

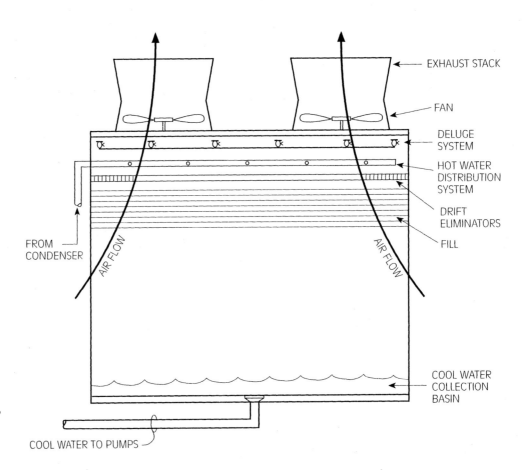

Figure 8-2 *Water cooling tower.*

NFPA 214

Standard on Water Cooling Towers

significant heat transfer occurs. By the time the water hits the floor of the cooling tower, it has dropped in temperature to a level where it may be repumped into the boiler. Combustible cooling towers may be primarily constructed of wood and plastic. The combination of the wood exterior and the plastic drift eliminators and fill can provide a scenario for a significant fire, especially when the tower is out of service, or down for repair. Welding and woodworking within the tower during periods of shutdown make the tower a high-hazard risk. Fire protection designers use **NFPA 214**, *Standard on Water Cooling Towers*, for the design of cooling tower fire protection systems.

The Rankine Cycle

Those who have taken a course in thermodynamics should recognize the process through which the water undergoes transformation in phase on Figure 8-1 as the Rankine cycle, a vapor power cycle used by most power plants that generate power for our homes. We will not discuss the intricacies of this elegant cycle, but it is important to recognize the beauty of obtaining power by taking advantage of the liquid and vapor phases of that most amazing fluid, water.

WATER SPRAY PROTECTION FOR TRANSFORMERS

The remainder of this chapter concentrates on transformer water spray system design. Transformers are among the most familiar and commonplace of water spray applications associated with power plants. Similar to our study of sprinkler systems, we take a methodical approach to the design of water spray systems for transformers. Transformer water spray protection is illustrated in Figure 8.3.

Transformer Hazards

transformer case
the transformer tank that contains oil to cool the electric coils

radiator
a unit on a power transformer in which hot oil is circulated for cooling by fans

mineral oil
an oil used in transformers with a flash point of about 300°F

A power transformer requires water spray fire protection because it contains oil within the **transformer case** to cool its electric coils. The oil is cooled by fans when circulated through **radiators**, not unlike the way your car's engine cools its water. An example of an oil used in transformers is **mineral oil**, with a flash point of about 300°F. A lightning strike, a broken seal or coil winding, or other source of overheat or metal failure, can initiate the leakage of mineral oil from the transformer onto hot transformer parts, resulting in an ignition and fire that could destroy the transformer and threaten adjacent buildings, personnel, routes of egress, or devices.

Figure 8-3 *Water spray protection for transformers. The electric detection system is the small conduit that can be seen directly above the transformer loop. (Courtesy Grinnell Fire Protection Systems Co.)*

Personnel and Other Hazards

Although transformer facilities are usually fenced and personnel are not usually present in a transformer yard, the fire could occur while someone is in the fenced area, or could grow to the point where personnel or adjacent facilities may be in danger. The transformer may be in a location where personnel are required to pass near the transformer in order to escape to a safe area. The loss of a transformer can result in an expensive, lengthy interruption of power-generating capability. By protecting the transformer with water spray, we are lessening the danger to adjacent areas and increasing the possibility of being able to reuse the transformer in the event of a fire.

Power Generation

A transformer receives power generated by a turbine in a power plant or other generating facility, and steps it up so that it is capable of overcoming the frictional resistance of the high voltage lines that deliver the power to your home. A local step-down transformer, either on a telephone pole for overhead lines or a box in your front yard for underground lines, brings the power to a reliable voltage of 120 volts AC for use in your home as shown in Figure 8-1.

Oil Recovery and Reclamation

New transformers are most often installed on foundations surrounded by gravel pits, designed to collect any oil that might drip or spray from a faulty transformer and the water that sprays onto the transformer surface from the water spray system would also be collected in the pit. An oil-water separator, as shown in Figure 8-4, makes it possible to reclaim the oil and reuse the water. A fire protection designer should be sensitive to the environmental hazard that could ensue if a procedure is not established for the recovery and reclamation of the oil. Policies of the Environmental Protection Agency may dictate that a procedure be implemented to recover all oil that leaks from a transformer, whether from a fire incident or other scenario.

Extinguishment/Control Mechanism

Extinguishment/control mechanisms applicable to water spray protection, as shown on Figure 8-5, include:

1. **Surface cooling.** The cooling of a combustible liquid can be obtained by water application if the combustible liquid has a flash point greater than 125°F. By cooling a surface, a likely source of ignition is removed or is cooled to a temperature where the propagation of flame cannot be sustained.

2. **Steam smothering.** The process of a fire turning droplets of water into steam yields another mechanism for cooling of the flame. During vaporization, the

---

surface cooling
the cooling of a flammable liquid that has a flash point greater than 125°F

steam smothering
the process of a fire turning droplets of water into steam, cooling the flame

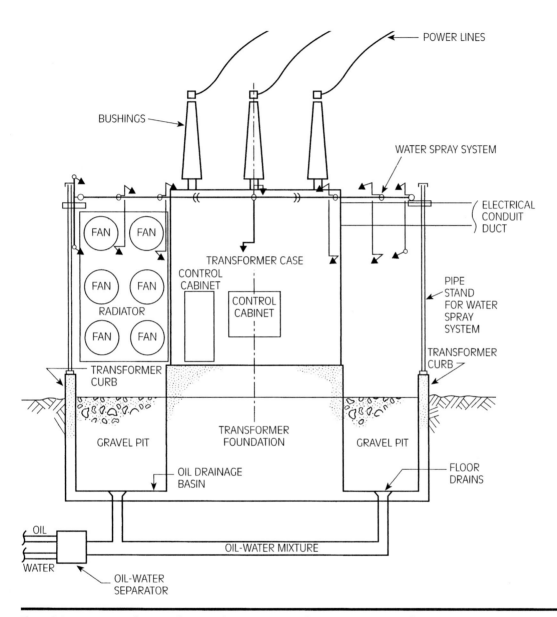

Figure 8-4 *A sectional view of a transformer. (Note: This view corresponds to section "B-B" of Figure 8-14.)*

dilution
reducing the concentration of a water-soluble flammable liquid by mixing it with water

steam is capable of absorbing 9.72 kilocalories per mole of heat from the fire, assuming that the combustible liquid boils at a temperature greater than 212°F, the boiling point of water.

3. **Dilution.** If the combustible liquid is water soluble, the water application mixes with and dilutes the combustible liquid.

redirection
sweeping standing oil from the surfaces of a transformer to a less hazardous area

coating
condition when water floats and provides a barrier between the surface of a flammable liquid and the combustible vapor above the surface

vapor exhaust
a ventilation system to remove vapors to a point 25% below the lower flammable limit of the vapor

transformer outline drawing
a dimensioned shop drawing of a transformer

4. **Redirection.** Water under pressure is capable of sweeping standing oil from the surfaces of a transformer to less hazardous areas, such as a predesigned drainage pit.

5. **Coating.** Water spray application provides a thin coating of water on vertical and horizontal surfaces of the transformer during system activation.

6. **Vapor exhaust.** If the transformer is indoors, or in an enclosed vault, a ventilation system is recommended or required to reduce vapor concentrations to a point 25% below the lower flammable limit of the vapor. In cases where the transformer is exposed to the elements, vapors are removed by the wind.

TRANSFORMER DESIGN PROCEDURE

Compile Reference Data

Acquiring the necessary reference information is the most prudent way to begin your layout of a water spray system, as discussed in Chapter 1. You should obtain the following information in order to perform a complete design:

   1. Obtain a set of **transformer outline drawings** from the manufacturer of the transformer. A transformer outline drawing, as shown in Figure 8-6, is a shop

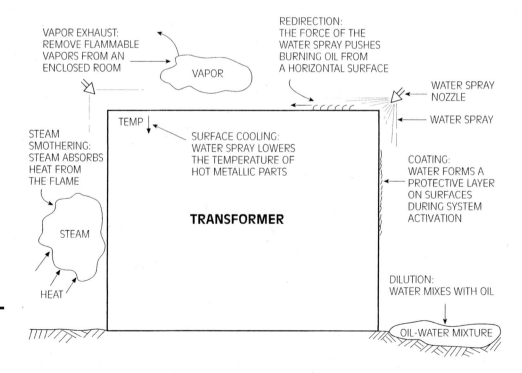

Figure 8-5
Mechanisms of extinguishment.

Chapter 8 Water Spray System Design for Transformers

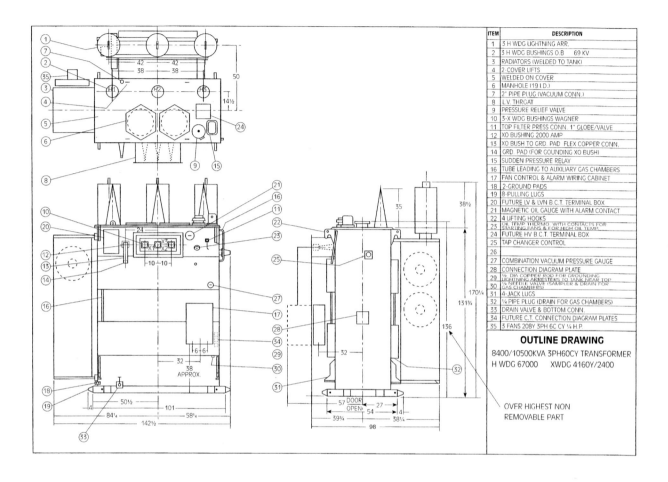

Figure 8-6 *A sample outline drawing for a transformer.*

drawing of a transformer, and is rarely to scale, since dimensions are added by the manufacturer to a generic layout that is modified for a variety of transformer power ratings. Do not trace an outline drawing.

The outline drawing shows:

- The dimensions of the transformer case, in inches. For example, for the transformer shown on Figure 8-6, the base is 54" wide by 101" long by 131 3/4" tall.

- Locations and ratings of the transformer bushings in kilovolts (Kv). For the transformer shown in Figure 8-6, the bushings are shown as item 2. In the description list in the upper right-hand corner, the rating is shown as 69 kilovolts (Kv).

- Sizes and locations of cooling fins or radiators, shown as item 3 on Figure 8-6. The radiators have fans, shown as item 35 on Figure 8-6, that blow cool air across the cooling fins in the radiator, cooling the hot oil.
- Location of the oil expansion tank, if applicable.

Working with outline drawings tends to be somewhat frustrating because dimensions are usually shown in inches, as opposed to feet, and because invariably one or more dimensions are needed that do not appear on the drawing. Outline drawings have always been this way, so work with outline drawings as best you can. Transformer outline drawings are not created for use by fire protection designers. Their primary purpose is to show the minimum necessary information for shipment of a transformer on a rail car or a truck from the factory to where it is to be installed, and to facilitate installation.

2. A set of transformer structural drawings shows the transformer foundation upon which the transformer is to sit, the oil drainage basin, and the transformer curb as shown on Figure 8-4. Be certain to compare the elevations of the pad and curb with the finished floor elevation of the deluge valve house. The structural drawing also shows any nonabsorbing ground surface that requires water spray in accordance with NFPA 15.

3. A civil drawing or plot plan is needed to reference the transformer to the deluge valve room when they are separated by extended distances. This plan is especially helpful in resolving underground conflicts if the pipe from the valve room to the transformer must be below grade.

4. An architectural plan or fire protection plan may be needed to show how the transformer is positioned and oriented on the transformer foundation pad.

Commence Layout

Once the necessary information is obtained, the layout of the transformer can begin.

1. Be certain to use a scale of 1/4" = 1'-0" or larger for the transformer layout. If the transformer pad structural drawing is in 1/4" scale, you can trace or scan it.

2. The manufacturer's outline drawing can be used to obtain the dimensions necessary to show the transformer on the foundation pad.

3. Make certain that sufficient space is available on your drawing to depict the deluge valve house location and to draw the feedmain from the valve house to the transformer. The valve house should be located such that a transformer fire will not damage or hinder access to the valvehouse.

4. A transformer key plan, with a minimum scale of 1/8" = 1'-0", may be needed if the valvehouse is an extended distance from the transformer, and should show the overhead or underground piping from the valvehouse to the transformer.

5. A minimum of one sectional view of the transformer and water spray system is required, with two sectional views highly recommended. An additional section is required for showing details of the riser and deluge valve.

Items that need to be shown on the plan and sections include:
- Electrical clearances. These clearances, shown on Figure 8-7, demonstrate electrical field and necessary clearances from system components in accordance with NFPA 15. Electrical clearance is determined by obtaining the voltage of the bushings, as previously discussed, and determining a radius of clearance from the bushings, using Table 8-1. BIL values used in Table 8-1 refer to Basic Insulation Level values given by the manufacturer of the transformer. The bushings shown on Figure 8-6 have a rating of 69kV and require 25 inches of clearance per Table 8-1.

Table 8-1 *Clearance from water spray equipment to live uninsulated electrical components*

	Nominal System Voltage (kV)	Maximum System Voltage (kV)	Design BIL (kV)	Minimum* Clearance (in.)	Minimum Clearance (mm)
To	13.8	14.5	110	7	178
	23	24.3	150	10	254
	34.5	36.5	200	13	330
	46	48.3	250	17	432
	69	72.5	350	25	635
	115	121	550	42	1067
	138	145	650	50	1270
	161	169	750	58	1473
	230	242	900	76	1930
			1050	84	2134
	345	362	1050	84	2134
			1300	104	2642
	500	550	1500	124	3150
			1800	144	3658
	765	800	2050	167	4242

Source: Reprinted with permission from NFPA 15, *Water Spray Fixed Systems for Fire Protection.* Copyright © 1996, National Fire Protection Association, Quincy, MA 02269. This reprinted material is not the complete and official position of the National Fire Protection Association on the referenced subject, which is represented only by the standard in its entirety.

*For voltages up to 161 kV the clearances are taken from NFPA 70, *National Electrical Code*®. For voltages 230 kV and above the clearances are taken from Table 124 of ANSI C-2, *National Electrical Safety Code.*

NOTE: BIL values are expressed as kilovolts (kV), the number being the crest value of the full wave impulse test that the electrical equipment is designed to withstand. For BIL values that are not listed in the table, clearances may be found by interpolation.

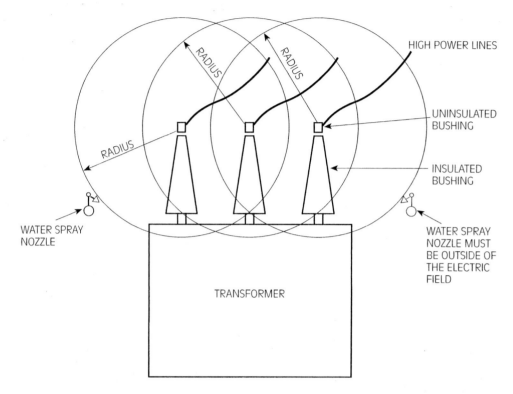

Figure 8-7 *Clearance from uninsulated bushings. For a given voltage of uninsulated bushing, NFPA gives a radius within which no portion of the water spray system may be installed.*

- Transformer control cabinets. Demonstrate proper water spray protection around cabinets and assure adequate clearances for door swing.
- Pipe stand locations. Denote these pipe stands as a circle on the plan view.
- Nozzle spray patterns. Specifically note the water spray direction and angle for each nozzle.

Transformer Loop Design

transformer loop
water spray piping designed for rigidity and enhanced hydraulic characteristics

1. Each transformer requires at least one **transformer loop** of fire protection piping surrounding it. Transformer piping is traditionally designed as a loop for better pipe rigidity when hung by pipestands, and for the improved hydraulic advantages involved in looped system design.

2. A good rule of thumb is to place the loops 2'-0" to 2'-6" from the vertical surface of the transformer following along its perimeter, as shown in Figure 8-8. This technique ensures the best water spray coverage for most nozzles. The spray pattern for the nozzle selected must be plotted on the plan to verify proper coverage.

3. Transformers up to 12 feet in height can be satisfactorily protected with one transformer loop. See Figure 8-9.

Chapter 8 Water Spray System Design for Transformers

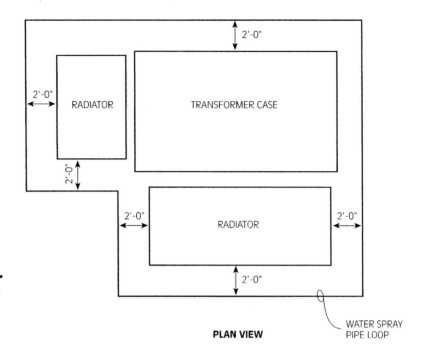

Figure 8-8 Pipe loop location with respect to sides of transformer.

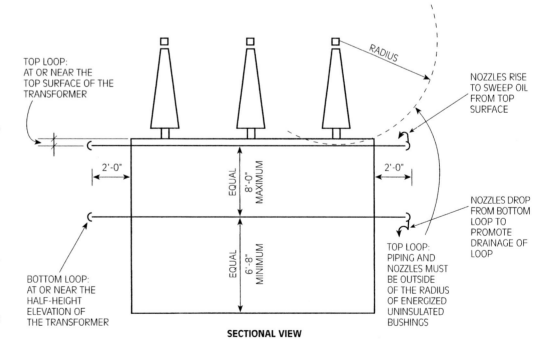

Figure 8-9
Transformer loop arrangement, sectional view. One loop is recommended if transformer is up to 16 feet, assuming no major water spray interferences exist. Two loops are recommended for transformers greater than 16 feet in height.

4. With transformers from 12 to 16 feet in height, one loop may satisfactorily protect the transformer, provided that the water spray is not obstructed by control cabinets or other elements.

5. If water spray obstructions prevent **direct impingement**, or direct application of water spray to all transformer surfaces on transformers between 12 and 16 feet, or if the transformer exceeds 16 feet in height, two loops are recommended. NFPA 15 requires direct impingement of water spray on all transformer surfaces, meaning that all surfaces should be struck directly by water spray from the system. An example of a surface that does not qualify as being directly impinged is a surface that is not being directly struck by water spray, but is being wetted by the water running down from surfaces being struck directly above.

6. In most cases, the top transformer loop should be no higher than the plane of the top of the transformer. Placing it higher may result in electrical clearance problems. Nozzles on the top loop must be positioned to be able to sweep oil from the top of the transformer yet remain outside of the electrical clearance of live uninsulated bushings.

7. The second loop should be at the half-height elevation of the transformer (i.e., elevation 8'-0" for a 16'-0" high transformer).

8. The bottom loop should be no lower than 6'-8" to provide clearance for personnel, and must be coordinated to avoid interference with control panels, access doors, and other operating features of the transformer.

9. Transformer loops should be spaced no more than 10'-0" apart vertically. Dry or unwetted spots on the transformer tank surface will occur if the loops are too far apart.

10. As a rule of thumb, to achieve direct impingement, transformer loops should be spaced 8 feet apart, measured vertically. Closer loop spacing may be required for some transformers where significant obstructions to water spray exist.

11. When a transformer is raised above the surface of the foundation pad, the bottom surface of the transformer must also be wetted by water spray.

Calculation of Transformer Surface Area

NFPA 15 requires a density of 0.25 gpm/sq. ft. for each square foot of **rectangular prism** of the transformer on all exposed surfaces of the transformer. A rectangular prism is an imaginary box, created by the waterspray system designer, that makes it unnecessary to calculate the surface area of minor discontinuities on the transformer tank and makes it possible to consider radiators whose fins are less than 12 inches apart as a box. The designer uses this prism to calculate the surface area of each major transformer component and to ensure that the minimum density is applied to all surfaces.

The transformer tank, consisting of four sides and a top, is a simple calculation when using the rectangular prism concept. If the bottom of the transformer case is raised above the transformer foundation, this bottom surface must be added to the transformer case calculation, and waterspray must be provided.

---

direct impingement
direct application of water spray to all transformer surfaces

rectangular prism
an imaginary box that simplifies any discontinuities, making it unnecessary to calculate minor irregularities on the transformer surface

The radiators are usually one or more appendages to the transformer case, with each radiator consisting of four sides, a top, and a bottom. The sides of these devices may be calculated as rectangles, provided that the radiator fins are spaced 12 inches or less apart. Radiators with fins spaced greater than 12 inches must have nozzles located between each fin, with the surface area of each fin calculated. As a word of caution, the outline drawing may not provide detail relative to the fin spacing. This information may need to be obtained in a survey. The bottoms of all radiators are raised above the gravel surface and must also receive water spray application.

Other appendages that require surface area calculation include control cabinets and expansion tanks. Although control cabinets are not an oil hazard, they should be considered for their ability to allow oil to form pools on the top of the cabinet surface, and for their ability to block water spray from reaching the transformer tank behind and below the cabinet.

Example 8.1 Given a transformer whose case measures:

$$(6'\text{-}0'' \text{ wide}) \times (12'\text{-}0'' \text{ long}) \times (16'\text{-}0'' \text{ high}),$$

with three radiators measuring:

$$(4'\text{-}0'' \text{ wide}) \times (4'\text{-}0'' \text{ deep}) \times (12'\text{-}0'' \text{ high}),$$

and with two control cabinets measuring:

$$(2'\text{-}0'' \text{ wide}) \times (1'\text{-}0'' \text{ deep}) \times (3'\text{-}0'' \text{ high}):$$

Calculate the surface of the rectangular prism of the transformer and determine the minimum gallonage that must be applied to the transformer surface in accordance with NFPA 15.

Solution

Transformer case area calculation:

Top:	(6' × 12') = 72 sq. ft.
Sides:	(6' × 16') × (2) = 192 sq. ft.
	(12' × 16') × (2) = 384 sq. ft.
Bottom:	case is flush to slab = 0 sq. ft.
Transformer case total:	648 square feet

Radiator area calculation:

Top and bottom:	(4' × 4') x (2) = 32 sq. ft.
Sides:	(4' × 12') × (4) = 192 sq. ft.
Radiator total:	224 square feet
Total 3 radiators:	672 square feet

Control cabinet area calculation:

Front:	(2' × 3') = 6 sq. ft.
Back:	Flush to case = 0 sq. ft.
Sides:	(1' × 3') × (2) = 6 sq. ft.
Top and bottom:	(1' × 2') × (2) = 4 sq. ft.
Cabinet total:	16 square feet
Total 2 cabinets:	32 square feet

Transformer case density calculation:

$$(648 \text{ sq. ft.}) \times (0.25 \text{ gpm/sq. ft.}) = 162 \text{ gpm}$$

Radiator density calculation:

$$(224 \text{ sq. ft.}) \times (0.25 \text{ gpm}) = 56 \text{ gpm}$$
$$3 \text{ radiators: } (56 \text{ gpm}) \times (3) = 168 \text{ gpm}$$

Control cabinet density calculation:

$$(16 \text{ sq. ft.}) \times (0.25 \text{ gpm/sq. ft.}) = 4 \text{ gpm}$$
$$2 \text{ cabinets: } (4 \text{ gpm}) \times (2) = 8 \text{ gpm}$$

Total density calculation:

$$(162 \text{ gpm}) + (168 \text{ gpm}) + (8 \text{ gpm}) = 338 \text{ gpm}$$

Conclusion: A minimum waterspray demand of 338 gpm is required to adequately protect this transformer. Note that this demand total does not consider any frictional or elevational losses. Therefore this calculated demand represents the absolute minimum gpm that can be applied to protect the transformer. As we learned in Chapter 6, friction loss increases the total flow of a fire protection system and the actual gpm figure would be greater when hydraulic calculations are performed. The preceding minimum gpm calculation is necessary to use as a basis for comparison to ensure that the hydraulic calculation of the transformer is providing the minimum gpm per NFPA 15.

Water Spray for Nonabsorbing Ground Surfaces

Most modern transformer installations are designed so that the transformer is surrounded by a gravel-filled drainage pit, as shown in Figure 8-4. One of the primary purposes for the water spray system is to push oil away from the transformer surfaces and into the drainage pit and drainage system where disposal can be facilitated. Oil separators collect oil for disposal or reuse.

In rare instances, particularly for older transformers or for temporary transformers, a transformer may be mounted onto nonabsorbing concrete slabs with no drainage pit. Without supplementary waterspray impingement protection of the concrete slab, the transformer could conceivably be surrounded by a ring of fire.

nonabsorbing ground area
concrete slabs not equipped with drainage pits onto which a transformer may be mounted

For this reason, NFPA 15 requires that a minimum density of 0.15 gpm/sq.ft. be applied to any **nonabsorbing ground area** to handle a potential for a pool fire around the transformer. While a fire should be controlled by this level of protection, an environmental or toxic hazard may ensue from any uncontained oil.

Recommendations for Water Spray System Layout

1. For exterior exposures, a minimum pressure of 20 psi is required by NFPA 15 for each nozzle on a water spray system to best preserve the velocity force of the water spray when exposed to wind conditions. The appendix of NFPA 15 strongly recommends 30 psi for nozzles with orifices less than 3/8" protecting exterior exposures. Many water spray systems for transformers are located next to rivers or other large bodies of water and the wind can be quite intense. Wind conditions can divert water spray from its intended target and render the water spray system to be considerably less effective. By having a high pressure available at the nozzle, wind effects can be minimized.

It is recommended that a minimum of 20 psi be calculated at each nozzle for interior waterspray exposures. To maintain spray integrity for exterior exposures, a maximum nozzle pressure of 50 psi is recommended. Pressures significantly greater than 50 psi break the water spray into smaller droplets that may be swept away by wind or evaporate before achieving their performance objective of coating and surface cooling.

2. All piping and piping supports used outside must be shop galvanized inside and outside the piping to protect against pipe corrosion and blockage of nozzles from loosened pipe corrosion. Another variety of galvanization, called hot-dipped galvanization, involves dipping black pipe into liquid galvanization. This process is not recommended because the galvanization does not always adhere to internal pipe surfaces, especially surfaces not cleaned of oil or other foreign matter. Flakes of loose galvanization may block the free flow of water to the nozzles.

3. Malleable iron screwed fittings should be used on water spray systems to prevent damage resulting from the shock of the water hitting the empty fittings and the movement of the system upon system actuation.

4. Sway braces or earthquake braces in accordance with NFPA 13 are recommended to minimize pipe movement and maintain piping rigidity and integrity during the discharge of the system. A sway brace is a piece of rod or pipe that more solidly attaches the pipe loop to the pipe stand. Upon system activation, the force of the water rushing into the pipe loop can create a swaying or minor seismic event. Sway braces help to minimize this pipe movement.

5. A minimum loop size of 2-inch diameter is recommended to enhance loop rigidity. Larger loop sizes (2½ inches or 3 inches) may be needed if hydraulic calculations dictate.

6. A strainer is required by NFPA 15 if the nozzles have clear openings for waterway passages less than 3/8-inch diameter.

7. Nozzles should ideally be no closer than 2 feet from the transformer in order to allow the development of a sufficient spray pattern on the transformer surface, and should not exceed 3 feet to ensure proper densities and to preserve the spray pattern in the presence of wind. Although "long-throw" nozzles are manufactured, their spray pattern is highly dependent on wind conditions.

8. Nozzles should not be extended further than 2 feet in any direction from the loop without a hanger or support.

9. Nozzles must be aimed to sweep water across the top surface of the transformer case without spraying the bushings above the case or into the electric field of uninsulated bushings. The sweeping action is intended to remove standing oil from the top surface of the transformer. Verification of this positioning must be demonstrated in a sectional view on the plan.

10. Pipe stand supports should not be spaced further than 14 feet apart.

11. Pipe stand diameter should be 2 inches for transformers up to 8 feet in height, 2½ inches for transformers up to 12 feet, 3 inches for transformers up to 22 feet, and 4 inches for transformers up to 30 feet.

12. Transformer removal: The owner must be contacted to determine the manner and direction in which the transformer would be removed for service or replacement. Flanges or grooved fittings should be placed on the corners of the loops to allow for expeditious removal of portions of the water spray system to allow for an orderly removal of the transformer and an orderly reconnection of the water spray system.

13. Screwed unions are susceptible to corrosion and are not recommended for exterior exposures.

14. Where rubber-gasketed grooved fittings are used, nozzles should be positioned to provide water spray coverage of the fittings.

15. The spacing of the nozzles along the loop is based on the:
- Spray angle of the nozzle,
- Area protected by the nozzle,
- Available pressure at the nozzle,
- Distance from the nozzle to the transformer surface,
- Gpm discharging from the nozzle.

16. A good design method is to commence nozzle spacing at the corners of the transformer and angle the nozzles toward the centerline of the transformer in a counteropposed fashion, as shown in Figure 8-10.

17. Additional nozzles are then added to provide overlapping spray patterns ensuring proper coverage and density application, as shown in Figure 8-11.

18. Densities must be reverified before completing the nozzle layout. It may not be the best strategy to use large nozzles with high flow rates to protect large areas. Although large nozzles may provide the density required, they may not provide direct impingement of all transformer surface areas that smaller orifice nozzles more closely spaced may provide. Vertical and horizontal obstructions on the transformers may necessitate using smaller nozzles spaced closer

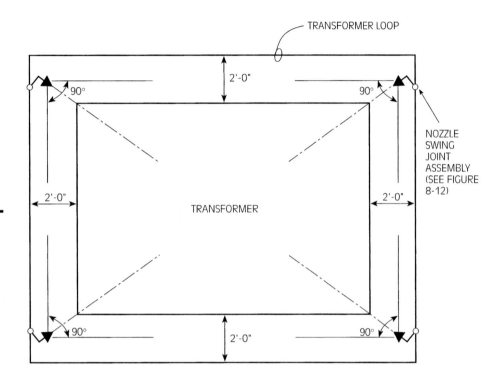

Figure 8-10 *Nozzle layout procedure (step 1). Transformer is shown in plan view. A good nozzle spacing procedure is to begin spacing with nozzles aimed at the corners.*

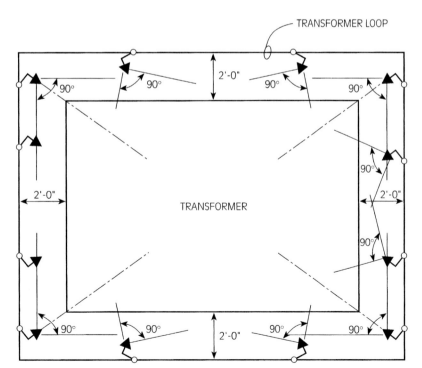

Figure 8-11 *Nozzle layout procedure (step 2). Transformer is shown in plan view. Additional nozzles are then added to provide overlapping spray patterns.*

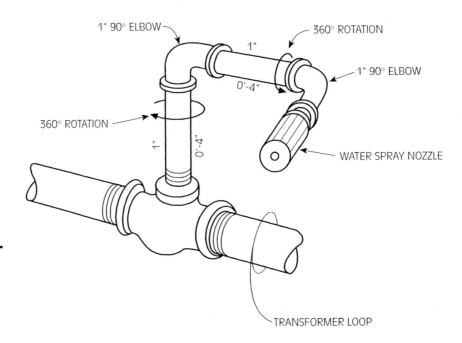

Figure 8-12 *Nozzle swing joint. Swing joints offer a wide range of field flexibility.*

swing joints

joints that allow a nozzle to be adjusted in all directions

together and overlapped, to better protect the obstructions that appear on transformer surfaces.

19. Nozzles should be installed on **swing joints**, as shown on Figure 8-12. Swing joints are adjustable piping assemblies that allow a nozzle to be adjusted in all directions and provide the ability to direct the precise application of water spray to specific areas without major field piping modifications.

20. The feedmain, shown in Figure 8-13, connects one or more loops to the riser. This main should be run no lower than 8'-0" above finished grade. When crossing roads, the minimum height of the feedmain must be coordinated with the tallest anticipated vehicle that would travel under it.

21. The riser, shown on Figure 8-13, consists of a vertical pipe emanating from the underground and contains the deluge valve with solenoid valve and system control valve with tamper switch or lock and chain. The riser must be in an area protected from freezing and mechanical injury, such as a valve house, which also protects the system control panel. The panel monitors the status of the detection system that initiates water spray system activation by sending a signal to the solenoid valve on the deluge valve.

22. The valve house, shown on Figure 8-13, should be located a safe distance from the transformer so that fire service access to the valves are not impeded by a transformer fire.

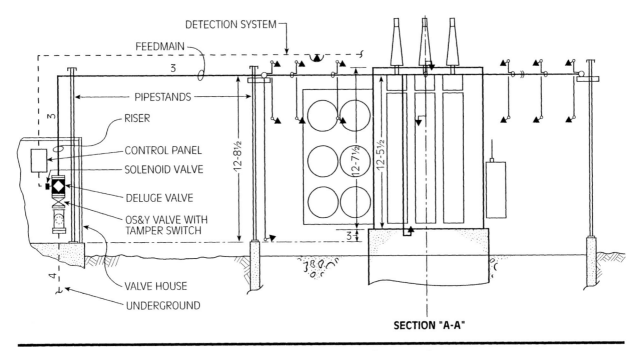

Figure 8-13 *Sectional view of the transformer water spray system plan view shown in Figure 8-14.*

Hydraulic Calculation Procedure for Transformers

1. Start by laying out the nozzle spray pattern, using plastic or cardboard templates which show the spray angle of the nozzle chosen.

2. Determine the minimum gpm of each nozzle as a function of the individual area protected by each nozzle, using a density of 0.25 gpm/sq. ft. per nozzle, in accordance with NFPA 15. Some authorities having jurisdiction may require a higher density. Where the area covered by each nozzle is approximately uniform, the gpm at the nozzle may be determined by taking the total minimum gpm, as calculated in example 6.1, and dividing it by the number of nozzles on the water spray system.

$$Q_{nozzle} = \frac{(\text{Total transformer surface area}) \times (0.25 \text{ gpm/sq.ft.})}{(\text{Number of nozzles})}$$

Where the area protected by each nozzle is not uniform, each nozzle must be calculated individually:

$$Q_{nozzle} = (A_{nozzle}) \times (0.25 \text{ gpm/sq. ft.})$$

3. Determine the minimum nozzle starting pressure. It is recommended that a minimum of 30 psi be used for small orifice nozzles protecting exterior exposures to help to counteract the effects of wind velocity on the water spray, and to enhance the ability of the water spray to sweep oil from the transformer surfaces. For interior water spray exposures, 20 psi is recommended. As a maximum nozzle pressure, 50 psi is recommended to preserve droplet size and maintain the spray pattern.

4. Having calculated Q_{nozzle}, and having selected a minimum starting pressure at the nozzle, the fire protection designer may then select a nozzle manufacturer and determine a nozzle K-factor, spray angle, and minimum flow equal to or greater than Q_{nozzle}.

5. Using the formula:

$$Q = (K) \times \sqrt{P}$$

where Q = gpm at nozzle
K = nozzle K-factor
P = minimum water pressure at the nozzle

Substituting the K-factor from item 4 above, and the minimum pressure from item 3 above, one can determine the minimum gpm of the most remote nozzle. It must equal or exceed Q_{nozzle} calculated above.

6. Reverify that every nozzle provides 0.25 gpm per square foot over the surface area protected by that nozzle.

7. The c-factor of galvanized piping on deluge systems is C-120, per Appendix B. Stainless steel or aluminum piping may be necessary in cases of unusual atmospheric corrosive conditions, such as salt spray or corrosive chemicals.

8. Use the Hazen-Williams formula to determine the pressure loss through each pipe segment.

$$P_t = \frac{(4.52)(Q^{1.85}) \times (L)}{(C^{1.85})(D^{4.87})}$$

where P = the pressure lost in the pipes (psi)
Q = the gpm flowing through the pipe
C = the roughness coefficient (C = 120 for galvanized)
D = the actual internal piping diameter, in inches.

orifice plate
a disk designed to restrict the flow of water in the water spray system

9. **Orifice plates** shall not be used to balance the pressures in the system. An orifice plate is a disk with a hole in it placed in the piping system, designed to restrict the flow of water in the waterspray system, such that the maximum nozzle pressure of 50 psi is not exceeded. In the days of manual or hand hydraulic calculations, orifice plates were once used to balance systems with significant water pressures. With the advent of computer calculation programs, we can now perform

accurate calculations and modify pipe sizes to ensure that pressures do not exceed 50 psi without the use of orifice plates.

10. Equivalent pressure loss through fittings is determined by using the equivalent loss tables in Appendix D.

11. Elevation pressure loss is calculated by the formula:

$$P_e = (0.433) \times (h)$$

where P_e = the pressure loss due to elevation (psi)
         h = the elevation, in feet.

12. Plot calculated system results with respect to the water supply curve per Chapter 6.

13. Add the outside hose requirement to system demand. This requirement is a function of your evaluation of the probability for the use of hydrants and other hose stream sources during system activation. If no hydrants or hose equipment exist within the proximity of the transformer, there is no requirement for hose demand.

Example 8.2 A transformer outline drawing shows a total surface area of a transformer and its radiators to be 1,700 square feet. A transformer layout depicting 33 water spray nozzles is shown on Figures 8-14, 8-15, and 8-16. Nozzles used on the system discharge 14 gpm at 30 psi, with a discharge spray angle of 120°. Hydraulically calculate the system and determine the system demand.
Solution

(1700 square feet of transformer surface area) × (0.25 gpm/sq. ft.) = 425 gpm minimum water spray system demand.

$$Q_{nozzle} = \frac{(425 \text{ gpm})}{33 \text{ nozzles}} = 12.9 \text{ gpm per nozzle}$$

A nozzle has been selected that is manufactured to discharge 14 gpm at 30 psi, providing the minimum acceptable flow.

Nozzles on the loop are divided into two segments of approximately equal flow. Figure 8-14 shows the dividing point at node B where nozzles at nodes 1 and 2 are connected. The hydraulic calculation in Figure 8-17 flows from node B in a clockwise direction to nodes C, D, E, F, G, H, I, J, and K. Then, on Figure 8-18, the calculation commences at node L and travels counterclockwise to nodes M, N, O, P, Q, R, S, T, and K. The flows of the two loop segments are then balanced at node K on Figure 8-17, and the combined flow is carried to the deluge valve.

The calculation on Figures 8-17 and 8-18 shows a system demand of 481.78 gpm at 62.51 psi, plus a hose allowance of 500 gpm, for a total of 981.78 gpm at 62.51 psi. Note that the actual calculated water spray demand of 481.78 gpm exceeds the estimated minimum gpm demand of 425 gpm that we had previous-

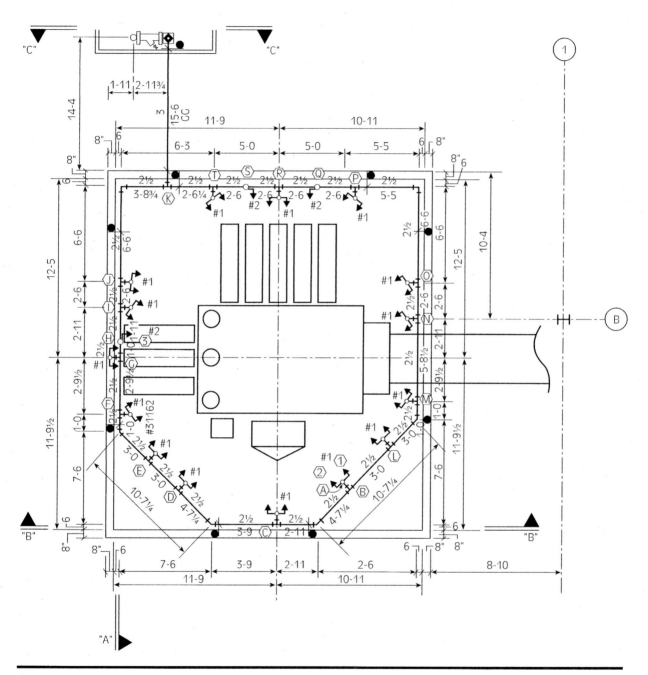

Figure 8-14 *Transformer described in example 8-2. Section A-A is shown on Figure 8-13. Section B-B is shown on Figure 8-4. Section C-C is shown on Figure 8-15. Detail of nozzle assemblies is shown on Figure 8-16.*

Chapter 8   Water Spray System Design for Transformers

detection system
a system that supervises the area being protected and is capable of actuating a fire protection system when a fire is detected

electric detection system
a system employing detectors that send an electronic signal to a control panel when a fire is detected

pilot head detection system
a system that consists of closed sprinklers connected to piping supervised by a pressure switch

ly calculated in this example. The difference is that the hydraulically calculated demand considers frictional and elevational pressure losses.

Detection Systems

Water spray systems differ from wet pipe and dry pipe systems because they use open nozzles and thus, cannot be actuated by a loss of pressure in the piping system. Water spray systems must therefore have a **detection system** that supervises the area being protected by water spray and that will actuate the system when a fire is detected.

Two types of detection systems can be considered. **Electric detection systems** send an electronic signal to a control panel when a fire is detected. The control panel then sends an electric signal to a solenoid switch on the deluge valve, electrically opening the valve and allowing water to flow into the water spray system. An electric detection system for a transformer water spray system is shown on Figure 8-3 and 8-13 on page 197 and 213 respectively.

The other type of detection system is the **pilot head detection system**. A pilot head system consists of closed sprinklers, called *pilot heads*, on a loop of 1/2-inch pilot piping supported from the transformer loop. The pilot piping is pressurized

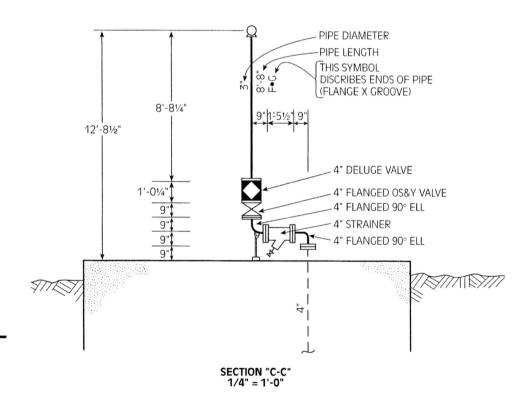

Figure 8-15 *Valve arrangement for Figure 8-14.*

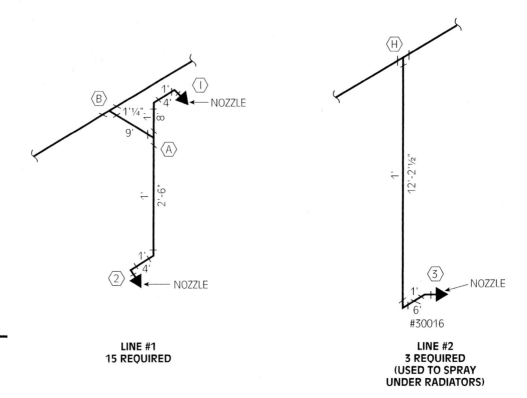

Figure 8-16 *Nozzle assemblies for Figure 8-14.*

heat detector
a device that sends a signal to a control panel when it detects a temperature at a predetermined level

rate compensated detector
a detector that measures the rate of temperature rise, sending a signal when the temperature rises a predetermined rate of temperature change within a predetermined time

with air and is connected to a low air switch on a control panel or on a deluge valve that actuates the system when a pilot head actuates. In areas not subject to freezing, a wet pilot system is used where the pilot piping is pressurized with water, and in areas where freezing could occur, the pilot line is pressurized with air. Pilot head detection systems use standard sprinklers with glass bulb or eutectic elements. When a sprinkler is connected to pilot piping, it is called a pilot head. It is recommended that quick response elements be used on pilot head systems for exterior exposures.

Electric detectors may be chosen from several types, each performing differing functions:

- **Heat detectors** measure the temperature of the air in the vicinity of the detector and send a signal to a control panel. Fixed-temperature detectors send a signal when the detector reaches a predetermined temperature.
- **Rate compensated detectors** measure the rate of temperature rise and send a signal to a control panel when the temperature rises a predetermined rate of tem-

Chapter 8   Water Spray System Design for Transformers

HYDRAULIC CALCULATION SHEET

CONTRACT NO. _____   System No. _TRANSFORMER_ Sheet ___1___

CONTRACT NAME EXAMPLE 8-2

NOTES   DATE

Calculated by _____

Plan Ref. Pt.	SPRINKLERS NOZZLES Total Heads	Type	Added Gpm	Total Gpm	Pipe Size In.	Pipe and Equivalent Length Fittings - Devices Feet	FRICTION LOSS Psi/Ft. C = 120	Total Psi	(Elevation) Static Plus or Minus Psi	Required Pressure Psi	CALCULATN. REFERENCE Sheet No.	Point No.
①	1	NOZZLE					PRESSURE AT HEAD →			30.0		
				14.0	1	BL=1,E=2,T=5 TOTAL 8	.067	.54	.29	30.83		Ⓐ
②	1	NOZZLE	—	14.0	1	BL=2.9,E=2,T=5 TOTAL 9.9	.067	.67	-1.12	30.37		Ⓐ
		ADJUST FLOW:		$Q_N = 14 \sqrt{30.83/30.37} = 14.10$						30.83		Ⓐ
Ⓐ	2	—	14.10	28.10	1¼	BL=.75,T3 TOTAL 3.75	.068	.26	—	31.09		Ⓑ
						LINE $K = 28.10/\sqrt{31.09} = 5.04$						
Ⓑ	2	—	—	28.10	2½	LOOP=7.8,45°=3; TOTAL=10.8	.004	.04	—	31.13		Ⓒ
Ⓒ	4	K = 5.04	28.12	56.22	2½	LOOP=8.8,45°=3; TOTAL=11.8	.015	.17	—	31.30		Ⓓ
Ⓓ	6	K = 5.04	28.19	84.41	2½	LOOP 3	.029	.09	—	31.39		Ⓔ
Ⓔ	8	K = 5.04	28.24	112.65	2½	LOOP=4, 45°=3; TOTAL=7	.0496	.35	—	31.74		Ⓕ
Ⓕ	10	K = 5.04	28.39	141.04	2½	LOOP 2.8	.0756	.21	—	31.95		Ⓖ
Ⓖ	12	K = 5.04	28.49	169.53	2½	LOOP 1	.1055	.11	—	32.06	2	Ⓗ
Ⓗ	13	K = 2.74	15.51	185.04	2½	LOOP 2	.1246	.25	—	32.31		Ⓘ
Ⓘ	15	K = 5.04	28.65	213.69	2½	LOOP 2.5	.1700	.43	—	32.74		Ⓙ
Ⓙ	17	K = 5.04	28.84	242.53	2½	LOOP=10.33,E=6,T=12;TOTAL=28.33	.2154	6.10	—	38.84		Ⓚ
Ⓚ	33	SEE FIG. 7.18	239.25	481.78	3	FEEDMAIN=24.5,E=7;TOTAL=31.5	.2597	8.18	3.9	50.92	TOP DELUGE VA.	
DEL VA.	33	—	—	481.78	4	2E=20,0SY=2 DV=18;TOTAL=40	.0691	2.76	1.7	55.38		
					4	STRAINER	—	3.18	—	58.56	FLANGED ENTRY	
			—	481.78	4	UNDG 12 } 17 LTE 3 GV 2 }	.0446	.76	2.2	61.52		
					8	UNDG 30, T30 TOTAL 60	.0015	.09	—	61.61		
					10	UNDG 480 } 610 2T 72 2LTE 14 2E 34 2GV 10	.0005	.31	.9	62.51		
		HOSE	500	981.78						62.51		

SYSTEM DEMAND AT PUMP DISCHARGE 981.78 @ 62.51 PSI

Figure 8-17   *Example 8-2 calculation.*

HYDRAULIC CALCULATION SHEET

CONTRACT NO. ______ System No. ______ Sheet 2

CONTRACT EXAMPLE 8-2

NOTES DATE

Calculated by ______

SPRINKLERS NOZZLES			(q) Added Gpm	(Q) Total Gpm	Pipe Size In.	Pipe and Equivalent Length Fittings - Devices Feet	FRICTION LOSS		(Elevation) Static Plus or Minus Psi	Required Pressure Psi	CALCULATN. REFERENCE	
Plan Ref. Pt.	Total Heads	Type					Psi/Ft. C = 120	Total Psi			Sheet No.	Point No.
⟨3⟩	1	NOZZLE					PRESSURE AT HEAD ⟶			30.0		
				14.0	1	BL=12.8,E=2,T=5; TOTAL=19.8	.067	1.33	-5.32	26.01		
						LINE K = 2.74						
		"L" SAME AS "B" SHEET 1								31.09		⟨L⟩
⟨L⟩	2	NOZZLE	—	28.10	$2\frac{1}{2}$	LOOP=4 ,45 = 3 TOTAL=7	.004	.03	—	31.12		⟨M⟩
⟨M⟩	4	K = 5.04	28.11	56.21	$2\frac{1}{2}$	LOOP 5.8	.0140	.08	—	31.20		⟨N⟩
⟨N⟩	6	K = 5.04	28.15	84.36	$2\frac{1}{2}$	LOOP 2.5	.0293	.07	—	31.27		⟨O⟩
⟨O⟩	8	K = 5.04	28.18	112.54	$2\frac{1}{2}$	LOOP=12, E6, TOTAL=18	.0496	.89	—	32.16		⟨P⟩
⟨P⟩	10	K = 5.04	28.58	141.13	$2\frac{1}{2}$	LOOP 2.5	.0756	.19	—	32.35		⟨Q⟩
⟨Q⟩	11	K = 2.74	15.58	156.71	$2\frac{1}{2}$	LOOP 2.5	.0911	.23	—	32.58		⟨R⟩
⟨R⟩	13	K = 5.04	28.77	185.48	$2\frac{1}{2}$	LOOP 2.5	.1234	.31	—	32.89		⟨S⟩
⟨S⟩	14	K = 2.74	15.71	201.19	$2\frac{1}{2}$	LOOP 2.5	.1560	.39	—	33.28		⟨T⟩
⟨T⟩	16	K = 5.04	29.08	230.27	$2\frac{1}{2}$	LOOP=2.6, T12, TOTAL=14.6	.1846	2.70	—	35.98	—	⟨K⟩
										38.84	1	⟨K⟩

$$Q_{ADJUST} = 230.27\sqrt{38.84/35.98} = 239.25$$

Figure 8-18 *Example 8-2 calculation.*

linear heat detection system
heat-sensitive electric wiring installed around a transformer loop

perature change within a predetermined time. These detectors are very reliable for exterior transformer and vessel protection since gradual fluctuations in temperature, such as the sun rising in the morning, would not affect the detection system.

- A **linear heat detection system** is heat-sensitive wiring that may be installed around a transformer loop. When the wiring senses heat, the current flow in the wiring changes. When the current flow changes a predetermined amount, the control panel sends a signal to a switch on the deluge valve that activates the water spray system. This detection system is especially effective on linear hazards, such as coal conveyors and cable tray protection, but have also been found to be effective for transformers and vessels.

Smoke detectors and beam detectors are also electric detectors, but are not suited for exterior exposures.

Summary

A water spray system protects specialized hazards with directional water spray application. Power plants offer a wide variety of opportunities for the use of water spray systems, with the protection of transformers being one of the most prominent. Water spray systems control and extinguish fires by surface cooling, steam smothering, dilution, redirection, coating, and vapor exhaust. An outline drawing is the foundation for the commencement of design of a water spray system. Also needed for transformer fire protection layouts are structural, civil, and architectural drawings of the area in which the transformer is located. Transformer water spray systems, because of the detail that must be shown, should be drawn to a minimum scale of 1/4" = 1'-0". By following the methodical design and calculation procedure in this chapter, the reader should be able to design and calculate a transformer water spray system.

Review Questions

1. List the applications for water spray systems in a power-generating plant.
2. Describe the differences between the extinguishment/control mechanism employed by water spray systems for interior and exterior transformer exposures.
3. Using the outline drawing shown on Figure 8-6 and Table 8-1, determine the electrical clearance radius for the live uninsulated bushings.
4. Using the outline drawing shown on Figure 8-6, determine the locations and elevations of the transformer water spray loops.
5. Using the outline drawing shown on Figure 8-6, design a water spray system for the transformer shown using nozzles with a 90° spray angle, and rated for a discharge of 18 gpm at a 30 psi starting pressure.
6. Using the outline drawing shown on Figure 8-6 and using a loop diameter of 2" and the nozzle assemblies shown on Figure 8-12, perform a hydraulic calculation of the water spray system designed in review question 5 above.
7. Discuss the similarities and differences between sprinkler system design and calculation, and water spray system design and calculation.
8. Using the computer hydraulic program in Appendix E, perform the calculation of the system described in review question 6 above.

Activities

1. Arrange to tour a power generating plant and meet the fire protection engineer or chief of the fire brigade for the plant. Determine whether the plant has any requirements that differ from NFPA standards.

2. Visit a sprinkler design or installation firm, and obtain a blueprint of a completed water spray system design. Note the level of detail required for plans that are required to be submitted to an authority having jurisdiction.

3. Arrange to witness a discharge test of a water spray system protecting a transformer. Note the wind speed and wind direction. Determine the nozzle pressure from the hydraulic calculations, and observe the spray pattern to see if it is being affected by the wind. Do you think that the spray pattern would be affected if the nozzle pressure were reduced by half its current value or if the wind was significantly stronger?

Chapter 9

Water Spray System Design for Flammable Liquid Storage Tanks or Vessels

Objectives

Upon completion of this chapter, you should be able to:

- Describe the modes of ignition and extinguishment for water spray protection of a flammable liquid storage tank or vessel.
- Calculate the surface area of a flammable liquid storage tank or vessel and determine the minimum water supply requirement.
- Perform a layout of a simple water spray system protecting a flammable liquid storage tank or vessel.
- Perform a hydraulic calculation of a water spray system protecting a flammable liquid storage tank or vessel.

Water spray protection of vessels is performed in accordance with NFPA 15, *Standard for Water Spray Fixed Systems for Fire Protection*. NFPA 15 is not a detailed design document, but a list of requirements without a methodology for the accomplishment of the design objectives. The purpose of this chapter is to help the reader meet the design objectives for water spray protection in accordance with NFPA 15, by following a step-by-step method.

HAZARDS ASSOCIATED WITH TANKS AND VESSELS

Piloted and Autoignition

A pressure vessel, a low pressure tank, or an atmospheric tank protected by water spray predominantly contains flammable or combustible liquids possessing a **flash point** conducive to **piloted ignition** or **autoignition**. A flash point is the temperature at which a flammable liquid ignites vapor from the fuel for a brief period. Piloted ignition is ignition and sustained combustion from an outside source, such as a spark or match, and autoignition is ignition and sustained combustion that could occur in cases where the fuel could ignite without an outside ignition source.

flash point
the temperature at which a flammable liquid ignites vapor from the fuel for a brief period

piloted ignition
ignition and sustained combustion from an outside source such as a spark or match

autoignition
ignition and sustained combustion that could occur in cases where the fuel could ignite without an outside ignition source

Pool Fires

The accidental spill of the flammable liquid from the tank and the ignition of that liquid on the floor or ground below the tank results in a pool fire. A pool fire is a pool of burning flammable liquid. The danger of a pool fire beneath a flammable or combustible liquid storage tank is the impingement of flame upon the storage tank surface, heating the flammable or combustible liquid contained within the tank.

Pool Fire Containment

The severity of a pool fire can be lessened by providing drains that are capable of removing the spilled flammable liquid to a safe location where it may be reprocessed and reused. A wall or dike around a flammable liquid storage tank contains the spilled liquid, and drains that have been specially designed to accept flaming flammable liquid are installed within the diked area. The containment is necessary to prevent a running spill fire that would carry flaming flammable liquid to other areas.

Pressure Spray Fires

A common ignition scenario involves a pump responsible for pumping flammable liquid to or from the tank. A pump might overheat or might burst a seal, spraying flammable liquid onto and beneath the tank. Ignition of the escaped liquid from an overheated pump could result in a spray of ignited flammable liquid that could be extremely damaging to the storage vessel or personnel, and could initiate a pool fire below the vessel, endangering the remaining contents of the vessel and of adjacent vessels.

Deflagration and Rupture

In a worst case scenario, the tank shell is heated by a fire beneath the tank, and the heated shell transfers heat to the liquid remaining within the tank, igniting it, creating a **deflagration**, or a fire that spreads at a velocity less than the speed of sound. A deflagration would pose a serious threat to adjacent personnel, tanks, or combustibles. A similar result could ensue if a tank support or tank shell is heated to its failure point, causing the shell to deform and burst, resulting in a **rupture** and a serious deflagration or an **explosion**. An explosion is the sudden overpressurization of a flammable liquids storage vessel beyond the maximum pressure potential of the vessel.

If a portion of a flammable liquid storage tank is unwetted by water spray, that portion of the tank surface can deform and lose strength, and eventually rupture, causing a **BLEVE** (boiling liquid-expanding vapor explosion). These explosions are extremely serious, because they can result in pieces of the tank being thrown considerable distances, possibly igniting all other tanks in the vicinity of the BLEVE, and placing fire service personnel in serious danger.

Such incidents result not only in extensive damage to storage and processing equipment, but could also pose a life safety threat to maintenance personnel trapped within a **tank farm**, a group of closely spaced tanks or vessels, sometimes located in pits or walled areas. A tank farm that is fully involved in flame could jeopardize all areas in its vicinity. Death, injury, and business interruption caused by the loss of storage capacity for an extended period of time could be the result of the loss of a tank farm.

VESSEL PROTECTION MECHANISMS

Exposure Protection

Exposure protection is the uniform application of a coating of water to the entire exposed exterior surface of a vessel that acts like an insulation blanket. To be successful, exposure protection must involve the uniform application of water over the entire surface area of the vessel, with the understanding that even small unprotected areas could result in deformation of the tank wall and the failure of the vessel at that point. Careful examination must be made to ensure that water spray patterns at least meet or overlap and that water spray application to the vessel successfully considers water spray obstructions on the vessel, such as tank supports.

---

deflagration
a fire in which the expanding flame front travels at a velocity less than the speed of sound

rupture
a situation in which a tank shell is heated to its failure point, causing the shell to deform and burst

explosion
the sudden overpressurization of a vessel or enclosure beyond its maximum pressure potential

BLEVE
a boiling liquid-expanding vapor explosion that may occur when a portion of a flammable liquid storage tank is unwetted by water spray, causing that portion of the tank surface to rupture

tank farm
a group of closely spaced flammable liquid tanks

Surface Cooling

Because the hazard to the vessel predominately involves an ignition of pooled liquid beneath the tank or vessel, the most critical application of water spray is to accomplish the performance objective of surface cooling to the underside of the tank or vessel. Water applied to this area absorbs a considerable amount of heat from a pool fire and should prevent the container from failing. Water spray application on the entire surface area is required to keep the vessel cool in the event of a pressurized flammable or combustible liquid fire impinging on adjacent areas other than the bottom of the vessel.

Rundown

For a vertically oriented vessel, water applied to the top of the vessel runs down the sides and supplements water spray applied to the bottom, resulting in densities applied to the bottom that exceed those applied to the top. The increase of density at the bottom of a vertical vessel is called **rundown**, the accumulation of water applied from nozzles at higher elevations of the tank. Rundown is a considerably more prominent phenomenon for vertically oriented vessels than it is for a horizontally oriented vessels, as shown on Figure 9-1.

The presence of rundown does not mean that direct application of water spray to the bottom of the vessel can be eliminated, and the probability of a fire below the vessel does not mean that direct application of water spray to the top of the vessel can be eliminated. The best way to be assured that adequate protection is provided is to design a water spray system that provides direct impingement of water spray to the entire exposed surface area of the vessel.

rundown
the accumulation of water at lower elevations of a vessel, applied from nozzles at higher elevations of the vessel

Prevention of Boiling Contents

Figures 9-2 and 9-3 illustrate the differences between a vessel protected with water spray and an unprotected vessel. Testing has established that a vessel without insulation or exposure protection can absorb up to 20,000 Btu of heat from a pool fire below the vessel, and that the application of water spray exposure protection to the vessel reduces this heat absorption to a considerably lower level of 6,000 Btu. Figure 9-2 can be used for estimating the time for the flammable liquid in the tank to reach a temperature of 100°F, assuming no water spray protection is provided for the vessel. Figure 9-3 shows the time for most liquids in a vessel to reach their boiling points, assuming that water spray protection is provided.

Example 9.1 For a vessel with a diameter of 10 feet, determine the estimated time for the contents to boil, with and without water spray.

Solution Without water spray, for a 10-foot diameter tank, the estimated time is 7 minutes, per Figure 9-2. For the same diameter tank with water spray the estimated time to boiling is 23 minutes, per Figure 9-3.

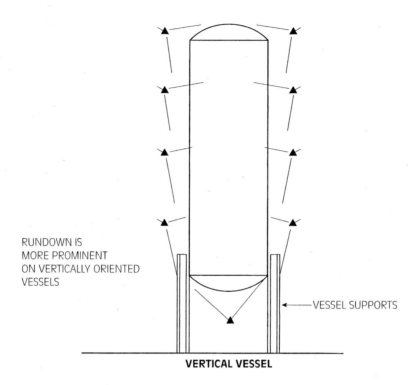

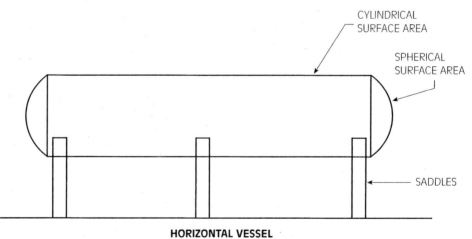

Figure 9-1 *Tank or vessel orientation.*

Assuming that a water spray system has been properly designed for total impingement for all surface areas of the vessel, the 23 minutes of protection from boiling should allow the system to provide adequate protection for the tank until extinguishment has been achieved or until such time as fire service personnel have intervened to extinguish the fire. Note that the time frame of 7 minutes to boiling

Figure 9-2 *Estimated time for liquids to boil (without water spray). (Reprinted with permission from NFPA 15, Water Spray Fixed Systems for Fire Protection. Copyright © 1996, National Fire Protection Association, Quincy, MA 02269. This reprinted material is not the complete and official position of the National Fire Protection Association on the referenced subject, which is represented only by the standard in its entirety.)*

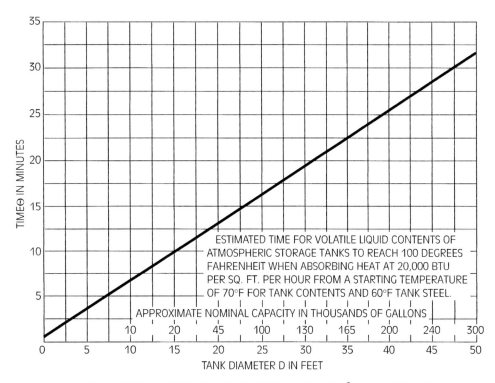

may not be sufficient to notify the fire service, deliver fire suppression equipment, set up equipment, and deliver effective hose streams to the tank.

Supplementary Protection for Vessels

In addition to exposure protection of the vessel, water spray protection should be considered for the nonabsorbing ground surface below the tank and for any tank supports or structural elements. It makes little sense to protect the vessel yet still lose it when the tank supports fail. It is possible to have a design that successfully cools the vessel but results in vessel failure when the supports buckle from the heat from a pool fire. Extra nozzles, aimed directly at the supports, providing complete water spray protection of the supports, may be necessary to ensure that the vessel remains erect.

Water spray protection of the vessel is greatly enhanced by applying water directly onto the pool fire below the vessel. Direct water spray on a nonabsorbing ground surface provides an extra level of protection for the vessel and its supports

Figure 9-3 *Estimated time for liquids to boil (with water spray). (Reprinted with permission from NFPA 15, Water Spray Fixed Systems for Fire Protection. Copyright © 1996, National Fire Protection Association, Quincy, MA 02269. This reprinted material is not the complete and official position of the National Fire Protection Association on the referenced subject, which is represented only by the standard in its entirety.)*

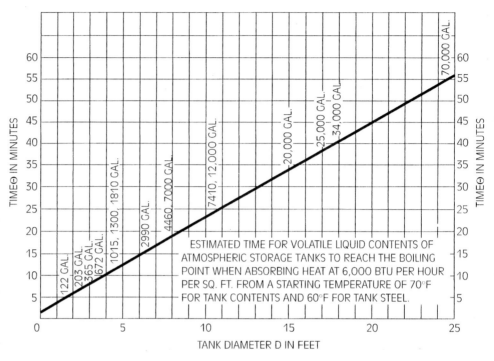

FOR SI UNITS: 1 BTU PER SQ. FT. PER HOUR = 3.155 W/m²; 1 GAL = 0.003 79 m³ = 3.785 L; °C = 5/9 (°F − 32); 1 FT = 0.305 m.

by directly attacking the source of the hazard to the vessel, by diluting the fuel, by surface cooling of the burning surface of the fuel, and by redirecting the fuel to a drain. Water spray protection of nonabsorbing ground area promotes extinguishment of the fire below the tank, which significantly benefits the exposure protection of the tanks.

VESSEL WATER SPRAY DESIGN PROCEDURE

Surface Area Calculation

NFPA 15 requires that a minimum density of 0.25 gpm of water spray per square foot be applied over all exposed surface areas of the vessel. The density requirement for tanks is identical to that required for transformers, because the same performance objective of exposure protection is being sought for both. In order to ensure proper exposure protection, accurate surface area calculations must be performed.

1. Cylindrical tank surface area, as shown in Figure 9-1, in square feet:

$$\text{Area of tank cylinder} = (\text{diameter}) \times (3.1416) \times (\text{length})$$

2. Spherical tank surface area (2 tank ends), as shown in Figure 9-1, in square feet. Note that this is the formula for a sphere, with half the sphere located at each end of the tank:

$$\text{Spherical end area} = (3.1416) \times (\text{diameter})^2$$

3. If tank has flat (nonspherical) ends, in square feet:

$$\text{Flat end area} = (1/4) \times [\,(3.1416) \times (\text{diameter})^2\,] \times (2 \text{ ends})$$

4. Total tank surface area, in square feet:

$$\text{Total tank area} = (\text{cylindrical area}) + (\text{spherical or flat end area})$$

Example 9.2 A horizontal cylindrical flammable liquid storage tank is 12'- 0" in diameter, 75'- 0" long, and has belled (spherical) ends with a radius of 6'-0". Calculate the total surface area of the tank.
Solution

$$\begin{aligned}
\text{Area of cylinder} &= (12\text{'-}0") \times (3.1416) \times (75\text{'-}0") \\
\text{Area of cylinder} &= 2{,}827 \text{ square feet} \\
\text{Area of two bells} &= (12\text{'-}0")^2 \times (3.1416) \\
\text{Area of two bells} &= 453 \text{ square feet} \\
\text{Total surface area} &= (2827 \text{ sq. ft.}) + (453 \text{ sq. ft.}) \\
\text{Total surface area} &= 3{,}280 \text{ square feet}
\end{aligned}$$

Estimating the Minimum Water Supply Requirement

The minimum density requirement of 0.25 gpm per square foot is applied to the total surface area of the vessel:

$$Q = (\text{total surface area}) \times (0.25 \text{ gpm per square foot})$$

Example 9.3 Continuing with the tank whose surface area was calculated in example 9.2, calculate the minimum water supply requirement for the tank.
Solution

$$\begin{aligned}
Q &= (\text{total surface area}) \times (0.25 \text{ gpm per square foot}) \\
&= (3280 \text{ square feet}) \times (0.25 \text{ gpm per square foot}) \\
&= 820 \text{ gpm.}
\end{aligned}$$

It must be emphasized that 820 gpm is the minimum flow required to wet the tank. It does not represent the minimum hydraulically calculated water supply required at the base of the riser.

Total Water Requirement

The total water requirement at the base of the riser is determined by hydraulic calculations, and the calculation must consider the following factors:

- Frictional losses in pipe, obtained by hydraulic calculation.
- Pressure imbalances in the system, obtained by hydraulic calculation.
- Elevation losses, obtained by hydraulic calculation.
- **Water wastage**, or water that flows from a water spray nozzle, but does not directly impinge upon the vessel surface. Wastage could take the form of **overspray**, or water that misses the vessel surface, or **wind wastage**, water droplets that have been swept away by wind. NFPA 15 assumes that wastage does not exceed 0.05 gpm per square foot, and includes this amount of wastage in the minimum densities specified by the standard. Wastage is reduced by selecting a nozzle spray angle that is narrow enough to provide direct impingement and pattern overlap, and by aiming the nozzles in such a way that most of the spray hits the tank.
- Additional water spray that might be required to protect vessel supports or nonabsorbing ground surface.

water wastage
water that flows from a water spray nozzle but does not directly impinge upon the target area

overspray
water that misses the target surface

wind wastage
water droplets that have been swept away by wind

Estimation of Water Supply Adequacy

Although the estimation method for determining the minimum water supply requirement needs to be used with some degree of caution, a preliminary evaluation of the water supply can be performed at this juncture for most water spray systems. By adding the anticipated flow for vessel supports and nonabsorbing ground quantity to the total flow required for protection of the vessel surface, and multiplying this total by a factor of 1.25 to account for frictional losses of 25%, an estimated water supply total can be obtained. Estimated gpm and psi totals are used as a basis of comparison to determine the accuracy of hydraulic calculations.

Example 9.4 In example 9.3, a water requirement of 820 gpm was calculated to provide exposure protection for the vessel surface. If the vessel has four supports protected by nozzles flowing 15 gpm each, and if a nonabsorbing surface of 1,000 square feet is protected with a minimum density of 0.15 gpm per square foot, estimate the minimum water supply that must be provided, assuming a 25% loss factor for frictional effects, and an allotment of 500 gpm for minimum hose stream requirements.

Vessel surface:
$$Q = 820 \text{ gpm}$$

Vessel supports:
$$Q = (4) \times (15.0 \text{ gpm}) = 60 \text{ gpm}$$

Nonabsorbing ground surface:

$$Q = (1{,}000 \text{ square feet}) \times (0.15 \text{ gpm/sq. ft.}) = 150 \text{ gpm}$$

Total quantity, neglecting friction:

$$Q = 820 \text{ gpm} + 60 \text{ gpm} + 150 \text{ gpm} = 1{,}030 \text{ gpm}$$

Estimated gallonage requirement, assuming 25% system frictional losses:

$$Q = (1{,}030 \text{ gpm}) \times (1.25) = 1{,}288 \text{ gpm}$$

Estimated gallonage requirement, including hose streams:

$$Q = (1{,}288 \text{ gpm}) + (500 \text{ gpm}) = 1{,}788 \text{ gpm}$$

Pressure Estimation

The pressure requirement can be similarly estimated by starting with the design pressure at the most remote nozzle, accounting for the elevation to the highest nozzle, and multiplying by a factor of 1.3 to provide an estimated 30% loss resulting from system pressure imbalances.

Example 9.5 The tank in example 9.4 has a minimum pressure requirement of 30 psi at the end nozzle which is 20 feet above the water supply source. Estimate the minimum system pressure requirement, assuming a 30% system pressure loss.
  Solution

End nozzle pressure:

$$P = 30.0 \text{ psi}$$

Elevation losses:

$$P = (20'\text{-}0") \times (.433 \text{ psi/ft.})$$
$$= 8.7 \text{ psi}$$

Total pressure, neglecting losses:

$$P = 38.7 \text{ psi}$$

Estimated system pressure, assuming a 30% system balancing loss:

$$P = (38.7 \text{ psi}) \times (1.30) = 50.3 \text{ psi}$$

The estimated water supply requirement for the example calculated is 1788 gpm at 50.3 psi. It is recommended that the estimated method presented above serve as the basis for water supply estimates of water spray systems, so that the flow and pressure values obtained in a thorough hydraulic calculation can be verified and placed into context.

Loop and Nozzle Spacing Procedure

1. Draw a scaled sectional view of the tank shell from the end view, showing its circumference, similar to Figure 9-4.

2. Plot nozzle locations around the circumference of the vessel, assuming a distance of 2'-0" from the vessel shell to the nozzle, plotting the nozzle spray angle selected for protection of the vessel. The following can be used as a guide:

Tank diameter less than 3'-0": 2 nozzles around circumference,

Tank diameter 3'-0" to 8'-0": 3 nozzles around circumference,

Tank diameter 8'-0" to 15'-0": 4 nozzles around circumference.

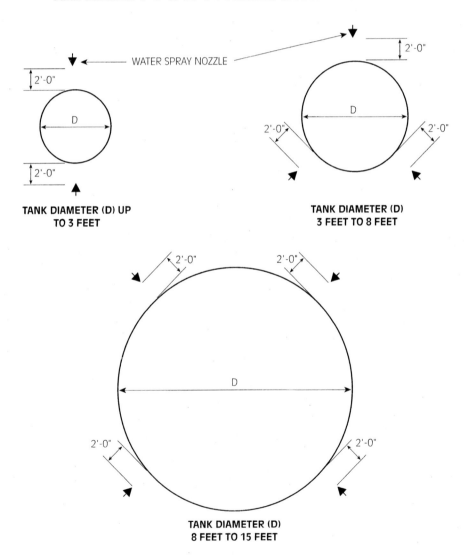

Figure 9-4 *End view of a horizontal tank, or top view of a vertical tank.*

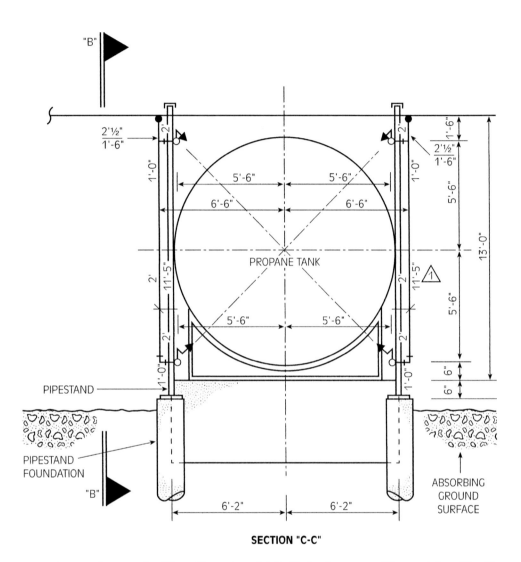

Figure 9-5 *Sectional view of an 11'0" diameter tank. Section B-B appears on Figure 9-6, and the plan view is shown on Figure 9-9.*

A sample vessel with a diameter of 11'-0" is shown on Figure 9-5, with all dimensions and details shown. Since it is 11 feet in diameter, four nozzles are required around its circumference.

3. Space nozzles along the length of the vessel, assuming a horizontal distance not to exceed 7'-0" between nozzles, with nozzles angled at 60° to the tank surface. For vertical tanks, the nozzles are angled down, and for horizontal tanks, the nozzles are angled toward the center line of the tank, as shown on Figure 9-6. Note that Figure 9-6 shows a side view of the tank detailed on Figure 9-5.

4. Adjust nozzle locations for obstructions and vessel supports, providing additional water spray for the vessel supports. On Figure 9-6, note how the nozzles on the bottom loop are spaced differently than nozzles on the top loop to avoid

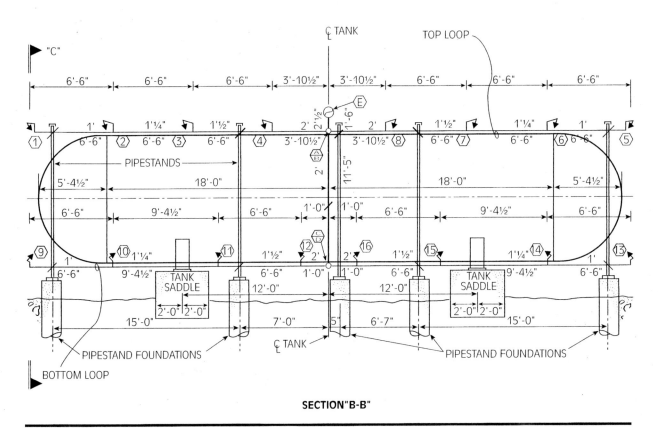

Figure 9-6 *Longitudinal sectional view of a horizontal cylindrical storage tank water spray system. Section C-C appears on Figure 9-5, and the plan view is shown on Figure 9-9.*

obstructions from the tank saddles. Tank saddles are steel or concrete foundations upon which a horizontal vessel is supported.

5. Determine the location of vessel relief valves, and make certain that water spray is provided for direct impingement of these devices. A relief valve is intended to relieve excessive internal pressure from a vessel that can build up during a fire. Keeping these valves cool is necessary to ensure the proper operation of these devices.

6. The points of connection for the flammable liquid supply pipe can cause water spray obstruction, and nozzle position may require adjustment to protect the vessel in the vicinity of these connections. It may be desirable to provide direct impingement water spray protection for the flammable liquid pipes entering and leaving the vessel.

7. Design protection for any nonabsorbing ground surface that may surround the vessel, using a density of 0.15 gpm per square foot of area protected, in a manner similar to what was done in Chapter 8.

8. Select a gallonage for each nozzle, assuming an application rate of 0.25 gpm per square foot per nozzle, commensurate to the area covered by the nozzle, and in accordance with rated nozzle flow rates listed by an approved testing laboratory, such as Underwriters Laboratory. Many layouts result in a uniform area of coverage and a uniform minimum gallonage for each nozzle. Where nozzles protect areas that are approximately equal on a tank, use the formula below to calculate the minimum gpm required at each nozzle:

$$Q_{nozzle} = \frac{(\text{Total vessel area}) \times (0.25 \text{ gpm/sq ft})}{(\text{Number of nozzles})}$$

9. Select a nozzle K-factor. Using Q_{nozzle} calculated above and a minimum psi requirement at the nozzle, one can determine the minimum nozzle K-factor required using the formula:

$$Q_{nozzle} = (K) \times \sqrt{P}$$

NFPA 15 requires a minimum of 20 psi at each nozzle, but recommends 30 psi for nozzles with orifices smaller than 3/8 inch.

Example 9.6 The vessel shown in Figures 9-5 and 9-6 is protected by 32 nozzles with a total surface area of 1,900 square feet. At a 30 psi starting pressure, what is the minimum flow at the most remote nozzle, and what is the minimum K-factor to achieve this flow at 30 psi?

Solution The minimum gpm at each nozzle, assuming a uniform coverage area per nozzle, is calculated by using the formula:

$$Q_{nozzle} = \frac{(\text{Total vessel area})(0.25 \text{ gpm/sq. ft.})}{(\text{Number of nozzles})}$$

$$Q_{nozzle} = \frac{(1,900 \text{ sq ft}) \times (0.25 \text{ gpm/sq. ft.})}{32 \text{ nozzles}}$$

$$Q_{nozzle} = 14.8 \text{ gpm per nozzle}$$

$$P_{nozzle} = 30 \text{ psi (given)}$$

$$Q_{nozzle} = (K) \times \sqrt{P}$$

$$14.8 \text{ gpm} = (K) \times \sqrt{30 \text{ psi}}$$

$$K = 2.7$$

When selecting a K-factor from a list of listed or approved nozzles, 2.7 is the minimum, or smallest K-factor to be used for any nozzle protecting the surface area of the tank in example 9.6. The nozzle must be capable of providing a minimum of 14.8 gpm at 30 psi.

10. Water spray protection of a vertical cylindrical storage tank consists of several horizontal loops, spaced 7'-0" apart. The advantage of rundown can result in a dimension between loops not to exceed 12'-0" at the lower levels of a tall vertical tank, as shown on Figure 9-7a. A photograph of the water spray protection of a vertical vessel is shown in Figure 9-7b.

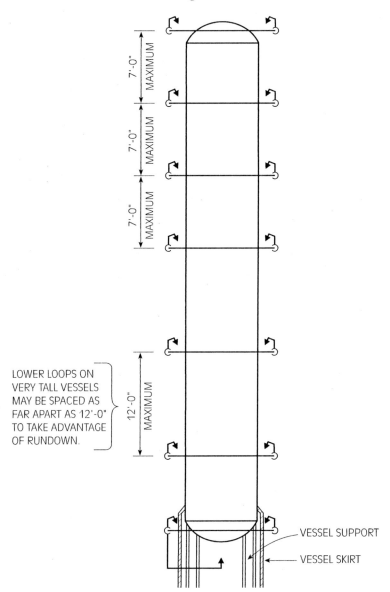

Figure 9-7a *Water spray protection of vertical vessels.*

Figure 9-7b *Water spray protection of vertical vessels. (Courtesy Grinnell Fire Protection Systems Co.)*

11. Protection of the vessel skirt and protection of the bottom of the vessel within the vessel skirt is required. A vessel skirt provides some protection for the vertical vessel supports and for the bottom of a vertical vessel. NFPA 15 requires water spray on uninsulated vessel skirts with a density of 0.10 gpm per square foot. It is recommended that protection be provided for insulated skirts as well.

12. A horizontal cylindrical storage tank requires two horizontal loops, one covering the top and one covering the bottom, running parallel to the length of the tank.

13. Spherical tanks are to have spherical water spray piping loops spaced at a maximum distance of 7 feet oriented in the horizontal plane, as shown on Figure 9-8. Pipe loop diameter should not be smaller than two inches for best loop rigidity. Loop spacing may be increased for the bottom loops for spherical tanks where rundown is considerable.

14. Draw the riser sectional view showing the deluge valve, control valve, strainer, and water supply feedmain or underground entry.

15. Reverify nozzle coverage and density requirements for all nozzles on the system before hydraulic calculations are performed.

16. Perform hydraulic calculations for the system.

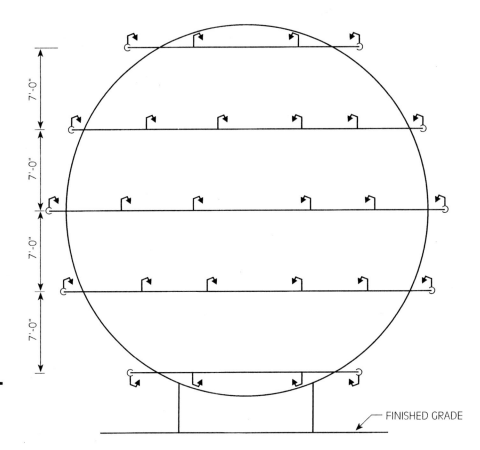

Figure 9-8 *Water spray protection of a spherical vessel.*

Example 9.7 Perform a hydraulic calculation of the water spray system for the six horizontal flammable liquid storage tanks shown on Figure 9-9.

Solution A sample hydraulic calculation, based on the six identical vessels shown on Figure 9-9, appears on Figure 9-10. Note Figures 9-5 and 9-6 are sectional views of the vessels shown on Figure 9-9, and that the nozzles, discharging 15 gpm at 30 psi, were based on the calculation performed in example 9.6. Figure 9-6 shows 16 nozzles on one side of the tank, with an identical number of nozzles on the other side of the tank, for a total of 32 nozzles per tank. The total number of nozzles for the six tanks is 192 nozzles. For these vessels, the designer opted not to loop the pipe around the tanks.

Completed calculations show 3,001.93 gpm at 55.10 psi for protection of the six vessels. Proper design of a water spray system for flammable liquid storage vessels provides an impressive distribution of water over the surface area of the vessels, as illustrated in Figure 9-11.

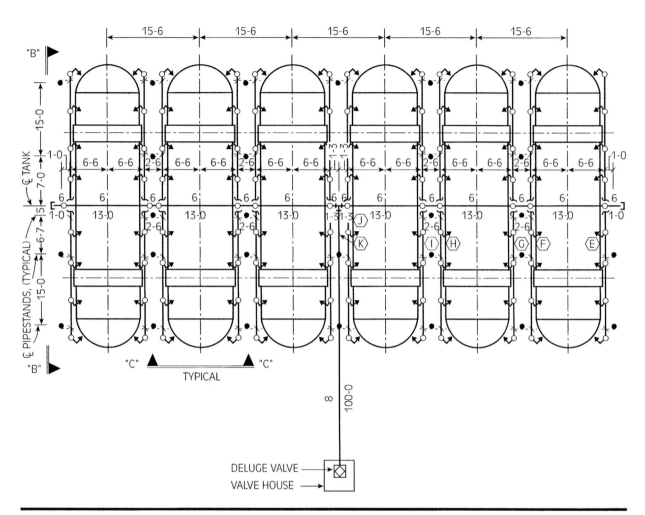

Figure 9-9 *Plan view of horizontal vessels. Section B-B is Figure 9-6. Section C-C is Figure 9-5.*

HYDRAULIC CALCULATION SHEET

CONTRACT NO. _____ System No. _____ Sheet __1__

CONTRACT NAME: CALCULATION OF SYSTEM SHOWN ON FIGURE 9-9

NOTES DATE

Calculated by _____

SPRINKLERS NOZZLES			(q) Added Gpm	(Q) Total Gpm	Pipe Size In.	Pipe and Equivalent Length Fittings - Devices Feet	FRICTION LOSS		(Elevation) Static Plus or Minus Psi	Required Pressure Psi	CALCULATN. REFERENCE	
Plan Ref. Pt.	Total Heads	Type					Psi/Ft. C = 120	Total Psi			Sheet No.	Point No.
1	1	NOZZLE					PRESSURE AT HEAD →			30.0		
				15.0	1	BL .75, E2,T5, TOTAL 7.75	.0764	.59	.22	30.81		1B
1B	1	—	—	15.0	1	BL 6.5	.0764	.50		31.31		2
2	2	LINE K = 2.70	15.11	30.11	1¼	BL 6.5	.0770	.50		31.81		3
3	3	LINE K = 2.70	15.23	45.34	1½	BL 6.5	.0754	.49		32.30		4
4	4	LINE K = 2.70	15.34	60.68	2	BL 3.9, T5, TOTAL 8.5	.0377	.32		32.62		A
A	8	—	60.68	121.36	2	BL=1', T=10', TOTAL 11'	.1357	1.49		34.11		B
B	16	SEE BELOW	127.71	249.07	2½	BL 1.5, T12, TOTAL 13.5	.2154	2.91	-.62	36.40		E
E	16	—	—	249.07	6	CM 13	.0027	.04		36.44		F
F	32	LINE K = 41.28	249.17	498.24	6	CM 2.5	.0098	.02		36.46		G
G	48	LINE K = 41.28	249.27	747.51	6	CM 3.5	.0206	.07		36.53		—
			—	747.51	6	CM 9.5	.0206	.02		36.73		H
H	64	LINE K = 41.28	250.16	997.68	6	CM 2.5	.0352	.09		36.82		I
I	80	LINE K = 41.28	250.48	1248.16	6	CM 8	.0531	.42		37.24		
			—	1248.16	6	CM 5	.0531	.27		37.51		J
J	96	LINE K = 41.28	252.81	1500.96	6	CM 1.25, T30, TOTAL 31.25	.0745	2.33		39.84		K
K	192	—	1500.96	3001.93	8	FM 100 RISER 10, 154 ELL 18 OS&Y 4 DELUGE VA = 22	.071	10.93	4.33	55.10	BASE OF RISER	

SYSTEM DEMAND AT BASE OF DELUGE VALVE
3001.93 GPM @ 55.10 PSI

										34.11		
D	8	HEADS 9-16	—	121.36	2	BL 11	.1357	1.49	-4.8	30.80	—	B
			$Q_N = 121.36 \sqrt{34.11/30.90} = 127.71$							34.11		B ←

Figure 9-10 *Calculation of system shown on Figure 9-9.*

Figure 9-11 *Water spray protection of vessels. Top: Water spray protection of a horizontal cylindrical flammable liquids storage tank. Bottom: Water spray protection of vertical vessels. (Courtesy Grinnell Fire Protection Systems Co.)*

Summary

Flammable and combustible liquids storage vessels require exposure protection to provide cooling to the entire surface area of the tank, and for the tank supports, relief valves, and flammable or combustible liquid piping entry. Vessels fail by deformation of the metal, rupture of the tank shell, failure of the tank supports, or overpressurization caused by boiling of the liquid. Vertically oriented vessels experience rundown from water spray applied to the upper surfaces. Consideration of rundown may allow for loop spacing up to 12 feet at the lower levels of a vertical vessel. A fire protection designer must consider water wastage in the fire protection design for a vessel. Water not striking the surface is providing less density to the surface of the vessel than is leaving the nozzle, and this water wastage must be considered in the hydraulic calculation.

Review Questions

1. Discuss the performance objectives of a water spray system for a flammable or combustible liquid storage tank or vessel.

2. Calculate the surface area of a vessel with a cylindrical length of 60 feet and a diameter of 10 feet, and with belled ends of 5-foot radius.

3. Using the method shown in example 9.4, estimate the minimum water supply required for the tank in review question 2, assuming 2,500 square feet of nonabsorbing ground surface below the tank, and four tank saddles requiring 30 gpm of water spray each.

4. Determine Q_{nozzle} if a vessel has a total surface area of 6,200 square feet and the vessel is protected by 200 nozzles. Determine the minimum nozzle K-factor if P_{nozzle} is 30 psi.

5. An authority having jurisdiction has determined that the water spray system protecting the six vessels shown on Figure 9-9 must be designed to deliver 0.5 gpm per square foot of tank surface area. Recalculate the water spray system to provide this density.

6. Calculate the system described in example 9.7 using the computer program included in this book.

Activities

1. Arrange a visit to a power plant or other facility that has water spray systems protecting flammable or combustible liquid storage tanks or vessels. Draw the systems and vessels shown, discuss vessel protection with the plant fire protection engineer, and take notes on your findings.

2. Visit a sprinkler company or fire protection design firm, and review some shop drawings of water spray systems. Were the systems designed in accordance with NFPA 15, or were different criteria used? If differing criteria were used, explain the reasons for the difference.

3. Interview a fire chief, a fire service officer, or a fire brigade commander. What actions are taken by the fire service when fighting flammable liquid storage tank fires with fixed water spray protection installed, and what actions are taken without fixed water spray protection installed?

4. Witness a discharge test of a water spray system protecting a flammable liquid storage tank. Note the effect of the wind on water sprays that protect the vessel.

Ultra High Speed Water Spray Systems

Objectives

Upon completion of this chapter, you should be able to:

- Describe the differences between an ultra high speed water spray system and an explosion suppression system.
- Discuss detection criteria for ultra high speed water spray systems, specifically with regard to ultraviolet and infrared detectors.
- Outline the differences in performance objectives between area protection and point protection.
- Describe the differences between a squib-actuated ultra high speed water spray system and a solenoid-operated ultra high speed water spray system.
- Determine the timing requirements for ultra high speed water spray systems.
- List the processes commonly involved with munitions manufacturing, and determine a fire protection scenario for each process.
- Know the components of squib-actuated ultra high speed water spray systems and solenoid-operated ultra high speed water spray systems, and be familiar with design guidelines for each.

Chapter 10 Ultra High Speed Water Spray Systems

ultra high speed water spray system
a system designed to discharge water within 100 milliseconds of detection for the protection of unconfined areas where overpressurization is not the primary concern

millisecond
a thousandth of a second

munitions
hazardous commodities used in ammunition

propellants
hazardous commodities used in rockets

pyrotechnics
hazardous commodities used in fireworks or flares

explosion suppression systems
systems designed in accordance with NFPA 69 for enclosed vessels

Ultra high speed deluge systems have had an impressive history of effective protection of explosive hazards. It is very rewarding to design such systems for facilities that have experienced numerous lost-time injuries, because the value of a well-designed system is evident from the moment it is placed in service. Historically, lost-time injuries are eliminated or dramatically reduced, and production dramatically increases as personnel gain an appreciation for the value of the system.

Since the ultra high speed water spray system is essentially a highly sophisticated water spray system, a review of the fundamentals of water spray system design in Chapters 8 and 9 may be of considerable assistance for users of this chapter.

This chapter concentrates primarily on ultra high speed water spray system design. For more detailed information on the design of a detection system for an ultra high speed water spray system and for more detail on system design recommendations, see Chapter 06-15 in the 18th edition of the *NFPA Fire Protection Handbook* and *Design of Special Hazard and Fire Alarm Systems* (Delmar, 1997).

ULTRA HIGH SPEED WATER SPRAY SYSTEMS

Definition

An **ultra high speed water spray system** is generally defined as a fire suppression system that can initiate discharge of water to a volatile commodity within 100 **milliseconds** of detection, measured from presentation of an energy source to the detector to the commencement of the flow of water from the nozzle being tested. A millisecond is a thousandth of a second.

A significantly heightened level of system response is generally required where materials of extremely high flame spread and heat release, such as **munitions**, **propellants**, and **pyrotechnic** materials, are involved. Propellants are used in rockets, munitions are used in ammunition, and pyrotechnics are fireworks or flares. Ultra high speed water spray systems were included in Chapter 9 of NFPA 15 for the first time in the 1996 edition, so fire protection professionals may consider such systems for applications where rapidly growing fires are anticipated.

Ultra High Speed Water Spray Systems and Explosion Suppression Systems

The NFPA Technical Committee on Water Spray Fixed Systems, the committee responsible for NFPA 15, has determined that a dividing line must exist between ultra high speed water spray systems, covered in NFPA 15, *Standard for Water Spray Fixed Systems for Fire Protection*, and **explosion suppression systems**, covered in **NFPA 69**, *Standard on Explosion Suppression Systems*:

- Explosion suppression systems are to be designed in accordance with NFPA 69 for enclosed vessels or other enclosed devices or conveyances where overpressurization or rupture is the primary concern.

Standard on Explosion Suppression Systems

- Ultra high speed water spray systems are predominately designed in accordance with NFPA 15 for rooms or open, unconfined areas where rupture resulting from overpressurization is not the primary concern.

It should be noted that a room may be subject to overpressurization, but by the time the room pressure approaches the failure limit, the time available for protection of personnel and equipment within a room has passed. It is therefore necessary to design an ultra high speed water spray system to stop a rapidly spreading fire in advance of any room overpressurization concerns that may exist. The primary performance objective of an ultra high speed water spray system would therefore be personnel and equipment protection, not room rupture.

Normally unoccupied, tightly sealed rooms and enclosed or partially enclosed process equipment are examples of applications that may be amenable for the design of either ultra high speed water spray systems or explosion suppression systems, depending on the configuration of the commodity and its associated manufacturing process.

A Venn diagram, or pictorial logic diagram, that details the respective uses for ultra high speed water spray systems and explosion suppression systems and depicts the areas of commonality where either system may be considered is shown in Figure 10-1.

Why Water?

Water has long been known to have many uses in fire protection and a broad array of uses in other fields. It is the agent of choice in the protection of explosive haz-

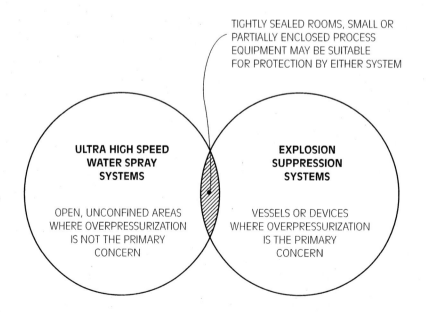

Figure 10-1
Applications for ultra high speed water spray systems and explosion suppression.

Chapter 10 Ultra High Speed Water Spray Systems

ards for its ability to wet and cool a hazardous combustible, insulate the exposed surfaces, dilute a hazardous material, and cool the expanding flame front.

Deflagrations and Detonations

Both ultra high speed water spray systems and explosion suppression systems can be satisfactorily designed for the suppression of a deflagration, a fire where the expanding flame front travels at a velocity less than the speed of sound. Neither system has been found to be effective for applications where a **detonation**, a fire where the expanding flame front travels at a velocity greater than the speed of sound, could occur. Test data or field testing with respect to the velocity of the flame front may be necessary before selecting a suppression system for an explosive hazard.

detonation
a fire in which the expanding flame front travels at a velocity greater than the speed of sound

Uses for Ultra High Speed Water Spray Systems

Examples of facilities in which ultra high speed water spray systems are advantageous include rocket fuel manufacturing or processing, solid propellant manufacturing or handling, ammunition manufacturing, pyrotechnics manufacturing, and the manufacture or handling of other volatile solids, chemicals, dusts, and gases.

Note the use of the word manufacturing in the preceding descriptions. An ultra high speed water spray system is most likely to be effective where limited and controlled amounts of hazardous combustibles are present in open areas, where an opportunity for wetting the combustibles is present. An ultra high speed suppression system is not likely to be appropriate for the protection of stored munitions, sealed within shells where manufacture is complete.

Munitions manufacturing is a prominent application for ultra high speed water spray systems, and is among the most hazardous fire protection challenges that a fire protection professional can encounter.

Reaction Time

reaction time
the time from activation of a detector to water flow at nozzles

The 100 millisecond maximum **reaction time** used by NFPA 15 for ultra high speed water spray systems considers the time from activation of the detector to water flow at the nozzles.

Since different detectors have different sensitivities, sensitive detectors must be chosen, detectors must be placed as close as possible to the target, and no timers or other system delay features are permitted.

Of great concern is the travel time for the water to hit the target. To minimize water travel time, nozzles are placed as close as physically possible to the target.

Human Senses

The process of detection, transfer of the signal to the control panel, interpretation of the signal, transmission of signal to the valve, and delivery of water to the hazard happens before human senses are aware of a problem. System actuation histories correspond quite closely to the frequency of actual fire and explosion incidents reported before installation. It is recommended that a diary of fire and explosion incidents be kept before and after system installation for analysis.

Detectors for Ultra High Speed Water Spray Systems

To achieve the required reaction time, optical flame detection is a necessity for ultra high speed water spray systems. Examples of optical detection devices include ultraviolet detectors and infrared detectors, placed in close proximity to the hazard, that instantaneously detect flashes of light associated with the wavelengths normally found in combustion.

Ultraviolet detectors, being insensitive to sunlight, are well suited for surveillance of an entire process area or a process requiring a wide angle of vision, whereas infrared detectors are better suited for close surveillance of a specific contact point within a processing machine or supervision of the interior of an enclosed vessel. Combination ultraviolet and infrared detectors are not suitable for ultra high speed applications because of the delay involved in verifying both wavelengths before a detection signal is sent.

Design Approaches

Two general design approaches, **point protection** and **area protection**, are used to design ultra high speed water spray systems.

Point protection, illustrated by Figure 10-2, is the application of concentrated high speed water spray onto a likely point of ignition, and is accomplished by providing two or more nozzles at positions as close as physically possible to the anticipated point of ignition.

Area protection, illustrated by Figure 10-3, is the application of high speed water spray throughout the surface area of a room or over the surface area of an object. In both cases, the nozzles must be as close to the area protected as possible, to minimize the time for the water to travel from the nozzle to the intended target. A system can be designed to meet the performance objectives of both point protection and area protection, as shown in Figure 10-4.

Types of Ultra High Speed Water Spray Systems

The piping methods used to get water to the hazard are as varied as the hazards they protect. The nozzles must be in close proximity to the hazard, with water filling all piping up to each nozzle, enhancing rapid delivery of water to the target. The remainder of this chapter concentrates on the two types of ultra high speed water spray systems.

---

point protection
the application of concentrated high speed water spray onto a likely point of ignition

area protection
the application of high speed water spray over the surface area of a room or over the surface area of an object

Chapter 10 Ultra High Speed Water Spray Systems

Figure 10-2 *Point protection (solenoid-operated system shown). (Courtesy Superior Automatic Sprinkler Corp.)*

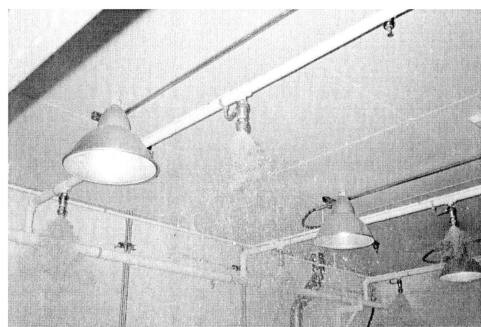

Figure 10-3 *Ultra high speed area application (solenoid-operated system shown). Note that standard sprinklers are installed at the ceiling. They are not connected to the ultra high speed water spray system. (Courtesy Superior Automatic Sprinkler Corp.)*

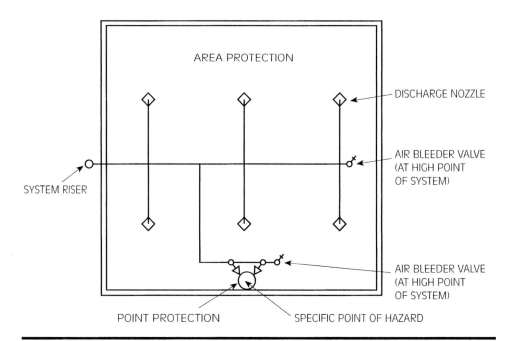

Figure 10-4 *Ultra high speed combination system. (Reprinted with permission from NFPA 15,* Water Spray Fixed Systems for Fire Protection. *Copyright © 1996, National Fire Protection Association, Quincy, MA 02269. This reprinted material is not the complete and official position of the National Fire Protection Association on the referenced subject, which is represented only by the standard in its entirety.)*

THE SQUIB-ACTUATED ULTRA HIGH SPEED WATER SPRAY SYSTEM

Definition

squib-actuated system
a preprimed ultra high speed water spray system with a squib-operated deluge valve; a system of piping preprimed with water, nozzles with rupture discs, and a flame detection system

The **squib-actuated system**, shown in Figures 10-5 and 10-6, is a preprimed ultra high speed water spray system with a squib-actuated deluge valve, a system of preprimed piping, nozzles with rupture discs, and a flame detection system. A squib is a small explosive charge that is attached to the deluge valve to enhance rapid opening of the valve.

Preprimed Piping System

The preprimed piping system, illustrated in Figure 10-6, is controlled by a priming bypass valve that provides water pressure to the nozzles. Allowing the water to be preprimed in the piping has a strong advantage over standard deluge systems. The time it would have taken for the water to travel from the deluge valve to the

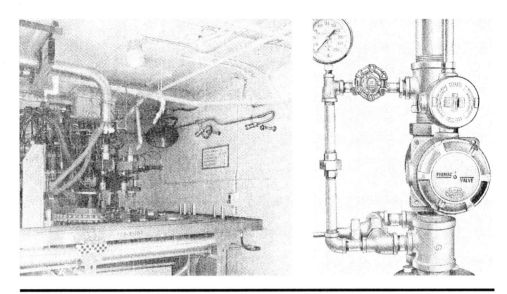

Figure 10-5 *Squib-actuated ultra high speed water spray system. Left: Propellant compacting operation protected by a squib-actuated ultra high speed water spray system. Detectors are mounted to the wall to the right of the photo. Right: Squib-actuated valve. (Courtesy Grinnell Fire Protection Systems Co.)*

nozzle has been eliminated. This difference can be significant, especially for systems requiring the deluge valve to be remote from the hazard. The preprimed aspect of this system requires that the hazard area and valve riser be heated.

Rupture discs capable of maintaining a minimum system static pressure of 50 psi are required. The system must be hydraulically designed to ensure a minimum residual pressure of 50 psi at each nozzle.

The Squib-Actuated Valve

The squib-actuated valve is an electrically actuated valve with two explosive primers, wired in parallel. The valve is intended to be installed vertically, but also functions, with a negative effect upon response time, in the horizontal position. The primers actuate when they receive an electric signal from the detection system. The rupture discs are broken once their inlet pressure exceeds their respective release pressure.

Increased system size significantly increases reaction time. A 500 gallon water-holding capacity is required by NFPA 15 as the maximum system size, but best results would be obtained for squib-actuated systems by subdividing the systems such that the number of nozzles served by one squib-actuated valve is between two and six nozzles.

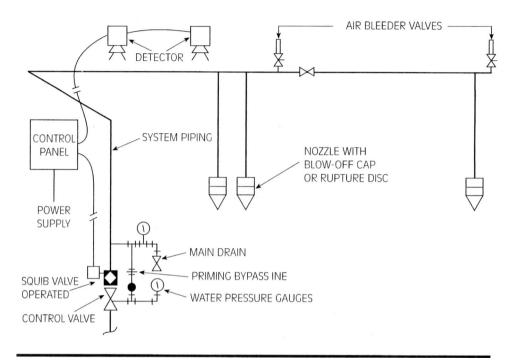

Figure 10-6 *Ultra high speed system using squib-actuated valve. (Reprinted with permission from NFPA 15, Water Spray Fixed Systems for Fire Protection. Copyright © 1996, National Fire Protection Association, Quincy, MA 02269. This reprinted material is not the complete and official position of the National Fire Protection Association on the referenced subject, which is represented only by the standard in its entirety.)*

System Response Time

The piping in an ultra high speed water spray system must be sloped, with air bleeder valves placed at all piping high points, to facilitate the removal of trapped air bubbles which dramatically increase response time if not removed.

Another major factor influencing response time is the water supply pressure. When all other variables are constant, the response time decreases when water supply pressure increases. For example, if one could double the supply pressure from 50 to 100 psi, the actuation time should decrease by about 20 to 25%.

Other factors influencing the effectiveness of a squib-actuated system are the number of nozzles on the system, orifice diameter of the nozzles, length of piping from the deluge valve to the nozzles, number of fittings in the piping system, rupture disc rupture pressure, and cross-sectional area of the supply pipe.

For squib-actuated ultra high speed water spray systems, a minimum static pressure of 50 psi should be maintained.

Squib-Actuated Ultra High Speed Water Spray System Design

1. Squib-actuated valves should be installed in the vertical position.
2. The pipe routing from squib-actuated valve to the nozzles must be as direct as possible, with as few fittings and bends as possible.
3. Slope all piping up to air bleeder valves at piping high points to facilitate the elimination of air from the system.
4. For point protection, nozzles should be counteropposed for efficient delivery of water to a likely source of ignition, as shown on Figure 10-2 on page 251.
5. A minimum operating pressure at the most remote nozzle of 50 psi is required by NFPA 15.
6. NFPA 15 requires that each nozzle discharge no less than 25 gpm per nozzle for point protection, nor less than 0.50 gpm per square foot for area protection.

THE SOLENOID-OPERATED ULTRA HIGH SPEED WATER SPRAY SYSTEM

System Arrangement

solenoid-operated ultra high speed water spray system
a system consisting of a preprimed discharge system with solenoid-actuated nozzles, a wet pilot system supervised by solenoid valves, a flame detection system, and a central control panel

A **solenoid-operated ultra high speed water spray system**, shown in Figure 10-7, consists of a preprimed discharge system with solenoid-operated nozzles, a wet pilot system supervised by solenoid valves, a flame detection system, and a control panel. Figures 10-8 and 10-9 (page 257 and 258 respectively) show a riser and control arrangement for a solenoid-operated ultra high speed water spray system. Although munitions provide the lion's share of current pilot-actuated system applications, this system has been used, with a preset timer to limit discharge, to protect commodities such as mail conveyors and chutes, and ovens.

The Discharge System

A solenoid-operated nozzle is connected to two systems of piping. The discharge system, normally filled with water to the nozzle, delivers the water for distribution to the hazard. The wet pilot system applies pressure so that water in the discharge system is prevented from flowing through the nozzle. Release of pilot pressure allows the deluge system to overcome the pressure differential, initiating instantaneous discharge to all nozzles monitored by the pilot system.

Sloping of the discharge system to air vents facilitates the purging of air from the discharge system, reducing system discharge time. NFPA 15 requires that all piping be sloped a minimum of 1 inch per 10 feet of pipe to air vents placed at all high points for removal of air from the pilot system.

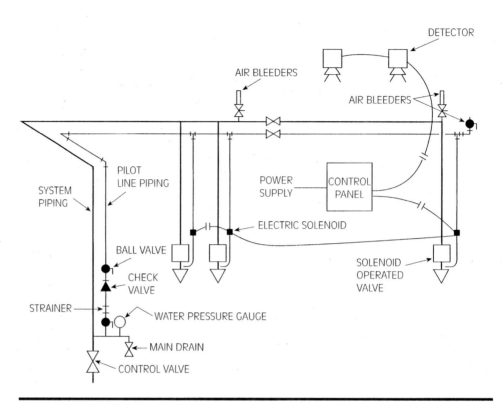

Figure 10-7 *Ultra high speed system using solenoid-operated valves. (Reprinted with permission from NFPA 15,* Water Spray Fixed Systems for Fire Protection. *Copyright © 1996, National Fire Protection Association, Quincy, MA 02269. This reprinted material is not the complete and official position of the National Fire Protection Association on the referenced subject, which is represented only by the standard in its entirety.*

The Pilot System

The pilot system, shown in Figure 10-10 on page 259, is monitored by normally closed two-way solenoid valves spaced throughout the system. Removing air in the pilot line is of paramount concern because of the smaller cross-sectional area of the pilot line, as compared to the calculated pipe sizes for the discharge piping system. To protect the small solenoid orifices, a strainer must be installed on the pilot line riser to filter sediment before it reaches the solenoids.

The two systems of piping are simultaneously pressurized by an excess pressure pump maintaining static system pressures between 170 to 180 psi, with a relief valve and pressure limiting switch ensuring this range. An orifice union in the pilot line maintains the pressure ratio necessary to keep the pilot-actuated nozzles closed.

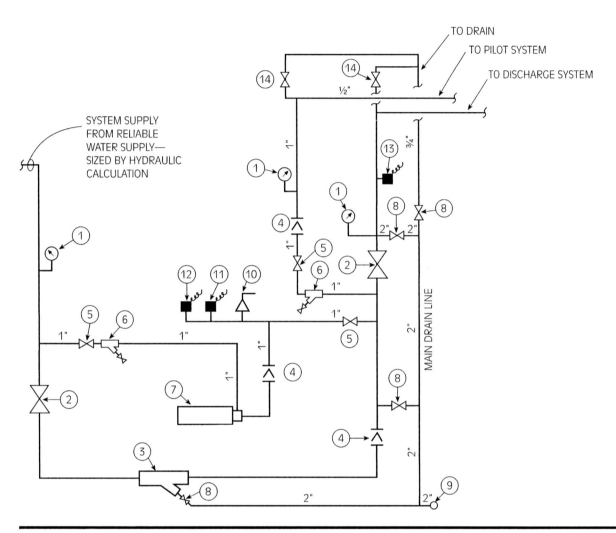

Figure 10-8 *Sectional view showing valving arrangement for an ultra high speed solenoid-operated water spray system. Symbol key: 1, pressure gauge; 2, control valve with tamper switch; 3, main line strainer with flushing connection; 4, check valve; 5, ball valve; 6, strainer with flushing connection; 7, excess pressure pump; 8, drain valve; 9, drain to outside; 10, relief valve; 11, pressure switch, adjustable settings 170–180 psi; 12, low pressure switch; 13, water flow switch; 14, manual or automatic air bleed valve at high points of system. (Reprinted with permission from* Fire Protection Handbook, *18th edition. Copyright © 1997, National Fire Protection Association, Quincy, MA 02269.)*

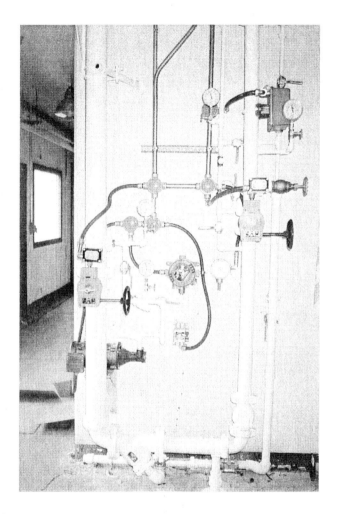

Figure 10-9
Solenoid-operated control apparatus. See Figure 10-8 for identification of parts. (Courtesy Superior Automatic Sprinkler Corp.)

A manual means of actuating the system can be provided by installing a pull ring to a ball valve at the end of a pilot line, as shown in Figure 10-11.

Sequence of Operation for a Pilot-Actuated System

Flame detectors are installed as close to the points of anticipated hazard as practicable and angled and spaced so that all areas where the hazard might be found are observed by the detectors. Flame detectors constantly view the hazard areas, searching for changes in light intensity, and the status is reported to the controller for analysis. Light intensities are screened by the ability of the detector to recognize common signals, such as house lighting and sunlight. Physical screening may be necessary to prevent detector signals from such intermittent light sources as welding torches, lightning, or sun reflecting from the windshield of a car.

Chapter 10 Ultra High Speed Water Spray Systems

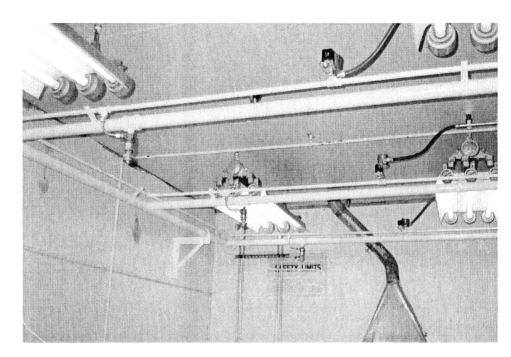

Figure 10-10 *Pilot and discharge system for a solenoid-operated water spray system. The pilot line is the small pipe running above the discharge line. Connections are made from the pilot line to each nozzle. Solenoids are installed on the pilot line, and are wired to the control panel. (Courtesy Superior Automatic Sprinkler Corp.)*

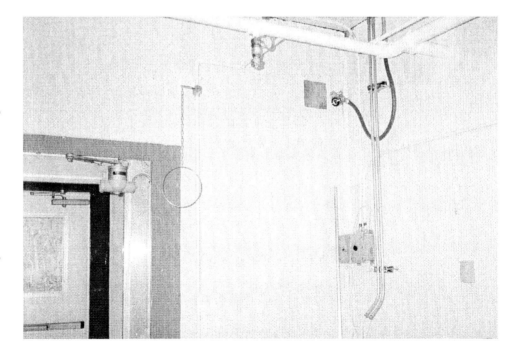

Figure 10-11 *Manual release for solenoid-operated system. An o-ring is connected to a ball valve. Pulling on the o-ring releases water pressure on the pilot line. Note the ultraviolet flame detector in the corner of the room. (Courtesy Superior Automatic Sprinkler Corp.)*

Upon detection, the signal is interpreted and relayed to the appropriate high speed module, and is sent simultaneously to all the high speed solenoids on the system.

The normally closed solenoids open, allowing water to flow through the solenoids, purging the pilot system of water and relieving the pilot pressure. The pressure drop allows the poppet valves in the pilot-actuated nozzles to move, allowing water to flow through the nozzle and onto the hazard.

Factors Influencing the Speed of a Pilot-Actuated System

Trapped air in the pilot line can be a significant hindrance to the hydraulic evacuation of the pilot line. To prevent this, the pilot line is sloped to air bleeder valves at all high points of the pilot system. It is a common practice to bleed air from the system after it is reset and repeat the process at the start of the next working day to ensure that all air is removed from the system.

Excess system volumetric capacity of a pilot-actuated system is a considerable contributor to increasing the response time of the system. As it was with squib-actuated systems, it is important that the system volume not exceed 500 gallons, that the systems be subdivided such that the maximum number of nozzles on a system is about 10, and that the number of fittings and length of the feedmain be minimized.

Since the distance between the solenoid valve and the connection to the pilot-operated nozzle can be a major factor in the speed of the evacuation of the pilot line, solenoids should be adjacent to each nozzle, releasing the pilot pressure to the nozzle instantaneously.

Pilot pressure is also a major factor in reaction time. An excess pressure pump should be provided for each system or group of systems, to maintain 170 to 180 pounds of static pressure on the system piping and pilot line. This pressure at each nozzle provides the impetus for rapid evacuation of pilot pressure and rapid delivery of water to the hazard.

As with squib-actuated systems, NFPA 15 requires that no less than 25 gpm per nozzle be designed for point protection and that no less than 0.50 gpm per square foot of exposed area be designed for area protection. Point protection should be accomplished with no fewer than two counteropposed nozzles per point of likely ignition, with nozzles aimed from differing directions at a common point. Area protection should be accomplished with a nozzle spacing of between 30 and 60 square feet, depending on the configuration of the hazardous commodity or process.

Response Time for Solenoid-Operated Systems

NFPA 15 mandates a response time not to exceed 100 milliseconds, measured from presentation of an energy source to the detector to water discharge from the most distant solenoid-operated nozzle. Whenever possible, travel time for the water to hit its intended target should be included in the total response time.

Summary

A differentiation was emphasized between ultra high speed water spray systems and explosion suppression systems. Even though both systems can suppress explosive hazards, ultra high speed water spray systems are used predominantly for open machine protection or room area protection, whereas ultra high speed explosion suppression systems primarily protect enclosed vessels.

Flash fire and explosion incidents involving damage and injury, recorded before the installation of a high speed system, correspond to post-installation tripping reports quite closely, clearly indicating successful actuations. Although field experience for these systems is impressive, the risks involved require careful consideration to ensure continued success.

Review Questions

1. Describe the differences between an ultra high speed water spray system and an explosion suppression system.
2. List and compare the features of a squib-actuated system and a pilot-actuated system.
3. Discuss the types of hazards protected by ultra high speed water spray systems, and differentiate these uses from other systems discussed previously in this book.
4. List the minimum requirements for ultra high speed water spray systems required by NFPA 15. Describe a situation where an ultra high speed water spray system would not be effective.
5. Discuss the difference between a deflagration and a detonation and give examples of each.
6. Why is water recommended for explosion protection?
7. Describe the sequence of operation for a squib-actuated system and a pilot-actuated system.
8. What are the differences between area protection, point protection, and a combination system?

Activities

1. Interview a fire protection engineer who has experience with the design of systems for explosive hazards. Discuss the unique hazards that have been protected by ultra high speed water spray systems.
2. Visit a facility where an ultra high speed water spray system has been installed. What has been the history of its use, including number of actuations and effectiveness?
3. Interview a fire protection professional to determine if this individual knows of standard sprinkler systems or water spray systems that have failed because their reaction time was too slow. Would an ultra high speed water spray system have been an improvement?
4. Contact a sprinkler company or fire protection engineering firm. What is the cost differential between a standard water spray system and an ultra high speed water spray system?

Chapter 11

Water Mist System Design

Objectives

Upon completion of this chapter, you should be able to:

- Discuss the applications for water mist systems.
- Compare the performance objectives of a water mist spray to the spray of a large drop sprinkler.
- List water mist system types and configurations.
- Discuss the reasons why zoning of water mist systems can increase system effectiveness for systems with limited water supplies.

Chapter 11 Water Mist System Design

NFPA 750

Standard for the Installation of Water Mist Fire Protection

water mist system
an automatic water-based fire protection system with nozzles capable of distributing water mist to a variety of hazards

With the introduction of the first NFPA standard on water mist systems in 1996, water mist technology reached a new and critical stage. **NFPA 750**, *Standard for the Installation of Water Mist Fire Protection Systems*, will create an atmosphere for the expansion of the use of water mist and the broadening of the number of applications for this unique system. A **water mist system** is an automatic water-based fire protection system with nozzles capable of distributing water mist to a variety of hazards. Water mist systems have the potential to solve problems that may be unsolvable or impractical with standard sprinkler systems.

This chapter covers the theory of water mist extinguishment, the limitations of water mist systems, and applications for which water mist systems have been tested, and in some cases, installed. Water mist systems are undergoing considerable research, but are clearly capable of being used with confidence to solve a wide array of fire protection problems when designed in accordance with NFPA 750 and the recommendations of the water mist equipment manufacturer.

WATER MIST PERFORMANCE OBJECTIVES

In selecting a water mist system, it is important to specify and design the system to meet one or more performance objectives specific to the hazard being protected. The NFPA is in the process of reorganizing its standards to emphasize performance objectives as a focal point for system selection and design. In the first edition of NFPA 750, five performance objectives are listed. Water mist has been found to be effective for:

- Fire extinguishment, where the fire is intended to be suppressed with no further burning combustibles.
- Fire suppression, where the rate of heat release is expected to be sharply reduced to the point that regrowth would not occur.
- Fire control, where the fire growth has been limited by prewetting combustibles in the vicinity of the fire.
- **Temperature control**, where the room temperatures are reduced during combustion to allow safe egress and reduced damage.
- Exposure protection, where combustibles adjacent to the fire are wetted to delay their ignition.

temperature control
the reduction of room temperature during combustion to allow safe egress and reduced damage

WATER MIST DROPLETS

Droplet Size

water mist
a fine water spray having droplets less than 1000 microns at a distance of 3.3 feet from the discharge nozzle

NFPA 750 defines a **water mist** as a fine water spray whose water droplets are less than 1000 microns at a distance of 3.3 feet from the discharge nozzle. Three classifications of water mist have been established with this document:

Class 1: 90% of the drops in the water spray are less than 200 microns.

Class 2: 90% of the drops in the water spray are less than 400 microns.

Class 3: A water mist meeting the definition above that is neither class 1 or class 2.

Early spray sprinkler research clearly indicated that water droplet size, as a function of the discharge pressure through an orifice of fixed diameter, was a key contributing factor in the ability of a water spray to evaporate and cool a flame for certain configurations of combustibles.

In an effort to learn the mechanics involved with this variable, Dundas (1974) reports that in 1974, Factory Mutual Research Corporation performed tests using a high-speed photographic technique to measure drop size and distribution to predict the drop size that would be delivered by discharge devices with various orifice diameters. It was confirmed that larger drops at given velocity are more likely to penetrate a plume, whereas smaller drops rapidly decrease in velocity as they approach a plume, and may be swept away by the plume. The total absorption of heat, however, greatly increases as drop size decreases, given a fixed water volume, because the available surface area for heat transfer increases, as illustrated in Figure 11-1.

 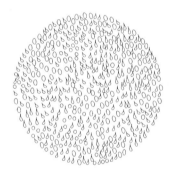

ONE GALLON OF WATER DIVIDED INTO VERY SMALL DROPLETS

- MORE DROPLETS
- LARGER WATER SURFACE AREA EXPOSED TO HEAT
- MORE DROPS WILL EVAPORATE AND TURN TO STEAM
- STEAM ABSORBS MORE HEAT FROM THE FLAME, REDUCING THE FLAME TEMPERATURE

ONE GALLON OF WATER DIVIDED INTO VERY LARGE DROPLETS

- FEWER DROPLETS
- SMALLER WATER SURFACE AREA EXPOSED TO HEAT
- MORE DROPS WILL REMAIN UNEVAPORATED AND COAT THE SURFACE OF THE COMBUSTILBES
- A COMBUSTIBLE COATED WITH WATER IS COOLED BY THE WATER

Figure 11-1 *Droplet size.*

Droplet Size Performance Objectives

It makes sense that if droplet sizes of differing sprinklers or discharge devices can be identified, a fire protection system designer armed with this information, could make important decisions relative to the performance objective of the system.

Two performance objectives are available to the designer:

1. Provide a discharge device that produces small water drops and create a scenario where most of the water turns to steam and absorbs heat from the flame.
2. Provide a discharge device that can produce large water drops that penetrate the fire plume and can coat and cool the flaming combustible and provide exposure protection for adjacent combustibles.

By controlling the pressure, deflector design, and orifice size, a designer can accomplish the performance objective chosen:

- Large drop sprinklers, previously discussed in Chapter 6, are used primarily for storage situations where coating of the combustible and the penetration of the fire plume are the primary performance objectives. Combustibles for which this tactic may be appropriate include rolled paper and rubber tires.
- Standard spray sprinklers, discussed in Chapter 5, provide much smaller droplets than large drop sprinklers, and provide considerable evaporative properties.
- Water mist may be selected where it is decided that the creation of heat-absorbing vapor is the primary performance objective, where water supplies are limited, or where water damage may be a concern. Combustibles that may be successfully protected by using this performance objective include ordinary combustibles, electronics or telecommunications equipment, and flammable liquids.

Droplet Size and the Fire Service

Tests of fire service fog nozzles have been performed to determine the amount of radiant energy transmitted to the firefighter using different fog orifices and angles. Reischl (1979) reports that for both 1-inch and 1½-inch fog nozzles discharging at 100 psi, the 90° fog angle absorbed more radiant energy than 60° and 30° angled streams, thus providing smaller droplets, more evaporation and cooling, and an increased level of safety for the firefighter holding the nozzle.

Droplet size decreases as the water spray angle is increased, and smaller drops absorb more heat energy and provide safer fire fighting conditions. Results of this study prompted a change in the manner in which many fire service personnel were trained to approach a fire with a fog nozzle.

Measuring Droplet Size

Early efforts to measure droplet size used shallow dishes coated in castor oil, distributed at various points in the spray. Scaled photographs were taken of the droplets as they contacted the plate and measurements were taken.

Other droplet sizing methods included optical, impact, thermal, shadowgraphic, and electrical, all very different from each other, but each measuring water mist effectiveness by the rate of heat transfer to the droplet and the rate of evaporation of the droplet.

The evaporation of a droplet involves a change in phase for the water, which occurs in the following steps:

1. Heating of the droplet,
2. Boiling of the droplet,
3. Evaporation of the droplet, turning the droplet to steam. The steam occupies a much larger volume than did the droplet, and absorbs considerably more heat than the droplet could have absorbed before evaporation.

Through testing, the heat absorbed by a droplet and the heat absorbed by the steam produced by the evaporation of that droplet can be predicted, given the initial droplet diameter. Testing results confirm that smaller droplets evaporate faster, and that droplets spaced far apart evaporate faster than droplets closely spaced. If efficient evaporation is the desired result, a sprinkler or nozzle at high pressure producing small drops over a wide angle is the distribution device of choice. Results of droplet measurement experiments and evaporation efficiency analysis were crucial to the development of an engineering approach to water mist applications.

Water Mist Pressure and Droplet Size

Pressure has been found to be a strong contributor to the size of a water mist droplet, given a nozzle of fixed diameter. In addition, the pressure at the nozzle is a significant contributor to the ability of a nozzle to project water mist droplets at extended distances from the nozzle. Generally, the lower the pressure at the nozzle, the closer the nozzle needs to be to the axis of the plume centerline. NFPA 750 has established three classifications for system pressurization:

- **Low-pressure water mist systems**, in which the pressures encountered by the system piping are 175 psi or less, the same pressure range used for most standard sprinkler systems.
- **Intermediate-pressure water mist systems**, in which the pressures encountered by the system piping are between 175 psi and 500 psi.
- **High-pressure water mist systems**, in which the pressures encountered by the system piping are 500 psi or greater.

---

low-pressure water mist systems
systems where the pressures encountered by the system piping are 175 psi or less

intermediate-pressure water mist systems
systems where the pressures encountered by the system piping are between 175 psi and 500 psi

high-pressure water mist systems
systems where the pressures encountered by the system piping are 500 psi or greater

WATER MIST SYSTEM DESIGN

local application
applying water mist to a specific object

total flooding
flooding an entire compartment with water mist

zoned application systems
systems where a volume is protected by several distinct zones, each with its own detection system

deluge water mist systems
systems with open nozzles that discharge water mist simultaneously from all nozzles on the system

wet pipe water mist systems
systems with nozzles having individual heat-responsive actuating devices and with piping filled with water in the ambient or normal condition

In a manner similar to gaseous agent suppression systems, water mist systems can be designed for **local application** of water mist onto a specific object, or designed for **total flooding** of an entire compartment with water mist, as shown in Figure 11-2. Local application systems may be considered in very large compartments where total flooding is not an achievable option and where the object of protection is immobile.

Systems in room volumes too large to be practically designed for a total compartment flooding application, such as in aircraft, may be considered as a candidate for a **zoned application system**, where a volume is protected by several distinct zones, each with its own detection system, as shown in Figure 11-3. A designer can determine the number of zones that should actuate upon receipt of a detection signal from any zone, based on the amount of water available and the expected duration of discharge. The zoned arrangement saves water and provides opportunities for the possible protection of aircraft in the future.

Water Mist System Types

Water mist systems differ greatly from standard sprinkler systems, but water mist system types are the same as we have already studied for sprinkler systems:

- **Deluge water mist systems**, with open nozzles that discharge water mist simultaneously from all nozzles on the system. Most current applications of water mist technology employ this type of system.
- **Wet pipe water mist systems**, possessing nozzles with an individual actuating device, where each nozzle actuates individually, and with piping filled with water.

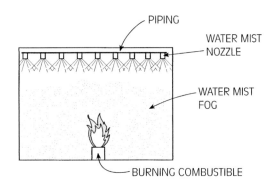
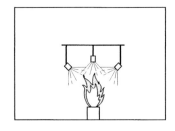

Figure 11-2 *Water mist system configuration.*

TOTAL COMPARTMENT APPLICATION
THE ENTIRE COMPARTMENT IS FLOODED WITH WATER MIST DROPLETS.

LOCAL APPLICATION SYSTEM
WATER MIST IS PROJECTED DIRECTLY ONTO AN EXPECTED FIRE.

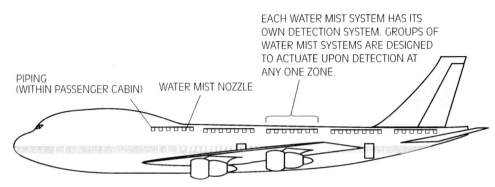

Figure 11-3 *Zoned application system.*

AIRCRAFT ARE BEING TESTED FOR A ZONED APPLICATION SCENARIO.

preaction water mist systems
systems having nozzles with individual heat-responsive actuating devices, closed heads, and piping filled with pressurized air

dry pipe water mist systems
systems with air-filled piping in the ambient condition and nozzles with individual heat-responsive actuating devices

- **Preaction water mist systems**, possessing nozzles with an individual actuating device, where each nozzle actuates individually, with piping filled with air. A preaction water mist system is designed to prevent water flow until a reliable detection signal is received.
- **Dry pipe water mist systems**, systems with air-filled piping, possessing nozzles with an individual actuating device, where each nozzle actuates individually. Water flows into the piping when a nozzle actuates.

Water Mist System Calculation Criteria

NFPA 750 recognizes that at the present time, no uniform design method or criteria exist for the design and calculation of water mist systems. Each manufacturer has a design method associated with the system, and the listing of the components is obtained for the system in accordance with the performance objectives for that system. When a water mist system is designed or specified, the water mist criteria for that system must be met in accordance with its listing.

Advice and reference material for the performance of flow calculations is included in NFPA 750, which may be used once the system is selected and the performance criteria of the manufacturer are determined.

WATER MIST APPLICATIONS

The potential uses for water mist technology are very diverse, and this diversity creates a solid basis for the specification and design of water mist systems for numerous unique problems, some of which are unsolvable by using standard sprinkler technology. For example, we all know that designing a standard sprinkler or water spray system that requires a huge water storage tank is impossible on a passenger aircraft, on the space shuttle, or for a circus that travels by rail and requires an automatic fixed fire protection system in order to remain insured. A water mist system is a possible solution that can be considered for many unique situations where a standard sprinkler system would be prohibitive.

Some applications, most notably shipboard applications, are well past the experimental stage and are developing an impressive record of reliable performance. Other applications, such as aircraft applications, remain experimental but provide an exciting potential for an increased level of safety.

Submarine Applications

Lugar (1979) reports that the U.S. Navy conducted numerous tests to determine the feasibility of water mist fire protection of main auxiliary machine spaces and hull insulation fires in submarines. Problems associated with fires on submarines are exacerbated by the inability of the vessel to exhaust smoke. Water mist was considered for fighting submarine fires when submerged because of severe weight restrictions, problems with evacuation of the crew, and a limited onboard water supply, as illustrated by Figure 11-4.

The tests were established under the following parameters:

- Hexane or wood cribs were used as the fuel.
- Nozzles with orifices sized from .020 inches to .066 inches in diameter were used.
- Nozzles discharged at a water pressure of 400 psi.
- A 1/2-inch piping network was used.
- A hydropneumatic tank containing a rechargeable water supply of 1-minute duration was provided.
- An application rate of 0.05 gpm per square foot was applied.
- Nozzle and pipe loop placement was tested at both deck level and at ceiling level.

Figure 11-4
Shipboard applications. Both vessels carry flammable fuels.

SUBMARINES
- VERY LIMITED ONBOARD WATER SUPPLY
- SMOKE EXHAUST IS A SERIOUS PROBLEM

SURFACE SHIPS
- ON LARGER SHIPS, STORED WATER SUPPLY IS LIMITED, BUT OF SUFFICIENT SIZE TO SUPPLY A WELL-DESIGNED SPRINKLER SYSTEM
- SMOKE EXHAUSTS TO THE OPEN AIR

- A heat sensor set at 350°F was employed as the system detector and actuator.

Test results showed that for heptane fires:

- All nozzle orifice sizes were capable of extinguishing the fire in 10 seconds at 400 psi.
- All orifice sizes could lower the room temperature from 350°F to 250°F within one minute.
- The .032-inch nozzle orifice was more efficient than the other nozzles tested. With the .032-inch nozzle, only 1.3 gpm was required for extinguishment, compared to 2.0 gpm for the .020-inch nozzle, 3.5 gpm for the .048-inch nozzle, and 5.0 gpm for the .066-inch nozzle.
- The .032-inch and .048-inch nozzles performed better at ceiling level than at deck level.

The Navy concluded from these preliminary tests that water mist extinguishment of flammable liquid fires at 400 psi was feasible with water mist protection and that compartment cooling could be accomplished with a minimum expenditure of water.

Additional tests were performed to determine whether a lowered pressure of 250 psi could result in an effective extinguishing system. These tests, using .032-inch nozzles at 250 psi, helped to define the limitations of water mist systems, particularly with respect to the distance from the nozzle to the fire. Results of these tests showed:

- If the fire occurred in the center of the test room, it was extinguished in all cases within 9 seconds using 3 gallons of water.
- Nozzles do not necessarily need to be directly over the fire, but extended distances between nozzle and test pan required more time for extinguishment and more gallons used.
- When a test fire was against a wall, the time and gallonage requirements were increased and piping arrangements remote to the fire could not accomplish extinguishment.
- When the spray angle of the nozzle is widened from 90° to 120°, the velocity of the water droplets was considerably reduced, hindering extinguishment.

It was concluded that .032-inch nozzles, using a lowered pressure of 250 psi with a spray angle of 90° and regularly spaced throughout an enclosure could effectively extinguish flammable liquid fires that could ordinarily be encountered in submarine engine room enclosures.

Surface Ship Applications

Incidents of shipboard fires, most notably involving a disaster on the passenger ship *Scandinavian Star* where 150 people died, have prompted the International

Maritime Organization to require sprinkler systems in passenger ship guest rooms by the year 2005. This action has spurred the development of water mist systems for the protection of cabins, public spaces, ship machinery spaces, and engine rooms. Water mist systems are currently replacing existing halon systems in some of these spaces. Halon is a gaseous extinguishing agent the production of which was prohibited because of its adverse effect on the earth's ozone layer. (More information is available in *Design of Special Hazard and Fire Alarm Systems*, Delmar, 1997.) The International Maritime Organization has prohibited the use of halon on new vessels since 1992.

Marioff (1993) reports that tests involving fires on mattresses in a passenger cabin were sufficient to result in a flashover condition without a fire protection system, and that a water mist system provided protection approximately equivalent to an onboard sprinkler system, when the cabin door was open. Tests on large open spaces, such as a passenger dining area on a ship, demonstrated that the systems needed to be arranged such that all nozzles in a minimum area of 1,076 square feet must actuate. This subdivision was needed to overcome the effects of air entrainment within the large space.

Halon system replacement concerns prompted several tests in shipboard engine machinery spaces of varying sizes. Water mist systems extinguished pool fires and full spray fires in 6 to 35 seconds using 6 to 34 liters of water, even when doors and hatches were left open. Tests on shipboard computers and energized electrical equipment demonstrated effective extinguishment and no disruptive electrical discharges when the water mist system was using deionized water, or when the water mist nozzles were placed strategically and tap water was used.

Aircraft Applications

Anyone who has flown in an aircraft may be cognizant of the hazards of onboard aircraft fires or postcrash aircraft fires where the ignition of spilled jet fuel threatens passenger safety and affects passenger egress from the aircraft. Fires originating within the passenger cabin provide a serious threat to passenger safety, but the Federal Aviation Administration (FAA) determined that postcrash fires were to be considered the worst case scenario.

Hill et al. (1993) and Spring et al. (1993) report that full-scale fire testing, using an onboard water mist fire suppression system, was undertaken at the Federal Aviation Administration Technical Center in Atlantic City, New Jersey. The tests were established to simulate the possibility of a large pool fire directly adjacent to the cabin caused by fuel escaping from ruptured wings after a crash landing. The primary objective of the tests was not extinguishment of the external flame, but provision of a water mist discharge within the cabin to create a survivable atmosphere. The performance objective of the system was to provide sufficient time for egress of all passengers.

With water being the primary weight burden, a zoned scenario was devised to reduce the total water storage requirement. Each zone was individually detected and actuated, with the narrow-body aircraft divided into zones 8 feet in length. The systems were designed with the assumption that five zones would simultaneously actuate, centered on the point of flame impingement of the fuselage. This scenario provided protection for one-third of the aircraft, with 24 gallons of water available for discharge.

Using nozzle orifices ranging from a low discharge of 0.23 gpm to a high of 0.50 gpm, it was concluded that the zoned scenario provided greater survival time than did the nonzoned scenario because the concentration of density onto the area of fuselage most severely affected by the fire provided more efficient use of a limited water supply.

Using low flow rate nozzles, a zoned water mist system, and an 8-gallon water supply provided increased survival time, making it clear that zoning not only provides weight feasibility, but increases survival times.

Zoning of the water mist systems provides a real opportunity for these systems to become a part of the aircraft of the future. The combination of increased survival time and the visibility improvement within the two-thirds of the craft in which the system is not discharging provides an exciting opportunity to fly with a higher degree of confidence.

Total Flooding Electronics and Telecommunications Applications

class A fires
fires involving ordinary combustibles

class B fires
fires involving flammable liquids

class C fires
fires involving electrical hazards

Applications mentioned previously in this book were primarily concerned with **class A fires**, involving ordinary combustibles, and **class B fires**, involving flammable liquids. A reluctance to provide water protection of **class C fires**, involving electrical hazards, exists among some fire protection professionals because of fears relative to water conductivity, some of which are unfounded. Most of the valid problems can be solved by judicious water application and an automatic shutdown of the power supply to the electrical equipment being protected by a fire detection system.

Halon systems or carbon dioxide systems provide the majority of current protection schemes for electrical equipment. Carbon dioxide, applied in concentrations of 34% or greater, may be dangerous in areas occupied by personnel, whereas halon systems are marked for replacement by the Montreal Protocol, an international agreement intended to reduce the depletion of the ozone layer caused by the release of agents such as halon into the atmosphere.

Hills et al. (1993) reports that a water mist system was designed to test protection within switchgear modules, as opposed to a ceiling-mounted sprinkler or water mist distribution. Ignition was initiated by a ribbon placed on a board in the center of a switchgear module. With no water mist protection, temperatures ranged in the 600–1,000°C range. Without water mist, fire spread was rapid vertically, then horizontal spread was eventually accomplished at peak temperature, with flames extending up to 4 meters above the module and smoke reaching 100% in the test room in 20 minutes. Damage to the switchgear was considerable.

Tests involving water mist protection reduced maximum temperatures to the 350–500°C range, and smoke obstruction was reduced after system actuation, with nozzles under high pressure providing effective results. Nozzles in the switchgear modules were significantly more effective than nozzles mounted at the ceiling. It was noticed that the water mist was able to negotiate some obstructions within the switchgear, depending on placement of the nozzle. Extinguishment was accomplished within 2 seconds, using less than 1 liter of water. The water mist proved to be less conductive than smoke encountered in the unsuppressed tests and did not damage any electrical equipment within the switchgear module.

The tests were encouraging enough to prompt further testing on other types of electrical equipment, with the intent of providing a water mist system that could be standardized among varying types of electronic equipment.

A study by the National Institute of Standards and Technology for the U.S. Navy showed that fresh water was not found to be a cause of the shorting of energized electrical equipment. Salt water or water heavily laden with rust was significantly more likely to be a potential problem. Fresh water in stainless steel piping, when used in conjunction with a water mist system, appears to be a promising fire protection scheme for electrical and switchgear equipment.

Commercial and Residential Applications

Water droplet efficiency in living quarters dates back to the earliest uses of sprinklers as a life and property saving tool. An awareness of drop size occurred in the early 1950s with the development of the standard spray sprinkler as a replacement for the "old style" sprinkler. It was noted during testing that the smaller droplets provided a greater available surface area for cooling and heat absorption. It was decided to make the "teeth" on the sprinkler deflector more closely spaced than the old style sprinkler, thus breaking the water spray into finer drops.

Liu (1977) reports that subsequent tests, related to sprinklers installed in corridors outside of a living unit, were conducted by the National Institute of Standards and Technology in 1977. Full scale tests, bench scale tests, and mathematical modeling were employed to determine the best nozzle placement, nozzle diameter, and water droplet size.

It was determined that the large droplets provided by old style sprinklers were an inefficient method of distributing water to a fire hazard in a living occupancy. The findings clearly showed that smaller droplets, as predicted by intuition and mathematical modeling, were significantly more efficient in absorbing heat, evaporation, and in cooling of the fire plume. Additional studies indicated that velocity and droplet trajectory were also significant factors in the rate of evaporation.

Full-scale studies confirmed the bench studies and resulted in the conclusion that smaller orifice nozzles, and their associated smaller droplet sizes, achieved a greater plume temperature reduction, resulting from more efficient evaporation.

Testing has been performed at the Maryland Fire and Rescue Institute of the University of Maryland (Figure 11-5) to test water mist systems for sleeping quarters. A 24-foot by 48-foot building was constructed that includes a control center for data collection, a 12-foot by 12-foot room with a table and ladder, and a 12-foot

Figure 11-5 *Water mist testing facility at the University of Maryland.*

by 12-foot room with an underfloor area. One room contained a residential sprinkler system installed in accordance with NFPA 13D, *Standard for Installation of Sprinkler Systems in One and Two Family Dwellings and Manufactured Homes*. The other room was equipped with a water mist system of varying piping configurations and with nozzle sizes of varying orifice diameters.

Tests featured a room with a table in the center and a ladder in the corner of the room. Four containers of heptane were situated at the quarter points of the room, with one additional container under the table and three containers at varying elevations on the ladder. The piping configurations that were tested included the cornice protection system, with nozzles surrounding the room at the ceiling perimeter, and the ceiling protection system, with two lines of nozzles installed at the ceiling, as shown on Figure 11-6. Nozzle outlets were about 6 inches on center, installed into pretapped outlets on stainless steel pipe. The ceiling-mounted system was mounted on the ceiling at a regular spacing. The cornice-mounted system was installed along the wall at the ceiling. Each system was tested individually. Only one system would be installed in a water mist application.

The system was pressurized by one of two pumps, shown in Figure 11-7, and monitored by a personal computer, shown in Figure 11-8. The pumps were greatly oversized to test a wide range of pressures.

Numerous tests were performed for varying applications, including tests to measure the effectiveness of water mist on underfloor applications as an alternate to halon, which had been the protection method of choice for this application in

Chapter 11 Water Mist System Design

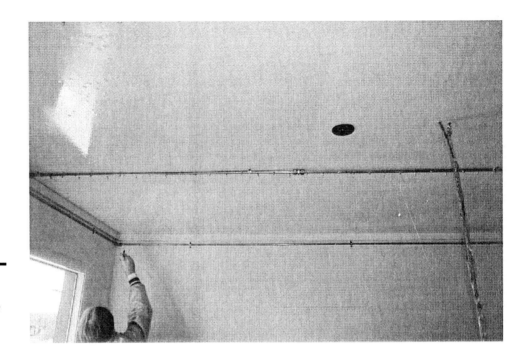

Figure 11-6 Test room showing ceiling and cornice protection water mist systems.

Figure 11-7 Test pumps for water mist system.

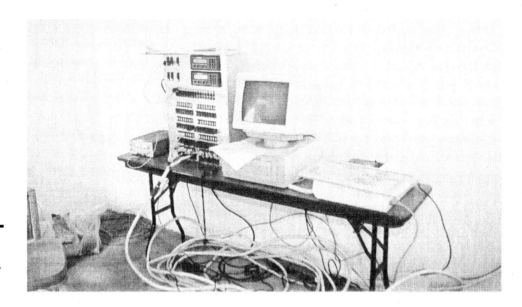

Figure 11-8 *Control and recording apparatus for water mist testing.*

the past. Protection scenarios include library stack protection and electronic equipment protection. Applications for water mist that have been traditionally associated with the use of halon as an agent are being conducted.

Tests were performed for both heptane flammable liquid fires and wood crib ordinary combustible fires using a wide variety of nozzle orifices, piping configurations, and design applications, with promising results.

Summary

The many successful research efforts with water mist are cause for encouragement and excitement, and the issuance of NFPA 750 is certain to draw attention to this new technology. Successful application of water mist for several passenger vessels in Europe, and approvals obtained by several maritime authorities demonstrates only a portion of the future of water mist. In the United States, water mist is making a fascinating transition from a research effort to a system with exciting possibilities for varied applications. For environmental protection, water mist is a promising potential substitute for halon, and for public passenger safety, water mist is a potential answer to the vexing problem of fire protection on aircraft and ships. Ongoing testing demonstrates that under certain conditions, water mist could serve as a potential option to traditional sprinkler systems in areas with limited water supplies.

Review Questions

1. List water mist applications that meet the following criteria:
   a. Applications that may replace existing halon systems.
   b. Applications for which a standard sprinkler system is impossible or impractical.
2. Why is it important that water mist droplets be very small? What performance objective is desired when small droplets are specified?
3. Discuss the advantage that a fire service person may have when approaching a fire with a fog nozzle opened to a wide angle.
4. Give an example of a situation in which a local application water mist system would be preferred, and one in which a total flooding water mist system would be preferred.
5. Describe the difference between wet pipe, dry pipe, deluge, and preaction water mist systems.
6. Would water mist droplets be larger or smaller if the pressure is increased for water flowing through a fixed orifice? Describe the reason for your answer.
7. For a large passenger aircraft or a very large room on a passenger ship, why would it be necessary to install a zoned application water mist system? For a system with a fixed amount of stored water, why would a zoned application water mist system be more effective than a total flooding water mist system?

References

Dundas, P. H., 1974, *The Scaling of Sprinkler Discharge Prediction of Drop Size.* Factory Mutual Research Corp., Norwood, Massachusetts.

Gagnon, R.M., 1997, *Design of Special Hazard and Fire Alarm Systems.* Delmar Publishers, Albany, New York.

Hill, R. G., Marker, T. M., and Sarkos, C. P., 1993, "Evaluation and Optimization of On-Board Water Spray Fire Suppression Systems in Aircraft," in *Proceedings of the Water Mist Fire Suppression Workshop,* March 1–2, 1993. National Institute of Standards and Technology, Gaithersburg, Maryland.

Hills, A. T., Simpson, T., and Smith, D. P., 1993, "Water Mist Fire Protection Systems for Telecommunications Switchgear and Other Electronic Facilities," in *Proceedings of the Water Mist Fire Suppression Workshop,* March 1–2, 1993. National Institute of Standards and Technology, Gaithersburg, Maryland.

Liu, S. T., 1977, *Analytical and Experimental Study of Evaporative Cooling and Room Fire Suppression by Corridor Sprinkler System.* National Institute for Standards and Technology, Gaithersburg, Maryland.

Lugar, J. R., 1979, *Water Mist Fire Protection.* David W. Taylor Naval Ship Research and Development Center, Bethesda, Maryland.

Marioff Hi-Fog Oy, 1993. Vantaa, Finland.

Reischl, U., 1979, "Water Fog Stream Heat Radiation," *Fire Technology,* vol. 15 (Nov.), pp. 262–270.

Spring, D. J., Simpson, T., Smith, D. P., and Ball, D. N., 1993, "New Applications of Aqueous Agents for Fire Suppression," in *Proceedings of the Halon Alternatives Technical Working Conference,* May 11–13, Albuquerque, NM.

Chapter 12

Quality Assurance and Coordination

Objectives

Upon completion of this chapter, you should be able to:

- Develop a procedure for the establishment of a quality assurance program for a fire protection firm.
- Discuss the reasons why coordination of mechanical trades is necessary.
- Explain why fire protection systems are prefabricated on many projects.
- Discuss the value of sanctioning a coordination meeting and the methodology for conducting a successful coordination meeting.
- Confirm and enact coordination meeting agreements.
- Develop a set of coordination plans.

A fire protection system does not stand alone in a building. Once designed and installed, it becomes one of many building components that must perform in accordance with its respective performance objectives and must function properly in conjunction with other mechanical systems. For this reason, a fire protection plan should not be completed until a quality assurance review of the drawing is conducted and a coordination meeting for all mechanical trades is successfully completed.

QUALITY ASSURANCE PROCEDURES

Once a fire protection system design and calculation has been completed using Chapters 1 through 11, the work should be thoroughly reviewed by an individual other than the person who designed it. This review process should become a part of a set of **quality assurance procedures** established by each contractor or design firm.

Quality assurance procedures can be individualized to suit the specific needs or organizational structure of each firm, but should consist of no less than an independent review of each fire protection design before it leaves the office for approval and again before construction.

The review must be performed by a **quality assurance administrator** who should be a senior level supervisory person who is an expert relative to the subject matter contained on the drawing. This process helps to increase the level of compliance to the contract documents and the requirements of the AHJ.

Quality Assurance Review

Some may argue that the AHJ should be responsible for quality assurance. Those thinking this way should seriously consider that an error made by a designer or installer of a fire protection system is not absolved of responsibility if an AHJ misses the error. Most governmental AHJs are immune from civil liability in cases of deaths caused by faulty fire protection systems, whereas designers and contractors are not.

A quality assurance review should consist of the following steps as a minimum:

- Verifying that the system selected is appropriate and adequate for the hazard described.
- Contacting the AHJ to ensure that the system selection is congruent to the requirements of the jurisdiction, as shown on Figure 12-1.
- Visiting the building in which the system is to be installed to make certain that occupancy selection and agent selection is appropriate. Noting areas of differing occupancy or areas requiring protection not shown on the plans.
- Using the appropriate standard as a checklist, review the plan for conformance to technical provisions of the standard.

quality assurance procedure
an independent review of each fire protection design before it leaves the office for approval and again before construction

quality assurance administrator
a senior level supervisory person who is an expert on the subject matter contained on a drawing

Figure 12-1 *The quality assurance administrator (right) reviewing requirements with the authority having jurisdiction. (Courtesy Grinnell Fire Protection Systems Co.)*

- When the quality assurance review has been completed, the plan should be signed by the reviewer, or the reviewer's name or initials should be added to the plan.
- Releasing the drawing to the AHJ for approval.
- Reviewing the design again before construction of the fire protection system.
- The completed installation of the fire protection system should be inspected by the quality assurance administrator to ensure conformance to the plans.

If the quality assurance review is correctly performed, the plan should return from the AHJ with few comments. It is strongly recommended that a quality assurance review of completed fire protection systems be performed by the fire protection system installation company before turning the system over to the owner, and that a record of this review be kept on file.

COORDINATION

Because buildings are constructed to provide the maximum amount of usable space for occupants, with a minimum amount of usable space taken up by mechanical systems, a fire protection system is usually installed in a building in very close proximity to other mechanical systems. Each system has specific performance objectives that it must meet independently of the others. To ensure that each mechanical system is installed so that the performance objectives of each system are met without interference, the locations of all mechanical systems in the building should be carefully coordinated.

Coordinating the fire protection plan with other mechanical systems, as shown in Figure 12-2, can be time consuming but rewarding. The cost of the coordination process may be recovered from the reduced costs of a smoother fire protection system installation. It is necessary to have the completed shop drawings of all other mechanical trades before the coordination process can be completed.

Figure 12-2 *A coordination meeting. (Courtesy Grinnell Fire Protection Systems Co.)*

Scheduling the Coordination Process

The shop drawings for each mechanical trade are usually created simultaneously, therefore it may not always be possible to schedule a complete coordination session before the fire protection plans are submitted for approval. If all of the required information is not available at the time of submitting the fire protection drawing for approval, try to make the most remote area as accurate as the available information allows. Significant changes affecting the most remote area, made to the drawing after the initial submission of the drawing to the AHJ, may require resubmission for approval.

    The coordination process begins by contacting the general contractor and obtaining all available updated shop drawings, including any HVAC shop drawings that may be completed, and a schedule for the completion of the remaining shop drawings. If shop drawings are not going to be performed for the major mechanical trades, a coordination meeting, involving all mechanical trades, must be scheduled.

Prefabrication of Fire Protection Systems

Many fire protection firms, in order to reduce the cost of installing a fire protection system, **prefabricate** the system before it is installed in the building. Prefab-

prefabrication
cutting all pipe, installing fittings, and, sometimes, cutting hanger rods in advance of installation

rication involves cutting all pipe to the length shown on the plans, installing a fitting to one end of each pipe, and sometimes precutting hanger rods, with hanger rings installed in advance of installation. The pipes are numbered and bundled, and references are placed on the fire protection plan relative to the pipe numbering sequence. By having the pipes and hangers precut, the installation time is dramatically reduced.

Because a fire protection system is one of the few prefabricated mechanical trades in a building and because it must meet a very narrow range of installation tolerances, the fire protection system designer may discover that the full responsibility for initiating and facilitating the coordination process may be the responsibility of the fire protection system designer on some projects.

The Coordination Meeting

Because a successful coordination meeting largely depends on cooperation between mechanical trades, fire protection engineers preparing contract documents for the installation of a fire protection system in a building should strongly consider establishing a coordination procedure and requiring one or more coordination meetings that contractually requires the cooperation of all mechanical trades.

Some mechanical trades may be uncooperative because the sprinkler system may be the only prefabricated mechanical system in the building and because some trades find that being on the job first and installing their work without consideration of other trades is, for the most part, simpler and more economical. Where this tactic is used, the installation of mechanical systems can become chaotic and sometimes violent, and serious conflicts can ensue that delay the construction of the building and may force installing contractors to make unwise field changes that violate the performance objectives of the system.

Sanctioning a Coordination Meeting

sanctioned coordination meeting
a meeting required by the contract documents

If a coordination meeting is a contractual requirement, the meeting is a **sanctioned coordination meeting**, authorized by the contract documents.

For projects where the coordination process is not sanctioned by the contract documents, it is imperative that a company installing a prefabricated, hydraulically calculated fire protection system obtain the cooperation of the general contractor in setting up and sanctioning the meeting. Sanctioning is crucial, because the completed results of a coordination meeting must be enforced by the building owner or general contractor in order to ensure that all parties taking part in the coordination process abide by their agreements.

The best probability of success for a sprinkler designer attending a sanctioned coordination meeting is to go to the meeting with a carefully arranged agenda and with a set of blueprints of the fire protection system that summarizes possible areas of conflict between the fire protection system and other mechanical systems.

During the coordination process, you will be thankful for having traced the HVAC ducts and other potential mechanical interferences lightly on your drawing, as suggested earlier in this book. By running a print of your drawing through the print machine very slowly, the ducts and other mechanical features will be seen on the print. All points where sprinkler piping crosses ductwork will be shown, and areas where possible conflicts are evident can be circled in red on the print. The coordination meeting can then focus on the resolution of these circled areas.

Coordination Negotiations

Being a reasonable negotiator is of invaluable assistance in a coordination meeting. Being a negotiator means being prepared to make reasonable changes where needed. Being reasonable means being receptive to making changes, but also means making other trades aware that the fire protection system:

- Must meet the requirements of the applicable codes and standards that restrict the options available to the positioning of pipes and sprinklers. Changes affecting an approved fire protection design may require a resubmittal to the AHJ.
- Must be designed as efficiently as possible, with changes in direction held to a minimum to best ensure the proper delivery of water to the hydraulically most demanding area.
- Must be recalculated and resubmitted for approval if the coordination process changes the hydraulic demand of the design area or creates an area more hydraulically demanding than the design area previously calculated.
- For buildings with a very limited water supply, it may not be possible to have a large number of coordination changes and meet the system performance objectives with respect to the water supply.

Putting the Coordination Agreements into Action

Knowing when to agree to make a change may take some experience. It is best for a fire protection system installation firm to establish a designer apprenticeship program to train design personnel in the art of coordination.

The results of the agreements made at such a meeting need to be confirmed. Some recommendations for enforcement of the agreements would include:

- Having each trade sign your coordination print, and issuing copies of the print to all attendees.
- Having the meeting tape recorded and issuing a transcript of the meeting to all attendees.
- Taking notes and issuing a written summary of the coordination meeting minutes.

- Issuing revised fire protection drawings to all trades, clearly marked as final coordinated plans, for construction.
- Exchanging business cards with each attendee and encouraging a continued dialogue to discuss problems that may arise.
- Reminding each attendee of their contractual responsibility if a meeting is sanctioned by the contract documents.
- Arranging for an enforcement authority to oversee the agreements if a meeting is not sanctioned by the contract documents.

Coordination Failures

Should it not be possible to arrange a sanctioned coordination meeting or should a sanctioned coordination meeting break down or fail, the fire protection installing company has two options, both of which could be very risky:

1. Trying to be the first contractor on the job and installing the fire protection system as quickly as possible. This is very risky because an HVAC duct that may be 4 feet deep may not have the space to change direction to avoid sprinkler branch lines and mains. In such a circumstance, the entire fire protection system may need to be completely removed to allow access for the duct.
2. Waiting until all other trades are installed, surveying the completed installation, then installing the system. This option is preferable to the first option and can work if installation schedules can be maintained, but it results in a very uncomfortable and possibly costly scenario for the fire protection contractor if schedules can not be maintained.

Coordination Drawings

Some contract documents mandate a sanctioned coordination procedure and mandate the generation of large-scale coordination drawings. These coordination drawings, if required, may require two separate sets of fire protection drawings:

A set of 1/8" = 1'- 0" plans for approval,

A set of 1/4" = 1'- 0" or 3/8" = 1'- 0" plans for coordination.

Coordination drawings are a valuable tool for the avoiding conflicts, but creating the plans can be expensive and time consuming. It may be necessary to transfer information from the approval set to the coordination set, then transferring coordination changes from the coordination set to the approval set. Two options exist that may reduce the design and layout time involved:

- Drawing the fire protection drawings to the same scale as that required for the coordination plans and combining the approval set and the coordination set onto one set of plans.

- It may be possible to have one set of large-scale coordination plans prepared by one trade, perhaps by the HVAC contractor, then have the other systems laid out on that same set, using color codes to identify the systems.

The Value of Coordination

One of the most impressive results of the coordination process is to create a set of fire protection drawings that best ensures that the installed system is representative of the hydraulic calculations performed, enhancing system performance.

If fire protection piping changes affecting the design area are needed after coordination is performed, the calculations may need to be revised in order to ensure that the minimum gallonage and pressure requirements of the applicable codes and standards are met.

Completion of the coordination process and receipt of approvals authorizes the plan to be released for prefabrication and installation. Resubmittal of the plan to the authorities having jurisdiction is necessary if the coordination process resulted in many changes to the drawing after submittal. Resubmittals of this nature are contractually required by most specifications calling for sanctioned coordination meetings.

Summary

A fire protection firm must, in the interest of reducing liability and serving the public, establish a quality assurance program for the design and installation of each system it installs. In the long run, quality assurance programs pay for themselves in reduced field changes required by authorities having jurisdiction.

Coordination of fire protection systems with other trades can result in a smoother, more efficient, and less expensive installation process.

The public protected by fire protection system expects that care has been exercised to ensure that the system functions properly and appropriately, and performs in proper conjunction with other mechanical systems.

Review Questions

1. Discuss the methodology for establishing a quality assurance review of fire protection designs and installations.
2. Why are fire protection systems prefabricated? What factors might determine whether a system is prefabricated or cut in the field? Why does prefabrication make coordination of mechanical trades necessary?
3. Discuss how a fire protection designer might react differently during the coordination process, depending on whether coordination is contractually sanctioned or not contractually sanctioned.

Activities

1. Contact an experienced fire protection design professional and determine the qualities that a good negotiator should have in the coordination process.
2. Review a set of coordination plans or a set of shop drawings that have undergone coordination and have been marked up to show coordination changes.
3. Attend a coordination meeting and observe how the attendees interact with each other.

Chapter 13

Ethics and Professional Development

Objectives

Upon completion of this chapter, you should be able to:

- Discuss the code of ethics applicable to your professional qualifications.
- Discuss the ethical responsibilities and obligations of students, technicians, and engineers.
- Evaluate a methodology for determining whether a fire protection professional should become involved in whistle-blowing.
- Compare and contrast the respective ethical responsibilities to the individual, the company, society, the client, and the fire protection profession.

The proper application of technical knowledge relative to the design of a fire protection system is essential to your success as a fire protection professional. Most of this text has been dedicated to helping you acquire or sharpen technical skills.

In addition to the minimum requisite technical skills, ethical factors—factors that are not covered in most technical books—may contribute as much or more to the success of a fire protection professional. This chapter concludes our introduction to fire protection system design with material intended to help each reader on the road to a successful career, armed with both technical knowledge and an introduction to the procedures of ethical practice.

Ethical practice of fire protection system design ensures that the designs for which you are responsible serve the best interests of the public. Active participation in professional organizations and continued education and study advance your knowledge and your career, and benefits the society we serve. Continuing education is essential to the proper performance of the fire protection system design.

GOOD VERSUS EVIL

Some writers, philosophers, and psychologists theorize that some degree of bad behavior or the potential for bad behavior resides within each individual. Perhaps we could visualize this complex dichotomy as the angel and devil syndrome, with the forces of good and evil perched on our opposite shoulders, competing for our attention.

Our ethical dilemma in fire protection design must be to overcome the inclination to exhibit bad behavior, so that the angel defeats the devil. Society demands that professions institute an ethical structure intended to assist an individual in being guided by the angel.

The Ethical Dilemma

ethical dilemma
a situation where pressure is placed upon an individual to do something that is not ethically proper, or where it is not immediately obvious which option would be ethically proper

Situations arise in every professional's life where it is tempting to stray from accepted ethical behavior. In such situations, the individual is faced with an **ethical dilemma**. For a fire protection professional, such temptations can occur in the following situations:

- Situations where significant pressure is placed on an individual by a company or other outside entity to do something other than what would be considered ethically proper.
- Situations where it is not immediately obvious which option would be ethically proper.

In our attempts to deal with such situations, we encounter ethical dilemmas that may be difficult to solve. Some knowledge of what the profession expects of its members can be helpful in properly resolving the dilemma.

RESPONSIBILITIES OF A FIRE PROTECTION PROFESSIONAL

When we perform an engineering design or layout, we have a responsibility to perform the design in accordance with the values of several complex and sometimes intermingled strata of ethical judgment:

- our personal values,
- our company's values,
- society's values,
- values of the client,
- values of the fire protection profession.

These values reflect the diverse loyalties that a professional must respect.

Personal Values

> **personal values**
> values obtained from our parents, peers, mentors, television, film, the media, the community, and schools

Our system of **personal values** comes primarily from our parents in the early years of our lives. Many people consider these early impressions to be of lasting and influential value throughout one's life.

As we get older and more sophisticated, our value capacity increases. We recognize the importance of other values, and the weight that we give to differing values changes. Many forces act to change and expand our value system. Our personal values expand to include the values of our peers and mentors, the values obtained from television, film, and the print media, the values of the religious or social community in which we live, and the values of the schools from which we obtain our technical and literary knowledge.

Our personal value system, therefore, becomes a fascinating patchwork quilt, fabricated from the totality of our varied personal experiences. The exercise of this complex personal value system is the testing of our collective conscience with respect to each encounter with an ethical dilemma.

Values often reflect responsibilities. In professional life you are not required to neglect these personal responsibilities. One difficulty in resolving ethical dilemmas is to properly balance these personal responsibilities with the responsibilities that a professional has toward the company, the client, the profession, and society.

Company Values

> **company values**
> values that require a fire protection professional to behave in a way that would meet the goals of, and bring credit to, the company

Your employer may expect or require, that as a representative of the company you adopt the **company values** and behave in a way that meets the goals of the company and brings credit to the company. For example, a company would not want you to either divulge proprietary information to a competitor or behave in a manner that would bring dishonor to the company. Proprietary information is unique data that is the sole property of the company.

Company values are rarely written down for an individual to read. However, they are assumed to be congruous to the personal values of each employee.

company manual
a written set of company regulations that requires your signature

Only in a few cases are they written as part of a **company manual**, a written set of company regulations that requires your signature. Most company manuals require that all inventions and written material become the property of the company. Some companies prohibit an employee from interacting with a client of the company for a certain period of time after an employee has ceased employment with the company, or may require a new employee to submit to a drug test before the employee is allowed to commence work.

Even where a company has a written policy, many other company values exist. A young professional may not recognize all of these responsibilities. This is one reason why every entry-level professional should seek a mentor. A mentor can provide guidance related to the many unwritten values a professional is expected to respect.

The considerable influence exerted by the company on an employee can be exercised in either a positive or a negative manner. A company that primarily allows a fire protection professional to resolve ethical concerns in accordance with personal, societal, and professional values will benefit from an enhanced reputation for ethical behavior.

Companies that pressure fire protection professionals to betray personal, societal, or professional values could create a very uncomfortable atmosphere for the employee. Such action could possibly place the lives of people in jeopardy and could, in time, place the welfare of the company in danger. An employee must decide whether working for such a company is in his or her best interest.

An extreme example of negative company influence might be the case of a company that convinces a fire protection system installer to glue phony sprinklers to a ceiling. Several eventualities could ensue from such a crime:

- If the company is not caught, people may die.
- If the company is caught, all individuals involved in the plot are in jeopardy of being convicted of a serious crime and going to jail.
- The company may need to be dissolved to pay damage claims and legal fees.
- Personal property may be confiscated to pay for damage claims related to personal responsibility.
- Company employees who were not directly related to the unprofessional conduct may have their professional reputation damaged by such conduct.
- Such action can seriously damage the reputation of all fire protection professionals, even those who do not work for the company.

Societal Values

When the public enters a sprinklered building, it has every right to expect that the system has been designed and installed in a manner that provides adequate fire protection, and that the system provides adequate opportunity for the safe egress

societal values
values consisting of the public expectation that all fire protection professionals competently discharge their responsibilities to assure the safety of all who enter a building protected by a fire protection system

of all occupants. In addition, the public reasonably expects that the system will not malfunction, resulting in injury or property damage. The **societal values** consist of the public expectation that all fire protection professionals competently discharge their responsibilities to ensure the safety of all who enter a building protected by a fire protection system. Failure to do so can cause the wrath of society and unleash a frenzied whirlwind of bad publicity, indictments, financial implications, and perhaps jail. The performance of fire protection design involves an element of public trust that must not be betrayed.

The public is often fickle when it comes to awareness of and appreciation for fire protection systems. Sometimes fire protection systems are treated with benign neglect when problems with the system do not exist, but the public can become righteously indignant and even vengeful when a system fails to perform as expected.

To ensure the protection of the public, society demands a system of checks and balances that requires the competent review of fire protection plans, the inspection and testing of fire protection systems, and the ongoing maintenance of all fire protection systems. A system, no matter how competently designed, fails in its primary performance objective if the main control valve is turned off. Society has little patience with such a failure and goes to great lengths to ensure that the individuals responsible are punished.

The NFPA is a very effective standards-making institution that enforces the societal demand by providing minimum standards for the safety and welfare of the public. This complex layered system of standards, review, and inspection seeks to establish a standard for comparison, to correct errors, and to prevent unethical behavior.

Values of the Client

The client may be the building owner or a firm that represents the owner of the building. In the case of a government-owned building, the client is the public. The **client's value system** demands:

- A code-compliant fire protection system, meeting or exceeding the minimum accepted standard of care.
- A fire protection system obtained for the least possible price.

client's value system
a system directing a client to demand a code-compliant fire protection system for the least possible price

These two goals of the client may be at cross-purposes if the installing or designing firm allows it to be so. For example, a fire protection design and construction company designated as the lowest bidder may discover subsequent to acceptance of the contract, that supplying a code-compliant system would render a cost for the system that exceeds the contracted amount. Such an instance might be where an estimator based the system cost on less than the correct number of floors in a multistory building. The manner in which the company reacts to such a situation cuts to the heart of the ethical behavior of a company. Two options could ensue:

- The company installs a code compliant fire protection system and absorbs the loss.
- The company attempts to install a system that it knows to be substandard in an effort to make a profit.

Although the client may be expecting to buy a limousine for the price of a motor scooter, the client can no more afford the societal cost of a major fire protection system failure than can the installing and design firm. A client might react to the situation described in two ways:

- Force the company to install the system as specified at a loss.
- Negotiate terms for an extra to the contract to have the job done correctly.

The client may not be contractually obligated to issue authorization for extra money to the contractor, but may opt to do so for fear of a major catastrophe resulting from an improper installation or for fear of problems that may ensue if the company declares bankruptcy before construction is completed.

Values of the Fire Protection Profession

Professionals who perform fire protection design are obliged to constantly exhibit the **values of the fire protection profession**. These values are summarized in the Code of Responsibility for NICET-Certified Engineering Technicians and Technologists, Figure 13-1, and the Code of Ethics for Engineers, Figure 13-2.

Notice that these codes are not a list of dos and don'ts. Instead, they are a series of value statements that reflect what the profession believes is important. It is easy to read such codes, but it is much more difficult to interpret and apply them when faced with a value dilemma. Thus, it is important for the professional to read these codes carefully, identify the values, and reflect on how they should be applied in professional practice.

PROFESSIONAL STANDARDS OF CONDUCT

Code of Responsibility for NICET-Certified Engineering Technicians and Technologists

Practitioners who work as fire protection technicians for sprinkler contracting firms or who work as technicians or technologists in architectural or engineering firms are strongly encouraged to seek certification with **NICET**, the National Institute for Certification in Engineering Technologies. NICET provides evaluations and examinations for fire protection technicians in the field of fire protection engineering technology by sponsoring examinations in the subfields of automatic sprinkler system layout, special hazards systems layout, fire alarm systems, and fire protection system inspections. These subfields of fire protection engineering technology comprise the principal areas where the application of technical knowledge must be performed in an ethical manner. Practitioners of fire protection

---

values of the fire protection profession
values summarized by the Code of Responsibility for NICET-Certified Engineering Technicians and Technologists and the Code of Ethics for Engineers

NICET
the National Institute for Certification in Engineering Technologies

Figure 13-1 *Code of responsibility for NICET-certified engineering technicians and technologists. (Courtesy National Institute for Certification in Engineering Technologies)*

> **CODE OF RESPONSIBILITY FOR NICET-CERTIFIED ENGINEERING TECHNICIANS AND TECHNOLOGISTS**
>
> NICET-certified engineering technicians and technologists recognize that the services they render have a significant impact on the quality of life for everyone. As they perform their duties and responsibilities on behalf of the public, employers, and clients, they should demonstrate personal integrity and competence. Accordingly, certificants should:
>
> 1. Have due regard for the physical environment and for public safety, health, and wellbeing. If their judgment is overruled under circumstances where the safety, health, property, or welfare of the public may be endangered, they should notify their employer, client, and/or such other authority as may be appropriate.
> 2. Undertake only those assignments for which they are competent by way of their education, training, and experience.
> 3. Discharge their duties to their employer or client in an efficient and competent manner with complete fidelity and honesty.
> 4. Conduct themselves in a dignified manner. They will admit and accept their own errors when proven wrong and refrain from distorting or altering the facts in an attempt to justify their decisions.
> 5. Disclose to their employers or clients, in writing, any conflict of interest or other interest which might create the appearance of a conflict of interest.
> 6. Not receive, either directly or indirectly, any gratuity, commission, or other financial benefit in connection with any work which they are performing unless such benefit has been authorized by their employer or client.
> 7. Neither personally nor through any other agency or persons improperly seek to obtain work or, by way of commission or otherwise, make or offer to make payment to clients or prospective clients for obtaining such work.
> 8. Strive to maintain their proficiency by updating their technical knowledge and skills in engineering technology.
> 9. Not misrepresent or permit misrepresentation of their own or their associate's academic or professional qualifications nor exaggerate their degree of responsibility for any work.
> 10. Not reveal facts, data, or information obtained in connection with services rendered without the prior consent of the client or employer except as authorized or required by law.

technology can become certified in one of four grades, depending upon their experience and their ability to pass a series of examinations. Many authorities having jurisdiction are requiring a minimum level of certification for an individual performing layout, or certification for an individual performing a quality assurance review of a drawing in advance of submission for approval.

Code Of Ethics for Engineers

Preamble
Engineering is an important and learned profession. The members of the profession recognize that their work has a direct and vital impact on the quality of life for all people. Accordingly, the services provided by engineers require honesty, impartiality, fairness and equity, and must be dedicated to the protection of the public health, safety and welfare. In the practice of their profession, engineers must perform under a standard of professional behavior which requires adherence to the highest principles of ethical conduct on behalf of the public, clients, employers and the profession.

I. Fundamental Canons
Engineers, in the fulfillment of their professional duties, shall:
1. Hold paramount the safety, health and welfare of the public in the performance of their professional duties.
2. Perform services only in areas of their competence.
3. Issue public statements only in an objective and truthful manner.
4. Act in professional matters for each employer or client as faithful agents or trustees.
5. Avoid deceptive acts in the solicitation of professional employment.

II. Rules of Practice
1. Engineers shall hold paramount the safety, health and welfare of the public in the performance of their professional duties.
    a. Engineers shall at all times recognize that their primary obligation is to protect the safety, health, property and welfare of the public. If their professional judgment is overruled under circumstances where the safety, health, property or welfare of the public are endangered, they shall notify their employer or client and such other authority as may be appropriate.
    b. Engineers shall approve only those engineering documents which are safe for public health, property and welfare in conformity with accepted standards.
    c. Engineers shall not reveal facts, data or information obtained in a professional capacity without the prior consent of the client or employer except as authorized or required by law or this Code.
    d. Engineers shall not permit the use of their name or firm name nor associate in business ventures with any person or firm which they have reason to believe is engaging in fraudulent or dishonest business or professional practices.
    e. Engineers having knowledge of any alleged violation of this Code shall cooperate with the proper authorities in furnishing such information or assistance as may be required.
2. Engineers shall perform services only in the areas of their competence.
    a. Engineers shall undertake assignments only when qualified by education or experience in the specific technical fields involved.
    b. Engineers shall not affix their signatures to any plans or documents dealing with subject matter in which they lack competence, nor to any plan or document not prepared under their direction and control.
    c. Engineers may accept assignments and assume responsibility for coordination of an entire project and sign and seal the engineering documents for the entire project, provided that each technical segment is signed and sealed only by the qualified engineers who prepared the segment.
3. Engineers shall issue public statements only in an objective and truthful manner.
    a. Engineers shall be objective and truthful in professional reports, statements or testimony. They shall include all relevant and pertinent information in such reports, statements or testimony.
    b. Engineers may express publicly a professional opinion on technical subjects only when that opinion is founded upon adequate knowledge of the facts and competence in the subject matter.
    c. Engineers shall issue no statements, criticisms or arguments on technical matters which are inspired or paid for by interested parties, unless they have prefaced their comments by explicitly identifying the interested parties on whose behalf they are speaking, and by revealing the existence of any interest the engineers may have in the matters.
4. Engineers shall act in professional matters for each employer or client as faithful agents or trustees.
    a. Engineers shall disclose all known or potential conflicts of interest to their employers or clients by promptly informing them of any business association, interest, or other circumstances which could influence or appear to influence their judgment or the quality of their services.
    b. Engineers shall not accept compensation, financial or otherwise, from more than one party for services on the same project, or for services pertaining to the same project, unless the circumstances are fully disclosed to, and agreed to by, all interested parties.
    c. Engineers shall not solicit or accept financial or other valuable consideration directly or indirectly, from contractors, their agents, or other parties in connection with work for employers or clients for which they are responsible.
    d. Engineers in public service as members, advisors or employees of a governmental or quasi-governmental body or department shall not participate in decisions with respect to professional services solicited or provided by them or their organizations in private or public engineering practice.
    e. Engineers shall not solicit or accept a professional contract from a governmental body on which a principal or officer of their organization serves as a member.
5. Engineers shall avoid deceptive acts in the solicitation of professional employment.
    a. Engineers shall not falsify or permit misrepresentation of their, or their associates', academic or professional qualifications. They shall not misrepresent or exaggerate their degree of responsibility in or for the subject matter of prior assignments. Brochures or other presentations incident to the solicitation of employment shall not misrepresent pertinent facts concerning employers, employees, associates, joint ventures or past accomplishments with the intent and purpose of enhancing their qualifications and their work.
    b. Engineers shall not offer, give, solicit or receive, either directly or indirectly, any political contribution in an amount intended to influence the award of a contract by public authority, or which may be reasonably construed by the public of having the effect or intent to influence the award of a contract. They shall not offer any gift, or other valuable consideration in order to secure work. They shall not pay a commission, percentage or brokerage fee in order to secure work except to a bona fide employee or bona fide established commercial or marketing agencies retained by them.

III. Professional Obligations
1. Engineers shall be guided in all their professional relations by the highest standards of integrity.
    a. Engineers shall admit and accept their own errors when proven wrong and refrain from distorting or altering the facts in an attempt to justify their decisions.
    b. Engineers shall advise their clients or employers when they believe a project will not be successful.
    c. Engineers shall not accept outside employment to the detriment of their regular work or interest. Before accepting any outside employment they will notify their employers.
    d. Engineers shall not attempt to attract an engineer from another employer by false or misleading pretenses.
    e. Engineers shall not actively participate in strikes, picket lines, or other collective coercive action.
    f. Engineers shall avoid any act tending to promote their own interest at the expense of the dignity and integrity of the profession.
2. Engineers shall at all times strive to serve the public interest.
    a. Engineers shall seek opportunities to be of constructive service in civic affairs and work for the advancement of the safety, health and well-being of their community.
    b. Engineers shall not complete, sign or seal plans and/or specifications that are not of a design safe to the public health and welfare and in conformity with accepted engineering standards. If the client or employer insists on such unprofessional conduct, they shall notify the proper authorities and withdraw from further service on the project.
    c. Engineers shall endeavor to extend public knowledge and appreciation of engineering and its achievements and to protect the engineering profession from misrepresentation and misunderstanding.
3. Engineers shall avoid all conduct or practice which is likely to discredit the profession or deceive the public.
    a. Engineers shall avoid the use of statements containing a material misrepresentation of fact or omitting a material fact necessary to keep statements from being misleading or intended or likely to create an unjustified expectation, or statements containing prediction of future success.
    b. Consistent with the foregoing, Engineers may advertise for recruitment of personnel.
    c. Consistent with the foregoing, Engineers may prepare articles for the lay or technical press, but such articles shall not imply credit to the author for work performed by others.
4. Engineers shall not disclose confidential information concerning the business affairs or technical processes of any present or former client or employer without his consent.

Figure 13-2 Code of ethics for engineers. (*Courtesy National Society of Professional Engineers*)

 a. Engineers in the employ of others shall not without the consent of all interested parties enter promotional efforts or negotiations for work or make arrangements for other employment as a principal or to practice in connection with a specific project for which the Engineer has gained particular and specialized knowledge.
 b. Engineers shall not, without the consent of all interested parties, participate in or represent an adversary interest in connection with a specific project or proceeding in which the Engineer has gained particular specialized knowledge on behalf of a former client or employer.

5. Engineers shall not be influenced in their professional duties by conflicting interests.
 a. Engineers shall not accept financial or other considerations, including free engineering designs, from material or equipment suppliers for specifying their product.
 b. Engineers shall not accept commissions or allowances, directly or indirectly, from contractors or other parties dealing with clients or employers of the Engineer in connection with work for which the Engineer is responsible.

6. Engineers shall uphold the principle of appropriate and adequate compensation for those engaged in engineering work.
 a. Engineers shall not accept remuneration from either an employee or employment agency for giving employment.
 b. Engineers, when employing other engineers, shall offer a salary according to professional qualifications.

7. Engineers shall not attempt to obtain employment or advancement or professional engagements by untruthfully criticizing other engineers, or by other improper or questionable methods.
 a. Engineers shall not request, propose, or accept a professional commission on a contingent basis under circumstances in which their professional judgment may be compromised.
 b. Engineers in salaried positions shall accept part-time engineering work only to the extent consistent with policies of the employer and in accordance with ethical considerations.
 c. Engineers shall not use equipment, supplies, laboratory, or office facilities of an employer to carry on outside private practice without consent.

8. Engineers shall not attempt to injure, maliciously or falsely, directly or indirectly, the professional reputation, prospects, practice or employment of other engineers, nor untruthfully criticize other engineers' work. Engineers who believe others are guilty of unethical or illegal practice shall present such information to the proper authority for action.
 a. Engineers in private practice shall not review the work of another engineer for the same client, except with the knowledge of such engineer, or unless the connection of such engineer with the work has been terminated.
 b. Engineers in governmental, industrial or educational employ are entitled to review and evaluate the work of other engineers when so required by their employment duties.
 c. Engineers in sales or industrial employ are entitled to make engineering comparisons of represented products with products of other suppliers.

9. Engineers shall accept personal responsibility for their professional activities; provided, however, that Engineers may seek indemnification for professional services arising out of their practice for other than gross negligence, where the Engineer's interests cannot otherwise be protected.
 a. Engineers shall conform with state registration laws in the practice of engineering.
 b. Engineers shall not use association with a nonengineer, a corporation, or partnership as a "cloak" for unethical acts, but must accept personal responsibility for all professional acts.

10. Engineers shall give credit for engineering work to those to whom credit is due, and will recognize the proprietary interests of others.
 a. Engineers shall, whenever possible, name the person or persons who may be individually responsible for designs, inventions, writings, or other accomplishments.
 b. Engineers using designs supplied by a client recognize that the designs remain the property of the client and may not be duplicated by the Engineer for others without express permission.
 c. Engineers, before undertaking work for others in connection with which the Engineer may make improvements, plans, designs, inventions, or other records which may justify copyrights or patents, should enter into a positive agreement regarding ownership.
 d. Engineers' designs, data, records, and notes referring exclusively to an employer's work are the employer's property.

11. Engineers shall cooperate in extending the effectiveness of the profession by interchanging information and experience with other engineers and students, and will endeavor to provide opportunity for the professional development and advancement of engineers under their supervision.
 a. Engineers shall encourage engineering employees' efforts to improve their education.
 b. Engineers shall encourage engineering employees to attend and present papers at professional and technical society meetings.
 c. Engineers shall urge engineering employees to become registered at the earliest possible date.
 d. Engineers shall assign a professional engineer duties of a nature to utilize full training and experience, insofar as possible, and delegate lesser functions to subprofessionals or to technicians.
 e. Engineers shall provide a prospective engineering employee with complete information on working conditions and proposed status of employment, and after employment will keep employees informed of any changes.

"By order of the United States District Court for the District of Columbia, former Section 11(c) of the NSPE Code of Ethics prohibiting competitive bidding, and all policy statements, opinions, rulings or other guidelines interpreting its scope, have been rescinded as unlawfully interfering with the legal right of engineers, protected under the antitrust laws, to provide price information to prospective clients; accordingly, nothing contained in the NSPE Code of Ethics, policy statements, opinions, rulings or other guidelines prohibits the submission of price quotations or competitive bids for engineering services at any time or in any amount."

Statement by NSPE Executive Committee

In order to correct misunderstandings which have been indicated in some instances since the issuance of the Supreme Court decision and the entry of the Final Judgment, it is noted that in its decision of April 25, 1978, the Supreme Court of the United States declared: "The Sherman Act does not require competitive bidding."

It is further noted that as made clear in the Supreme Court decision:

1. Engineers and firms may individually refuse to bid for engineering services.
2. Clients are not required to seek bids for engineering services.
3. Federal, state, and local laws governing procedures to procure engineering services are not affected, and remain in full force and effect.
4. State societies and local chapters are free to actively and aggressively seek legislation for professional selection and negotiation procedures by public agencies.
5. State registration board rules of professional conduct, including rules prohibiting competitive bidding for engineering services, are not affected and remain in full force and effect. State registration boards with authority to adopt rules of professional conduct may adopt rules governing procedures to obtain engineering services.
6. As noted by the Supreme Court, "nothing in the judgment prevents NSPE and its members from attempting to influence governmental action . . . ."

Note:
 In regard to the question of application of the Code to corporations vis-a-vis real persons, business form or type should not negate nor influence conformance of individuals to the Code. The Code deals with professional services, which services must be performed by real persons. Real persons in turn establish and implement policies within business structures. The Code is clearly written to apply to the Engineer and it is incumbent on a member of NSPE to endeavor to live up to its provisions. This applies to all pertinent sections of the Code.

National Society of Professional Engineers
1420 King Street
Alexandria, Virginia 22314-2794
703/684-2800 • FAX: 703/836-4875
Publication date as revised: July 1993 • Publication #1102

Figure 13-2 *(Cont.)*

The NICET Code of Responsibility has a broad similarity to many other codes of ethics that have been devised. The code advises:

1. Public safety is paramount and takes precedence over other pressures that may be brought to bear on a fire protection designer. If a technician is ordered or advised to do something that violates the Code of Responsibility, the technician must provide timely notification and be prepared to take appropriate action.
2. Work only within the subfield of your expertise. If you are certified in automatic sprinkler system layout, your practice of fire protection engineering technology should be limited exclusively to that field. Conversely, you should obtain certification for all subfields applicable to your education and experience.
3. Loyalty, honesty, dignity, and a lack of a conflict of interest is an employee's responsibility to an employer.
4. Reporting of all gifts received as part of one's job is required. Accepting gifts presents the appearance that a professional's judgment has been influenced by factors other than the important technical details.
5. Don't undercut your employer by using its clients for personal gain. Such action reflects a conflict of interest and poor professional judgment.
6. Always keep current with fire protection technology through education, training, reading, and vigilance on the job. Failure to maintain competency implies that you are not providing the client or society with the best engineering judgment. NICET has instituted a requirement for verification of continuing professional education and competency, and several states have begun to require the same of licensed professional engineers.
7. Represent your qualifications accurately, and maintain the confidentiality of your employer's information.

The Code of Responsibility is intended to express the values that are important to the profession and to help a professional to resolve ethical dilemmas. NICET staff members may be able to assist the technician with the resolution of specific ethical problems that are not directly addressed by the code. Failure to comply with the Code of Responsibility may result in loss of certification.

Code of Ethics for Engineers
a compilation of rules and obligations published by the National Society of Professional Engineers used by engineers to resolve ethical conflicts in fire protection system design

The Code of Ethics for Engineers

Professional engineers are bound by the **Code of Ethics for Engineers**, published by the National Society of Professional Engineers (NSPE), which expands on the NICET Code of Responsibility, and can be used by engineers to resolve ethical conflicts in fire protection system design. Registration as a professional engineer is required of all principal fire protection engineers performing private practice, and is becoming increasingly important for engineers in government, business, and industry. The Code of Ethics is a legal requirement, and differs from the Code of Responsibility, which is administered and enforced by the NICET internal board of review.

Many states have adopted codes of ethics similar to the Code of Ethics for Engineers. As a part of the licensing procedure for engineers, they are legally required to become familiar with and follow the code of ethics for each state in which they practice. Each state has a board of ethical review that schedules hearings to evaluate claims of unethical conduct and levies penalties for conduct judged to be unethical, which includes fines and revocation of licenses.

The fundamental canons of the Code of Ethics reproduce the primary responsibilities required of fire protection technicians listed in the Code of Responsibility. The rules of practice and the professional obligations expand on the fundamental canons and provide detailed advice for the resolution of several specific situations. Some of the specific instances addressed by the Code of Ethics include:

1. Guilt by association. Don't become involved in business ventures with firms or individuals known to be violators of ethical procedures.
2. Engineers must seal or sign only that portion of the work for which they have expertise and have supplied supervision.
3. Engineers must be very careful in dealing with political representatives, especially in the realm of contributions and payments.
4. An engineer must give an honest evaluation of a project in a timely manner and cannot accept being guided into an inaccurate conclusion.
5. An engineer must protect the profession by avoiding criticism of others and other behavior that may discredit the profession.

The Profession of Engineering

Engineering is consistently one of the most widely respected professions in the world. The public willingly extends blind trust to engineers to protect the public from harm through the application of specialized technical knowledge, not unlike the trust given by the public to the pilot of a commercial airliner.

The early railroad era was a significant benchmark in the development of the profession of engineering. As our nation expanded and rail service followed, railroad bridge failures became commonplace. Society demanded that an analytical and methodological approach to railroad trestle design be performed. The profession instituted a requirement that the safety of railroad bridges be determined in advance of construction to demonstrate their ability to perform properly. Tau Beta Pi, the National Engineering Honor Society, uses a railroad trestle as its symbol to commemorate this significant development to the profession.

It is commonly thought that engineers got their name from locomotive engineers or that engineers are primarily designers or operators of machinery. The term *engineer* dates a thousand or more years before the colorful locomotive era, and is based loosely on the word *ingenuity*. The responsibility of engineers goes far beyond the design of machinery. Although all engineering professions have specific ethical responsibilities, fire protection engineers have the overall responsibility of protecting the public from being killed or injured by buildings and machines designed by other engineers. This added responsibility is what makes fire protection engineers unique among all engineers.

LEGAL/ETHICAL CONFLICTS

If something is legal, is it by necessity ethical, and vice versa? Does ethical behavior supersede the law? Does the law supersede ethical behavior? **Legal/ethical conflicts** are conflicts related to the congruence between our system of laws and our system of ethical values, and have historically been among the more difficult areas of ethical behavior to interpret.

legal/ethical conflicts
conflicts related to the congruence between our system of laws and our system of ethical values

Examples of legal behavior being subsequently judged to be unethical are historically numerous. Racial segregation and the denial of voting rights to women are examples of behavior that was once legal, but has subsequently become illegal, primarily because these practices were judged to be unethical.

Other behaviors, such as speeding or running a red light to deliver a severely injured person to the hospital, are clearly illegal, but perhaps may be morally justified in certain cases, on ethical grounds. Other behavior, such as selling items deemed unsafe in the United States to countries whose safety laws are less restrictive, are clearly legal, but may be viewed as being unethical.

The rapid and seemingly unbridled growth of technological development and innovation make it increasingly more difficult for our laws to remain precisely congruent with our system of perceived ethical behavior. Conflicts related to the gap between our laws and our code of ethics are becoming more numerous and more noteworthy, making it increasingly difficult for a fire protection professional to determine how to behave ethically. Picking up the newspaper on any given day would present the reader with an array of legal/ethical conflicts that could include international piracy of recordings and computer programs, child labor, and trade quotas.

A legal/ethical conflict related to fire protection could be the example of a fire inspector who inspects a nightclub for occupancy violations during normal daytime working hours. Some may consider that the letter of the law is being met, but the spirit of the law is lacking when the worst case scenario is not viewed. Be assured that in the final analysis, the public will expect that the spirit, not the letter of the law is enforced.

RESOLVING ETHICAL DILEMMAS

Rationalization

rationalization
an attempt to justify one's own action that is known to be wrong

In almost every case of unethical behavior, **rationalization** is evident. We have all used this word in our personal lives, usually without a comprehensive understanding of its meaning. Rationalization is an attempt to justify one's own action that is known to be wrong. Individuals use rationalization to avoid modifying their internal value system, therefore allowing repeated unethical conduct in the future.

Some examples of rationalization might include:

"I included some inaccurate information on my resume because others are doing it."

"I falsified my expense statement because the company owes me."

"If I hadn't signed the falsified report, someone else would have."

"I made an illegal copy of the software program because the company is making too much profit off of it."

Whistle-Blowing

whistle-blowing
an external action where an individual exposes a situation perceived to be unethical

Whistle-blowing is an external action in which an individual exposes a situation perceived to be unethical. Whistle-blowing may violate the Code of Ethics relative to an employee's responsibility to an employer and the responsibility of an employee to maintain a company's proprietary information. If the company is involved in illegal behavior, the employee is placed in an ethical dilemma that might result in whistle-blowing. For this reason, a procedure should be followed before an employee undertakes the extremely serious steps that result in whistle-blowing:

Internal appeal preparation

Evaluate the problem clinically.

Look at the problem from more than one point of view.

Collect evidence and deal with factual data only.

Develop a synopsis of the problem.

Explore and identify alternative courses of action and alternate solutions.

Examine the firm's internal appeal process.

Evaluate the problem with respect to the code of ethics.

Determine the outcome of a negotiated appeal with which you would be satisfied.

Commence the Appeal Process

Bring the problem and proposed solutions to an immediate supervisor in an informal manner.

Make a formal appeal.

Continue the appeal to the highest level within the company.

External Appeal Options

If possible, seek additional professional opinions in a confidential manner.

Make a list of available alternatives.

Protect yourself: Cover your options and keep detailed records.

Seek legal advice: Check local, state, and federal laws pertaining to the problem.

Contact a professional society and if possible obtain support or an endorsement.

If additional evidence or support is obtained, try to revive an internal appeal.

Contact the client and reveal the problem.

If all else fails, go public with the support of a lawyer as a last resort.

Whistle-blowing can result in severe repercussions for the employee, such as demotion, firing, legal problems, harassment, and financial implications. In the case of exposing a situation where sprinklers were glued to a ceiling, whistle-blowing is clearly ethically justified, since failure to expose such a serious situation could lead to death. A person who exposes such a situation can be rewarded with the knowledge that a public service has been performed and may eventually triumph over any negative repercussions.

RESPONSIBILITIES OF A STUDENT

Students should join and actively participate in student organizations and societies related to their intended profession. This activity not only gives students a practical idea of how professional societies work, but may help them to get jobs by networking with alumni and practicing professionals. A student should seek leadership positions in student organizations to obtain executive skills, experience in public speaking, organizational skills, and the development of self-confidence.

It is also essential that a student seek out opportunities to obtain relevant work experience before graduation. The development of professional behavior is enhanced by the combination of scholastic and professional experience.

Students should seek out a mentor or role model and pattern themselves after the positive qualities of this individual. A mentor can help to positively guide a student's career long after graduation.

RESPONSIBILITIES OF A PROFESSIONAL

Fire protection professionals must not only behave ethically, but they must become active in the advancement of the profession by participating in as many professional organizations as time allows, and by updating their skills by attending seminars and reading applicable publications.

Professional Development

Many organizations for fire protection professionals not only provide regular educational programs for their members, but publish technical literature to refresh and advance knowledge. Time is a valuable commodity for a fire protection professional, but it is strongly recommended that serious consideration be given to actively participating in as many of the following applicable organizations associated with fire protection system design as possible:

National Fire Protection Association
1 Batterymarch Park
P.O. Box 9101
Quincy, Massachusetts 02269
(617) 770-3000
   NFPA Publications include: NFPA Journal; Fire Technology; National Fire Codes; Fire Protection Handbook; Automatic Sprinkler and Standpipe Systems; Automatic Sprinkler Systems Handbook; Building Construction for the Fire Service.

Society of Fire Protection Engineers
One Liberty Square
Boston, Massachusetts 02109
(617) 482-0686
   SFPE Publications include: SFPE Bulletin; SFPE Handbook of Fire Protection Engineering; Journal of Fire Protection Engineering.

National Institute for Certification in Engineering Technologies
1420 King Street
Alexandria, Virginia 22314
(703) 684-2835
   NICET Publication: NICET News.

National Fire Sprinkler Association
P.O. Box 1000
Patterson, New York 1256
(914) 878-4200
   NFSA Publications include: Sprinkler Quarterly; Tech Notes; Grass Roots; Regional Report.

American Fire Sprinkler Association
Suite 142
12959 Jupiter Road
Dallas, Texas 75238
(214) 349-5965
   AFSA Publications include: Sprinkler Age; Membergram.

National Society of Professional Engineers
1420 King Street
Alexandria, Virginia 22314
(703) 684-2800
   NSPE Publications include: Engineering Times; State News Letter.

The SFPE, NFSA, AFSA, and NSPE have local chapters that provide the opportunity for the development of leadership skills and provide an exposure to valuable seminars, industry tours, and noted speakers. Readers of this book are encouraged to join and participate actively.

Publications that can be of assistance to the fire protection professional include:

Fire Protection Contractor Magazine, 12972 Earhart Avenue, Suite 302, Auburn, California 95602.

Consulting-Specifying Engineer, P.O. Box 7525, Highlands Ranch, Colorado 80126-9325.

Summary

Fire protection design professionals must be vigilant in keeping up with the current technology and using the appropriate code to determine accepted ethical methods of practice. Active participation in organizations related to the design of fire protection systems sharpens skills and helps to develop a working set of ethical practice procedures. Reading design-related fire protection publications keeps a designer current with the rapidly expanding technology of fire protection system design.

Review Questions

1. Using the Code of Ethics for Engineers as a basis, determine which of the following, when received as a gift from a client or other person with whom one has a professional relationship, must be reported to an employer. Identify items that, if accepted, could create a conflict of interest, and in each case, state the justification for your determination:
   a. A pen with promotional advertising printed on it.
   b. A promotional videotape.
   c. A book of engineering data.
   d. An engraved briefcase.
   e. A car.
   f. A calendar.
   g. A sterling silver pen with your name engraved on it.
   h. An encyclopedia.

2. Using the Code of Ethics for Engineers as a basis, determine which activities are a violation of the Code of Ethics. Provide the Code reference for each item.
   a. Performing outside design work for your employer's client without your employer's knowledge.
   b. Allowing a nonregistered or noncertified individual to use your registration or certification number.
   c. Accepting a free computer in exchange for specifying a product from a company.
   d. Claiming credit for the work of another individual.
   e. Paying a public official to get a design contract.
   f. Designing a bridge with a degree in fire protection engineering and professional registration in fire protection engineering.
   g. Performing part-time work for another firm with your employer's permission.

3. Describe a situation related to fire protection system design not cited in this chapter in which whistle-blowing may be justified.

4. Using the Code of Ethics for Engineers, identify the fundamental values, and explain how these relate to the profession and to society.

5. Discuss why cheating on an examination in a fire protection course as a student can be interpreted as being unprofessional conduct by the profession.

6. List some rationalizations that students give for cheating on a test and explain why they represent rationalization.

7. Would you turn in a classmate who had cheated on an exam? Would you turn in a colleague who falsified his or her resume on a proposal to a client? If your responses are different explain.

8. Obtain a copy of a recent newspaper and select an article related to ethical or unethical conduct. Write an evaluation of the impact that the conduct has had or will have with respect to a particular individual, a company, society, and the profession. Apply the Code of Ethics or the Code of Responsibility to the situation, and evaluate ethical courses of action.

Appendix A

Reference Tables

Inches converted to decimals of a foot.

Inches		Decimal of a Foot	Inches		Decimal of a Foot
0	1/8	.010416	6	1/8	.510416
	1/4	.020833	.50	1/4	.520833
	3/8	.031250		3/8	.531250
	1/2	.041666		1/2	.541666
	5/8	.052083		5/8	.552083
	3/4	.062500		3/4	.562500
	7/8	.072916		7/8	.572916
1	1/8	.093750	7	1/8	.593750
.083333	1/4	.104166	.583333	1/4	.604166
	3/8	.114583		3/8	.614583
	1/2	.125000		1/2	.625000
	5/8	.135416		5/8	.635416
	3/4	.145833		3/4	.645833
	7/8	.156250		7/8	.656250
2	1/8	.177083	8	1/8	.677083
.166666	1/4	.187500	.666666	1/4	.687500
	3/8	.197916		3/8	.697916
	1/2	.208333		1/2	.708333
	5/8	.218750		5/8	.718750
	3/4	.229166		3/4	.729166
	7/8	.239583		7/8	.739583
3	1/8	.260416	9	1/8	.760416
.250	1/4	.270833	.750	1/4	.770833
	3/8	.281250		3/8	.781250
	1/2	.291666		1/2	.791666
	5/8	.302083		5/8	.802083
	3/4	.312500		3/4	.812500
	7/8	.322916		7/8	.822916
4	1/8	.343750	10	1/8	.843750
.333333	1/4	.354166	.833333	1/4	.854166
	3/8	.364583		3/8	.864583
	1/2	.375000		1/2	.875000
	5/8	.385416		5/8	.885416
	3/4	.395833		3/4	.895833
	7/8	.406250		7/8	.906250
5	1/8	.427083	11	1/8	.927083
.416666	1/4	.437500	.916666	1/4	.937500
	3/8	.447916		3/8	.947916
	1/2	.458333		1/2	.958333
	5/8	.468750		5/8	.968750
	3/4	.479166		3/4	.979166
	7/8	.489583		7/8	.989583

Appendix A Reference Tables

Decimal equivalents of fractions.

Inches	Decimal of an Inch	Inches	Decimal of an Inch
1/64	.015625	33/64	.515625
1/32	.03125	17/32	.53125
3/64	.046875	35/64	.546875
1/16	.0625	9/16	.5625
5/64	.078125	37/64	.578125
3/32	.09375	19/32	.59375
7/64	.109375	39/64	.609375
1/8	.125	5/8	.625
9/64	.140625	41/64	.640625
5/32	.15625	21/32	.65625
11/64	.171875	43/64	.671875
3/16	.1875	11/16	.6875
13/64	.203125	45/64	.703125
7/32	.21875	23/32	.71875
15/64	.234375	47/64	.734375
1/4	.25	3/4	.75
17/64	.265625	49/64	.765625
9/32	.28125	25/32	.78125
19/64	.296875	51/64	.796875
5/16	.3125	13/16	.8125
21/64	.328125	53/64	.828125
1/3	.333	27/32	.84375
11/32	.34375	55/64	.859375
23/64	.359375	7/8	.875
3/8	.375	57/64	.890625
25/64	.390625	29/32	.90625
13/32	.40625	59/64	.921875
27/64	.421875	15/16	.9375
7/16	.4375	61/64	.953125
29/64	.453125	31/32	.96875
15/32	.46875	63/64	.984375
31/64	.484375	1	1.
1/2	.5		

Minutes converted to decimals of a degree.

Min.	Deg.	Min.	Deg.	Min.	Deg.	Min.	Deg.	Min.	Deg.	Min.	Deg.
1	.0166	11	.1833	21	.3500	31	.5166	41	.6833	51	.8500
2	.0333	12	.2000	22	.3666	32	.5333	42	.7000	52	.8666
3	.0500	13	.2166	23	.3833	33	.5500	43	.7166	53	.8833
4	.0666	14	.2333	24	.4000	34	.5666	44	.7333	54	.9000
5	.0833	15	.2500	25	.4166	35	.5833	45	.7500	55	.9166
6	.1000	16	.2666	26	.4333	36	.6000	46	.7666	56	.9333
7	.1166	17	.2833	27	.4500	37	.6166	47	.7833	57	.9500
8	.1333	18	.3000	28	.4666	38	.6333	48	.8000	58	.9666
9	.1500	19	.3166	29	.4833	39	.6500	49	.8166	59	.9833
10	.1666	20	.3333	30	.5000	40	.6666	50	.8333	60	1.0000

Standard conversions.

To Change	To	Multiply By	To Change	To	Multiply by
Inches	Feet	0.0833	Inches of mercury	Inches of water	13.6
Inches	Millimeters	25.4	Inches of mercury	Feet of water	1.1333
Feet	Inches	12	Inches of mercury	Pounds per square inch	0.4914
Feet	Yards	0.3333	Ounces per square inch	Inches of mercury	0.127
Yards	Feet	3	Ounces per square inch	Inches of water	1.733
Square inches	Square feet	0.00694	Pounds per square inch	Inches of water	27.72
Square feet	Square inches	144	Pounds per square inch	Feet of water	2.310
Square feet	Square yards	0.11111	Pounds per square inch	Inches of mercury	2.04
Square yards	Square feet	9	Pounds per square inch	Atmospheres	0.0681
Cubic inches	Cubic feet	0.00058	Feet of water	Pounds per square inch	0.434
Cubic feet	Cubic inches	1728	Feet of water	Pounds per square foot	62.5
Cubic feet	Cubic yards	0.03703	Feet of water	Inches of mercury	0.8824
Cubic yards	Cubic feet	27	Atmospheres	Pounds per square inch	14.696
Cubic inches	Gallons	0.00433	Atmospheres	Inches of mercury	29.92
Cubic feet	Gallons	7.48	Atmospheres	Feet of water	34
Gallons	Cubic inches	231	Long tons	Pounds	2240
Gallons	Cubic feet	0.1337	Short tons	Pounds	2000
Gallons	Pounds of water	8.33	Short tons	Long tons	0.89285
Pounds of water	Gallons	0.12004			
Ounces	Pounds	0.0625			
Pounds	Ounces	16			
Inches of water	Pounds per square inch	0.0361			
Inches of water	Inches of mercury	0.0735			
Inches of water	Ounces per square inch	0.578			
Inches of water	Pounds per square foot	5.2			

Appendix A Reference Tables

Conversion factors for water.

---

1 U.S. gallon = 8.3356 pounds
1 U.S. gallon = 0.1337 cubic feet
1 U.S. gallon = 231 cubic inches
1 U.S. gallon = 0.83356 Imperial gallons
1 U.S. gallon = 3.7854 liters

---

1 Imperial gallon = 10.00 pounds
1 Imperial gallon = 0.16037 cubic feet
1 Imperial gallon = 277.12 cubic inches
1 Imperial gallon = 1.1997 U.S. gallons
1 Imperial gallon = 4.5413 liters

---

1 liter = 2.202 pounds
1 liter = 0.0353 cubic feet
1 liter = 61.023 cubic inches
1 liter = 0.2642 U.S. gallons
1 liter = 0.2202 Imperial gallons

---

1 cubic foot of water = 62.355 pounds
1 cubic foot of water = 1728.00 cubic inches
1 cubic foot of water = 7.4805 U.S. gallons
1 cubic foot of water = 6.2355 Imperial gallons
1 cubic foot of water = 28.317 liters

---

1 pound of water = .01604 cubic feet
1 pound of water = 27.712 cubic inches
1 pound of water = 0.11997 U.S. gallons
1 pound of water = 0.100 Imperial gallons
1 pound of water = 0.45413 liters

---

1 cubic inch of water = 0.0361 pounds
1 gallon of water = 8.33 pounds

---

1 inch of water = 0.0361 pounds per square inch
1 foot of water = 0.4334 pounds per square inch
1 pound per square inch = 2.310 feet of water
1 pound per square inch = 2.04 inches of mercury
1 atmosphere = 14.696 pounds per square inch

Geometric formulae for area (A), volume (V), and circumference (C).

Circle
$A = 3.142 \times R^2$
$C = 3.142 \times D$
$R = \dfrac{D}{2}$
$D = 2 \times R$

Cone
$A_1 = 3.142 \times R \times S + 3.142 \times R^2$
$V = 1.047 \times R^2 \times H$

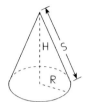

Elipse
$A = 3.142 \times A \times B$
$C = 6.283 \times \dfrac{\sqrt{A^2 + B^2}}{2}$

Cylinder
Area of curved surface =
$(2) \times (3.146) \times (R) \times (H)$
$V = 3.142 \times R^2 \times H$

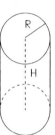

Parallelogram
$A = H \times L$

Rectangle
$A = W \times L$

Sector of circle
$A = \dfrac{3.142 \times R^2 \times \alpha}{360}$
$L = .01745 \times R \times \alpha$
$\alpha = \dfrac{L}{.01745 \times R}$
$R = \dfrac{L}{.01745 \times \alpha}$

Elliptical Tanks
$V = 3.142 \times A \times B \times H$
$A_1 = 6.283 \times \dfrac{\sqrt{A^2 + B^2}}{2} \times H + 6.283 \times A \times B$

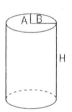

Rectangular solid
$A_1 = 2[W \times L + L \times H + H \times W]$
$V = W \times L \times H$

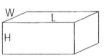

Trapezoid
$A = H \times \dfrac{L_1 + L_2}{2}$

Sphere
$A_1 = 12.56 \times R^2$
$V = 4.188 \times R^3$

Triangle
$A = \dfrac{W \times H}{2}$
$C^2 = H^2 + \left(\dfrac{W}{2}\right)^2$

For above containers:

Capacity in gallons = $\dfrac{V}{231}$ when V is in cubic inches

Capacity in gallons = $7.48 \times V$ when V is in cubic feet

Appendix A Reference Tables

Sizes and capacity of steel pipe.

Nominal Pipe Size (inches)	Outside Diameter (inches)	Inside Diameter (inches)	Capacity of One Foot Length of Pipe		Nominal Pipe Size (inches)	Outside Diameter (inches)	Inside Diameter (inches)	Capacity of One Foot Length of Pipe	
			Cubic Feet	Gallons				Cubic Feet	Gallons
		Schedule 40 Pipe					Schedule 80 Pipe		
½	0.840	0.622	0.0021	0.0158	½	0.840	0.546	0.0016	0.012
¾	1.050	0.824	0.0037	0.0276	¾	1.050	0.742	0.0030	0.022
1	1.315	1.049	0.0060	0.0449	1	1.315	0.957	0.0050	0.037
1¼	1.660	1.380	0.0104	0.0774	1¼	1.660	1.278	0.0089	0.066
1½	1.900	1.610	0.0142	0.106	1½	1.900	1.500	0.0123	0.092
2	2.375	2.067	0.0233	0.174	2	2.375	1.939	0.0205	0.153
2½	2.875	2.469	0.0332	0.248	2½	2.875	2.323	0.0294	0.220
3	3.500	3.068	0.0513	0.383	3	3.500	2.900	0.0548	0.344
3½	4.000	3.548	0.0686	0.513	3½	4.000	3.364	0.0617	0.458
4	4.500	4.026	0.0883	0.660	4	4.500	3.826	0.0798	0.597
5	5.563	5.047	0.139	1.04	5	5.563	4.813	0.126	0.947
6	6.625	6.065	0.200	1.50	6	6.625	5.761	0.181	1.35
8	8.625	7.981	0.3474	2.60	8	8.625	7.625	0.3171	2.38
10	10.75	10.020	0.5475	4.10	10	10.75	9.564	0.4989	3.74
12	12.75	11.938	0.7773	5.82	12	12.75	11.376	0.7058	5.28
14	14.0	13.126	0.9397	7.03	14	14.0	12.500	0.8522	6.38
16	16.0	15.000	1.2272	9.16	16	16.0	14.314	1.1175	8.36
18	18.0	16.876	1.5533	11.61	18	18.0	16.126	1.4183	10.61
20	20.0	18.814	1.9305	14.44	20	20.0	17.938	1.7550	13.13
24	24.0	22.626	2.7920	20.87	24	24.0	21.564	2.536	19.0

Conversion factors.

Length

1 inch = 0.08333 foot, 1,000 mils, 25.40 millimeters.
1 foot = 0.3333 yard, 12 inches, 0.3048 meter, 304.8 millimeters.
1 yard = 3 feet, 36 inches, 0.9144 meter.
1 rod = 16.5 feet, 5.5 yards, 5.029 meters.
1 mile (U.S. and British) = 5,280 feet, 1.609 kilometers, 0.8684 nautical mile.
1 millimeter = 0.03937 inch, 39.37 mils, 0.001 meter, 0.1 centimeter, 100 microns.
1 meter = 1.094 yards, 3.281 feet, 39.37 inches, 1,000 millimeters.
1 kilometer = 0.6214 mile, 1.094 yards, 3,281 feet, 1,000 meters.
1 nautical mile = 1.152 miles (statute), 1.853 kilometers.
1 micron = 0.03937 mil, 0.00003937 inch.
1 mil = 0.001 inch, 0.0254 millimeters, 25.40 microns.
1 degree = 1/360 circumference of a circle, 60 minutes, 3,600 seconds.
1 minute = 1/60 degree, 60 seconds.
1 second = 1/60 minute, 1/3600 degree.

Area

1 square inch = 0.006944 square foot, 1,273,000 circular mils, 645.2 square millimeters.
1 square foot = 0.1111 square yard, 144 square inches, 0.09290 square meter, 92,900 square millimeters.
1 square yard = 9 square feet, 1,296 square inches, 0.8361 square meter.
1 acre = 43,560 square feet, 4,840 square yards, 0.001563 square mile, 4,047 square meters, 160 square rods.
1 square mile = 640 acres, 102,400 square rods, 3,097,600 square yards, 2.590 square kilometers.
1 square millimeter = 0.001550 square inch, 1,974 circular mils.
1 square meter = 1.196 square yards, 10.76 square feet, 1,550 square inches, 1,000,000 square millimeters.
1 square kilometer = 0.3861 square mile, 247.1 acres, 1,196,000 square yards, 1,000,000 square meters.
1 circular mil = 0.7854 square mil, 0.0005067 square millimeter, 0.0000007854 square inch.

Volume (Capacity)

1 fluid ounce = 1.805 cubic inches, 29.57 milliliters, 0.03125 quarts (U.S.) liquid measure.
1 cubic inch = 0.5541 fluid ounce, 16.39 milliliters.
1 cubic foot = 7.481 gallons (U.S.) 6.229 gallons (British), 1,728 cubic inches, 0.02832 cubic meter, 28.32 liters.
1 cubic yard = 27 cubic feet, 46,656 cubic inches, 0.7646 cubic meter, 746.6 liters, 202.2 gallons (U.S.), 168.4 gallons (British).
1 gill = 0.03125 gallon, 0.125 quart, 4 ounces, 7.219 cubic inches, 118.3 milliliters.
1 pint = 0.01671 cubic foot, 28.88 cubic inches, 0.125 gallon, 4 gills, 16 fluid ounces, 473.2 milliliters.
1 quart = 2 pints, 32 fluid ounces, 0.9464 liter, 946.4 milliliters, 8 gills, 57.75 cubic inches.
1 U.S. gallon = 4 quarts, 128 fluid ounces, 231.0 cubic inches, 0.1337 cubic foot, 3.785 liters (cubic decimeters), 3,785 milliliters, 0.8327 Imperial gallon.
1 Imperial (British and Canadian) gallon = 1.201 U.S. gallons, 0.1605 cubic foot, 277.3 cubic inches, 4.546 liters (cubic decimeters), 4,546 milliliters.
1 U.S. bushel = 2,150 cubic inches, 0.9694 British bushel, 35.24 liters.
1 barrel (U.S. liquid) = 31.5 gallons (various industries have special definitions of a barrel).
1 barrel (petroleum) = 42.0 gallons.
1 millimeter = 0.03381 fluid ounce, 0.06102 cubic inch, 0.001 liter.
1 liter (cubic decimeter) = 0.2642 gallon, 0.03532 cubic foot, 1.057 quarts, 33.81 fluid ounces, 61.03 cubic inches, 1,000 milliliters.
1 cubic meter (kiloliter) = 1.308 cubic yards, 35.32 cubic feet, 264.2 gallons, 1,000 liters.
1 cord = 128 cubic feet, 8 feet × 4 feet × 4 feet, 3.625 cubic meters.

Weight

1 grain = 0.0001428 pound.
1 ounce (avoirdupois) = 0.06250 pound (avoirdupois) = 0.06250 pound (avoirdupois), 28.35 grams, 437.5 grains.
1 pound (avoirdupois) = the mass of 27.69 cubic inches of water weighed in air at 4°C (39.2°F) and 760 millimeters of mercury (atmospheric pressure), 16 ounces (avoirdupois), 0.4536 kilogram, 453.6 grams, 7,000 grains.
1 long ton (U.S. and British) = 1.120 short tons, 2,240 pounds, 1.016 metric tons, 1016 kilograms.
1 short ton (U.S. and British) = 0.8929 long ton, 2,000 pounds, 0.9072 metric ton, 907.2 kilograms.
1 milligram = 0.001 gram, 0.000002205 pound (avoirdupois).

Conversion factors (Cont.)

1 gram = 0.002205 pound (avoirdupois), 0.03527 ounce, 0.001 kilogram, 15.43 grains.

1 kilogram = the mass of 1 liter of water in air at 4°C and 760 millimeters of mercury (atmospheric pressure), 2.205 pounds (avoirdupois), 35.27 ounces (avoirdupois), 1,000 grams.

1 metric ton = 0.9842 long ton, 1.1023 short tons, 2,205 pounds, 1,000 kilograms.

Velocity

1 foot per second = 0.6818 mile per hour, 18.29 meters per minute, 0.3048 meters per second.

1 mile per hour = 1.467 feet per second, 1.609 kilometers per hour, 26.82 meters per minute, 0.4470 meters per second.

1 kilometer per hour = 0.2778 meter per second, 0.5396 knot per hour, 0.6214 mile per hour, 54.68 feet per minute.

1 meter per minute = 0.03728 mile per hour, 0.05468 foot per second, 0.06 kilometer per hour, 16.67 millimeters per second, 3.281 feet per minute.

1 knot per hour = 1.152 miles per hour, 1.689 feet per second, 1.853 kilometers per hour.

1 revolution per minute = 0.01667 revolution per second, 6 degrees per second.

1 revolution per second = 60 revolutions per minute, 360 degrees per second.

Acceleration

Standard gravity = 32.17 feet per second per second, 9.807 meters per second per second.

Density

1 gram per millimeter = 0.03613 pound per cubic inch, 8.345 pounds per gallon, 62.43 pounds per cubic foot, 998.9 ounces per cubic foot.

Mercury at 0°C = 0.1360 grams per millimeter, basic value used in expressing pressures in terms of columns of mercury.

1 pound per cubic foot = 16.02 kilograms per cubic meter.

1 pound per gallon = 0.1198 grams per milliliter.

Flow

1 cubic foot per minute = 0.1247 gallon per second, 0.4720 liter per second, 472.0 milliliters per second.

1 gallon per minute = 0.06308 liter per second, 1,440 gallons per day, 0.002228 cubic foot per second.

1 liter per second = 2.119 cubic feet per minute, 15.85 gallons (U.S.) per minute.

1 liter per minute = 0.0005885 cubic foot per second, 0.004403 gallon per second.

1 gallon per minute per square foot = 40.746 liters per minute per square meter.

Pressure

Absolute pressure = the sum of the gage pressure and the barometric pressure.

1 atmosphere = pressure exerted by 760 millimeters of mercury of standard density of 0°C, 14.70 pounds per square inch, 29.92 inches of mercury at 32°F, 33.90 feet of water at 39.2°F, 101.3 kilopascal.

1 millimeter of mercury (at 0°C) = 0.001316 atmosphere, 0.01934 pound per square inch, 0.04460 foot of water (4°C or 39.2°F), 0.0193 pound per square inch, 0.1333 kilopascal.

1 inch of water (at 39.2°F) = 0.00246 atmosphere, 0.0361 pound per square inch, 0.0736 inch of mercury (at 32°F). 0.2491 kilopascal.

1 foot of water (at 39.2°F) = 0.02950 atmosphere, 0.4335 pound per square inch, 0.8827 inch of mercury (at 32°F), 22.42 millimeters of mercury, 2.989 kilopascal.

1 inch of mercury (at 32°F) = 0.03342 atmosphere, 0.4912 pound per square inch, 1.133 feet of water, 13.60 inches of water (at 39.2°F), 3.386 kilopascal.

1 millibar (1/1000 bar) = 0.02953 inch of mercury. A bar is the pressure exerted by a force of one million dynes on a square centimeter of surface.

1 pound per square inch = 0.06805 atmosphere, 2.036 inches of mercury, 2.307 feet of water, 51.72 millimeters of mercury, 27.67 inches of water (at 39.2°F), 144 pounds per square foot, 2,304 ounces per square foot, 6.895 kilopascal.

1 pound per square foot = 0.00047 atmosphere, 0.00694 pound per square inch, 0.0160 foot of water, 0.391 millimeter of mercury, 0.04788 kilopascal.

1 ton (short) per square foot = 0.9451 atmosphere, 13.89 pounds per square inch, 9.765 kilograms per square meter.

1 Torr = 0.013 pascal.

Power

1 British thermal unit per hour = 0.293 watts.

1 British chemical unit per minute = 1.054 kW.

1 British chemical unit per second per square foot = 11.33 kilowatts per square meter.

Conversion factors (Cont.)

---

1 horsepower = 746 watts, 1.014 metric horsepower, 10.69 kilograms-calories per minute, 42.42 British thermal units per minute, 550 pound-feet per second, 33,000 pound-feet per minute.

1 kilowatt = 1.341 horsepower, 1.360 metric horsepower, 14.33 kilogram-calories per minute, 56.90 British thermal units per minute, 1,000 watts.

Heat (Mean Values)

1 British thermal unit = 0.2520 kilogram-calorie, 1,055 joules (absolute).

1 kilogram-calorie = 3.969 British thermal units, 4,187 joules.

1 British thermal unit per pound = 0.5556 kilogram-calorie per kilogram, 2.325 joules per gram.

1 gram-calorie per gram = 1.8 British thermal units per pound, 4.187 joules per gram.

Electrical

1 volt = potential required to produce current flow of 1 ampere through a resistance or impedance of 1 ohm, or current flow of 2 amperes through resistance of ½ ohm, etc.

1 ampere = current flow through a resistance or impedance of 1 ohm produced by a potential of 1 volt, or current flow through a resistance of 100 ohms produced by a potential of 100 volts, etc.

1 milliampere = 0.001 ampere.

1 ohm = resistance or impedance through which current of 1 ampere will flow under a potential of 1 volt.

1 microhm = 0.000001 ohm.

mho = Unit of conductance. In a direct current, circuit conductance in mhos is the reciprocal of number of ohms resistance.

1 watt = power developed by current flow of 1 ampere under potential of 1 volt. (DC, or AC with power factor unity.) See also Power.

1 joule = 1 watt second. A flow of 1 ampere through a resistance of 1 ohm for 1 second (see also Heat).

1 millijoule = 0.001 joule.

SI Units

Included in the conversion tables are the appropriate SI units. SI is an abbreviation for the International System of Units. Most of the industrialized nations of the world have already adopted this system. In all probability, the United States will adopt this system in the near future. SI units consist of seven base units and two supplementary units. Some of the units are derivations of these base and supplementary units.

Base Units

Quantity	Unit	SI Symbol
length	meter	m
mass	kilogram	kg
time	second	s
electric current	ampere	A
thermodynamic temperature	kelvin	K
amount of substance	mole	mol
luminous intensity	candela	cd

Supplementary Units

Quantity	Unit	SI Symbol
plane angle	radian	rad
solid angle	steradian	sr

Number Grouping

To facilitate the reading of numbers having four or more digits, the digits should be placed in groups of three separated by a space instead of commas—counting both to the left and to the right of the decimal point. In the case of four digits, the spacing is optional.

Prefixes

The usual practice is to use not more than four digits and then use the next higher prefix.

Name	Symbol	Multiplication Factor
tera	T	10^{12}
giga	G	10^{9}
mega	M	10^{6}
kilo	k	10^{3}
milli	m	10^{-3}
micro	μ	10^{-6}
nano	n	10^{-9}
pico	p	10^{-12}
femto	f	10^{-15}
atto	a	10^{-18}

Source: Reprinted with permission from the *Fire Protection Handbook*, 17th edition. Copyright © 1991, National Fire Protection Association, Quincy, MA 02269.

Appendix A Reference Tables

Inches to millimeters.

in.	mm	in.	mm	in.	mm	in.	mm
1	25.4	26	660.4	51	1295.4	76	1930.4
2	50.8	27	685.8	52	1320.8	77	1955.8
3	76.2	28	711.2	53	1346.2	78	1981.2
4	101.6	29	736.6	54	1371.6	79	2006.6
5	127.0	30	762.0	55	1397.0	80	2032.0
6	152.4	31	787.4	56	1422.4	81	2057.4
7	177.8	32	812.8	57	1447.8	82	2082.8
8	203.2	33	838.2	58	1473.2	83	2108.2
9	228.6	34	863.6	59	1498.6	84	2133.6
10	254.0	35	889.0	60	1524.0	85	2159.0
11	279.4	36	914.4	61	1549.4	86	2184.4
12	304.8	37	939.8	62	1574.8	87	2209.8
13	330.2	38	965.2	63	1600.2	88	2235.2
14	355.6	39	990.6	64	1625.6	89	2260.6
15	381.0	40	1016.0	65	1651.0	90	2286.0
16	406.4	41	1041.4	66	1676.4	91	2311.4
17	431.8	42	1066.8	67	1701.8	92	2336.8
18	457.2	43	1092.2	68	1727.2	93	2362.2
19	482.6	44	1117.6	69	1752.6	94	2387.6
20	508.0	45	1143.0	70	1778.0	95	2413.0
21	533.4	46	1168.4	71	1803.4	96	2438.4
22	558.8	47	1193.8	72	1828.8	97	2463.8
23	584.2	48	1219.2	73	1854.2	98	2489.2
24	609.6	49	1244.6	74	1879.6	99	2514.6
25	635.0	50	1270.0	75	1905.0	100	2540.0

The above table is exact on the basis: 1 in. = 25.4 mm

Millimeters to inches.

mm	in.	mm	in.	mm	in.	mm	in.
1	0.039370	26	1.023622	51	2.007874	76	2.992126
2	0.078740	27	1.062992	52	2.047244	77	3.031496
3	0.118110	28	1.102362	53	2.086614	78	3.070866
4	0.157480	29	1.141732	54	2.125984	79	3.110236
5	0.196850	30	1.181102	55	2.165354	80	3.149606
6	0.236220	31	1.220472	56	2.204724	81	3.188976
7	0.275591	32	1.259843	57	2.244094	82	3.228346
8	0.314961	33	1.299213	58	2.283465	83	3.267717
9	0.354331	34	1.338583	59	2.322835	84	3.307087
10	0.393701	35	1.377953	60	2.362205	85	3.346457
11	0.433071	36	1.417323	61	2.401575	86	3.385827
12	0.472441	37	1.456693	62	2.440945	87	3.425197
13	0.511811	38	1.496063	63	2.480315	88	3.464567
14	0.551181	39	1.535433	64	2.519685	89	3.503937
15	0.590551	40	1.574803	65	2.559055	90	3.543307
16	0.629921	41	1.614173	66	2.598425	91	3.582677
17	0.669291	42	1.653543	67	2.637795	92	3.622047
18	0.708661	43	1.692913	68	2.677165	93	3.661417
19	0.748031	44	1.732283	69	2.716535	94	3.700787
20	0.787402	45	1.771654	70	2.755906	95	3.740157
21	0.826772	46	1.811024	71	2.795276	96	3.779528
22	0.866142	47	1.850394	72	2.834646	97	3.818898
23	0.905512	48	1.889764	73	2.874016	98	3.858268
24	0.944882	49	1.929134	74	2.913386	99	3.897638
25	0.984252	50	1.968504	75	2.952756	100	3.937008

The above table is approximate on the basis: 1 in. = 25.4 mm, 1/25.4 = 0.039370078740+

Appendix A Reference Tables

Inch/Metric—equivalents.

Fraction	Decimal Equivalent Customary (in.)	Metric (mm)	Fraction	Decimal Equivalent Customary (in.)	Metric(mm)
1/64	.015625	0.3969	33/64	.515625	13.0969
1/32	.03125	0.7938	17/32	.53125	13.4938
3/64	.046875	1.1906	35/64	.546875	13.8906
1/16	.0625	1.5875	9/16	.5625	14.2875
5/64	.078125	1.9844	37/64	.578125	14.6844
3/32	.09375	2.3813	19/32	.59375	15.0813
7/64	.109375	2.7781	39/64	.609375	15.4781
1/8	.1250	3.1750	5/8	.6250	15.8750
9/64	.140625	3.5719	41/64	.640625	16.2719
5/32	.15625	3.9688	21/32	.65625	16.6688
11/64	.171875	4.3656	43/64	.671875	17.0656
3/16	.1875	4.7625	11/16	.6875	17.4625
13/64	.203125	5.1594	45/64	.703125	17.8594
7/32	.21875	5.5563	23/32	.71875	18.2563
15/64	.234375	5.9531	47/64	.734375	18.6531
1/4	.250	6.3500	3/4	.750	19.0500
17/64	.265625	6.7469	49/64	.765625	19.4469
9/32	.28125	7.1438	25/32	.78125	19.8438
19/64	.296875	7.5406	51/64	.796875	20.2406
5/16	.3125	7.9375	13/16	.8125	20.6375
21/64	.328125	8.3384	53/64	.828125	21.0344
11/32	.34375	8.7313	27/32	.84375	21.4313
23/64	.359375	9.1281	55/64	.859375	21.8281
3/8	.3750	9.5250	7/8	.8750	22.2250
25/64	.390625	9.9219	57/64	.890625	22.6219
13/32	.40625	10.3188	29/32	.90625	23.0188
27/64	.421875	10.7156	59/64	.921875	23.4156
7/16	.4375	11.1125	15/16	.9375	23.8125
29/64	.453125	11.5094	61/64	.953125	24.2094
15/32	.46875	11.9063	31/32	.96875	24.6063
31/64	.484375	12.3031	63/64	.984375	25.0031
1/2	.500	12.7000	1	1.000	25.4000

Metric equivalents.

Length

U.S. to Metric	Metric to U.S.
1 inch = 2.540 centimeters	1 millimeter = .039 inch
1 foot = .305 meter	1 centimeter = .394 inch
1 yard = .914 meter	1 meter = 3.281 feet or 1.094 yards
1 mile = 1.609 kilometers	1 kilometer = .621 mile

Area

1 inch^2 = 6.451 centimeter^2	1 millimeter^2 = .00155 inch^2
1 foot^2 = .093 meter^2	1 centimeter^2 = .155 inch^2
1 yard^2 = .836 meter^2	1 meter^2 = 10.764 foot^2 or 1.196 yard^2
1 acre^2 = 4,046.873 meter^2	1 kilometer^2 = .386 mile^2 or 247.04 acre^2

Volume

1 inch^3 = 16.387 centimeter^3	1 centimeter^3 = 0.61 inch^3
1 foot^3 = .028 meter^3	1 meter^3 = 35.314 foot^3 or 1.308 yard^3
1 yard^3 = .764 meter^3	1 liter = .2642 gallons
1 quart = .946 liter	1 liter = 1.057 quarts
1 gallon = .003785 meter^3	1 meter^3 = 264.02 gallons

Weight

1 ounce = 28.349 grams	1 gram = .035 ounce
1 pound = .454 kilogram	1 kilogram = 2.205 pounds
1 ton = .907 metric ton	1 metric ton = 1.102 tons

Velocity

1 foot/second = .305 meter/second	1 meter/second = 3.281 feet/second
1 mile/hour = .447 meter/second	1 kilometer/hour = .621 mile/second

Acceleration

1 inch/second^2 = .0254 meter/second^2	1 meter/second^2 = 3.278 feet/second^2
1 foot/second^2 = .305 meter/second^2	

Force

N (newton) = basic unit of force, kg-m/s^2. A mass of one kilogram (1 kg) exerts a gravitational force of 9.8 N (theoretically 9.80665 N) at mean sea level.

Temperature conversion, Celsius—Fahrenheit.

Temp. Celsius = 5/9 (Temp. F − 32 deg.)
Rankine (Fahrenheit Absolute) = Temp. F + 459.67 deg.
Freezing point of water: Celsius = 0 deg.; Fahr. = 32 deg.
Absolute zero: Celsius = −273.15 deg.; Fahr. = −459.67 deg.

Temp. Fahrenheit = 9/5 × Temp. C + 32 deg.
Kelvin (Celsius Absolute) = Temp. C + 273.15 deg.
Boiling point of water: Celsius = 100 deg.; Fahr. = 212 deg.

Celsius	Fahrenheit	Celsius	Fahrenheit	Celsius	Fahrenheit	Celsius	Fahrenheit
−273.15	−459.67	21	69.8	48.9	120	98	208.4
−200	−328	21.1	70	49	120.2	98.9	210
−100	−148	21.7	71	50	122	99	210.2
0	32	22	71.6	51	123.8	100	212
0.56	33	22.2	72	52	125.6	120	248
1	33.8	22.8	73	53	127.4	121.1	250
1.11	34	23	73.4	54	129.2	140	284
1.67	35	23.3	74	54.4	130	148.9	300
2	35.6	23.9	75	55	131	160	320
2.22	36	24	75.2	56	132.8	176.7	350
2.78	37	24.4	76	57	134.6	180	356
3	37.4	25	77	58	136.4	200	392
3.33	38	25.6	78	59	138.2	204.4	400
3.89	39	26	78.8	60	140	250	482
4	39.2	26.1	79	61	141.8	260	500
4.44	40	26.7	80	62	143.6	300	572
5	41	27	80.6	63	145.4	315.8	600
5.56	42	27.2	81	64	147.2	350	662
6	42.8	27.8	82	65	149	371.1	700
6.11	43	28	82.4	65.6	150	400	752
6.67	44	28.3	83	66	150.8	426.7	800
7	44.6	28.9	84	67	152.6	450	842
7.22	45	29	84.2	68	154.4	482.2	900
7.78	46	29.4	85	69	156.2	500	932
8	46.4	30	86	70	158	537.8	1000
8.33	47	30.6	87	71	159.8	600	1112
8.89	48	31	87.8	71.1	160	648.9	1200
9	48.2	31.1	88	72	161.6	700	1292
9.44	49	31.7	89	73	163.4	760	1400
10	50	32	89.6	74	165.2	800	1472
10.6	51	32.2	90	75	167	871.1	1600
11	51.8	32.8	91	76	168.8	900	1652
11.1	52	33	91.4	76.7	170	982.2	1800
11.7	53	33.3	92	77	171.6	1000	1832
12	53.6	33.9	93	78	172.4	1093.3	2000
12.2	54	34	93.2	79	174.2	1100	2012
12.8	55	34.4	94	80	176	1200	2192
13	55.4	35	95	81	177.8	1204.4	2200
13.3	56	35.6	96	82	179.6	1300	2372
13.9	57	36	96.8	82.2	180	1315.6	2400
14	57.2	36.1	97	83	181.4	1400	2552
14.4	58	36.7	98	84	183.2	1428	2600
15	59	37	98.6	85	185	1500	2732
15.6	60	37.2	99	86	186.8	1537.8	2800
16	60.8	37.8	100	87	188.6	1600	2912
16.1	61	38	100.4	87.8	190	1648.9	3000
16.7	62	39	102.2	88	190.4	1700	3092
17	62.6	40	104	89	192.2	1760	3200
17.2	63	41	105.8	90	194	1800	3272
17.8	64	42	107.6	91	195.8	1871.1	3400
18	64.4	43	109.4	92	197.6	1900	3452
18.3	65	43.3	110	93	199.4	1982	3600
18.9	66	44	111.2	93.3	200	2000	3632
19	66.2	45	113	94	201.2	2204.4	4000
19.4	67	46	114.8	95	203	2500	4532
20	68	47	116.6	96	204.8	2760	5000
20.6	69	48	118.4	97	206.6	3000	5432

Source: Reprinted with permission from the *Fire Protection Handbook*, 17th edition. Copyright © 1991, National Fire Protection Association, Quincy, MA 02269.

GPM flow tables using the formula $Q = 29.83 \, cd^2 \sqrt{P}$.

Pitot Pressures (psi)	Nozzle Orifice Diameters					Pitot Pressures (psi)	Nozzle Orifice Diameters				
	1"	1⅛"	1¼"	1½"	1¾"		2⅜"	2½"	2⅝"	4"	4½"
6	73	93	114	164	224	6	412	457	500	1170	1480
8	84	107	132	190	259	8	475	528	575	1351	1710
10	94	119	148	212	289	10	538	590	650	1510	1910
12	103	131	162	233	317	12	588	646	712	1655	2100
14	112	141	175	251	342	14	625	698	775	1787	2260
16	120	151	187	269	366	16	662	746	825	1910	2420
18	127	160	198	285	388	18	712	791	875	2026	2570
20	134	169	209	300	409	20	750	834	925	2136	2710
22	140	177	219	315	429	22	788	875	962	2240	2840
24	146	185	229	329	448	24	825	914	1012	2340	2970
26	152	193	238	343	466	26	850	951	1050	2435	3090
28	158	200	247	356	484	28	888	987	1088	2527	3210
30	164	207	256	368	501	30	925	1022	1125	2616	3320
35	177	223	276	398	541	32	950	1055	1062	2702	3430
40	189	239	295	425	578	34	975	1088	1200	2785	3540
45	201	254	314	450	613	36	1012	1119	1238	2866	3640
50	211	267	330	475	646	48	1038	1150	1262	2944	3740
60	231	293	362	520	708	40	1062	1180	1300	3021	3840
70	250	316	391	562	765	42	1088	1209	1338	3095	3935
80	267	338	418	601	818	44	1112	1237	1362	3168	4030
90	283	358	443	637	867	46	1137	1265	1400	3239	4120
100	299	378	467	672	914	48	1162	1293	1425	3309	4205
						50	1188	1319	1450	3377	4290

This table assumes a discharge coefficient (c) of 1.00. Refer to Chapter 4 for modifiers that apply to the fire hydrant outlet coefficient used. The hydrant outlet coefficient is multiplied by the value obtained from this table.

Appendix A Reference Tables

Water pressure to feet head

Pounds Per Square Inch	Feet Head	Pounds Per Square Inch	Feet Head
1	2.31	100	230.90
2	4.62	110	253.99
3	6.93	120	277.08
4	9.24	130	300.17
5	11.55	140	323.26
6	13.85	150	346.35
7	16.16	160	369.44
8	18.47	170	392.53
9	20.78	180	415.62
10	23.09	200	461.80
15	34.64	250	577.25
20	46.18	300	692.70
25	57.73	350	808.15
30	69.27	400	923.60
40	92.36	500	1154.50
50	115.45	600	1385.40
60	138.54	700	1616.30
70	161.63	800	1847.20
80	184.72	900	2078.10
90	207.81	1000	2309.00

NOTE: One pound of pressure per square inch of water equals 2.309 feet of water at 62° Fahrenheit. Therefore, to find the feet head of water for any pressure not given in the table above, multiply the pressure pounds per square inch by 2.309.

Feet head of water to psi.

Feet Head	Pounds Per Square Inch	Feet Head	Pounds Per Square Inch
1	.43	100	43.31
2	.87	110	47.64
3	1.30	120	51.97
4	1.73	130	56.30
5	2.17	140	60.63
6	2.60	150	64.97
7	3.03	160	69.30
8	3.46	170	73.63
9	3.90	180	77.96
10	4.33	200	86.62
15	6.50	250	108.28
20	8.66	300	129.93
25	10.83	350	151.59
30	12.99	400	173.24
40	17.32	500	216.55
50	21.65	600	259.86
60	25.99	700	303.17
70	30.32	800	346.48
80	34.65	900	389.79
90	38.98	1000	433.00

NOTE: One foot of water at 62° Fahrenheit equals .433 pound pressure per square inch. To find the pressure per square inch for any feet head not given in the table above, multiply the feet head by .433.

c-Factors

Table B-1 *c-Factors required for hydraulic calculations of sprinkler systems per NFPA 13.*

Kind of Pipe	c-Factor
Unlined Cast or Ductile Iron	100
Black Steel (Dry systems including preaction)	100
Black Steel (Wet and deluge systems)	120
Galvanized (All systems)	120
Listed Plastic	150
Cement Lined Cast or Ductile Iron	140
Copper Tube or Stainless Steel	150

Source: Reprinted with permission from NFPA 13-1996, *Installation of Sprinkler Systems*, Copyright © 1996, National Fire Protection Association, Quincy, MA 02269. This reprinted material is not the official position of the National Fire Protection Association, which is represented only by the standard in its entirety.

Table B-2 *Guide for estimating the Hazen-Williams coefficient of roughness, c, for situations involving moderately corrosive water.*

Kind of Pipe	Value of c
Cast Iron, Unlined:	
10 years old	90
15 years old	75
20 years old	65
30 years old	55
50 years old	50
Cast Iron, Unlined, New	120
Cast Iron, Cement Lined	140
Cast Iron, Bitumastic Enamel Lined	140
Average Steel, New	110
Asbestos—Cement	140
Reinforced Concrete	140
Plastic	150

Source: Reprinted with permission from the *Fire Protection Handbook*, 17th edition. Copyright © 1991, National Fire Protection Association, Quincy, MA 02269.

Table B-3 *Values of c for underground piping.*

Kind of Pipe	c-Factor
Unlined Cast or Ductile Iron	100
Asbestos Cement, Cement Lined Cast or Ductile, Iron, and Steel	140
Fiberglass Filament Wound Epoxy, Polyethylene and Polyvinyl Chloride (PVC)	150

Source: Reprinted with permission from NFPA 24-1995, *Installation of Private Fire Service Mains*, Copyright © 1995, National Fire Protection Association, Quincy, MA 02269. This reprinted material is not the official position of the National Fire Protection Association, which is represented only by the standard in its entirety.

Pipe Diameters

Appendix C Pipe Diameters

Table C-1 *Aboveground piping diameters (in inches).*

	Nominal Diameter											
	.75"	1"	1.25"	1.5"	2"	2.5"	3"	3.5"	4"	5"	6"	8"
Steel Pipe Diameter												
Schedule 10		1.097	1.442	1.682	2.157	2.635	3.26	3.76	4.26	5.295	6.357	8.249
Schedule 30												8.071
Schedule 40		1.049	1.38	1.61	2.067	2.469	3.068	3.548	4.026	5.047	6.065	7.981
Copper Pipe Diameter												
Type K	.745	.995	1.245	1.481	1.959	2.435	2.907	3.385	3.857	4.805	5.741	7.583
Type L	.785	1.025	1.265	1.505	1.985	2.465	2.945	3.425	3.905	4.875	5.845	7.725
Type M	.811	1.055	1.291	1.527	2.009	2.495	2.981	3.459	3.935	4.907	5.881	7.785
CPVC Plastic Pipe Diameter												
	.884	1.109	1.400	1.602	2.003	2.423	2.951					

Table C-2 *Underground Piping Diameters*

	Nominal Diameter									
	4"	6"	8"	10"	12"	14"	16"	18"	20"	24"
Ductile Iron Pipe Diameter										
Class 50		6.4	8.51	10.52	12.58	14.64	16.72	18.8	20.88	24.04
Class 51	4.28	6.34	8.45	10.46	12.52	14.58	16.66	18.74	20.82	24.78
Class 52	4.155	6.275	8.385	10.4	12.46	14.52	16.4	18.68	20.76	24.72
Class 54		6.16	8.27	10.28	12.34					
Cast-Iron Pipe Class 150 Diameter										
Unlined	4.1	6.14	8.23	10.22	12.24	14.28	16.32	18.34	20.36	24.34
Enamel-Lined	3.98	6.02	8.11	10.1	12.12	14.09	16.13			
Cement-Lined	3.85	5.89	7.98	9.97	11.99	13.9	15.94			
Plastic Underground Pipe Diameter										
PVC Class 150	4.24	6.09	7.98	9.79	11.65					
PVC Class 200	4.08	5.86	7.68	9.42	11.2					

Friction Loss Tables

Source: Reprinted with permission from the Fire Sprinkler Advisory Board of Southern California.

Appendix D Friction Loss Tables

Friction Loss Table For 1" Sch. 40 Steel Pipe   I.D. = 1.049

Equiv. Feet of Pipe

	LONG EL	90 ELL	45 ELL	TEE	GV	CV
C–120	2	2	1	5		5
C–100	1.43	1.43	0.71	3.57		3.57

Q	P	P	V	Q	P	P	V	Q	P	P	V
	PSI/FT	PSI/FT	FT/SEC		PSI/FT	PSI/FT	FT/SEC		PSI/FT	PSI/FT	FT/SEC
GPM	C–100	C–120		GPM	C–100	C–120		GPM	C–100	C–120	
7	0.0261	0.0187	2.6	42	0.7194	0.5134	15.6	77	2.2079	1.5757	28.6
8	0.0335	0.0239	3.0	43	0.7514	0.5363	16.0	78	2.2612	1.6138	29.0
9	0.0416	0.0297	3.3	44	0.7841	0.5596	16.3	79	2.3151	1.6523	29.3
10	0.0506	0.0361	3.7	45	0.8173	0.5833	16.7	80	2.3696	1.6912	29.7
11	0.0603	0.0431	4.1	46	0.8513	0.6075	17.1	81	2.4247	1.7305	30.1
12	0.0709	0.0506	4.5	47	0.8858	0.6322	17.5	82	2.4804	1.7702	30.5
13	0.0822	0.0587	4.8	48	0.9210	0.6573	17.8	83	2.5366	1.8104	30.8
14	0.0943	0.0673	5.2	49	0.9568	0.6829	18.2	84	2.5935	1.8509	31.2
15	0.1071	0.0764	5.6	50	0.9932	0.7089	18.6	85	2.6509	1.8919	31.6
16	0.1207	0.0861	5.9	51	1.0303	0.7353	18.9	86	2.7088	1.9333	31.9
17	0.1350	0.0963	6.3	52	1.0680	0.7622	19.3	87	2.7674	1.9751	32.3
18	0.1500	0.1071	6.7	53	1.1063	0.7896	19.7	88	2.8265	2.0173	32.7
19	0.1658	0.1184	7.1	54	1.1452	0.8173	20.1	89	2.8862	2.0599	33.1
20	0.1823	0.1301	7.4	55	1.1848	0.8456	20.4				
21	0.1996	0.1424	7.8	56	1.2249	0.8742	20.8				
22	0.2175	0.1552	8.2	57	1.2657	0.9033	21.2				
23	0.2361	0.1685	8.5	58	1.3071	0.9329	21.5				
24	0.2555	0.1823	8.9	59	1.3491	0.9628	21.9				
25	0.2755	0.1966	9.3	60	1.3917	0.9932	22.3				
26	0.2963	0.2114	9.7	61	1.4349	1.0241	22.7				
27	0.3177	0.2267	10.0	62	1.4787	1.0554	23.0				
28	0.3398	0.2425	10.4	63	1.5231	1.0871	23.4				
29	0.3626	0.2588	10.8	64	1.5682	1.1192	23.8				
30	0.3860	0.2755	11.1	65	1.6138	1.1518	24.1				
31	0.4102	0.2927	11.5	66	1.6600	1.1848	24.5				
32	0.4350	0.3105	11.9	67	1.7069	1.2182	24.9				
33	0.4605	0.3286	12.3	68	1.7543	1.2520	25.3				
34	0.4866	0.3473	12.6	69	1.8023	1.2863	25.6				
35	0.5134	0.3664	13.0	70	1.8509	1.3210	26.0				
36	0.5409	0.3860	13.4	71	1.9002	1.3561	26.4				
37	0.5690	0.4061	13.7	72	1.9500	1.3917	26.7				
38	0.5978	0.4267	14.1	73	2.0004	1.4277	27.1				
39	0.6272	0.4477	14.5	74	2.0514	1.4640	27.5				
40	0.6573	0.4691	14.9	75	2.1029	1.5009	27.9				
41	0.6880	0.4911	15.2	76	2.1551	1.5381	28.2				

Friction Loss Table For 1¼" Sch. 40 Steel Pipe I.D. = 1.38

Equiv. Feet of Pipe

	LONG EL	90 ELL	45 ELL	TEE	GV	CV
C-120	2	3	1	6		7
C-100	1.43	2.14	0.71	4.28		4.99

Q	P	P	V	Q	P	P	V	Q	P	P	V
	PSI/FT	PSI/FT	FT/SEC		PSI/FT	PSI/FT	FT/SEC		PSI/FT	PSI/FT	FT/SEC
GPM	C-100	C-120		GPM	C-100	C-120		GPM	C-100	C-120	
7	0.0069	0.0049	1.5	42	0.1892	0.1350	9.0	77	0.5807	0.4144	16.5
8	0.0088	0.0063	1.7	43	0.1976	0.1410	9.2	78	0.5947	0.4244	16.7
9	0.0109	0.0078	1.9	44	0.2062	0.1472	9.4	79	0.6089	0.4346	17.0
10	0.0133	0.0095	2.1	45	0.2150	0.1534	9.7	80	0.6232	0.4448	17.2
11	0.0159	0.0113	2.4	46	0.2239	0.1598	9.9	81	0.6377	0.4551	17.4
12	0.0186	0.0133	2.6	47	0.2330	0.1663	10.1	82	0.6524	0.4656	17.6
13	0.0216	0.0154	2.8	48	0.2422	0.1729	10.3	83	0.6671	0.4761	17.8
14	0.0248	0.0177	3.0	49	0.2516	0.1796	10.5	84	0.6821	0.4868	18.0
15	0.0282	0.0201	3.2	50	0.2612	0.1864	10.7	85	0.6972	0.4976	18.2
16	0.0317	0.0226	3.4	51	0.2710	0.1934	10.9	86	0.7124	0.5085	18.5
17	0.0355	0.0253	3.6	52	0.2809	0.2005	11.2	87	0.7278	0.5195	18.7
18	0.0395	0.0282	3.9	53	0.2910	0.2077	11.4	88	0.7434	0.5306	18.9
19	0.0436	0.0311	4.1	54	0.3012	0.2150	11.6	89	0.7591	0.5418	19.1
20	0.0480	0.0342	4.3	55	0.3116	0.2224	11.8	90	0.7750	0.5531	19.3
21	0.0525	0.0375	4.5	56	0.3222	0.2299	12.0	91	0.7910	0.5645	19.5
22	0.0572	0.0408	4.7	57	0.3329	0.2376	12.2	92	0.8071	0.5760	19.7
23	0.0621	0.0443	4.9	58	0.3438	0.2453	12.4	93	0.8234	0.5877	20.0
24	0.0672	0.0480	5.2	59	0.3548	0.2532	12.7	94	0.8399	0.5994	20.2
25	0.0725	0.0517	5.4	60	0.3660	0.2612	12.9	95	0.8565	0.6113	20.4
26	0.0779	0.0556	5.6	61	0.3774	0.2693	13.1	96	0.8732	0.6232	20.6
27	0.0836	0.0596	5.8	62	0.3889	0.2776	13.3	97	0.8901	0.6353	20.8
28	0.0894	0.0638	6.0	63	0.4006	0.2859	13.5	98	0.9072	0.6475	21.0
29	0.0954	0.0681	6.2	64	0.4124	0.2944	13.7	99	0.9244	0.6597	21.2
30	0.1015	0.0725	6.4	65	0.4244	0.3029	14.0	100	0.9417	0.6721	21.5
31	0.1079	0.0770	6.7	66	0.4366	0.3116	14.2	105	1.0307	0.7356	22.5
32	0.1144	0.0817	6.9	67	0.4489	0.3204	14.4	107	1.0673	0.7617	23.0
33	0.1211	0.0864	7.1	68	0.4614	0.3293	14.6	109	1.1045	0.7883	23.4
34	0.1280	0.0913	7.3	69	0.4740	0.3383	14.8	111	1.1423	0.8152	23.8
35	0.1350	0.0964	7.5	70	0.4868	0.3474	15.0	113	1.1807	0.8426	24.3
36	0.1423	0.1015	7.7	71	0.4998	0.3567	15.2	115	1.2196	0.8704	24.7
37	0.1497	0.1068	7.9	72	0.5129	0.3660	15.5	120	1.3195	0.9417	25.8
38	0.1572	0.1122	8.2	73	0.5261	0.3755	15.7	130	1.5301	1.0920	27.9
39	0.1650	0.1177	8.4	74	0.5395	0.3851	15.9	140	1.7549	1.2525	30.0
40	0.1729	0.1234	8.6	75	0.5531	0.3947	16.1	150	1.9939	1.4230	32.2
41	0.1810	0.1291	8.8	76	0.5668	0.4045	16.3	160	2.2467	1.6035	34.3

Appendix D Friction Loss Tables

Friction Loss Table For 1½" Sch. 40 Steel Pipe   I.D. = 1.61

Equiv. Feet of Pipe

	LONG EL	90 ELL	45 ELL	TEE	GV	CV
C–120	2	4	2	8		9
C–100	1.43	2.85	1.43	5.70		6.42

Q GPM	P PSI/FT C–100	P PSI/FT C–120	V FT/SEC	Q GPM	P PSI/FT C–100	P PSI/FT C–120	V FT/SEC	Q GPM	P PSI/FT C–100	P PSI/FT C–120	V FT/SEC
15	0.0133	0.0095	2.4	50	0.1233	0.0880	7.9	85	0.3291	0.2349	13.4
16	0.0150	0.0107	2.5	51	0.1279	0.0913	8.0	86	0.3363	0.2400	13.6
17	0.0168	0.0120	2.7	52	0.1326	0.0946	8.2	87	0.3436	0.2452	13.7
18	0.0186	0.0133	2.8	53	0.1373	0.0980	8.4	88	0.3509	0.2504	13.9
19	0.0206	0.0147	3.0	54	0.1422	0.1015	8.5	89	0.3583	0.2557	14.0
20	0.0226	0.0162	3.2	55	0.1471	0.1050	8.7	90	0.3658	0.2611	14.2
21	0.0248	0.0177	3.3	56	0.1521	0.1085	8.8	91	0.3734	0.2665	14.3
22	0.0270	0.0193	3.5	57	0.1571	0.1121	9.0	92	0.3810	0.2719	14.5
23	0.0293	0.0209	3.6	58	0.1623	0.1158	9.1	93	0.3887	0.2774	14.7
24	0.0317	0.0226	3.8	59	0.1675	0.1195	9.3	94	0.3964	0.2829	14.8
25	0.0342	0.0244	3.9	60	0.1728	0.1233	9.5	95	0.4043	0.2885	15.0
26	0.0368	0.0262	4.1	61	0.1781	0.1271	9.6	96	0.4122	0.2942	15.1
27	0.0394	0.0281	4.3	62	0.1836	0.1310	9.8	97	0.4202	0.2999	15.3
28	0.0422	0.0301	4.4	63	0.1891	0.1350	9.9	98	0.4282	0.3056	15.5
29	0.0450	0.0321	4.6	64	0.1947	0.1389	10.1	99	0.4363	0.3114	15.6
30	0.0479	0.0342	4.7	65	0.2003	0.1430	10.2	100	0.4445	0.3173	15.8
31	0.0509	0.0363	4.9	66	0.2061	0.1471	10.4	101	0.4528	0.3232	15.9
32	0.0540	0.0385	5.0	67	0.2119	0.1512	10.6	106	0.4951	0.3534	16.7
33	0.0572	0.0408	5.2	68	0.2178	0.1554	10.7	111	0.5392	0.3848	17.5
34	0.0604	0.0431	5.4	69	0.2238	0.1597	10.9	116	0.5850	0.4175	18.3
35	0.0637	0.0455	5.5	70	0.2298	0.1640	11.0	121	0.6325	0.4514	19.1
36	0.0672	0.0479	5.7	71	0.2359	0.1684	11.2	126	0.6817	0.4865	19.9
37	0.0706	0.0504	5.8	72	0.2421	0.1728	11.4	131	0.7326	0.5228	20.7
38	0.0742	0.0530	6.0	73	0.2483	0.1772	11.5	136	0.7851	0.5603	21.4
39	0.0779	0.0556	6.1	74	0.2547	0.1818	11.7	141	0.8394	0.5991	22.2
40	0.0816	0.0582	6.3	75	0.2611	0.1863	11.8	146	0.8953	0.6389	23.0
41	0.0854	0.0610	6.5	76	0.2675	0.1909	12.0	151	0.9528	0.6800	23.8
42	0.0893	0.0637	6.6	77	0.2741	0.1956	12.1	156	1.0120	0.7223	24.6
43	0.0933	0.0666	6.8	78	0.2807	0.2003	12.3	161	1.0728	0.7657	25.4
44	0.0973	0.0695	6.9	79	0.2874	0.2051	12.5	166	1.1353	0.8102	26.2
45	0.1015	0.0724	7.1	80	0.2942	0.2100	12.6	171	1.1993	0.8560	27.0
46	0.1057	0.0754	7.3	81	0.3010	0.2148	12.8	181	1.3323	0.9509	28.5
47	0.1100	0.0785	7.4	82	0.3079	0.2198	12.9	191	1.4717	1.0503	30.1
48	0.1143	0.0816	7.6	83	0.3149	0.2248	13.1	201	1.6174	1.1543	31.7
49	0.1188	0.0848	7.7	84	0.3220	0.2298	13.2	203	1.6473	1.1757	32.0

Friction Loss Table For 2" Sch. 40 Steel Pipe   I.D. = 2.067

Equiv. Feet of Pipe

	LONG EL	90 ELL	45 ELL	TEE	GV	CV	BV
C-120	3	5	2	10	1	11	6
C-100	2.14	3.57	1.43	7.13	0.71	7.84	4.28

Q GPM	P PSI/FT C-100	P PSI/FT C-120	V FT/SEC	Q GPM	P PSI/FT C-100	P PSI/FT C-120	V FT/SEC	Q GPM	P PSI/FT C-100	P PSI/FT C-120	V FT/SEC
30	0.0142	0.0101	2.9	100	0.1317	0.0940	9.6	170	0.3514	0.2508	16.3
32	0.0160	0.0114	3.1	102	0.1366	0.0975	9.8	172	0.3591	0.2563	16.5
34	0.0179	0.0128	3.3	104	0.1416	0.1010	9.9	174	0.3668	0.2618	16.6
36	0.0199	0.0142	3.4	106	0.1466	0.1047	10.1	176	0.3747	0.2674	16.8
38	0.0220	0.0157	3.6	108	0.1518	0.1083	10.3	178	0.3826	0.2730	17.0
40	0.0242	0.0172	3.8	110	0.1570	0.1121	10.5	180	0.3906	0.2787	17.2
42	0.0265	0.0189	4.0	112	0.1624	0.1159	10.7	182	0.3986	0.2845	17.4
44	0.0288	0.0206	4.2	114	0.1678	0.1197	10.9	184	0.4068	0.2903	17.6
46	0.0313	0.0223	4.4	116	0.1733	0.1236	11.1	186	0.4150	0.2962	17.8
48	0.0339	0.0242	4.6	118	0.1788	0.1276	11.3	188	0.4233	0.3021	18.0
50	0.0365	0.0261	4.8	120	0.1845	0.1317	11.5	190	0.4316	0.3081	18.2
52	0.0393	0.0280	5.0	122	0.1902	0.1357	11.7	195	0.4529	0.3232	18.7
54	0.0421	0.0301	5.2	124	0.1960	0.1399	11.9	200	0.4746	0.3387	19.1
56	0.0450	0.0321	5.4	126	0.2019	0.1441	12.1	205	0.4968	0.3546	19.6
58	0.0481	0.0343	5.5	128	0.2079	0.1483	12.2	210	0.5194	0.3707	20.1
60	0.0512	0.0365	5.7	130	0.2139	0.1527	12.4	215	0.5426	0.3872	20.6
62	0.0544	0.0388	5.9	132	0.2200	0.1570	12.6	220	0.5661	0.4040	21.0
64	0.0577	0.0412	6.1	134	0.2262	0.1615	12.8	225	0.5902	0.4212	21.5
66	0.0610	0.0436	6.3	136	0.2325	0.1660	13.0	230	0.6146	0.4387	22.0
68	0.0645	0.0460	6.5	138	0.2389	0.1705	13.2	235	0.6396	0.4565	22.5
70	0.0681	0.0486	6.7	140	0.2453	0.1751	13.4	240	0.6650	0.4746	23.0
72	0.0717	0.0512	6.9	142	0.2519	0.1798	13.6	245	0.6909	0.4931	23.4
74	0.0754	0.0538	7.1	144	0.2585	0.1845	13.8	250	0.7172	0.5118	23.9
76	0.0792	0.0566	7.3	146	0.2651	0.1892	14.0	255	0.7439	0.5309	24.4
78	0.0831	0.0593	7.5	148	0.2719	0.1941	14.2	260	0.7711	0.5504	24.9
80	0.0871	0.0622	7.7	150	0.2787	0.1989	14.3	265	0.7988	0.5701	25.4
82	0.0912	0.0651	7.8	152	0.2857	0.2039	14.5	270	0.8269	0.5902	25.8
84	0.0954	0.0681	8.0	154	0.2926	0.2089	14.7	275	0.8555	0.6105	26.3
86	0.0996	0.0711	8.2	156	0.2997	0.2139	14.9	280	0.8844	0.6312	26.8
88	0.1039	0.0742	8.4	158	0.3069	0.2190	15.1	285	0.9139	0.6522	27.3
90	0.1083	0.0773	8.6	160	0.3141	0.2242	15.3	295	0.9741	0.6952	28.2
92	0.1128	0.0805	8.8	162	0.3214	0.2294	15.5	305	1.0361	0.7394	29.2
94	0.1174	0.0838	9.0	164	0.3288	0.2346	15.7	315	1.0998	0.7849	30.1
96	0.1221	0.0871	9.2	166	0.3362	0.2400	15.9	325	1.1652	0.8316	31.1
98	0.1268	0.0905	9.4	168	0.3438	0.2453	16.1	335	1.2324	0.8796	32.0

Appendix D Friction Loss Tables

Friction Loss Table For 2½" Sch. 40 Steel Pipe   I.D. = 2.469"

Equiv. Feet of Pipe

	LONG EL	90 ELL	45 ELL	TEE	GV	CV	ALM VA	BV
C-120	4	6	3	12	1	14	10	7
C-100	2.85	4.28	2.14	8.56	0.71	9.98	7.13	4.99

Q GPM	P PSI/FT C-100	P PSI/FT C-120	V FT/SEC	Q GPM	P PSI/FT C-100	P PSI/FT C-120	V FT/SEC	Q GPM	P PSI/FT C-100	P PSI/FT C-120	V FT/SEC
40	0.0102	0.0073	2.7	145	0.1102	0.0786	9.7	250	0.3018	0.2154	16.8
43	0.0116	0.0083	2.9	148	0.1144	0.0817	9.9	255	0.3131	0.2234	17.1
46	0.0132	0.0094	3.1	151	0.1188	0.0848	10.1	260	0.3245	0.2316	17.4
49	0.0148	0.0106	3.3	154	0.1232	0.0879	10.3	265	0.3362	0.2399	17.8
52	0.0165	0.0118	3.5	157	0.1276	0.0911	10.5	270	0.3480	0.2484	18.1
55	0.0183	0.0131	3.7	160	0.1322	0.0943	10.7	275	0.3600	0.2569	18.4
58	0.0202	0.0144	3.9	163	0.1368	0.0976	10.9	280	0.3722	0.2657	18.8
61	0.0222	0.0158	4.1	166	0.1415	0.1010	11.1	285	0.3846	0.2745	19.1
64	0.0243	0.0173	4.3	169	0.1463	0.1044	11.3	290	0.3972	0.2835	19.4
67	0.0264	0.0189	4.5	172	0.1511	0.1078	11.5	295	0.4100	0.2926	19.8
70	0.0286	0.0204	4.7	175	0.1560	0.1114	11.7	300	0.4229	0.3018	20.1
73	0.0310	0.0221	4.9	178	0.1610	0.1149	11.9	305	0.4360	0.3112	20.5
76	0.0333	0.0238	5.1	181	0.1661	0.1185	12.1	310	0.4493	0.3207	20.8
79	0.0358	0.0256	5.3	184	0.1712	0.1222	12.3	315	0.4628	0.3303	21.1
82	0.0384	0.0274	5.5	187	0.1764	0.1259	12.5	320	0.4765	0.3401	21.5
85	0.0410	0.0293	5.7	190	0.1817	0.1296	12.7	325	0.4904	0.3500	21.8
88	0.0437	0.0312	5.9	193	0.1870	0.1335	12.9	330	0.5044	0.3600	22.1
91	0.0465	0.0332	6.1	196	0.1924	0.1373	13.1	335	0.5187	0.3702	22.5
94	0.0494	0.0353	6.3	199	0.1979	0.1412	13.3	340	0.5331	0.3805	22.8
97	0.0524	0.0374	6.5	202	0.2035	0.1452	13.5	345	0.5477	0.3909	23.1
100	0.0554	0.0395	6.7	205	0.2091	0.1492	13.7	350	0.5625	0.4014	23.5
103	0.0585	0.0418	6.9	208	0.2148	0.1533	13.9	355	0.5774	0.4121	23.8
106	0.0617	0.0440	7.1	211	0.2205	0.1574	14.1	360	0.5925	0.4229	24.1
109	0.0650	0.0464	7.3	214	0.2264	0.1616	14.3	365	0.6079	0.4338	24.5
112	0.0683	0.0488	7.5	217	0.2323	0.1658	14.5	370	0.6234	0.4449	24.8
115	0.0718	0.0512	7.7	220	0.2383	0.1700	14.8	380	0.6549	0.4674	25.5
118	0.0753	0.0537	7.9	223	0.2443	0.1744	15.0	390	0.6871	0.4904	26.1
121	0.0788	0.0563	8.1	226	0.2504	0.1787	15.2	400	0.7201	0.5139	26.8
124	0.0825	0.0589	8.3	229	0.2566	0.1831	15.4	410	0.7537	0.5379	27.5
127	0.0862	0.0615	8.5	232	0.2629	0.1876	15.6	420	0.7881	0.5625	28.2
130	0.0900	0.0642	8.7	235	0.2692	0.1921	15.8	430	0.8231	0.5875	28.8
133	0.0939	0.0670	8.9	238	0.2756	0.1967	16.0	440	0.8589	0.6130	29.5
136	0.0979	0.0698	9.1	241	0.2820	0.2013	16.2	450	0.8954	0.6390	30.2
139	0.1019	0.0727	9.3	244	0.2886	0.2059	16.4	460	0.9325	0.6655	30.8
142	0.1060	0.0756	9.5	247	0.2952	0.2107	16.6	477	0.9973	0.7118	32.0

Friction Loss Table For 3" Sch. 40 Steel Pipe   I.D. = 3.068"

Equiv. Feet of Pipe

	LONG EL	90 ELL	45 ELL	TEE	GV	CV	BV
C–120	5	7	3	15	1	16	10
C–100	3.57	4.99	2.14	10.70	0.71	11.41	7.13	0.00

Q	P	P	V	Q	P	P	V	Q	P	P	V
	PSI/FT	PSI/FT	FT/SEC		PSI/FT	PSI/FT	FT/SEC		PSI/FT	PSI/FT	FT/SEC
GPM	C–100	C–120		GPM	C–100	C–120		GPM	C–100	C–120	
50	0.0053	0.0038	2.2	223	0.0848	0.0605	9.7	396	0.2454	0.1751	17.2
55	0.0064	0.0045	2.4	228	0.0884	0.0631	9.9	406	0.2570	0.1834	17.6
60	0.0075	0.0053	2.6	233	0.0920	0.0657	10.1	416	0.2688	0.1919	18.1
65	0.0087	0.0062	2.8	238	0.0957	0.0683	10.3	426	0.2809	0.2005	18.5
70	0.0099	0.0071	3.0	243	0.0994	0.0710	10.6	436	0.2932	0.2093	18.9
75	0.0113	0.0081	3.3	248	0.1032	0.0737	10.8	446	0.3058	0.2182	19.4
80	0.0127	0.0091	3.5	253	0.1071	0.0765	11.0	456	0.3186	0.2274	19.8
85	0.0142	0.0102	3.7	258	0.1111	0.0793	11.2	466	0.3316	0.2367	20.2
90	0.0158	0.0113	3.9	263	0.1151	0.0821	11.4	476	0.3449	0.2462	20.7
95	0.0175	0.0125	4.1	268	0.1192	0.0851	11.6	486	0.3585	0.2558	21.1
100	0.0192	0.0137	4.3	273	0.1233	0.0880	11.9	496	0.3722	0.2656	21.5
105	0.0211	0.0150	4.6	278	0.1275	0.0910	12.1	506	0.3862	0.2756	22.0
110	0.0229	0.0164	4.8	283	0.1318	0.0941	12.3	516	0.4005	0.2858	22.4
115	0.0249	0.0178	5.0	288	0.1362	0.0972	12.5	526	0.4149	0.2961	22.8
120	0.0270	0.0192	5.2	293	0.1406	0.1003	12.7	536	0.4296	0.3066	23.3
125	0.0291	0.0207	5.4	298	0.1450	0.1035	12.9	546	0.4446	0.3173	23.7
130	0.0313	0.0223	5.6	303	0.1496	0.1067	13.2	556	0.4598	0.3281	24.1
135	0.0335	0.0239	5.9	308	0.1542	0.1100	13.4	566	0.4752	0.3391	24.6
140	0.0359	0.0256	6.1	313	0.1588	0.1134	13.6	576	0.4908	0.3503	25.0
145	0.0383	0.0273	6.3	318	0.1635	0.1167	13.8	586	0.5067	0.3616	25.4
150	0.0407	0.0291	6.5	323	0.1683	0.1201	14.0	596	0.5228	0.3731	25.9
155	0.0433	0.0309	6.7	328	0.1732	0.1236	14.2	606	0.5392	0.3848	26.3
160	0.0459	0.0328	6.9	333	0.1781	0.1271	14.5	616	0.5557	0.3966	26.7
165	0.0486	0.0347	7.2	338	0.1831	0.1307	14.7	626	0.5726	0.4086	27.2
170	0.0513	0.0366	7.4	343	0.1881	0.1343	14.9	636	0.5896	0.4208	27.6
175	0.0542	0.0387	7.6	348	0.1932	0.1379	15.1	646	0.6069	0.4331	28.1
180	0.0571	0.0407	7.8	353	0.1984	0.1416	15.3	656	0.6243	0.4456	28.5
185	0.0600	0.0428	8.0	358	0.2036	0.1453	15.5	666	0.6421	0.4582	28.9
190	0.0631	0.0450	8.3	363	0.2089	0.1491	15.8	676	0.6600	0.4711	29.4
195	0.0662	0.0472	8.5	368	0.2143	0.1529	16.0	686	0.6782	0.4840	29.8
200	0.0694	0.0495	8.7	373	0.2197	0.1568	16.2	696	0.6966	0.4972	30.2
205	0.0726	0.0518	8.9	378	0.2252	0.1607	16.4	706	0.7152	0.5105	30.7
210	0.0759	0.0542	9.1	383	0.2307	0.1647	16.6	716	0.7341	0.5239	31.1
215	0.0793	0.0566	9.3	388	0.2363	0.1687	16.8	726	0.7532	0.5375	31.5
220	0.0827	0.0590	9.6	393	0.2420	0.1727	17.1	736	0.7725	0.5513	32.0

Friction Loss Table For 4" Sch. 40 Steel Pipe I.D. = 4.026"

Equiv. Feet of Pipe

	LONG ELL	90 ELL	45 ELL	TEE	GV	CV	ALM VA	DRY VA
C-120	6	10	4	20	2	22	20	20
C-100	4.28	7.13	2.85	14.26	1.43	15.69	14.26	14.26

Q	P	P	V	Q	P	P	V	Q	P	P	V
	PSI/FT	PSI/FT	FT/SEC		PSI/FT	PSI/FT	FT/SEC		PSI/FT	PSI/FT	FT/SEC
GPM	C-100	C-120		GPM	C-100	C-120		GPM	C-100	C-120	
100	0.0051	0.0037	2.5	450	0.0828	0.0591	11.3	800	0.2399	0.1713	20.2
110	0.0061	0.0044	2.8	460	0.0862	0.0615	11.6	815	0.2483	0.1772	20.6
120	0.0072	0.0051	3.0	470	0.0897	0.0640	11.9	830	0.2569	0.1833	20.9
130	0.0083	0.0059	3.3	480	0.0933	0.0666	12.1	845	0.2655	0.1895	21.3
140	0.0095	0.0068	3.5	490	0.0969	0.0691	12.4	860	0.2743	0.1958	21.7
150	0.0108	0.0077	3.8	500	0.1006	0.0718	12.6	875	0.2832	0.2021	22.1
160	0.0122	0.0087	4.0	510	0.1043	0.0745	12.9	890	0.2923	0.2086	22.4
170	0.0137	0.0098	4.3	520	0.1081	0.0772	13.1	905	0.3014	0.2151	22.8
180	0.0152	0.0108	4.5	530	0.1120	0.0800	13.4	920	0.3107	0.2218	23.2
190	0.0168	0.0120	4.8	540	0.1160	0.0828	13.6	935	0.3202	0.2285	23.6
200	0.0185	0.0132	5.0	550	0.1200	0.0856	13.9	950	0.3298	0.2353	24.0
210	0.0202	0.0144	5.3	560	0.1240	0.0885	14.1	965	0.3395	0.2423	24.3
220	0.0220	0.0157	5.5	570	0.1282	0.0915	14.4	980	0.3493	0.2493	24.7
230	0.0239	0.0171	5.8	580	0.1324	0.0945	14.6	995	0.3592	0.2564	25.1
240	0.0259	0.0185	6.1	590	0.1366	0.0975	14.9	1010	0.3693	0.2636	25.5
250	0.0279	0.0199	6.3	600	0.1409	0.1006	15.1	1025	0.3795	0.2709	25.8
260	0.0300	0.0214	6.6	610	0.1453	0.1037	15.4	1040	0.3899	0.2782	26.2
270	0.0322	0.0230	6.8	620	0.1497	0.1069	15.6	1055	0.4003	0.2857	26.6
280	0.0344	0.0246	7.1	630	0.1542	0.1101	15.9	1070	0.4109	0.2933	27.0
290	0.0367	0.0262	7.3	640	0.1588	0.1133	16.1	1085	0.4216	0.3009	27.4
300	0.0391	0.0279	7.6	650	0.1634	0.1166	16.4	1100	0.4325	0.3087	27.7
310	0.0415	0.0296	7.8	660	0.1681	0.1200	16.6	1115	0.4435	0.3165	28.1
320	0.0440	0.0314	8.1	670	0.1728	0.1234	16.9	1130	0.4546	0.3244	28.5
330	0.0466	0.0333	8.3	680	0.1776	0.1268	17.1	1145	0.4658	0.3324	28.9
340	0.0493	0.0352	8.6	690	0.1825	0.1303	17.4	1160	0.4771	0.3405	29.3
350	0.0520	0.0371	8.8	700	0.1874	0.1338	17.7	1175	0.4886	0.3487	29.6
360	0.0548	0.0391	9.1	710	0.1924	0.1373	17.9	1190	0.5002	0.3570	30.0
370	0.0576	0.0411	9.3	720	0.1975	0.1409	18.2	1205	0.5120	0.3654	30.4
380	0.0605	0.0432	9.6	730	0.2026	0.1446	18.4	1220	0.5238	0.3738	30.8
390	0.0635	0.0453	9.8	740	0.2077	0.1483	18.7	1235	0.5358	0.3824	31.1
400	0.0666	0.0475	10.1	750	0.2129	0.1520	18.9	1250	0.5479	0.3910	31.5
410	0.0697	0.0497	10.3	760	0.2182	0.1557	19.2	1265	0.5601	0.3997	31.9
420	0.0728	0.0520	10.6	770	0.2236	0.1596	19.4	1280	0.5725	0.4086	32.3
430	0.0761	0.0543	10.8	780	0.2290	0.1634	19.7	1295	0.5849	0.4175	32.7
440	0.0794	0.0567	11.1	790	0.2344	0.1673	19.9	1310	0.5975	0.4265	33.0

Friction Loss Table For 6" Sch. 40 Steel Pipe   I.D. = 6.065

Equiv. Feet of Pipe

	LONG EL	90 ELL	45 ELL	TEE	GV	CV	ALM VA	DRY VA	BV
C-120	9	14	7	30	3	32	18	19	10
C-100	6.42	9.98	4.99	21.39	2.14	22.82	12.83	13.55	7.13

Q	P	P	V	Q	P	P	V	Q	P	P	V
	PSI/FT	PSI/FT	FT/SEC		PSI/FT	PSI/FT	FT/SEC		PSI/FT	PSI/FT	FT/SEC
GPM	C-100	C-120		GPM	C-100	C-120		GPM	C-100	C-120	
400	0.0090	0.0065	4.4	1100	0.0588	0.0420	12.2	1800	0.1462	0.1044	20.0
420	0.0099	0.0071	4.7	1120	0.0608	0.0434	12.4	1830	0.1508	0.1076	20.3
440	0.0108	0.0077	4.9	1140	0.0628	0.0448	12.7	1860	0.1554	0.1109	20.7
460	0.0117	0.0084	5.1	1160	0.0649	0.0463	12.9	1890	0.1600	0.1142	21.0
480	0.0127	0.0090	5.3	1180	0.0669	0.0478	13.1	1920	0.1648	0.1176	21.3
500	0.0137	0.0098	5.6	1200	0.0691	0.0493	13.3	1950	0.1696	0.1210	21.7
520	0.0147	0.0105	5.8	1220	0.0712	0.0508	13.6	1980	0.1744	0.1245	22.0
540	0.0158	0.0113	6.0	1240	0.0734	0.0524	13.8	2010	0.1793	0.1280	22.3
560	0.0169	0.0120	6.2	1260	0.0756	0.0539	14.0	2040	0.1843	0.1315	22.7
580	0.0180	0.0128	6.4	1280	0.0778	0.0555	14.2	2070	0.1894	0.1351	23.0
600	0.0192	0.0137	6.7	1300	0.0801	0.0572	14.4	2100	0.1945	0.1388	23.3
620	0.0204	0.0145	6.9	1320	0.0824	0.0588	14.7	2130	0.1996	0.1425	23.7
640	0.0216	0.0154	7.1	1340	0.0847	0.0605	14.9	2160	0.2049	0.1462	24.0
660	0.0229	0.0163	7.3	1360	0.0871	0.0621	15.1	2190	0.2102	0.1500	24.3
680	0.0241	0.0172	7.6	1380	0.0894	0.0638	15.3	2220	0.2155	0.1538	24.7
700	0.0255	0.0182	7.8	1400	0.0919	0.0656	15.6	2250	0.2209	0.1577	25.0
720	0.0268	0.0192	8.0	1420	0.0943	0.0673	15.8	2280	0.2264	0.1616	25.3
740	0.0282	0.0202	8.2	1440	0.0968	0.0691	16.0	2310	0.2320	0.1656	25.7
760	0.0297	0.0212	8.4	1460	0.0993	0.0708	16.2	2340	0.2376	0.1696	26.0
780	0.0311	0.0222	8.7	1480	0.1018	0.0727	16.4	2370	0.2432	0.1736	26.3
800	0.0326	0.0233	8.9	1500	0.1044	0.0745	16.7	2400	0.2490	0.1777	26.7
820	0.0341	0.0244	9.1	1520	0.1069	0.0763	16.9	2430	0.2548	0.1818	27.0
840	0.0357	0.0255	9.3	1540	0.1096	0.0782	17.1	2460	0.2606	0.1860	27.3
860	0.0373	0.0266	9.6	1560	0.1122	0.0801	17.3	2490	0.2665	0.1902	27.7
880	0.0389	0.0278	9.8	1580	0.1149	0.0820	17.6	2520	0.2725	0.1945	28.0
900	0.0406	0.0289	10.0	1600	0.1176	0.0839	17.8	2550	0.2785	0.1988	28.3
920	0.0422	0.0301	10.2	1620	0.1203	0.0859	18.0	2580	0.2846	0.2031	28.7
940	0.0440	0.0314	10.4	1640	0.1231	0.0878	18.2	2610	0.2908	0.2075	29.0
960	0.0457	0.0326	10.7	1660	0.1259	0.0898	18.4	2640	0.2970	0.2120	29.3
980	0.0475	0.0339	10.9	1680	0.1287	0.0919	18.7	2670	0.3032	0.2164	29.7
1000	0.0493	0.0352	11.1	1700	0.1315	0.0939	18.9	2700	0.3096	0.2209	30.0
1020	0.0511	0.0365	11.3	1720	0.1344	0.0959	19.1	2730	0.3160	0.2255	30.3
1040	0.0530	0.0378	11.6	1740	0.1373	0.0980	19.3	2760	0.3224	0.2301	30.7
1060	0.0549	0.0392	11.8	1760	0.1403	0.1001	19.6	2790	0.3289	0.2348	31.0
1080	0.0568	0.0406	12.0	1780	0.1432	0.1022	19.8	2880	0.3488	0.2490	32.0

Appendix D Friction Loss Tables

Friction Loss Table For 8" Sch. 30 Steel Pipe   I.D. = 8.071

Equiv. Feet of Pipe

	LONG ELL	90 ELL	45 ELL	TEE	GV	CV	ALM VA	DRY VA	BV
C-120	13	18	9	35	4	45	35	27	12
C-100	9.27	12.83	6.42	24.96	2.85	32.09	24.96	19.25	8.56

Q	P	P	V	Q	P	P	V	Q	P	P	V
	PSI/FT	PSI/FT	FT/SEC		PSI/FT	PSI/FT	FT/SEC		PSI/FT	PSI/FT	FT/SEC
GPM	C-100	C-120		GPM	C-100	C-120		GPM	C-100	C-120	
500	0.0034	0.0024	3.1	1890	0.0398	0.0284	11.9	3280	0.1103	0.0788	20.6
540	0.0039	0.0028	3.4	1930	0.0414	0.0295	12.1	3330	0.1135	0.0810	20.9
580	0.0045	0.0032	3.6	1970	0.0430	0.0307	12.4	3380	0.1167	0.0833	21.2
620	0.0051	0.0036	3.9	2010	0.0446	0.0318	12.6	3430	0.1199	0.0855	21.5
660	0.0057	0.0041	4.1	2050	0.0463	0.0330	12.9	3480	0.1231	0.0879	21.8
700	0.0063	0.0045	4.4	2090	0.0479	0.0342	13.1	3530	0.1264	0.0902	22.1
740	0.0070	0.0050	4.6	2130	0.0496	0.0354	13.4	3580	0.1297	0.0926	22.5
780	0.0077	0.0055	4.9	2170	0.0514	0.0367	13.6	3630	0.1331	0.0950	22.8
820	0.0085	0.0061	5.1	2210	0.0532	0.0379	13.9	3680	0.1365	0.0974	23.1
860	0.0093	0.0066	5.4	2250	0.0549	0.0392	14.1	3730	0.1400	0.0999	23.4
900	0.0101	0.0072	5.6	2290	0.0568	0.0405	14.4	3780	0.1435	0.1024	23.7
940	0.0109	0.0078	5.9	2330	0.0586	0.0418	14.6	3830	0.1470	0.1049	24.0
980	0.0118	0.0084	6.1	2370	0.0605	0.0432	14.9	3880	0.1506	0.1075	24.3
1020	0.0127	0.0091	6.4	2410	0.0624	0.0445	15.1	3930	0.1542	0.1100	24.7
1060	0.0137	0.0097	6.7	2450	0.0643	0.0459	15.4	3980	0.1578	0.1126	25.0
1100	0.0146	0.0104	6.9	2490	0.0663	0.0473	15.6	4030	0.1615	0.1153	25.3
1140	0.0156	0.0111	7.2	2530	0.0683	0.0487	15.9	4080	0.1652	0.1179	25.6
1180	0.0166	0.0119	7.4	2570	0.0703	0.0502	16.1	4130	0.1690	0.1206	25.9
1220	0.0177	0.0126	7.7	2610	0.0723	0.0516	16.4	4180	0.1728	0.1233	26.2
1260	0.0188	0.0134	7.9	2650	0.0744	0.0531	16.6	4230	0.1767	0.1261	26.5
1300	0.0199	0.0142	8.2	2690	0.0765	0.0546	16.9	4280	0.1805	0.1289	26.9
1340	0.0211	0.0150	8.4	2730	0.0786	0.0561	17.1	4330	0.1845	0.1316	27.2
1380	0.0222	0.0159	8.7	2770	0.0807	0.0576	17.4	4380	0.1884	0.1345	27.5
1420	0.0234	0.0167	8.9	2810	0.0829	0.0592	17.6	4430	0.1924	0.1373	27.8
1460	0.0247	0.0176	9.2	2850	0.0851	0.0607	17.9	4480	0.1965	0.1402	28.1
1500	0.0260	0.0185	9.4	2890	0.0873	0.0623	18.1	4530	0.2005	0.1431	28.4
1540	0.0272	0.0194	9.7	2930	0.0896	0.0639	18.4	4580	0.2046	0.1461	28.7
1580	0.0286	0.0204	9.9	2970	0.0918	0.0655	18.6	4630	0.2088	0.1490	29.1
1620	0.0299	0.0214	10.2	3010	0.0941	0.0672	18.9	4680	0.2130	0.1520	29.4
1660	0.0313	0.0223	10.4	3050	0.0965	0.0688	19.1	4730	0.2172	0.1550	29.7
1700	0.0327	0.0233	10.7	3090	0.0988	0.0705	19.4	4780	0.2215	0.1581	30.0
1740	0.0342	0.0244	10.9	3130	0.1012	0.0722	19.6	4830	0.2258	0.1611	30.3
1780	0.0356	0.0254	11.2	3170	0.1036	0.0739	19.9	4880	0.2301	0.1642	30.6
1820	0.0371	0.0265	11.4	3210	0.1060	0.0757	20.1	4930	0.2345	0.1674	30.9
1860	0.0386	0.0276	11.7	3250	0.1085	0.0774	20.4	5100	0.2497	0.1782	32.0

Friction Loss Table For 1¼" Sch. 10 Steel Pipe   I.D. = 1.442

Equiv. Feet of Pipe

	LONG ELL	90 ELL	45 ELL	TEE	GV	CV				
C−120	2	3	1	6		7				
C−100	1.43	2.14	0.71	4.28	0.00	4.99	0.00	0.00		

Q	P	P	V	Q	P	P	V	Q	P	P	V
	PSI/FT	PSI/FT	FT/SEC		PSI/FT	PSI/FT	FT/SEC		PSI/FT	PSI/FT	FT/SEC
GPM	C−100	C−120		GPM	C−100	C−120		GPM	C−100	C−120	
7	0.0056	0.0040	1.4	42	0.1528	0.1090	8.3	77	0.4688	0.3346	15.1
8	0.0071	0.0051	1.6	43	0.1595	0.1139	8.5	80	0.5031	0.3591	15.7
9	0.0088	0.0063	1.8	44	0.1665	0.1188	8.6	83	0.5386	0.3844	16.3
10	0.0107	0.0077	2.0	45	0.1735	0.1239	8.8	86	0.5752	0.4105	16.9
11	0.0128	0.0091	2.2	46	0.1808	0.1290	9.0	89	0.6128	0.4374	17.5
12	0.0150	0.0107	2.4	47	0.1881	0.1342	9.2	92	0.6516	0.4651	18.1
13	0.0174	0.0125	2.6	48	0.1956	0.1396	9.4	95	0.6915	0.4935	18.7
14	0.0200	0.0143	2.8	49	0.2032	0.1450	9.6	98	0.7324	0.5227	19.3
15	0.0227	0.0162	2.9	50	0.2109	0.1505	9.8	101	0.7744	0.5527	19.9
16	0.0256	0.0183	3.1	51	0.2188	0.1561	10.0	104	0.8175	0.5835	20.4
17	0.0287	0.0205	3.3	52	0.2268	0.1618	10.2	107	0.8617	0.6150	21.0
18	0.0319	0.0227	3.5	53	0.2349	0.1676	10.4	110	0.9069	0.6472	21.6
19	0.0352	0.0251	3.7	54	0.2432	0.1735	10.6	113	0.9532	0.6803	22.2
20	0.0387	0.0276	3.9	55	0.2516	0.1795	10.8	116	1.0005	0.7141	22.8
21	0.0424	0.0302	4.1	56	0.2601	0.1856	11.0	119	1.0489	0.7486	23.4
22	0.0462	0.0330	4.3	57	0.2687	0.1918	11.2	122	1.0984	0.7839	24.0
23	0.0501	0.0358	4.5	58	0.2775	0.1981	11.4	125	1.1488	0.8199	24.6
24	0.0542	0.0387	4.7	59	0.2865	0.2044	11.6	128	1.2004	0.8567	25.2
25	0.0585	0.0418	4.9	60	0.2955	0.2109	11.8	131	1.2529	0.8942	25.8
26	0.0629	0.0449	5.1	61	0.3047	0.2174	12.0	134	1.3065	0.9325	26.3
27	0.0675	0.0481	5.3	62	0.3140	0.2241	12.2	137	1.3612	0.9715	26.9
28	0.0721	0.0515	5.5	63	0.3234	0.2308	12.4	140	1.4168	1.0112	27.5
29	0.0770	0.0549	5.7	64	0.3330	0.2376	12.6	143	1.4735	1.0516	28.1
30	0.0820	0.0585	5.9	65	0.3427	0.2446	12.8	146	1.5312	1.0928	28.7
31	0.0871	0.0622	6.1	66	0.3525	0.2516	13.0	149	1.5899	1.1347	29.3
32	0.0924	0.0659	6.3	67	0.3624	0.2587	13.2	152	1.6496	1.1773	29.9
33	0.0978	0.0698	6.5	68	0.3725	0.2658	13.4	155	1.7104	1.2207	30.5
34	0.1033	0.0737	6.7	69	0.3827	0.2731	13.6	158	1.7721	1.2648	31.1
35	0.1090	0.0778	6.9	70	0.3930	0.2805	13.8	161	1.8349	1.3095	31.6
36	0.1149	0.0820	7.1	71	0.4035	0.2880	14.0	164	1.8986	1.3550	32.2
37	0.1208	0.0862	7.3	72	0.4140	0.2955	14.2	167	1.9634	1.4013	32.8
38	0.1269	0.0906	7.5	73	0.4247	0.3031	14.3	170	2.0291	1.4482	33.4
39	0.1332	0.0951	7.7	74	0.4356	0.3109	14.5	173	2.0959	1.4958	34.0
40	0.1396	0.0996	7.9	75	0.4465	0.3187	14.7	176	2.1636	1.5442	34.6
41	0.1461	0.1043	8.1	76	0.4576	0.3266	14.9	179	2.2323	1.5932	35.2

Friction Loss Table For 1½" Sch. 10 Steel Pipe   I.D. = 1.682

Equiv. Feet of Pipe

	LONG EL	90 ELL	45 ELL	TEE	GV	CV		
C-120	2	4	2	8		9		
C-100	1.43	2.85	1.43	5.70	0.00	6.42	0.00	0.00

Q GPM	P PSI/FT C-100	P PSI/FT C-120	V FT/SEC	Q GPM	P PSI/FT C-100	P PSI/FT C-120	V FT/SEC	Q GPM	P PSI/FT C-100	P PSI/FT C-120	V FT/SEC
7	0.0026	0.0019	1.0	76	0.2162	0.1543	11.0	145	0.7143	0.5098	20.9
9	0.0042	0.0030	1.3	78	0.2269	0.1619	11.3	148	0.7419	0.5295	21.4
11	0.0061	0.0043	1.6	80	0.2377	0.1697	11.6	151	0.7700	0.5495	21.8
13	0.0082	0.0059	1.9	82	0.2488	0.1776	11.8	154	0.7985	0.5699	22.2
15	0.0107	0.0077	2.2	84	0.2602	0.1857	12.1	157	0.8275	0.5906	22.7
17	0.0135	0.0097	2.5	86	0.2718	0.1940	12.4	160	0.8570	0.6116	23.1
19	0.0166	0.0119	2.7	88	0.2836	0.2024	12.7	163	0.8870	0.6330	23.5
21	0.0200	0.0143	3.0	90	0.2956	0.2110	13.0	166	0.9174	0.6548	24.0
23	0.0237	0.0169	3.3	92	0.3079	0.2197	13.3	169	0.9483	0.6768	24.4
25	0.0276	0.0197	3.6	94	0.3204	0.2286	13.6	172	0.9797	0.6992	24.8
27	0.0319	0.0227	3.9	96	0.3331	0.2377	13.9	175	1.0115	0.7219	25.3
29	0.0364	0.0260	4.2	98	0.3460	0.2470	14.2	178	1.0439	0.7450	25.7
31	0.0412	0.0294	4.5	100	0.3592	0.2564	14.4	181	1.0766	0.7684	26.1
33	0.0462	0.0330	4.8	102	0.3726	0.2659	14.7	184	1.1099	0.7921	26.6
35	0.0515	0.0368	5.1	104	0.3863	0.2757	15.0	187	1.1436	0.8162	27.0
37	0.0571	0.0407	5.3	106	0.4001	0.2856	15.3	190	1.1778	0.8406	27.4
39	0.0629	0.0449	5.6	108	0.4142	0.2956	15.6	193	1.2124	0.8653	27.9
41	0.0690	0.0493	5.9	110	0.4285	0.3058	15.9	196	1.2475	0.8903	28.3
43	0.0754	0.0538	6.2	112	0.4430	0.3162	16.2	199	1.2831	0.9157	28.8
45	0.0820	0.0585	6.5	114	0.4578	0.3267	16.5	202	1.3191	0.9414	29.2
47	0.0889	0.0634	6.8	116	0.4727	0.3374	16.8	205	1.3555	0.9674	29.6
49	0.0960	0.0685	7.1	118	0.4879	0.3482	17.0	208	1.3925	0.9938	30.1
51	0.1034	0.0738	7.4	120	0.5033	0.3592	17.3	211	1.4298	1.0205	30.5
53	0.1110	0.0792	7.7	122	0.5190	0.3704	17.6	214	1.4677	1.0475	30.9
55	0.1189	0.0848	7.9	124	0.5348	0.3817	17.9	217	1.5060	1.0748	31.4
57	0.1270	0.0906	8.2	126	0.5509	0.3932	18.2	220	1.5447	1.1025	31.8
59	0.1353	0.0966	8.5	128	0.5672	0.4048	18.5	223	1.5839	1.1304	32.2
61	0.1440	0.1027	8.8	130	0.5837	0.4166	18.8	226	1.6236	1.1587	32.7
63	0.1528	0.1091	9.1	132	0.6004	0.4285	19.1	229	1.6636	1.1873	33.1
65	0.1619	0.1156	9.4	134	0.6173	0.4406	19.4	232	1.7042	1.2163	33.5
67	0.1712	0.1222	9.7	136	0.6345	0.4528	19.6	235	1.7452	1.2455	34.0
69	0.1808	0.1290	10.0	138	0.6518	0.4652	19.9	238	1.7866	1.2751	34.4
71	0.1906	0.1361	10.3	140	0.6694	0.4778	20.2	241	1.8285	1.3050	34.8
73	0.2007	0.1432	10.5	142	0.6872	0.4905	20.5	244	1.8708	1.3352	35.3
75	0.2110	0.1506	10.8	144	0.7052	0.5033	20.8	247	1.9136	1.3657	35.7

Friction Loss Table For 2" Sch. 10 Steel Pipe   I.D. = 2.157

Equiv. Feet of Pipe

	LONG EL	90 ELL	45 ELL	TEE	GV	CV	BV	DRY VA		
C-120	3	5	2	10	1	11	6			
C-100	2.14	3.57	1.43	7.13	0.71	7.84	4.28	0.00		

Q	P	P	V	Q	P	P	V	Q	P	P	V
	PSI/FT	PSI/FT	FT/SEC		PSI/FT	PSI/FT	FT/SEC		PSI/FT	PSI/FT	FT/SEC
GPM	C-100	C-120		GPM	C-100	C-120		GPM	C-100	C-120	
7	0.0008	0.0006	0.6	111	0.1298	0.0926	9.8	215	0.4409	0.3146	18.9
10	0.0015	0.0011	0.9	114	0.1363	0.0973	10.0	219	0.4562	0.3256	19.2
13	0.0025	0.0018	1.1	117	0.1430	0.1021	10.3	223	0.4717	0.3366	19.6
16	0.0036	0.0026	1.4	120	0.1499	0.1070	10.5	227	0.4875	0.3479	19.9
19	0.0050	0.0035	1.7	123	0.1569	0.1120	10.8	231	0.5035	0.3593	20.3
22	0.0065	0.0046	1.9	126	0.1640	0.1171	11.1	235	0.5197	0.3709	20.6
25	0.0082	0.0059	2.2	129	0.1713	0.1223	11.3	239	0.5362	0.3827	21.0
28	0.0102	0.0072	2.5	132	0.1788	0.1276	11.6	243	0.5529	0.3946	21.3
31	0.0123	0.0087	2.7	135	0.1864	0.1330	11.9	247	0.5699	0.4067	21.7
34	0.0145	0.0104	3.0	138	0.1941	0.1385	12.1	251	0.5871	0.4190	22.1
37	0.0170	0.0121	3.3	141	0.2020	0.1442	12.4	255	0.6045	0.4314	22.4
40	0.0196	0.0140	3.5	144	0.2100	0.1499	12.7	259	0.6221	0.4440	22.8
43	0.0224	0.0160	3.8	147	0.2182	0.1557	12.9	263	0.6400	0.4568	23.1
46	0.0254	0.0182	4.0	150	0.2265	0.1616	13.2	267	0.6582	0.4697	23.5
49	0.0286	0.0204	4.3	153	0.2349	0.1677	13.4	271	0.6765	0.4828	23.8
52	0.0319	0.0228	4.6	156	0.2435	0.1738	13.7	275	0.6951	0.4961	24.2
55	0.0354	0.0253	4.8	159	0.2523	0.1800	14.0	279	0.7139	0.5095	24.5
58	0.0391	0.0279	5.1	162	0.2612	0.1864	14.2	283	0.7330	0.5231	24.9
61	0.0429	0.0306	5.4	165	0.2702	0.1928	14.5	287	0.7523	0.5369	25.2
64	0.0469	0.0334	5.6	168	0.2793	0.1994	14.8	291	0.7718	0.5508	25.6
67	0.0510	0.0364	5.9	171	0.2886	0.2060	15.0	295	0.7915	0.5649	25.9
70	0.0553	0.0395	6.1	174	0.2981	0.2127	15.3	299	0.8115	0.5792	26.3
73	0.0598	0.0427	6.4	177	0.3076	0.2196	15.5	303	0.8317	0.5936	26.6
76	0.0644	0.0460	6.7	180	0.3174	0.2265	15.8	307	0.8521	0.6081	27.0
79	0.0692	0.0494	6.9	183	0.3272	0.2335	16.1	311	0.8728	0.6229	27.3
82	0.0741	0.0529	7.2	186	0.3372	0.2407	16.3	315	0.8936	0.6378	27.7
85	0.0792	0.0565	7.5	189	0.3473	0.2479	16.6	319	0.9147	0.6529	28.0
88	0.0844	0.0603	7.7	192	0.3576	0.2552	16.9	323	0.9361	0.6681	28.4
91	0.0898	0.0641	8.0	195	0.3680	0.2626	17.1	327	0.9576	0.6835	28.7
94	0.0954	0.0681	8.3	198	0.3785	0.2702	17.4	331	0.9794	0.6990	29.1
97	0.1011	0.0722	8.5	201	0.3892	0.2778	17.7	335	1.0014	0.7147	29.4
100	0.1070	0.0763	8.8	204	0.4000	0.2855	17.9	339	1.0237	0.7306	29.8
103	0.1130	0.0806	9.0	207	0.4110	0.2933	18.2	343	1.0461	0.7466	30.1
106	0.1192	0.0850	9.3	210	0.4221	0.3012	18.4	347	1.0688	0.7628	30.5
109	0.1255	0.0895	9.6	213	0.4333	0.3092	18.7	364	1.1677	0.8334	32.0

Appendix D Friction Loss Tables

Friction Loss Table For 2½" Sch. 10 Steel Pipe   I.D. = 2.635

Equiv. Feet of Pipe

	LONG EL	90 ELL	45 ELL	TEE	GV	CV	BV			
C-120	4	6	3	12	1	14	7			
C-100	2.85	4.28	2.14	8.56	0.71	9.98	4.99	0.00		

Q	P	P	V	Q	P	P	V	Q	P	P	V
	PSI/FT	PSI/FT	FT/SEC		PSI/FT	PSI/FT	FT/SEC		PSI/FT	PSI/FT	FT/SEC
GPM	C-100	C-120		GPM	C-100	C-120		GPM	C-100	C-120	
15	0.0012	0.0009	0.9	188	0.1298	0.0926	11.1	361	0.4338	0.3096	21.3
20	0.0021	0.0015	1.2	193	0.1362	0.0972	11.4	367	0.4473	0.3192	21.6
25	0.0031	0.0022	1.5	198	0.1428	0.1019	11.7	373	0.4609	0.3289	22.0
30	0.0044	0.0031	1.8	203	0.1496	0.1067	12.0	379	0.4747	0.3388	22.3
35	0.0058	0.0041	2.1	208	0.1564	0.1117	12.2	385	0.4887	0.3488	22.7
40	0.0074	0.0053	2.4	213	0.1635	0.1167	12.5	391	0.5029	0.3589	23.0
45	0.0092	0.0066	2.6	218	0.1706	0.1218	12.8	397	0.5172	0.3692	23.4
50	0.0112	0.0080	2.9	223	0.1780	0.1270	13.1	403	0.5318	0.3795	23.7
55	0.0134	0.0095	3.2	228	0.1854	0.1323	13.4	409	0.5465	0.3901	24.1
60	0.0157	0.0112	3.5	233	0.1930	0.1377	13.7	415	0.5615	0.4007	24.4
65	0.0182	0.0130	3.8	238	0.2007	0.1433	14.0	421	0.5766	0.4115	24.8
70	0.0209	0.0149	4.1	243	0.2086	0.1489	14.3	427	0.5919	0.4224	25.1
75	0.0237	0.0169	4.4	248	0.2166	0.1546	14.6	433	0.6074	0.4335	25.5
80	0.0267	0.0191	4.7	253	0.2248	0.1604	14.9	439	0.6230	0.4446	25.8
85	0.0299	0.0213	5.0	258	0.2330	0.1663	15.2	445	0.6389	0.4560	26.2
90	0.0332	0.0237	5.3	263	0.2415	0.1723	15.5	451	0.6549	0.4674	26.5
95	0.0367	0.0262	5.6	268	0.2500	0.1784	15.8	457	0.6711	0.4790	26.9
100	0.0404	0.0288	5.9	273	0.2587	0.1847	16.1	463	0.6875	0.4907	27.3
105	0.0442	0.0315	6.2	278	0.2676	0.1910	16.4	469	0.7041	0.5025	27.6
110	0.0481	0.0344	6.5	283	0.2765	0.1974	16.7	475	0.7208	0.5144	28.0
115	0.0523	0.0373	6.8	288	0.2856	0.2039	17.0	481	0.7377	0.5265	28.3
120	0.0565	0.0404	7.1	293	0.2949	0.2105	17.2	487	0.7549	0.5387	28.7
125	0.0610	0.0435	7.4	298	0.3043	0.2171	17.5	493	0.7722	0.5511	29.0
130	0.0656	0.0468	7.7	303	0.3138	0.2239	17.8	499	0.7896	0.5636	29.4
135	0.0703	0.0502	7.9	308	0.3234	0.2308	18.1	505	0.8073	0.5762	29.7
140	0.0752	0.0537	8.2	313	0.3332	0.2378	18.4	511	0.8251	0.5889	30.1
145	0.0803	0.0573	8.5	318	0.3431	0.2449	18.7	517	0.8431	0.6017	30.4
150	0.0854	0.0610	8.8	323	0.3532	0.2520	19.0	523	0.8613	0.6147	30.8
155	0.0908	0.0648	9.1	328	0.3633	0.2593	19.3	529	0.8797	0.6278	31.1
160	0.0963	0.0687	9.4	333	0.3736	0.2667	19.6	535	0.8982	0.6411	31.5
165	0.1019	0.0727	9.7	338	0.3841	0.2741	19.9	541	0.9170	0.6544	31.8
170	0.1077	0.0769	10.0	343	0.3947	0.2817	20.2	547	0.9359	0.6679	32.2
175	0.1136	0.0811	10.3	348	0.4054	0.2893	20.5	553	0.9549	0.6815	32.6
180	0.1197	0.0854	10.6	353	0.4162	0.2971	20.8	559	0.9742	0.6953	32.9
185	0.1260	0.0899	10.9	358	0.4272	0.3049	21.1	565	0.9936	0.7092	33.3

Friction Loss Table For 3" Sch. 10 Steel Pipe I.D. = 3.26

Equiv. Feet of Pipe

	LONG ELL	90 ELL	45 ELL	TEE	GV	CV	BV			
C-120	5	7	3	15	1	16	10			
C-100	3.57	4.99	2.14	10.70	0.71	11.41	7.13	0.00		

Q	P	P	V	Q	P	P	V	Q	P	P	V
	PSI/FT	PSI/FT	FT/SEC		PSI/FT	PSI/FT	FT/SEC		PSI/FT	PSI/FT	FT/SEC
GPM	C-100	C-120		GPM	C-100	C-120		GPM	C-100	C-120	
20	0.0007	0.0005	0.8	297	0.1072	0.0765	11.4	574	0.3629	0.2590	22.1
28	0.0014	0.0010	1.1	305	0.1126	0.0804	11.7	582	0.3723	0.2657	22.4
36	0.0022	0.0015	1.4	313	0.1182	0.0843	12.0	590	0.3818	0.2725	22.7
44	0.0031	0.0022	1.7	321	0.1238	0.0884	12.3	598	0.3914	0.2794	23.0
52	0.0043	0.0030	2.0	329	0.1296	0.0925	12.7	606	0.4012	0.2863	23.3
60	0.0056	0.0040	2.3	337	0.1355	0.0967	13.0	614	0.4110	0.2934	23.6
68	0.0070	0.0050	2.6	345	0.1415	0.1010	13.3	622	0.4210	0.3005	23.9
76	0.0086	0.0061	2.9	353	0.1476	0.1054	13.6	630	0.4311	0.3077	24.2
84	0.0104	0.0074	3.2	361	0.1539	0.1098	13.9	638	0.4413	0.3149	24.5
92	0.0123	0.0088	3.5	369	0.1602	0.1144	14.2	646	0.4515	0.3223	24.8
100	0.0143	0.0102	3.8	377	0.1667	0.1190	14.5	654	0.4619	0.3297	25.2
108	0.0165	0.0118	4.2	385	0.1733	0.1237	14.8	662	0.4725	0.3372	25.5
116	0.0188	0.0134	4.5	393	0.1801	0.1285	15.1	670	0.4831	0.3448	25.8
124	0.0213	0.0152	4.8	401	0.1869	0.1334	15.4	678	0.4938	0.3524	26.1
132	0.0239	0.0171	5.1	409	0.1938	0.1383	15.7	686	0.5046	0.3601	26.4
140	0.0267	0.0190	5.4	417	0.2009	0.1434	16.0	694	0.5156	0.3680	26.7
148	0.0296	0.0211	5.7	425	0.2081	0.1485	16.3	702	0.5266	0.3758	27.0
156	0.0326	0.0233	6.0	433	0.2154	0.1537	16.7	710	0.5378	0.3838	27.3
164	0.0357	0.0255	6.3	441	0.2228	0.1590	17.0	718	0.5490	0.3918	27.6
172	0.0390	0.0279	6.6	449	0.2304	0.1644	17.3	726	0.5604	0.4000	27.9
180	0.0425	0.0303	6.9	457	0.2380	0.1699	17.6	734	0.5719	0.4082	28.2
188	0.0460	0.0328	7.2	465	0.2458	0.1754	17.9	742	0.5835	0.4164	28.5
196	0.0497	0.0355	7.5	473	0.2537	0.1810	18.2	750	0.5952	0.4248	28.8
204	0.0535	0.0382	7.8	481	0.2617	0.1867	18.5	758	0.6070	0.4332	29.2
212	0.0575	0.0410	8.2	489	0.2698	0.1925	18.8	766	0.6189	0.4417	29.5
220	0.0616	0.0439	8.5	497	0.2780	0.1984	19.1	774	0.6309	0.4503	29.8
228	0.0658	0.0469	8.8	505	0.2863	0.2043	19.4	782	0.6430	0.4589	30.1
236	0.0701	0.0500	9.1	513	0.2948	0.2104	19.7	790	0.6552	0.4676	30.4
244	0.0745	0.0532	9.4	521	0.3033	0.2165	20.0	798	0.6675	0.4764	30.7
252	0.0791	0.0565	9.7	529	0.3120	0.2227	20.3	806	0.6800	0.4853	31.0
260	0.0838	0.0598	10.0	537	0.3208	0.2289	20.7	814	0.6925	0.4942	31.3
268	0.0887	0.0633	10.3	545	0.3297	0.2353	21.0	822	0.7052	0.5033	31.6
276	0.0936	0.0668	10.6	553	0.3387	0.2417	21.3	830	0.7179	0.5124	31.9
284	0.0987	0.0705	10.9	561	0.3478	0.2482	21.6	838	0.7308	0.5215	32.2
292	0.1039	0.0742	11.2	569	0.3570	0.2548	21.9	846	0.7437	0.5308	32.5

Friction Loss Table For 4" Sch. 10 Steel Pipe I.D. = 4.26

Equiv. Feet of Pipe

	LONG EL	90 ELL	45 ELL	TEE	GV	CV	ALM VA	DRY VA	BV
C-120	6	10	4	20	2	22	20	20	12
C-100	4.28	7.13	2.85	14.26	1.43	15.69	14.26	14.26	8.56

Q	P	P	V	Q	P	P	V	Q	P	P	V
	PSI/FT	PSI/FT	FT/SEC		PSI/FT	PSI/FT	FT/SEC		PSI/FT	PSI/FT	FT/SEC
GPM	C-100	C-120		GPM	C-100	C-120		GPM	C-100	C-120	
30	0.0004	0.0003	0.7	550	0.0911	0.0650	12.4	1070	0.3121	0.2227	24.1
45	0.0009	0.0006	1.0	565	0.0958	0.0683	12.7	1085	0.3202	0.2285	24.4
60	0.0015	0.0011	1.4	580	0.1005	0.0717	13.1	1100	0.3285	0.2344	24.8
75	0.0023	0.0016	1.7	595	0.1054	0.0752	13.4	1115	0.3368	0.2404	25.1
90	0.0032	0.0023	2.0	610	0.1104	0.0788	13.7	1130	0.3452	0.2464	25.5
105	0.0043	0.0030	2.4	625	0.1154	0.0824	14.1	1145	0.3538	0.2525	25.8
120	0.0055	0.0039	2.7	640	0.1206	0.0861	14.4	1160	0.3624	0.2586	26.1
135	0.0068	0.0048	3.0	655	0.1259	0.0898	14.8	1175	0.3711	0.2648	26.5
150	0.0082	0.0059	3.4	670	0.1313	0.0937	15.1	1190	0.3799	0.2711	26.8
165	0.0098	0.0070	3.7	685	0.1368	0.0976	15.4	1205	0.3888	0.2775	27.1
180	0.0115	0.0082	4.1	700	0.1423	0.1016	15.8	1220	0.3978	0.2839	27.5
195	0.0134	0.0095	4.4	715	0.1480	0.1057	16.1	1235	0.4069	0.2904	27.8
210	0.0153	0.0110	4.7	730	0.1538	0.1098	16.4	1250	0.4161	0.2970	28.2
225	0.0174	0.0124	5.1	745	0.1597	0.1140	16.8	1265	0.4254	0.3036	28.5
240	0.0196	0.0140	5.4	760	0.1657	0.1183	17.1	1280	0.4348	0.3103	28.8
255	0.0220	0.0157	5.7	775	0.1718	0.1226	17.5	1295	0.4442	0.3170	29.2
270	0.0244	0.0174	6.1	790	0.1780	0.1271	17.8	1310	0.4538	0.3239	29.5
285	0.0270	0.0193	6.4	805	0.1843	0.1316	18.1	1325	0.4635	0.3308	29.8
300	0.0297	0.0212	6.8	820	0.1908	0.1361	18.5	1340	0.4732	0.3377	30.2
315	0.0325	0.0232	7.1	835	0.1973	0.1408	18.8	1355	0.4831	0.3448	30.5
330	0.0354	0.0253	7.4	850	0.2039	0.1455	19.1	1370	0.4930	0.3519	30.9
345	0.0384	0.0274	7.8	865	0.2106	0.1503	19.5	1385	0.5030	0.3590	31.2
360	0.0416	0.0297	8.1	880	0.2174	0.1551	19.8	1400	0.5132	0.3662	31.5
375	0.0449	0.0320	8.4	895	0.2243	0.1601	20.2	1415	0.5234	0.3735	31.9
390	0.0482	0.0344	8.8	910	0.2313	0.1651	20.5	1430	0.5337	0.3809	32.2
405	0.0517	0.0369	9.1	925	0.2384	0.1701	20.8	1445	0.5441	0.3883	32.5
420	0.0553	0.0395	9.5	940	0.2456	0.1753	21.2	1460	0.5546	0.3958	32.9
435	0.0590	0.0421	9.8	955	0.2529	0.1805	21.5	1475	0.5652	0.4034	33.2
450	0.0629	0.0449	10.1	970	0.2603	0.1858	21.8	1490	0.5758	0.4110	33.6
465	0.0668	0.0477	10.5	985	0.2678	0.1911	22.2	1505	0.5866	0.4187	33.9
480	0.0708	0.0506	10.8	1000	0.2754	0.1965	22.5	1520	0.5975	0.4264	34.2
495	0.0750	0.0535	11.1	1015	0.2831	0.2020	22.9	1535	0.6084	0.4342	34.6
510	0.0792	0.0566	11.5	1030	0.2908	0.2076	23.2	1550	0.6195	0.4421	34.9
525	0.0836	0.0597	11.8	1045	0.2987	0.2132	23.5	1565	0.6306	0.4501	35.2
540	0.0881	0.0629	12.2	1060	0.3067	0.2189	23.9	1580	0.6418	0.4581	35.6

Friction Loss Table For 5" Sch. 10 Steel Pipe   I.D. = 5.295

Equiv. Feet of Pipe

	LONG EL	90 ELL	45 ELL	TEE	GV	CV	ALM VA	DRY VA	BV
C-120	8	12	5	25	2	27			9
C-100	5.70	8.56	3.57	17.83	1.43	19.25	0.00	0.00	6.42

Q GPM	P PSI/FT C-100	P PSI/FT C-120	V FT/SEC	Q GPM	P PSI/FT C-100	P PSI/FT C-120	V FT/SEC	Q GPM	P PSI/FT C-100	P PSI/FT C-120	V FT/SEC
50	0.0004	0.0003	0.7	745	0.0554	0.0395	10.9	1440	0.1874	0.1338	21.0
70	0.0007	0.0005	1.0	765	0.0582	0.0415	11.2	1465	0.1935	0.1381	21.4
90	0.0011	0.0008	1.3	785	0.0610	0.0435	11.4	1490	0.1997	0.1425	21.7
110	0.0016	0.0011	1.6	805	0.0639	0.0456	11.7	1515	0.2059	0.1470	22.1
130	0.0022	0.0016	1.9	825	0.0669	0.0477	12.0	1540	0.2122	0.1515	22.5
150	0.0029	0.0020	2.2	845	0.0699	0.0499	12.3	1565	0.2187	0.1561	22.8
170	0.0036	0.0026	2.5	865	0.0730	0.0521	12.6	1590	0.2252	0.1607	23.2
190	0.0044	0.0032	2.8	885	0.0762	0.0544	12.9	1615	0.2318	0.1654	23.5
210	0.0053	0.0038	3.1	905	0.0794	0.0567	13.2	1640	0.2384	0.1702	23.9
230	0.0063	0.0045	3.4	925	0.0827	0.0590	13.5	1665	0.2452	0.1750	24.3
250	0.0073	0.0052	3.6	945	0.0860	0.0614	13.8	1690	0.2521	0.1799	24.6
270	0.0085	0.0060	3.9	965	0.0894	0.0638	14.1	1715	0.2590	0.1848	25.0
290	0.0097	0.0069	4.2	985	0.0928	0.0663	14.4	1740	0.2660	0.1899	25.4
310	0.0109	0.0078	4.5	1005	0.0964	0.0688	14.7	1765	0.2731	0.1949	25.7
330	0.0123	0.0088	4.8	1025	0.0999	0.0713	14.9	1790	0.2803	0.2001	26.1
350	0.0137	0.0098	5.1	1045	0.1036	0.0739	15.2	1815	0.2876	0.2053	26.5
370	0.0152	0.0108	5.4	1065	0.1073	0.0766	15.5	1840	0.2950	0.2105	26.8
390	0.0167	0.0119	5.7	1085	0.1110	0.0792	15.8	1865	0.3025	0.2159	27.2
410	0.0183	0.0131	6.0	1105	0.1149	0.0820	16.1	1890	0.3100	0.2212	27.6
430	0.0200	0.0143	6.3	1125	0.1187	0.0847	16.4	1915	0.3176	0.2267	27.9
450	0.0218	0.0156	6.6	1145	0.1227	0.0875	16.7	1940	0.3253	0.2322	28.3
470	0.0236	0.0169	6.9	1165	0.1267	0.0904	17.0	1965	0.3331	0.2378	28.6
490	0.0255	0.0182	7.1	1185	0.1307	0.0933	17.3	1990	0.3410	0.2434	29.0
510	0.0275	0.0196	7.4	1205	0.1348	0.0962	17.6	2015	0.3490	0.2491	29.4
530	0.0295	0.0211	7.7	1225	0.1390	0.0992	17.9	2040	0.3570	0.2548	29.7
550	0.0316	0.0225	8.0	1245	0.1432	0.1022	18.2	2065	0.3652	0.2606	30.1
570	0.0338	0.0241	8.3	1265	0.1475	0.1053	18.4	2090	0.3734	0.2665	30.5
590	0.0360	0.0257	8.6	1285	0.1518	0.1084	18.7	2115	0.3817	0.2724	30.8
610	0.0383	0.0273	8.9	1305	0.1562	0.1115	19.0	2140	0.3901	0.2784	31.2
630	0.0406	0.0290	9.2	1325	0.1607	0.1147	19.3	2165	0.3986	0.2845	31.6
650	0.0430	0.0307	9.5	1345	0.1652	0.1179	19.6	2190	0.4071	0.2906	31.9
670	0.0455	0.0325	9.8	1365	0.1698	0.1212	19.9	2215	0.4158	0.2967	32.3
690	0.0481	0.0343	10.1	1385	0.1744	0.1245	20.2	2240	0.4245	0.3030	32.7
710	0.0507	0.0362	10.4	1405	0.1791	0.1278	20.5	2265	0.4333	0.3092	33.0
730	0.0533	0.0381	10.6	1425	0.1839	0.1312	20.8	2290	0.4422	0.3156	33.4

Friction Loss Table For 6" Sch. 10 Steel Pipe   I.D. = 6.357

Equiv. Feet of Pipe

	LONG EL	90 ELL	45 ELL	TEE	GV	CV	ALM VA	DRY VA	BV
C-120	9	14	7	30	3	32	18	19	10
C-100	6.42	9.98	4.99	21.39	2.14	22.82	12.83	13.55	7.13

Q GPM	P PSI/FT C-100	P PSI/FT C-120	V FT/SEC	Q GPM	P PSI/FT C-100	P PSI/FT C-120	V FT/SEC	Q GPM	P PSI/FT C-100	P PSI/FT C-120	V FT/SEC
50	0.0002	0.0001	0.5	920	0.0336	0.0240	9.3	1790	0.1151	0.0821	18.1
75	0.0003	0.0002	0.8	945	0.0353	0.0252	9.6	1825	0.1193	0.0851	18.5
100	0.0006	0.0004	1.0	970	0.0371	0.0264	9.8	1860	0.1236	0.0882	18.8
125	0.0008	0.0006	1.3	995	0.0388	0.0277	10.1	1895	0.1279	0.0913	19.2
150	0.0012	0.0008	1.5	1020	0.0407	0.0290	10.3	1930	0.1323	0.0944	19.5
175	0.0016	0.0011	1.8	1045	0.0425	0.0304	10.6	1965	0.1368	0.0976	19.9
200	0.0020	0.0014	2.0	1070	0.0444	0.0317	10.8	2000	0.1413	0.1009	20.2
225	0.0025	0.0018	2.3	1095	0.0464	0.0331	11.1	2035	0.1459	0.1041	20.6
250	0.0030	0.0022	2.5	1120	0.0483	0.0345	11.3	2070	0.1506	0.1075	20.9
275	0.0036	0.0026	2.8	1145	0.0504	0.0359	11.6	2105	0.1554	0.1109	21.3
300	0.0042	0.0030	3.0	1170	0.0524	0.0374	11.8	2140	0.1602	0.1143	21.6
325	0.0049	0.0035	3.3	1195	0.0545	0.0389	12.1	2175	0.1650	0.1178	22.0
350	0.0056	0.0040	3.5	1220	0.0566	0.0404	12.3	2210	0.1700	0.1213	22.4
375	0.0064	0.0046	3.8	1245	0.0588	0.0420	12.6	2245	0.1750	0.1249	22.7
400	0.0072	0.0051	4.0	1270	0.0610	0.0435	12.8	2280	0.1801	0.1285	23.1
425	0.0081	0.0057	4.3	1295	0.0632	0.0451	13.1	2315	0.1852	0.1322	23.4
450	0.0089	0.0064	4.6	1320	0.0655	0.0468	13.4	2350	0.1904	0.1359	23.8
475	0.0099	0.0071	4.8	1345	0.0678	0.0484	13.6	2385	0.1957	0.1397	24.1
500	0.0109	0.0078	5.1	1370	0.0702	0.0501	13.9	2420	0.2011	0.1435	24.5
525	0.0119	0.0085	5.3	1395	0.0726	0.0518	14.1	2455	0.2065	0.1474	24.8
550	0.0130	0.0093	5.6	1420	0.0750	0.0535	14.4	2490	0.2120	0.1513	25.2
575	0.0141	0.0101	5.8	1445	0.0775	0.0553	14.6	2525	0.2175	0.1552	25.5
600	0.0152	0.0109	6.1	1470	0.0800	0.0571	14.9	2560	0.2231	0.1592	25.9
625	0.0164	0.0117	6.3	1495	0.0825	0.0589	15.1	2595	0.2288	0.1633	26.2
650	0.0177	0.0126	6.6	1520	0.0851	0.0607	15.4	2630	0.2345	0.1674	26.6
675	0.0189	0.0135	6.8	1545	0.0877	0.0626	15.6	2665	0.2403	0.1715	27.0
700	0.0203	0.0145	7.1	1570	0.0903	0.0645	15.9	2700	0.2462	0.1757	27.3
725	0.0216	0.0154	7.3	1595	0.0930	0.0664	16.1	2735	0.2522	0.1800	27.7
750	0.0230	0.0164	7.6	1620	0.0957	0.0683	16.4	2770	0.2582	0.1842	28.0
775	0.0245	0.0175	7.8	1645	0.0984	0.0703	16.6	2805	0.2642	0.1886	28.4
800	0.0259	0.0185	8.1	1670	0.1012	0.0723	16.9	2840	0.2704	0.1930	28.7
825	0.0275	0.0196	8.3	1695	0.1041	0.0743	17.1	2875	0.2766	0.1974	29.1
850	0.0290	0.0207	8.6	1720	0.1069	0.0763	17.4	2910	0.2828	0.2018	29.4
875	0.0306	0.0219	8.9	1745	0.1098	0.0784	17.6	2945	0.2891	0.2064	29.8
900	0.0323	0.0230	9.1	1770	0.1127	0.0805	17.9	3165	0.3304	0.2358	32.0

Friction Loss Table For 8" Sch. 10 Steel Pipe   I.D. = 8.249

Equiv. Feet of Pipe

	LONG ELL	90 ELL	45 ELL	TEE	GV	CV	ALM VA	DRY VA	BV
C-120	13	18	9	35	4	45	35	27	12
C-100	9.27	12.83	6.42	24.96	2.85	32.09	24.96	19.25	8.56

Q	P	P	V	Q	P	P	V	Q	P	P	V
	PSI/FT	PSI/FT	FT/SEC		PSI/FT	PSI/FT	FT/SEC		PSI/FT	PSI/FT	FT/SEC
GPM	C-100	C-120		GPM	C-100	C-120		GPM	C-100	C-120	
100	0.0002	0.0001	0.6	1825	0.0335	0.0239	11.0	3550	0.1149	0.0820	21.3
150	0.0003	0.0002	0.9	1875	0.0353	0.0252	11.3	3600	0.1179	0.0841	21.6
200	0.0006	0.0004	1.2	1925	0.0370	0.0264	11.6	3650	0.1209	0.0863	21.9
250	0.0008	0.0006	1.5	1975	0.0388	0.0277	11.9	3700	0.1240	0.0885	22.2
300	0.0012	0.0008	1.8	2025	0.0407	0.0290	12.2	3750	0.1271	0.0907	22.5
350	0.0016	0.0011	2.1	2075	0.0425	0.0304	12.5	3800	0.1303	0.0930	22.8
400	0.0020	0.0014	2.4	2125	0.0445	0.0317	12.8	3850	0.1335	0.0953	23.1
450	0.0025	0.0018	2.7	2175	0.0464	0.0331	13.1	3900	0.1367	0.0976	23.4
500	0.0031	0.0022	3.0	2225	0.0484	0.0345	13.4	3950	0.1399	0.0999	23.7
550	0.0036	0.0026	3.3	2275	0.0504	0.0360	13.7	4000	0.1432	0.1022	24.0
600	0.0043	0.0031	3.6	2325	0.0525	0.0375	14.0	4050	0.1466	0.1046	24.3
650	0.0050	0.0035	3.9	2375	0.0546	0.0390	14.3	4100	0.1499	0.1070	24.6
700	0.0057	0.0041	4.2	2425	0.0568	0.0405	14.6	4150	0.1533	0.1094	24.9
750	0.0065	0.0046	4.5	2475	0.0589	0.0421	14.9	4200	0.1568	0.1119	25.2
800	0.0073	0.0052	4.8	2525	0.0612	0.0436	15.2	4250	0.1602	0.1144	25.5
850	0.0082	0.0058	5.1	2575	0.0634	0.0453	15.5	4300	0.1637	0.1169	25.8
900	0.0091	0.0065	5.4	2625	0.0657	0.0469	15.8	4350	0.1673	0.1194	26.1
950	0.0100	0.0072	5.7	2675	0.0680	0.0486	16.1	4400	0.1709	0.1219	26.4
1000	0.0110	0.0079	6.0	2725	0.0704	0.0503	16.4	4450	0.1745	0.1245	26.7
1050	0.0121	0.0086	6.3	2775	0.0728	0.0520	16.7	4500	0.1781	0.1271	27.0
1100	0.0131	0.0094	6.6	2825	0.0753	0.0537	17.0	4550	0.1818	0.1297	27.3
1150	0.0143	0.0102	6.9	2875	0.0778	0.0555	17.3	4600	0.1855	0.1324	27.6
1200	0.0154	0.0110	7.2	2925	0.0803	0.0573	17.6	4650	0.1893	0.1351	27.9
1250	0.0167	0.0119	7.5	2975	0.0828	0.0591	17.9	4700	0.1930	0.1378	28.2
1300	0.0179	0.0128	7.8	3025	0.0854	0.0610	18.2	4750	0.1969	0.1405	28.5
1350	0.0192	0.0137	8.1	3075	0.0881	0.0628	18.5	4800	0.2007	0.1432	28.8
1400	0.0205	0.0147	8.4	3125	0.0907	0.0648	18.8	4850	0.2046	0.1460	29.1
1450	0.0219	0.0156	8.7	3175	0.0934	0.0667	19.1	4900	0.2085	0.1488	29.4
1500	0.0233	0.0167	9.0	3225	0.0962	0.0686	19.4	4950	0.2125	0.1516	29.7
1550	0.0248	0.0177	9.3	3275	0.0989	0.0706	19.7	5000	0.2165	0.1545	30.0
1600	0.0263	0.0188	9.6	3325	0.1018	0.0726	20.0	5050	0.2205	0.1574	30.3
1650	0.0278	0.0199	9.9	3375	0.1046	0.0747	20.3	5100	0.2245	0.1602	30.6
1700	0.0294	0.0210	10.2	3425	0.1075	0.0767	20.6	5150	0.2286	0.1632	30.9
1750	0.0310	0.0222	10.5	3475	0.1104	0.0788	20.9	5200	0.2327	0.1661	31.2
1800	0.0327	0.0233	10.8	3525	0.1134	0.0809	21.2	5320	0.2428	0.1733	32.0

Appendix D Friction Loss Tables

Friction Loss Table For 10" Sch. 10 Steel Pipe I.D. = 10.374

Equiv. Feet of Pipe

	LONG EL	90 ELL	45 ELL	TEE	GV	CV	ALM VA	DRY VA
C-120	16	22	11	50	5			
C-100	11.41	15.69	7.84	35.65	3.57	0.00	0.00	0.00

Q	P	P	V	Q	P	P	V	Q	P	P	V
	PSI/FT	PSI/FT	FT/SEC		PSI/FT	PSI/FT	FT/SEC		PSI/FT	PSI/FT	FT/SEC
GPM	C-100	C-120		GPM	C-100	C-120		GPM	C-100	C-120	
500	0.0010	0.0007	1.9	3100	0.0293	0.0209	11.8	5700	0.0903	0.0645	21.6
575	0.0013	0.0009	2.2	3175	0.0306	0.0218	12.1	5775	0.0925	0.0660	21.9
650	0.0016	0.0012	2.5	3250	0.0319	0.0228	12.3	5850	0.0948	0.0676	22.2
725	0.0020	0.0014	2.8	3325	0.0333	0.0238	12.6	5925	0.0970	0.0693	22.5
800	0.0024	0.0017	3.0	3400	0.0347	0.0248	12.9	6000	0.0993	0.0709	22.8
875	0.0028	0.0020	3.3	3475	0.0362	0.0258	13.2	6075	0.1016	0.0725	23.1
950	0.0033	0.0023	3.6	3550	0.0376	0.0268	13.5	6150	0.1040	0.0742	23.4
1025	0.0038	0.0027	3.9	3625	0.0391	0.0279	13.8	6225	0.1063	0.0759	23.6
1100	0.0043	0.0031	4.2	3700	0.0406	0.0290	14.1	6300	0.1087	0.0776	23.9
1175	0.0049	0.0035	4.5	3775	0.0421	0.0301	14.3	6375	0.1111	0.0793	24.2
1250	0.0055	0.0039	4.7	3850	0.0437	0.0312	14.6	6450	0.1135	0.0810	24.5
1325	0.0061	0.0043	5.0	3925	0.0453	0.0323	14.9	6525	0.1160	0.0828	24.8
1400	0.0067	0.0048	5.3	4000	0.0469	0.0335	15.2	6600	0.1185	0.0846	25.1
1475	0.0074	0.0053	5.6	4075	0.0486	0.0347	15.5	6675	0.1210	0.0863	25.4
1550	0.0081	0.0058	5.9	4150	0.0502	0.0358	15.8	6750	0.1235	0.0881	25.6
1625	0.0089	0.0063	6.2	4225	0.0519	0.0370	16.0	6825	0.1261	0.0900	25.9
1700	0.0096	0.0069	6.5	4300	0.0536	0.0383	16.3	6900	0.1286	0.0918	26.2
1775	0.0104	0.0074	6.7	4375	0.0554	0.0395	16.6	6975	0.1312	0.0937	26.5
1850	0.0113	0.0080	7.0	4450	0.0571	0.0408	16.9	7050	0.1339	0.0955	26.8
1925	0.0121	0.0087	7.3	4525	0.0589	0.0421	17.2	7125	0.1365	0.0974	27.1
2000	0.0130	0.0093	7.6	4600	0.0608	0.0434	17.5	7200	0.1392	0.0993	27.3
2075	0.0139	0.0099	7.9	4675	0.0626	0.0447	17.8	7275	0.1419	0.1012	27.6
2150	0.0149	0.0106	8.2	4750	0.0645	0.0460	18.0	7350	0.1446	0.1032	27.9
2225	0.0159	0.0113	8.5	4825	0.0664	0.0474	18.3	7425	0.1473	0.1051	28.2
2300	0.0169	0.0120	8.7	4900	0.0683	0.0487	18.6	7500	0.1501	0.1071	28.5
2375	0.0179	0.0128	9.0	4975	0.0702	0.0501	18.9	7575	0.1529	0.1091	28.8
2450	0.0189	0.0135	9.3	5050	0.0722	0.0515	19.2	7650	0.1557	0.1111	29.1
2525	0.0200	0.0143	9.6	5125	0.0742	0.0530	19.5	7725	0.1585	0.1131	29.3
2600	0.0211	0.0151	9.9	5200	0.0762	0.0544	19.7	7800	0.1614	0.1152	29.6
2675	0.0223	0.0159	10.2	5275	0.0783	0.0559	20.0	7875	0.1643	0.1172	29.9
2750	0.0235	0.0167	10.4	5350	0.0803	0.0573	20.3	7950	0.1672	0.1193	30.2
2825	0.0247	0.0176	10.7	5425	0.0824	0.0588	20.6	8025	0.1701	0.1214	30.5
2900	0.0259	0.0185	11.0	5500	0.0846	0.0603	20.9	8100	0.1731	0.1235	30.8
2975	0.0271	0.0194	11.3	5575	0.0867	0.0619	21.2	8175	0.1760	0.1256	31.0
3050	0.0284	0.0203	11.6	5650	0.0889	0.0634	21.5	8420	0.1859	0.1327	32.0

Friction Loss Table For 3/4" Type L Copper   I.D. = .785

Equiv. Feet of Pipe

	LONG EL	90 ELL	45 ELL	TEE	GV	CV
C-120	1	2	1	3		
C-150	1.51	3.02	1.51	4.53		

Q	P	P	V	Q	P	P	V	Q	P	P	V
	PSI/FT	PSI/FT	FT/SEC		PSI/FT	PSI/FT	FT/SEC		PSI/FT	PSI/FT	FT/SEC
GPM	C-150			GPM	C-150			GPM	C-150		
7	0.0507		4.6	42	1.3943		27.9				
8	0.0649		5.3	43	1.4564		28.5				
9	0.0807		6.0	44	1.5196		29.2				
10	0.0980		6.6	45	1.5842		29.8				
11	0.1169		7.3	46	1.6499		30.5				
12	0.1374		8.0	47	1.7169		31.2				
13	0.1593		8.6	48	1.7851		31.8				
14	0.1827		9.3	49	1.8545		32.5				
15	0.2076		9.9								
16	0.2339		10.6								
17	0.2616		11.3								
18	0.2908		11.9								
19	0.3214		12.6								
20	0.3534		13.3								
21	0.3868		13.9								
22	0.4215		14.6								
23	0.4577		15.3								
24	0.4952		15.9								
25	0.5340		16.6								
26	0.5742		17.2								
27	0.6157		17.9								
28	0.6586		18.6								
29	0.7027		19.2								
30	0.7482		19.9								
31	0.7950		20.6								
32	0.8431		21.2								
33	0.8925		21.9								
34	0.9432		22.6								
35	0.9951		23.2								
36	1.0484		23.9								
37	1.1029		24.5								
38	1.1587		25.2								
39	1.2157		25.9								
40	1.2740		26.5								
41	1.3335		27.2								

Appendix D Friction Loss Tables

Friction Loss Table For 1" Type L Copper   I.D. = 1.025

Equiv. Feet of Pipe

	LONG EL	90 ELL	45 ELL	TEE	GV	CV
C–120	2	2	1	5		
C–150	3.02	3.02	1.51	7.55		

Q	P	P	V	Q	P	P	V	Q	P	P	V
	PSI/FT	PSI/FT	FT/SEC		PSI/FT	PSI/FT	FT/SEC		PSI/FT	PSI/FT	FT/SEC
GPM	C–150			GPM	C–150			GPM	C–150		
7	0.0138		2.7	42	0.3803		16.3	77	1.1672		30.0
8	0.0177		3.1	43	0.3972		16.7	78	1.1954		30.3
9	0.0220		3.5	44	0.4145		17.1	79	1.2239		30.7
10	0.0267		3.9	45	0.4321		17.5	80	1.2527		31.1
11	0.0319		4.3	46	0.4500		17.9	81	1.2819		31.5
12	0.0375		4.7	47	0.4683		18.3	82	1.3113		31.9
13	0.0434		5.1	48	0.4869		18.7	83	1.3410		32.3
14	0.0498		5.4	49	0.5058		19.1				
15	0.0566		5.8	50	0.5251		19.5				
16	0.0638		6.2	51	0.5447		19.8				
17	0.0714		6.6	52	0.5646		20.2				
18	0.0793		7.0	53	0.5849		20.6				
19	0.0877		7.4	54	0.6054		21.0				
20	0.0964		7.8	55	0.6263		21.4				
21	0.1055		8.2	56	0.6476		21.8				
22	0.1150		8.6	57	0.6691		22.2				
23	0.1248		8.9	58	0.6910		22.6				
24	0.1351		9.3	59	0.7132		23.0				
25	0.1457		9.7	60	0.7357		23.3				
26	0.1566		10.1	61	0.7586		23.7				
27	0.1679		10.5	62	0.7818		24.1				
28	0.1796		10.9	63	0.8052		24.5				
29	0.1917		11.3	64	0.8290		24.9				
30	0.2041		11.7	65	0.8532		25.3				
31	0.2169		12.1	66	0.8776		25.7				
32	0.2300		12.4	67	0.9024		26.1				
33	0.2434		12.8	68	0.9274		26.5				
34	0.2573		13.2	69	0.9528		26.8				
35	0.2714		13.6	70	0.9785		27.2				
36	0.2860		14.0	71	1.0046		27.6				
37	0.3008		14.4	72	1.0309		28.0				
38	0.3160		14.8	73	1.0575		28.4				
39	0.3316		15.2	74	1.0845		28.8				
40	0.3475		15.6	75	1.1118		29.2				
41	0.3637		16.0	76	1.1393		29.6				

Friction Loss Table For 1¼" Type L Copper   I.D. = 1.265

Equiv. Feet of Pipe

	LONG EL	90 ELL	45 ELL	TEE	GV	CV
C-120	2	3	1	6		
C-150	3.02	4.53	1.51	9.06		

Q GPM	P PSI/FT C-150	P PSI/FT	V FT/SEC	Q GPM	P PSI/FT C-150	P PSI/FT	V FT/SEC	Q GPM	P PSI/FT C-150	P PSI/FT	V FT/SEC
7	0.0050		1.8	42	0.1365		10.7	77	0.4190		19.7
8	0.0064		2.0	43	0.1426		11.0	79	0.4393		20.2
9	0.0079		2.3	44	0.1488		11.2	81	0.4601		20.7
10	0.0096		2.6	45	0.1551		11.5	83	0.4814		21.2
11	0.0114		2.8	46	0.1615		11.7	85	0.5031		21.7
12	0.0134		3.1	47	0.1681		12.0	87	0.5252		22.2
13	0.0156		3.3	48	0.1748		12.3	89	0.5477		22.7
14	0.0179		3.6	49	0.1816		12.5	91	0.5707		23.2
15	0.0203		3.8	50	0.1885		12.8	93	0.5941		23.8
16	0.0229		4.1	51	0.1955		13.0	95	0.6180		24.3
17	0.0256		4.3	52	0.2027		13.3	97	0.6423		24.8
18	0.0285		4.6	53	0.2099		13.5	99	0.6670		25.3
19	0.0315		4.9	54	0.2173		13.8	101	0.6921		25.8
20	0.0346		5.1	55	0.2248		14.0	103	0.7177		26.3
21	0.0379		5.4	56	0.2325		14.3	105	0.7437		26.8
22	0.0413		5.6	57	0.2402		14.6	107	0.7701		27.3
23	0.0448		5.9	58	0.2480		14.8	109	0.7969		27.8
24	0.0485		6.1	59	0.2560		15.1	111	0.8242		28.4
25	0.0523		6.4	60	0.2641		15.3	113	0.8519		28.9
26	0.0562		6.6	61	0.2723		15.6	115	0.8800		29.4
27	0.0603		6.9	62	0.2806		15.8	117	0.9085		29.9
28	0.0645		7.2	63	0.2890		16.1	119	0.9375		30.4
29	0.0688		7.4	64	0.2976		16.3	121	0.9668		30.9
30	0.0733		7.7	65	0.3063		16.6	123	0.9966		31.4
31	0.0778		7.9	66	0.3150		16.9	125	1.0268		31.9
32	0.0825		8.2	67	0.3239		17.1	127	1.0574		32.4
33	0.0874		8.4	68	0.3329		17.4	129	1.0884		32.9
34	0.0923		8.7	69	0.3420		17.6	131	1.1198		33.5
35	0.0974		8.9	70	0.3513		17.9	133	1.1516		34.0
36	0.1026		9.2	71	0.3606		18.1	135	1.1839		34.5
37	0.1080		9.5	72	0.3700		18.4	137	1.2165		35.0
38	0.1134		9.7	73	0.3796		18.6	139	1.2496		35.5
39	0.1190		10.0	74	0.3893		18.9	141	1.2831		36.0
40	0.1247		10.2	75	0.3991		19.2	143	1.3169		36.5
41	0.1306		10.5	76	0.4090		19.4	145	1.3512		37.0

Friction Loss Table For 1½" Type L Copper I.D. = 1.505

Equiv. Feet of Pipe

	LONG EL	90 ELL	45 ELL	TEE	GV	CV
C-120	2	4	2	8		9
C-150	3.02	6.04	3.02	12.08		13.59

Q	P	P	V	Q	P	P	V	Q	P	P	V
	PSI/FT	PSI/FT	FT/SEC		PSI/FT	PSI/FT	FT/SEC		PSI/FT	PSI/FT	FT/SEC
GPM	C-150			GPM	C-150			GPM	C-150		
7	0.0021		1.3	76	0.1755		13.7	111	0.3537		20.0
9	0.0034		1.6	77	0.1798		13.9	113	0.3656		20.4
11	0.0049		2.0	78	0.1841		14.1	115	0.3776		20.8
13	0.0067		2.3	79	0.1885		14.3	117	0.3899		21.1
15	0.0087		2.7	80	0.1930		14.4	119	0.4023		21.5
17	0.0110		3.1	81	0.1975		14.6	121	0.4149		21.8
19	0.0135		3.4	82	0.2020		14.8	123	0.4277		22.2
21	0.0163		3.8	83	0.2066		15.0	125	0.4406		22.6
23	0.0192		4.2	84	0.2112		15.2	127	0.4537		22.9
25	0.0224		4.5	85	0.2159		15.3	129	0.4670		23.3
27	0.0259		4.9	86	0.2206		15.5	131	0.4805		23.6
29	0.0295		5.2	87	0.2254		15.7	133	0.4942		24.0
31	0.0334		5.6	88	0.2302		15.9	135	0.5080		24.4
33	0.0375		6.0	89	0.2350		16.1	137	0.5220		24.7
35	0.0418		6.3	90	0.2399		16.2	139	0.5362		25.1
37	0.0463		6.7	91	0.2449		16.4	141	0.5506		25.4
39	0.0511		7.0	92	0.2499		16.6	143	0.5651		25.8
41	0.0560		7.4	93	0.2550		16.8	145	0.5798		26.2
43	0.0612		7.8	94	0.2600		17.0	147	0.5947		26.5
45	0.0666		8.1	95	0.2652		17.1	149	0.6098		26.9
47	0.0721		8.5	96	0.2704		17.3	151	0.6250		27.2
49	0.0779		8.8	97	0.2756		17.5	153	0.6404		27.6
51	0.0839		9.2	98	0.2809		17.7	155	0.6560		28.0
53	0.0901		9.6	99	0.2862		17.9	157	0.6717		28.3
55	0.0965		9.9	100	0.2916		18.0	159	0.6876		28.7
57	0.1031		10.3	101	0.2970		18.2	161	0.7037		29.1
59	0.1099		10.6	102	0.3025		18.4	163	0.7200		29.4
61	0.1168		11.0	103	0.3080		18.6	165	0.7364		29.8
63	0.1240		11.4	104	0.3135		18.8	167	0.7530		30.1
65	0.1314		11.7	105	0.3191		18.9	169	0.7698		30.5
67	0.1390		12.1	106	0.3248		19.1	171	0.7867		30.9
69	0.1468		12.5	107	0.3305		19.3	173	0.8038		31.2
71	0.1547		12.8	108	0.3362		19.5	175	0.8211		31.6
73	0.1629		13.2	109	0.3420		19.7	177	0.8385		31.9
75	0.1712		13.5	110	0.3478		19.8	179	0.8561		32.3

Friction Loss Table For 2" Type L Copper I.D. = 1.985

Equiv. Feet of Pipe

	LONG EL	90 ELL	45 ELL	TEE	GV	CV
C-120	3	5	2	10	1	11
C-150	4.53	7.55	3.02	15.10	1.51	16.61

Q GPM	P PSI/FT C-150	P PSI/FT	V FT/SEC	Q GPM	P PSI/FT C-150	P PSI/FT	V FT/SEC	Q GPM	P PSI/FT C-150	P PSI/FT	V FT/SEC
7	0.0006		0.7	110	0.0903		11.4	145	0.1506		15.0
10	0.0011		1.0	111	0.0919		11.5	150	0.1603		15.6
13	0.0017		1.3	112	0.0934		11.6	155	0.1704		16.1
16	0.0026		1.7	113	0.0949		11.7	160	0.1807		16.6
19	0.0035		2.0	114	0.0965		11.8	165	0.1913		17.1
22	0.0046		2.3	115	0.0981		11.9	170	0.2021		17.6
25	0.0058		2.6	116	0.0997		12.0	175	0.2133		18.2
28	0.0072		2.9	117	0.1013		12.1	180	0.2247		18.7
31	0.0087		3.2	118	0.1029		12.2	185	0.2363		19.2
34	0.0103		3.5	119	0.1045		12.3	190	0.2483		19.7
37	0.0120		3.8	120	0.1061		12.4	195	0.2605		20.2
40	0.0139		4.1	121	0.1078		12.6	200	0.2730		20.7
43	0.0159		4.5	122	0.1094		12.7	205	0.2858		21.3
46	0.0180		4.8	123	0.1111		12.8	210	0.2988		21.8
49	0.0202		5.1	124	0.1127		12.9	215	0.3121		22.3
52	0.0226		5.4	125	0.1144		13.0	220	0.3257		22.8
55	0.0251		5.7	126	0.1161		13.1	225	0.3395		23.3
58	0.0276		6.0	127	0.1178		13.2	230	0.3536		23.9
61	0.0303		6.3	128	0.1196		13.3	235	0.3679		24.4
64	0.0332		6.6	129	0.1213		13.4	240	0.3825		24.9
67	0.0361		7.0	130	0.1230		13.5	245	0.3974		25.4
70	0.0391		7.3	131	0.1248		13.6	250	0.4125		25.9
73	0.0423		7.6	132	0.1266		13.7	255	0.4279		26.5
76	0.0456		7.9	133	0.1284		13.8	260	0.4436		27.0
79	0.0490		8.2	134	0.1301		13.9	265	0.4595		27.5
82	0.0525		8.5	135	0.1319		14.0	270	0.4757		28.0
85	0.0561		8.8	136	0.1338		14.1	275	0.4921		28.5
88	0.0598		9.1	137	0.1356		14.2	280	0.5088		29.0
91	0.0636		9.4	138	0.1374		14.3	285	0.5257		29.6
94	0.0675		9.8	139	0.1393		14.4	290	0.5429		30.1
97	0.0716		10.1	140	0.1411		14.5	295	0.5603		30.6
100	0.0757		10.4	141	0.1430		14.6	300	0.5780		31.1
103	0.0800		10.7	142	0.1449		14.7	305	0.5960		31.6
106	0.0844		11.0	143	0.1468		14.8	310	0.6142		32.2
109	0.0888		11.3	144	0.1487		14.9	315	0.6326		32.7

Appendix D Friction Loss Tables

Friction Loss Table For 2½" Type L Copper   I.D. = 2.465

Equiv. Feet of Pipe

	LONG EL	90 ELL	45 ELL	TEE	GV	CV
C–120	4	6	3	12	1	14
C–150	6.04	9.06	4.53	18.12	1.51	21.14

Q	P	P	V	Q	P	P	V	Q	P	P	V
	PSI/FT	PSI/FT	FT/SEC		PSI/FT	PSI/FT	FT/SEC		PSI/FT	PSI/FT	FT/SEC
GPM	C–150			GPM	C–150			GPM	C–150		
15	0.0008		1.0	190	0.0865		12.8	365	0.2894		24.6
20	0.0013		1.3	195	0.0907		13.1	370	0.2968		24.9
25	0.0020		1.7	200	0.0951		13.5	375	0.3042		25.2
30	0.0028		2.0	205	0.0995		13.8	380	0.3118		25.6
35	0.0038		2.4	210	0.1041		14.1	385	0.3194		25.9
40	0.0048		2.7	215	0.1087		14.5	390	0.3271		26.2
45	0.0060		3.0	220	0.1134		14.8	395	0.3349		26.6
50	0.0073		3.4	225	0.1182		15.1	400	0.3428		26.9
55	0.0087		3.7	230	0.1231		15.5	405	0.3508		27.2
60	0.0103		4.0	235	0.1281		15.8	410	0.3588		27.6
65	0.0119		4.4	240	0.1332		16.1	415	0.3670		27.9
70	0.0136		4.7	245	0.1384		16.5	420	0.3752		28.3
75	0.0155		5.0	250	0.1437		16.8	425	0.3835		28.6
80	0.0175		5.4	255	0.1490		17.2	430	0.3919		28.9
85	0.0195		5.7	260	0.1545		17.5	435	0.4003		29.3
90	0.0217		6.1	265	0.1600		17.8	440	0.4089		29.6
95	0.0240		6.4	270	0.1657		18.2	445	0.4175		29.9
100	0.0264		6.7	275	0.1714		18.5	450	0.4262		30.3
105	0.0289		7.1	280	0.1772		18.8	455	0.4351		30.6
110	0.0315		7.4	285	0.1831		19.2	460	0.4439		30.9
115	0.0342		7.7	290	0.1891		19.5	465	0.4529		31.3
120	0.0370		8.1	295	0.1952		19.8	470	0.4620		31.6
125	0.0399		8.4	300	0.2013		20.2	475	0.4711		32.0
130	0.0429		8.7	305	0.2076		20.5	480	0.4803		32.3
135	0.0460		9.1	310	0.2139		20.9	485	0.4896		32.6
140	0.0492		9.4	315	0.2203		21.2	490	0.4990		33.0
145	0.0525		9.8	320	0.2269		21.5	495	0.5084		33.3
150	0.0558		10.1	325	0.2335		21.9	500	0.5180		33.6
155	0.0593		10.4	330	0.2401		22.2	505	0.5276		34.0
160	0.0629		10.8	335	0.2469		22.5	510	0.5373		34.3
165	0.0666		11.1	340	0.2538		22.9	515	0.5471		34.6
170	0.0704		11.4	345	0.2607		23.2	520	0.5570		35.0
175	0.0743		11.8	350	0.2678		23.5	525	0.5669		35.3
180	0.0782		12.1	355	0.2749		23.9	530	0.5769		35.7
185	0.0823		12.4	360	0.2821		24.2	535	0.5871		36.0

Friction Loss Table For 3" Type L Copper   I.D. = 2.945

Equiv. Feet of Pipe

	LONG EL	90 ELL	45 ELL	TEE	GV	CV
C-120	5	7	3	15	1	16
C-150	7.55	10.57	4.53	22.65	1.51	24.16

Q GPM	P PSI/FT C-150	P PSI/FT	V FT/SEC	Q GPM	P PSI/FT C-150	P PSI/FT	V FT/SEC	Q GPM	P PSI/FT C-150	P PSI/FT	V FT/SEC
20	0.0006		0.9	195	0.0381		9.2	375	0.1279		17.7
25	0.0009		1.2	200	0.0400		9.4	385	0.1343		18.1
30	0.0012		1.4	205	0.0418		9.7	395	0.1408		18.6
35	0.0016		1.6	210	0.0438		9.9	405	0.1475		19.1
40	0.0020		1.9	215	0.0457		10.1	415	0.1543		19.6
45	0.0025		2.1	220	0.0477		10.4	425	0.1612		20.0
50	0.0031		2.4	225	0.0497		10.6	435	0.1683		20.5
55	0.0037		2.6	230	0.0518		10.8	445	0.1755		21.0
60	0.0043		2.8	235	0.0539		11.1	455	0.1829		21.4
65	0.0050		3.1	240	0.0560		11.3	465	0.1904		21.9
70	0.0057		3.3	245	0.0582		11.5	475	0.1981		22.4
75	0.0065		3.5	250	0.0604		11.8	485	0.2058		22.9
80	0.0073		3.8	255	0.0627		12.0	495	0.2138		23.3
85	0.0082		4.0	260	0.0650		12.3	505	0.2218		23.8
90	0.0091		4.2	265	0.0673		12.5	515	0.2300		24.3
95	0.0101		4.5	270	0.0697		12.7	525	0.2384		24.7
100	0.0111		4.7	275	0.0721		13.0	535	0.2468		25.2
105	0.0121		4.9	280	0.0745		13.2	545	0.2554		25.7
110	0.0132		5.2	285	0.0770		13.4	555	0.2642		26.2
115	0.0144		5.4	290	0.0795		13.7	565	0.2730		26.6
120	0.0155		5.7	295	0.0821		13.9	575	0.2820		27.1
125	0.0168		5.9	300	0.0846		14.1	585	0.2912		27.6
130	0.0180		6.1	305	0.0873		14.4	595	0.3005		28.0
135	0.0193		6.4	310	0.0899		14.6	605	0.3099		28.5
140	0.0207		6.6	315	0.0926		14.8	615	0.3194		29.0
145	0.0221		6.8	320	0.0954		15.1	625	0.3291		29.5
150	0.0235		7.1	325	0.0982		15.3	635	0.3389		29.9
155	0.0249		7.3	330	0.1010		15.6	645	0.3488		30.4
160	0.0265		7.5	335	0.1038		15.8	655	0.3589		30.9
165	0.0280		7.8	340	0.1067		16.0	665	0.3691		31.3
170	0.0296		8.0	345	0.1096		16.3	675	0.3794		31.8
175	0.0312		8.2	350	0.1126		16.5	685	0.3899		32.3
180	0.0329		8.5	355	0.1156		16.7	695	0.4005		32.8
185	0.0346		8.7	360	0.1186		17.0	705	0.4112		33.2
190	0.0364		9.0	365	0.1217		17.2	715	0.4221		33.7

Friction Loss Table For 4" Type L Copper   I.D. = 3.905

Equiv. Feet of Pipe

	LONG EL	90 ELL	45 ELL	TEE	GV	CV
C-120	6	10	4	20	2	22
C-150	9.06	15.10	6.04	30.20	3.02	33.22

Q GPM	P PSI/FT C-150	P PSI/FT	V FT/SEC	Q GPM	P PSI/FT C-150	P PSI/FT	V FT/SEC	Q GPM	P PSI/FT C-150	P PSI/FT	V FT/SEC
30	0.0003		0.8	380	0.0332		10.2	735	0.1124		19.7
40	0.0005		1.1	390	0.0348		10.5	750	0.1167		20.1
50	0.0008		1.3	400	0.0365		10.7	765	0.1210		20.5
60	0.0011		1.6	410	0.0382		11.0	780	0.1255		20.9
70	0.0015		1.9	420	0.0399		11.3	795	0.1300		21.3
80	0.0019		2.1	430	0.0417		11.5	810	0.1345		21.7
90	0.0023		2.4	440	0.0435		11.8	825	0.1392		22.1
100	0.0028		2.7	450	0.0454		12.1	840	0.1439		22.5
110	0.0033		2.9	460	0.0472		12.3	855	0.1487		22.9
120	0.0039		3.2	470	0.0492		12.6	870	0.1536		23.3
130	0.0046		3.5	480	0.0511		12.9	885	0.1585		23.7
140	0.0052		3.8	490	0.0531		13.1	900	0.1635		24.1
150	0.0059		4.0	500	0.0551		13.4	915	0.1686		24.5
160	0.0067		4.3	510	0.0572		13.7	930	0.1737		24.9
170	0.0075		4.6	520	0.0593		13.9	945	0.1789		25.3
180	0.0083		4.8	530	0.0614		14.2	960	0.1842		25.7
190	0.0092		5.1	540	0.0635		14.5	975	0.1896		26.1
200	0.0101		5.4	550	0.0657		14.7	990	0.1950		26.5
210	0.0111		5.6	560	0.0680		15.0	1005	0.2005		26.9
220	0.0121		5.9	570	0.0702		15.3	1020	0.2061		27.3
230	0.0131		6.2	580	0.0725		15.5	1035	0.2117		27.7
240	0.0142		6.4	590	0.0749		15.8	1050	0.2175		28.1
250	0.0153		6.7	600	0.0772		16.1	1065	0.2232		28.5
260	0.0164		7.0	610	0.0796		16.4	1080	0.2291		28.9
270	0.0176		7.2	620	0.0821		16.6	1095	0.2350		29.4
280	0.0189		7.5	630	0.0845		16.9	1110	0.2410		29.8
290	0.0201		7.8	640	0.0870		17.2	1125	0.2471		30.2
300	0.0214		8.0	650	0.0895		17.4	1140	0.2532		30.6
310	0.0228		8.3	660	0.0921		17.7	1155	0.2594		31.0
320	0.0241		8.6	670	0.0947		18.0	1170	0.2657		31.4
330	0.0256		8.8	680	0.0973		18.2	1185	0.2720		31.8
340	0.0270		9.1	690	0.1000		18.5	1200	0.2784		32.2
350	0.0285		9.4	700	0.1027		18.8	1215	0.2849		32.6
360	0.0300		9.6	710	0.1054		19.0	1230	0.2914		33.0
370	0.0316		9.9	720	0.1082		19.3	1245	0.2980		33.4

Friction Loss Table For 6" Type L Copper   I.D. = 5.845

Equiv. Feet of Pipe

	LONG EL	90 ELL	45 ELL	TEE	GV	CV
C-120	9	14	7	30	3	32
C-150	13.59	21.14	10.57	45.30	4.53	48.32

Q GPM	P PSI/FT C-150	P PSI/FT	V FT/SEC	Q GPM	P PSI/FT C-150	P PSI/FT	V FT/SEC	Q GPM	P PSI/FT C-150	P PSI/FT	V FT/SEC
50	0.0001		0.6	925	0.0241		11.1	1810	0.0835		21.7
75	0.0002		0.9	950	0.0253		11.4	1845	0.0865		22.1
100	0.0004		1.2	975	0.0266		11.7	1880	0.0896		22.5
125	0.0006		1.5	1000	0.0279		12.0	1915	0.0927		22.9
150	0.0008		1.8	1025	0.0292		12.3	1950	0.0959		23.3
175	0.0011		2.1	1050	0.0305		12.6	1985	0.0991		23.7
200	0.0014		2.4	1075	0.0319		12.9	2020	0.1023		24.2
225	0.0018		2.7	1100	0.0332		13.2	2055	0.1056		24.6
250	0.0021		3.0	1125	0.0347		13.5	2090	0.1090		25.0
275	0.0026		3.3	1150	0.0361		13.8	2125	0.1124		25.4
300	0.0030		3.6	1175	0.0376		14.1	2160	0.1158		25.8
325	0.0035		3.9	1200	0.0390		14.4	2195	0.1193		26.3
350	0.0040		4.2	1225	0.0406		14.7	2230	0.1229		26.7
375	0.0045		4.5	1250	0.0421		15.0	2265	0.1265		27.1
400	0.0051		4.8	1275	0.0437		15.3	2300	0.1301		27.5
425	0.0057		5.1	1300	0.0453		15.6	2335	0.1338		27.9
450	0.0064		5.4	1325	0.0469		15.9	2370	0.1375		28.4
475	0.0070		5.7	1350	0.0486		16.2	2405	0.1413		28.8
500	0.0077		6.0	1375	0.0502		16.5	2440	0.1451		29.2
525	0.0085		6.3	1400	0.0519		16.7	2475	0.1490		29.6
550	0.0092		6.6	1425	0.0537		17.0	2510	0.1529		30.0
575	0.0100		6.9	1450	0.0554		17.3	2545	0.1569		30.4
600	0.0108		7.2	1475	0.0572		17.6	2580	0.1609		30.9
625	0.0117		7.5	1500	0.0590		17.9	2615	0.1650		31.3
650	0.0126		7.8	1525	0.0608		18.2	2650	0.1691		31.7
675	0.0135		8.1	1550	0.0627		18.5	2685	0.1733		32.1
700	0.0144		8.4	1575	0.0646		18.8	2720	0.1775		32.5
725	0.0154		8.7	1600	0.0665		19.1	2755	0.1817		33.0
750	0.0164		9.0	1625	0.0684		19.4	2790	0.1860		33.4
775	0.0174		9.3	1650	0.0704		19.7	2825	0.1903		33.8
800	0.0184		9.6	1675	0.0724		20.0	2860	0.1947		34.2
825	0.0195		9.9	1700	0.0744		20.3	2895	0.1991		34.6
850	0.0206		10.2	1725	0.0764		20.6	2930	0.2036		35.1
875	0.0218		10.5	1750	0.0785		20.9	2965	0.2081		35.5
900	0.0229		10.8	1775	0.0806		21.2	3000	0.2127		35.9

Friction Loss Table For   8" Type L Copper   I.D. = 7.725

Equiv. Feet of Pipe

	LONG ELL	90 ELL	45 ELL	TEE	GV	CV
C-120	13	18	9	35	4	45
C-150	19.63	27.18	13.59	52.85	6.04	67.95

Q	P	P	V	Q	P	P	V	Q	P	P	V
	PSI/FT	PSI/FT	FT/SEC		PSI/FT	PSI/FT	FT/SEC		PSI/FT	PSI/FT	FT/SEC
GPM	C-150			GPM	C-150			GPM	C-150		
100	0.0001		0.7	1850	0.0224		12.7	3600	0.0766		24.7
150	0.0002		1.0	1900	0.0235		13.0	3650	0.0786		25.0
200	0.0004		1.4	1950	0.0247		13.4	3700	0.0806		25.3
250	0.0006		1.7	2000	0.0258		13.7	3750	0.0827		25.7
300	0.0008		2.1	2050	0.0270		14.0	3800	0.0847		26.0
350	0.0010		2.4	2100	0.0283		14.4	3850	0.0868		26.4
400	0.0013		2.7	2150	0.0295		14.7	3900	0.0889		26.7
450	0.0016		3.1	2200	0.0308		15.1	3950	0.0910		27.1
500	0.0020		3.4	2250	0.0321		15.4	4000	0.0931		27.4
550	0.0024		3.8	2300	0.0335		15.8	4050	0.0953		27.7
600	0.0028		4.1	2350	0.0348		16.1	4100	0.0975		28.1
650	0.0032		4.5	2400	0.0362		16.4	4150	0.0997		28.4
700	0.0037		4.8	2450	0.0376		16.8	4200	0.1019		28.8
750	0.0042		5.1	2500	0.0390		17.1	4250	0.1042		29.1
800	0.0047		5.5	2550	0.0405		17.5	4300	0.1065		29.5
850	0.0053		5.8	2600	0.0420		17.8	4350	0.1088		29.8
900	0.0059		6.2	2650	0.0435		18.2	4400	0.1111		30.1
950	0.0065		6.5	2700	0.0450		18.5	4450	0.1134		30.5
1000	0.0072		6.8	2750	0.0466		18.8	4500	0.1158		30.8
1050	0.0078		7.2	2800	0.0481		19.2	4550	0.1182		31.2
1100	0.0085		7.5	2850	0.0497		19.5	4600	0.1206		31.5
1150	0.0093		7.9	2900	0.0514		19.9	4650	0.1231		31.8
1200	0.0100		8.2	2950	0.0530		20.2	4700	0.1255		32.2
1250	0.0108		8.6	3000	0.0547		20.5	4750	0.1280		32.5
1300	0.0116		8.9	3050	0.0564		20.9	4800	0.1305		32.9
1350	0.0125		9.2	3100	0.0581		21.2	4850	0.1330		33.2
1400	0.0134		9.6	3150	0.0599		21.6	4900	0.1356		33.6
1450	0.0143		9.9	3200	0.0616		21.9	4950	0.1381		33.9
1500	0.0152		10.3	3250	0.0634		22.3	5000	0.1407		34.2
1550	0.0161		10.6	3300	0.0652		22.6	5050	0.1433		34.6
1600	0.0171		11.0	3350	0.0671		22.9	5100	0.1460		34.9
1650	0.0181		11.3	3400	0.0690		23.3	5150	0.1486		35.3
1700	0.0191		11.6	3450	0.0708		23.6	5200	0.1513		35.6
1750	0.0202		12.0	3500	0.0727		24.0	5250	0.1540		36.0
1800	0.0213		12.3	3550	0.0747		24.3	5300	0.1568		36.3

Friction Loss Table For 3/4" Type M Copper   I.D. = .811

Equiv. Feet of Pipe

	LONG EL	90 ELL	45 ELL	TEE	GV	CV
C-120	1	2	1	3		
C-150	1.51	3.02	1.51	4.53		

Q GPM	P PSI/FT C-150	P PSI/FT	V FT/SEC	Q GPM	P PSI/FT C-150	P PSI/FT	V FT/SEC	Q GPM	P PSI/FT C-150	P PSI/FT	V FT/SEC
7	0.0432		4.4	42	1.1897		26.1				
8	0.0554		5.0	43	1.2427		26.7				
9	0.0688		5.6	44	1.2967		27.3				
10	0.0836		6.2	45	1.3517		28.0				
11	0.0998		6.8	46	1.4078		28.6				
12	0.1172		7.5	47	1.4649		29.2				
13	0.1359		8.1	48	1.5231		29.8				
14	0.1559		8.7	49	1.5823		30.5				
15	0.1771		9.3								
16	0.1996		9.9								
17	0.2232		10.6								
18	0.2481		11.2								
19	0.2742		11.8								
20	0.3015		12.4								
21	0.3300		13.1								
22	0.3597		13.7								
23	0.3905		14.3								
24	0.4225		14.9								
25	0.4556		15.5								
26	0.4899		16.2								
27	0.5254		16.8								
28	0.5619		17.4								
29	0.5996		18.0								
30	0.6384		18.6								
31	0.6784		19.3								
32	0.7194		19.9								
33	0.7615		20.5								
34	0.8048		21.1								
35	0.8491		21.8								
36	0.8945		22.4								
37	0.9410		23.0								
38	0.9886		23.6								
39	1.0373		24.2								
40	1.0870		24.9								
41	1.1379		25.5								

Friction Loss Table For 1" Type M Copper I.D. = 1.055

Equiv. Feet of Pipe

	LONG EL	90 ELL	45 ELL	TEE	GV	CV
C-120	2	2	1	5		
C-150	3.02	3.02	1.51	7.55		

Q	P	P	V	Q	P	P	V	Q	P	P	V
	PSI/FT	PSI/FT	FT/SEC		PSI/FT	PSI/FT	FT/SEC		PSI/FT	PSI/FT	FT/SEC
GPM	C-150			GPM	C-150			GPM	C-150		
7	0.0120		2.6	42	0.3305		15.4	77	1.0142		28.3
8	0.0154		2.9	43	0.3452		15.8	79	1.0635		29.0
9	0.0191		3.3	44	0.3602		16.2	81	1.1139		29.7
10	0.0232		3.7	45	0.3755		16.5	83	1.1653		30.5
11	0.0277		4.0	46	0.3910		16.9	85	1.2177		31.2
12	0.0326		4.4	47	0.4069		17.3	87	1.2713		31.9
13	0.0378		4.8	48	0.4231		17.6	89	1.3259		32.7
14	0.0433		5.1	49	0.4395		18.0				
15	0.0492		5.5	50	0.4563		18.4				
16	0.0554		5.9	51	0.4733		18.7				
17	0.0620		6.2	52	0.4906		19.1				
18	0.0689		6.6	53	0.5082		19.5				
19	0.0762		7.0	54	0.5261		19.8				
20	0.0838		7.3	55	0.5443		20.2				
21	0.0917		7.7	56	0.5627		20.6				
22	0.0999		8.1	57	0.5814		20.9				
23	0.1085		8.4	58	0.6004		21.3				
24	0.1174		8.8	59	0.6197		21.7				
25	0.1266		9.2	60	0.6393		22.0				
26	0.1361		9.5	61	0.6592		22.4				
27	0.1459		9.9	62	0.6793		22.8				
28	0.1561		10.3	63	0.6997		23.1				
29	0.1666		10.6	64	0.7204		23.5				
30	0.1773		11.0	65	0.7413		23.9				
31	0.1884		11.4	66	0.7626		24.2				
32	0.1998		11.8	67	0.7841		24.6				
33	0.2115		12.1	68	0.8059		25.0				
34	0.2235		12.5	69	0.8279		25.3				
35	0.2359		12.9	70	0.8503		25.7				
36	0.2485		13.2	71	0.8729		26.1				
37	0.2614		13.6	72	0.8958		26.4				
38	0.2746		14.0	73	0.9189		26.8				
39	0.2881		14.3	74	0.9423		27.2				
40	0.3020		14.7	75	0.9660		27.5				
41	0.3161		15.1	76	0.9900		27.9				

Friction Loss Table For 1¼" Type M Copper   I.D. = 1.291

Equiv. Feet of Pipe

	LONG EL	90 ELL	45 ELL	TEE	GV	CV
C-120	2	3	1	6		
C-150	3.02	4.53	1.51	9.06		

Q GPM	P PSI/FT C-150	P PSI/FT	V FT/SEC	Q GPM	P PSI/FT C-150	P PSI/FT	V FT/SEC	Q GPM	P PSI/FT C-150	P PSI/FT	V FT/SEC
7	0.0045		1.7	42	0.1236		10.3	77	0.3795		18.9
8	0.0058		2.0	43	0.1291		10.5	79	0.3979		19.4
9	0.0072		2.2	44	0.1348		10.8	81	0.4167		19.9
10	0.0087		2.5	45	0.1405		11.0	83	0.4360		20.4
11	0.0104		2.7	46	0.1463		11.3	85	0.4556		20.8
12	0.0122		2.9	47	0.1522		11.5	87	0.4756		21.3
13	0.0141		3.2	48	0.1583		11.8	89	0.4961		21.8
14	0.0162		3.4	49	0.1644		12.0	91	0.5169		22.3
15	0.0184		3.7	50	0.1707		12.3	93	0.5381		22.8
16	0.0207		3.9	51	0.1771		12.5	95	0.5597		23.3
17	0.0232		4.2	52	0.1836		12.8	97	0.5817		23.8
18	0.0258		4.4	53	0.1901		13.0	99	0.6041		24.3
19	0.0285		4.7	54	0.1968		13.2	101	0.6268		24.8
20	0.0313		4.9	55	0.2036		13.5	103	0.6500		25.3
21	0.0343		5.1	56	0.2105		13.7	105	0.6735		25.7
22	0.0374		5.4	57	0.2175		14.0	107	0.6975		26.2
23	0.0406		5.6	58	0.2246		14.2	109	0.7218		26.7
24	0.0439		5.9	59	0.2319		14.5	111	0.7465		27.2
25	0.0474		6.1	60	0.2392		14.7	113	0.7715		27.7
26	0.0509		6.4	61	0.2466		15.0	115	0.7970		28.2
27	0.0546		6.6	62	0.2541		15.2	117	0.8228		28.7
28	0.0584		6.9	63	0.2618		15.4	119	0.8490		29.2
29	0.0623		7.1	64	0.2695		15.7	121	0.8756		29.7
30	0.0663		7.4	65	0.2774		15.9	123	0.9026		30.2
31	0.0705		7.6	66	0.2853		16.2	125	0.9299		30.7
32	0.0748		7.8	67	0.2934		16.4	127	0.9576		31.1
33	0.0791		8.1	68	0.3015		16.7	129	0.9857		31.6
34	0.0836		8.3	69	0.3098		16.9	131	1.0142		32.1
35	0.0882		8.6	70	0.3181		17.2	133	1.0430		32.6
36	0.0930		8.8	71	0.3266		17.4	135	1.0722		33.1
37	0.0978		9.1	72	0.3351		17.7	137	1.1018		33.6
38	0.1027		9.3	73	0.3438		17.9	139	1.1317		34.1
39	0.1078		9.6	74	0.3526		18.1	141	1.1620		34.6
40	0.1130		9.8	75	0.3614		18.4	143	1.1927		35.1
41	0.1183		10.1	76	0.3704		18.6	145	1.2237		35.6

Friction Loss Table For 1½" Type M Copper   I.D. = 1.527

Equiv. Feet of Pipe

	LONG EL	90 ELL	45 ELL	TEE	GV	CV
C–120	2	4	2	8		9
C–150	3.02	6.04	3.02	12.08	0.00	13.59

Q	P	P	V	Q	P	P	V	Q	P	P	V
	PSI/FT	PSI/FT	FT/SEC		PSI/FT	PSI/FT	FT/SEC		PSI/FT	PSI/FT	FT/SEC
GPM	C–150			GPM	C–150			GPM	C–150		
7	0.0020		1.2	76	0.1635		13.3	111	0.3295		19.5
9	0.0032		1.6	77	0.1675		13.5	113	0.3406		19.8
11	0.0046		1.9	78	0.1716		13.7	115	0.3519		20.2
13	0.0062		2.3	79	0.1757		13.8	117	0.3633		20.5
15	0.0081		2.6	80	0.1798		14.0	119	0.3748		20.9
17	0.0102		3.0	81	0.1840		14.2	121	0.3866		21.2
19	0.0126		3.3	82	0.1882		14.4	123	0.3985		21.6
21	0.0151		3.7	83	0.1925		14.5	125	0.4105		21.9
23	0.0179		4.0	84	0.1968		14.7	127	0.4228		22.3
25	0.0209		4.4	85	0.2011		14.9	129	0.4352		22.6
27	0.0241		4.7	86	0.2055		15.1	131	0.4477		23.0
29	0.0275		5.1	87	0.2100		15.3	133	0.4605		23.3
31	0.0311		5.4	88	0.2145		15.4	135	0.4734		23.7
33	0.0349		5.8	89	0.2190		15.6	137	0.4864		24.0
35	0.0390		6.1	90	0.2236		15.8	139	0.4996		24.4
37	0.0432		6.5	91	0.2282		16.0	141	0.5130		24.7
39	0.0476		6.8	92	0.2329		16.1	143	0.5266		25.1
41	0.0522		7.2	93	0.2376		16.3	145	0.5403		25.4
43	0.0570		7.5	94	0.2423		16.5	147	0.5541		25.8
45	0.0620		7.9	95	0.2471		16.7	149	0.5682		26.1
47	0.0672		8.2	96	0.2519		16.8	151	0.5823		26.5
49	0.0726		8.6	97	0.2568		17.0	153	0.5967		26.8
51	0.0782		8.9	98	0.2617		17.2	155	0.6112		27.2
53	0.0839		9.3	99	0.2667		17.4	157	0.6259		27.5
55	0.0899		9.6	100	0.2717		17.5	159	0.6407		27.9
57	0.0960		10.0	101	0.2767		17.7	161	0.6557		28.2
59	0.1024		10.3	102	0.2818		17.9	163	0.6708		28.6
61	0.1089		10.7	103	0.2870		18.1	165	0.6861		28.9
63	0.1156		11.0	104	0.2921		18.2	167	0.7016		29.3
65	0.1225		11.4	105	0.2974		18.4	169	0.7172		29.6
67	0.1295		11.7	106	0.3026		18.6	171	0.7330		30.0
69	0.1368		12.1	107	0.3079		18.8	173	0.7490		30.3
71	0.1442		12.4	108	0.3133		18.9	175	0.7651		30.7
73	0.1518		12.8	109	0.3186		19.1	177	0.7813		31.0
75	0.1596		13.1	110	0.3241		19.3	179	0.7977		31.4

Friction Loss Table For 2" Type M Copper   I.D. = 2.009

Equiv. Feet of Pipe

	LONG EL	90 ELL	45 ELL	TEE	GV	CV
C-120	3	5	2	10	1	11
C-150	4.53	7.55	3.02	15.10	1.51	16.61

Q	P	P	V	Q	P	P	V	Q	P	P	V
	PSI/FT	PSI/FT	FT/SEC		PSI/FT	PSI/FT	FT/SEC		PSI/FT	PSI/FT	FT/SEC
GPM	C-150			GPM	C-150			GPM	C-150		
7	0.0005		0.7	110	0.0852		11.1	145	0.1420		14.7
10	0.0010		1.0	111	0.0866		11.2	150	0.1512		15.2
13	0.0016		1.3	112	0.0881		11.3	155	0.1607		15.7
16	0.0024		1.6	113	0.0895		11.4	160	0.1704		16.2
19	0.0033		1.9	114	0.0910		11.5	165	0.1804		16.7
22	0.0043		2.2	115	0.0925		11.6	170	0.1906		17.2
25	0.0055		2.5	116	0.0940		11.7	175	0.2011		17.7
28	0.0068		2.8	117	0.0955		11.8	180	0.2119		18.2
31	0.0082		3.1	118	0.0970		11.9	185	0.2229		18.7
34	0.0097		3.4	119	0.0985		12.1	190	0.2342		19.2
37	0.0114		3.7	120	0.1001		12.2	195	0.2457		19.7
40	0.0131		4.1	121	0.1016		12.3	200	0.2575		20.3
43	0.0150		4.4	122	0.1032		12.4	205	0.2695		20.8
46	0.0170		4.7	123	0.1048		12.5	210	0.2818		21.3
49	0.0191		5.0	124	0.1063		12.6	215	0.2944		21.8
52	0.0213		5.3	125	0.1079		12.7	220	0.3071		22.3
55	0.0236		5.6	126	0.1095		12.8	225	0.3202		22.8
58	0.0261		5.9	127	0.1111		12.9	230	0.3335		23.3
61	0.0286		6.2	128	0.1128		13.0	235	0.3470		23.8
64	0.0313		6.5	129	0.1144		13.1	240	0.3608		24.3
67	0.0340		6.8	130	0.1161		13.2	245	0.3748		24.8
70	0.0369		7.1	131	0.1177		13.3	250	0.3891		25.3
73	0.0399		7.4	132	0.1194		13.4	255	0.4036		25.8
76	0.0430		7.7	133	0.1211		13.5	260	0.4184		26.3
79	0.0462		8.0	134	0.1227		13.6	265	0.4334		26.8
82	0.0495		8.3	135	0.1244		13.7	270	0.4486		27.3
85	0.0529		8.6	136	0.1262		13.8	275	0.4641		27.8
88	0.0564		8.9	137	0.1279		13.9	280	0.4798		28.4
91	0.0600		9.2	138	0.1296		14.0	285	0.4958		28.9
94	0.0637		9.5	139	0.1314		14.1	290	0.5120		29.4
97	0.0675		9.8	140	0.1331		14.2	295	0.5285		29.9
100	0.0714		10.1	141	0.1349		14.3	300	0.5452		30.4
103	0.0754		10.4	142	0.1366		14.4	305	0.5621		30.9
106	0.0796		10.7	143	0.1384		14.5	310	0.5793		31.4
109	0.0838		11.0	144	0.1402		14.6	315	0.5967		31.9

Appendix D Friction Loss Tables

Friction Loss Table For 2½" Type M Copper I.D. = 2.495

Equiv. Feet of Pipe

	LONG EL	90 ELL	45 ELL	TEE	GV	CV
C-120	4	6	3	12	1	14
C-150	6.04	9.06	4.53	18.12	1.51	21.14

Q	P	P	V	Q	P	P	V	Q	P	P	V
	PSI/FT	PSI/FT	FT/SEC		PSI/FT	PSI/FT	FT/SEC		PSI/FT	PSI/FT	FT/SEC
GPM	C-150			GPM	C-150			GPM	C-150		
15	0.0007		1.0	190	0.0815		12.5	365	0.2728		24.0
20	0.0013		1.3	195	0.0855		12.8	370	0.2798		24.3
25	0.0019		1.6	200	0.0896		13.1	375	0.2868		24.6
30	0.0027		2.0	205	0.0938		13.5	380	0.2939		25.0
35	0.0036		2.3	210	0.0981		13.8	385	0.3011		25.3
40	0.0046		2.6	215	0.1025		14.1	390	0.3084		25.6
45	0.0057		3.0	220	0.1069		14.4	395	0.3157		25.9
50	0.0069		3.3	225	0.1115		14.8	400	0.3232		26.3
55	0.0082		3.6	230	0.1161		15.1	405	0.3307		26.6
60	0.0097		3.9	235	0.1208		15.4	410	0.3383		26.9
65	0.0112		4.3	240	0.1256		15.8	415	0.3460		27.2
70	0.0129		4.6	245	0.1305		16.1	420	0.3537		27.6
75	0.0146		4.9	250	0.1355		16.4	425	0.3615		27.9
80	0.0165		5.3	255	0.1405		16.7	430	0.3694		28.2
85	0.0184		5.6	260	0.1457		17.1	435	0.3774		28.6
90	0.0205		5.9	265	0.1509		17.4	440	0.3855		28.9
95	0.0226		6.2	270	0.1562		17.7	445	0.3936		29.2
100	0.0249		6.6	275	0.1616		18.1	450	0.4019		29.5
105	0.0272		6.9	280	0.1671		18.4	455	0.4102		29.9
110	0.0297		7.2	285	0.1726		18.7	460	0.4185		30.2
115	0.0322		7.6	290	0.1783		19.0	465	0.4270		30.5
120	0.0348		7.9	295	0.1840		19.4	470	0.4355		30.9
125	0.0376		8.2	300	0.1898		19.7	475	0.4441		31.2
130	0.0404		8.5	305	0.1957		20.0	480	0.4528		31.5
135	0.0433		8.9	310	0.2017		20.4	485	0.4616		31.8
140	0.0463		9.2	315	0.2077		20.7	490	0.4704		32.2
145	0.0495		9.5	320	0.2139		21.0	495	0.4794		32.5
150	0.0527		9.8	325	0.2201		21.3	500	0.4883		32.8
155	0.0559		10.2	330	0.2264		21.7	505	0.4974		33.2
160	0.0593		10.5	335	0.2328		22.0	510	0.5066		33.5
165	0.0628		10.8	340	0.2393		22.3	515	0.5158		33.8
170	0.0664		11.2	345	0.2458		22.7	520	0.5251		34.1
175	0.0700		11.5	350	0.2524		23.0	525	0.5345		34.5
180	0.0738		11.8	355	0.2592		23.3	530	0.5439		34.8
185	0.0776		12.1	360	0.2659		23.6	535	0.5535		35.1

Friction Loss Table For 3" Type M Copper   I.D. = 2.981

Equiv. Feet of Pipe

	LONG EL	90 ELL	45 ELL	TEE	GV	CV
C-120	5	7	3	15	1	16
C-150	7.55	10.57	4.53	22.65	1.51	24.16

Q	P	P	V	Q	P	P	V	Q	P	P	V
	PSI/FT	PSI/FT	FT/SEC		PSI/FT	PSI/FT	FT/SEC		PSI/FT	PSI/FT	FT/SEC
GPM	C-150			GPM	C-150			GPM	C-150		
20	0.0005		0.9	195	0.0360		9.0	375	0.1206		17.2
25	0.0008		1.1	200	0.0377		9.2	385	0.1266		17.7
30	0.0011		1.4	205	0.0394		9.4	395	0.1327		18.2
35	0.0015		1.6	210	0.0412		9.7	405	0.1390		18.6
40	0.0019		1.8	215	0.0431		9.9	415	0.1454		19.1
45	0.0024		2.1	220	0.0449		10.1	425	0.1520		19.5
50	0.0029		2.3	225	0.0469		10.3	435	0.1586		20.0
55	0.0035		2.5	230	0.0488		10.6	445	0.1655		20.5
60	0.0041		2.8	235	0.0508		10.8	455	0.1724		20.9
65	0.0047		3.0	240	0.0528		11.0	465	0.1795		21.4
70	0.0054		3.2	245	0.0549		11.3	475	0.1867		21.8
75	0.0061		3.4	250	0.0569		11.5	485	0.1940		22.3
80	0.0069		3.7	255	0.0591		11.7	495	0.2015		22.8
85	0.0077		3.9	260	0.0612		12.0	505	0.2091		23.2
90	0.0086		4.1	265	0.0634		12.2	515	0.2168		23.7
95	0.0095		4.4	270	0.0657		12.4	525	0.2247		24.1
100	0.0105		4.6	275	0.0679		12.6	535	0.2326		24.6
105	0.0114		4.8	280	0.0702		12.9	545	0.2407		25.1
110	0.0125		5.1	285	0.0726		13.1	555	0.2490		25.5
115	0.0135		5.3	290	0.0749		13.3	565	0.2573		26.0
120	0.0146		5.5	295	0.0773		13.6	575	0.2658		26.4
125	0.0158		5.7	300	0.0798		13.8	585	0.2745		26.9
130	0.0170		6.0	305	0.0823		14.0	595	0.2832		27.4
135	0.0182		6.2	310	0.0848		14.3	605	0.2921		27.8
140	0.0195		6.4	315	0.0873		14.5	615	0.3011		28.3
145	0.0208		6.7	320	0.0899		14.7	625	0.3102		28.7
150	0.0221		6.9	325	0.0925		14.9	635	0.3194		29.2
155	0.0235		7.1	330	0.0952		15.2	645	0.3288		29.7
160	0.0249		7.4	335	0.0978		15.4	655	0.3383		30.1
165	0.0264		7.6	340	0.1006		15.6	665	0.3479		30.6
170	0.0279		7.8	345	0.1033		15.9	675	0.3576		31.0
175	0.0294		8.0	350	0.1061		16.1	685	0.3675		31.5
180	0.0310		8.3	355	0.1089		16.3	695	0.3775		32.0
185	0.0326		8.5	360	0.1118		16.6	705	0.3876		32.4
190	0.0343		8.7	365	0.1147		16.8	715	0.3978		32.9

Friction Loss Table For 4" Type M Copper I.D. = 3.935

Equiv. Feet of Pipe

	LONG EL	90 ELL	45 ELL	TEE	GV	CV
C-120	6	10	4	20	2	22
C-150	9.06	15.10	6.04	30.20	3.02	33.22

Q	P	P	V	Q	P	P	V	Q	P	P	V
	PSI/FT	PSI/FT	FT/SEC		PSI/FT	PSI/FT	FT/SEC		PSI/FT	PSI/FT	FT/SEC
GPM	C-150			GPM	C-150			GPM	C-150		
30	0.0003		0.8	380	0.0320		10.0	735	0.1083		19.4
40	0.0005		1.1	390	0.0335		10.3	750	0.1124		19.8
50	0.0008		1.3	400	0.0351		10.6	765	0.1166		20.2
60	0.0011		1.6	410	0.0368		10.8	780	0.1209		20.6
70	0.0014		1.8	420	0.0385		11.1	795	0.1252		21.0
80	0.0018		2.1	430	0.0402		11.4	810	0.1296		21.4
90	0.0022		2.4	440	0.0419		11.6	825	0.1341		21.8
100	0.0027		2.6	450	0.0437		11.9	840	0.1386		22.2
110	0.0032		2.9	460	0.0455		12.1	855	0.1433		22.6
120	0.0038		3.2	470	0.0474		12.4	870	0.1479		23.0
130	0.0044		3.4	480	0.0492		12.7	885	0.1527		23.4
140	0.0050		3.7	490	0.0512		12.9	900	0.1575		23.8
150	0.0057		4.0	500	0.0531		13.2	915	0.1624		24.2
160	0.0065		4.2	510	0.0551		13.5	930	0.1674		24.5
170	0.0072		4.5	520	0.0571		13.7	945	0.1724		24.9
180	0.0080		4.8	530	0.0591		14.0	960	0.1775		25.3
190	0.0089		5.0	540	0.0612		14.3	975	0.1827		25.7
200	0.0097		5.3	550	0.0633		14.5	990	0.1879		26.1
210	0.0107		5.5	560	0.0655		14.8	1005	0.1932		26.5
220	0.0116		5.8	570	0.0677		15.0	1020	0.1986		26.9
230	0.0126		6.1	580	0.0699		15.3	1035	0.2040		27.3
240	0.0137		6.3	590	0.0721		15.6	1050	0.2095		27.7
250	0.0147		6.6	600	0.0744		15.8	1065	0.2151		28.1
260	0.0158		6.9	610	0.0767		16.1	1080	0.2207		28.5
270	0.0170		7.1	620	0.0791		16.4	1095	0.2264		28.9
280	0.0182		7.4	630	0.0814		16.6	1110	0.2322		29.3
290	0.0194		7.7	640	0.0838		16.9	1125	0.2380		29.7
300	0.0206		7.9	650	0.0863		17.2	1140	0.2439		30.1
310	0.0219		8.2	660	0.0887		17.4	1155	0.2499		30.5
320	0.0233		8.4	670	0.0912		17.7	1170	0.2559		30.9
330	0.0246		8.7	680	0.0938		17.9	1185	0.2620		31.3
340	0.0260		9.0	690	0.0964		18.2	1200	0.2682		31.7
350	0.0274		9.2	700	0.0990		18.5	1215	0.2744		32.1
360	0.0289		9.5	710	0.1016		18.7	1230	0.2807		32.5
370	0.0304		9.8	720	0.1042		19.0	1245	0.2871		32.9

Friction Loss Table For 6" Type M Copper   I.D. = 5.881

Equiv. Feet of Pipe

	LONG EL	90 ELL	45 ELL	TEE	GV	CV
C-120	9	14	7	30	3	32
C-150	13.59	21.14	10.57	45.30	4.53	48.32

Q GPM	P PSI/FT C-150	P PSI/FT	V FT/SEC	Q GPM	P PSI/FT C-150	P PSI/FT	V FT/SEC	Q GPM	P PSI/FT C-150	P PSI/FT	V FT/SEC
50	0.0001		0.6	925	0.0234		10.9	1810	0.0811		21.4
75	0.0002		0.9	950	0.0246		11.2	1845	0.0840		21.8
100	0.0004		1.2	975	0.0258		11.5	1880	0.0870		22.2
125	0.0006		1.5	1000	0.0270		11.8	1915	0.0900		22.6
150	0.0008		1.8	1025	0.0283		12.1	1950	0.0930		23.0
175	0.0011		2.1	1050	0.0296		12.4	1985	0.0962		23.5
200	0.0014		2.4	1075	0.0309		12.7	2020	0.0993		23.9
225	0.0017		2.7	1100	0.0323		13.0	2055	0.1025		24.3
250	0.0021		3.0	1125	0.0336		13.3	2090	0.1058		24.7
275	0.0025		3.2	1150	0.0350		13.6	2125	0.1091		25.1
300	0.0029		3.5	1175	0.0365		13.9	2160	0.1124		25.5
325	0.0034		3.8	1200	0.0379		14.2	2195	0.1158		25.9
350	0.0039		4.1	1225	0.0394		14.5	2230	0.1193		26.4
375	0.0044		4.4	1250	0.0409		14.8	2265	0.1227		26.8
400	0.0050		4.7	1275	0.0424		15.1	2300	0.1263		27.2
425	0.0056		5.0	1300	0.0439		15.4	2335	0.1299		27.6
450	0.0062		5.3	1325	0.0455		15.7	2370	0.1335		28.0
475	0.0068		5.6	1350	0.0471		16.0	2405	0.1372		28.4
500	0.0075		5.9	1375	0.0488		16.2	2440	0.1409		28.8
525	0.0082		6.2	1400	0.0504		16.5	2475	0.1446		29.2
550	0.0089		6.5	1425	0.0521		16.8	2510	0.1484		29.7
575	0.0097		6.8	1450	0.0538		17.1	2545	0.1523		30.1
600	0.0105		7.1	1475	0.0555		17.4	2580	0.1562		30.5
625	0.0113		7.4	1500	0.0573		17.7	2615	0.1601		30.9
650	0.0122		7.7	1525	0.0590		18.0	2650	0.1641		31.3
675	0.0131		8.0	1550	0.0608		18.3	2685	0.1681		31.7
700	0.0140		8.3	1575	0.0627		18.6	2720	0.1722		32.1
725	0.0149		8.6	1600	0.0645		18.9	2755	0.1763		32.6
750	0.0159		8.9	1625	0.0664		19.2	2790	0.1805		33.0
775	0.0169		9.2	1650	0.0683		19.5	2825	0.1847		33.4
800	0.0179		9.5	1675	0.0702		19.8	2860	0.1890		33.8
825	0.0189		9.7	1700	0.0722		20.1	2895	0.1933		34.2
850	0.0200		10.0	1725	0.0742		20.4	2930	0.1976		34.6
875	0.0211		10.3	1750	0.0762		20.7	2965	0.2020		35.0
900	0.0223		10.6	1775	0.0782		21.0	3000	0.2065		35.5

Friction Loss Table For 8" Type M Copper   I.D. = 7.785

Equiv. Feet of Pipe

	LONG EL	90 ELL	45 ELL	TEE	GV	CV
C-120	13	18	9	35	4	45
C-150	19.63	27.18	13.59	52.85	6.04	67.95

Q	P	P	V	Q	P	P	V	Q	P	P	V
	PSI/FT	PSI/FT	FT/SEC		PSI/FT	PSI/FT	FT/SEC		PSI/FT	PSI/FT	FT/SEC
GPM	C-150			GPM	C-150			GPM	C-150		
100	0.0001		0.7	1850	0.0215		12.5	3600	0.0738		24.3
150	0.0002		1.0	1900	0.0226		12.8	3650	0.0757		24.6
200	0.0004		1.3	1950	0.0237		13.2	3700	0.0776		25.0
250	0.0005		1.7	2000	0.0249		13.5	3750	0.0796		25.3
300	0.0007		2.0	2050	0.0260		13.8	3800	0.0816		25.6
350	0.0010		2.4	2100	0.0272		14.2	3850	0.0836		26.0
400	0.0013		2.7	2150	0.0284		14.5	3900	0.0856		26.3
450	0.0016		3.0	2200	0.0297		14.8	3950	0.0876		26.6
500	0.0019		3.4	2250	0.0309		15.2	4000	0.0897		27.0
550	0.0023		3.7	2300	0.0322		15.5	4050	0.0918		27.3
600	0.0027		4.0	2350	0.0335		15.8	4100	0.0939		27.7
650	0.0031		4.4	2400	0.0349		16.2	4150	0.0960		28.0
700	0.0036		4.7	2450	0.0362		16.5	4200	0.0982		28.3
750	0.0041		5.1	2500	0.0376		16.9	4250	0.1003		28.7
800	0.0046		5.4	2550	0.0390		17.2	4300	0.1025		29.0
850	0.0051		5.7	2600	0.0404		17.5	4350	0.1047		29.3
900	0.0057		6.1	2650	0.0419		17.9	4400	0.1070		29.7
950	0.0063		6.4	2700	0.0433		18.2	4450	0.1092		30.0
1000	0.0069		6.7	2750	0.0448		18.5	4500	0.1115		30.3
1050	0.0076		7.1	2800	0.0464		18.9	4550	0.1138		30.7
1100	0.0082		7.4	2850	0.0479		19.2	4600	0.1162		31.0
1150	0.0089		7.8	2900	0.0495		19.6	4650	0.1185		31.4
1200	0.0097		8.1	2950	0.0511		19.9	4700	0.1209		31.7
1250	0.0104		8.4	3000	0.0527		20.2	4750	0.1233		32.0
1300	0.0112		8.8	3050	0.0543		20.6	4800	0.1257		32.4
1350	0.0120		9.1	3100	0.0560		20.9	4850	0.1281		32.7
1400	0.0129		9.4	3150	0.0577		21.2	4900	0.1306		33.0
1450	0.0137		9.8	3200	0.0594		21.6	4950	0.1330		33.4
1500	0.0146		10.1	3250	0.0611		21.9	5000	0.1355		33.7
1550	0.0155		10.5	3300	0.0628		22.3	5050	0.1380		34.1
1600	0.0165		10.8	3350	0.0646		22.6	5100	0.1406		34.4
1650	0.0174		11.1	3400	0.0664		22.9	5150	0.1431		34.7
1700	0.0184		11.5	3450	0.0682		23.3	5200	0.1457		35.1
1750	0.0194		11.8	3500	0.0701		23.6	5250	0.1483		35.4
1800	0.0205		12.1	3550	0.0719		23.9	5300	0.1510		35.7

Friction Loss Table For 1" XL Steel Pipe   I.D. = 1.104

Equiv. Feet of Pipe

	LONG EL	90 ELL	45 ELL	TEE	GV	CV		
C-120	2	2	1	5		5		
C-100	1.43	1.43	0.71	3.57	0.00	3.57	0.00	0.00

Q GPM	P PSI/FT C-100	P PSI/FT C-120	V FT/SEC	Q GPM	P PSI/FT C-100	P PSI/FT C-120	V FT/SEC	Q GPM	P PSI/FT C-100	P PSI/FT C-120	V FT/SEC
7	0.0204	0.0145	2.3	42	0.5609	0.4003	14.1	77	1.7214	1.2286	25.8
8	0.0261	0.0186	2.7	43	0.5859	0.4181	14.4	78	1.7630	1.2583	26.2
9	0.0325	0.0232	3.0	44	0.6113	0.4363	14.8	79	1.8050	1.2883	26.5
10	0.0394	0.0281	3.4	45	0.6373	0.4548	15.1	80	1.8475	1.3186	26.8
11	0.0470	0.0336	3.7	46	0.6637	0.4737	15.4	81	1.8905	1.3492	27.2
12	0.0553	0.0394	4.0	47	0.6907	0.4929	15.8	82	1.9339	1.3802	27.5
13	0.0641	0.0457	4.4	48	0.7181	0.5125	16.1	83	1.9778	1.4115	27.8
14	0.0735	0.0524	4.7	49	0.7460	0.5324	16.4	84	2.0221	1.4431	28.2
15	0.0835	0.0596	5.0	50	0.7744	0.5527	16.8	85	2.0668	1.4751	28.5
16	0.0941	0.0671	5.4	51	0.8033	0.5733	17.1	86	2.1120	1.5074	28.8
17	0.1052	0.0751	5.7	52	0.8327	0.5943	17.4	87	2.1577	1.5399	29.2
18	0.1170	0.0835	6.0	53	0.8626	0.6156	17.8	88	2.2038	1.5728	29.5
19	0.1293	0.0923	6.4	54	0.8929	0.6373	18.1	89	2.2504	1.6061	29.8
20	0.1422	0.1015	6.7	55	0.9237	0.6593	18.4	90	2.2974	1.6396	30.2
21	0.1556	0.1110	7.0	56	0.9551	0.6816	18.8	91	2.3448	1.6735	30.5
22	0.1696	0.1210	7.4	57	0.9868	0.7043	19.1	92	2.3927	1.7077	30.9
23	0.1841	0.1314	7.7	58	1.0191	0.7273	19.5	93	2.4410	1.7422	31.2
24	0.1992	0.1422	8.0	59	1.0519	0.7507	19.8	94	2.4898	1.7770	31.5
25	0.2148	0.1533	8.4	60	1.0851	0.7744	20.1	95	2.5390	1.8121	31.9
26	0.2310	0.1649	8.7	61	1.1188	0.7985	20.5	96	2.5887	1.8475	32.2
27	0.2477	0.1768	9.1	62	1.1529	0.8228	20.8	97	2.6388	1.8833	32.5
28	0.2649	0.1891	9.4	63	1.1876	0.8476	21.1	98	2.6893	1.9194	32.9
29	0.2827	0.2018	9.7	64	1.2227	0.8726	21.5	99	2.7403	1.9558	33.2
30	0.3010	0.2148	10.1	65	1.2583	0.8980	21.8	100	2.7918	1.9925	33.5
31	0.3198	0.2283	10.4	66	1.2943	0.9237	22.1	101	2.8436	2.0295	33.9
32	0.3392	0.2421	10.7	67	1.3308	0.9498	22.5	102	2.8959	2.0668	34.2
33	0.3590	0.2562	11.1	68	1.3678	0.9762	22.8	103	2.9487	2.1045	34.5
34	0.3794	0.2708	11.4	69	1.4052	1.0029	23.1	104	3.0019	2.1424	34.9
35	0.4003	0.2857	11.7	70	1.4431	1.0300	23.5	105	3.0555	2.1807	35.2
36	0.4217	0.3010	12.1	71	1.4815	1.0574	23.8	106	3.1095	2.2193	35.5
37	0.4437	0.3166	12.4	72	1.5204	1.0851	24.1	107	3.1640	2.2582	35.9
38	0.4661	0.3327	12.7	73	1.5596	1.1131	24.5	108	3.2189	2.2974	36.2
39	0.4890	0.3490	13.1	74	1.5994	1.1415	24.8	109	3.2743	2.3369	36.6
40	0.5125	0.3658	13.4	75	1.6396	1.1702	25.2	110	3.3301	2.3767	36.9
41	0.5364	0.3829	13.7	76	1.6803	1.1992	25.5	111	3.3863	2.4168	37.2

Friction Loss Table For 1¼" XL Steel Pipe   I.D. = 1.452

Equiv. Feet of Pipe

	LONG EL	90 ELL	45 ELL	TEE	GV	CV				
C-120	2	3	1	6		7				
C-100	1.43	2.14	0.71	4.28	0.00	4.99	0.00	0.00		

Q	P	P	V	Q	P	P	V	Q	P	P	V
	PSI/FT	PSI/FT	FT/SEC		PSI/FT	PSI/FT	FT/SEC		PSI/FT	PSI/FT	FT/SEC
GPM	C-100	C-120		GPM	C-100	C-120		GPM	C-100	C-120	
7	0.0054	0.0038	1.4	42	0.1477	0.1054	8.1	77	0.4533	0.3235	14.9
8	0.0069	0.0049	1.6	43	0.1543	0.1101	8.3	80	0.4865	0.3472	15.5
9	0.0085	0.0061	1.7	44	0.1610	0.1149	8.5	83	0.5208	0.3717	16.1
10	0.0104	0.0074	1.9	45	0.1678	0.1198	8.7	86	0.5561	0.3969	16.7
11	0.0124	0.0088	2.1	46	0.1748	0.1247	8.9	89	0.5926	0.4229	17.3
12	0.0145	0.0104	2.3	47	0.1819	0.1298	9.1	92	0.6300	0.4497	17.8
13	0.0169	0.0120	2.5	48	0.1891	0.1349	9.3	95	0.6686	0.4772	18.4
14	0.0194	0.0138	2.7	49	0.1964	0.1402	9.5	98	0.7082	0.5054	19.0
15	0.0220	0.0157	2.9	50	0.2039	0.1455	9.7	101	0.7488	0.5344	19.6
16	0.0248	0.0177	3.1	51	0.2115	0.1510	9.9	104	0.7904	0.5641	20.2
17	0.0277	0.0198	3.3	52	0.2193	0.1565	10.1	107	0.8331	0.5946	20.7
18	0.0308	0.0220	3.5	53	0.2271	0.1621	10.3	110	0.8769	0.6258	21.3
19	0.0340	0.0243	3.7	54	0.2351	0.1678	10.5	113	0.9216	0.6578	21.9
20	0.0374	0.0267	3.9	55	0.2432	0.1736	10.7	116	0.9674	0.6904	22.5
21	0.0410	0.0292	4.1	56	0.2515	0.1795	10.9	119	1.0142	0.7238	23.1
22	0.0447	0.0319	4.3	57	0.2599	0.1855	11.1	122	1.0620	0.7580	23.7
23	0.0485	0.0346	4.5	58	0.2684	0.1915	11.2	125	1.1108	0.7928	24.2
24	0.0525	0.0374	4.7	59	0.2770	0.1977	11.4	128	1.1606	0.8284	24.8
25	0.0566	0.0404	4.8	60	0.2857	0.2039	11.6	131	1.2115	0.8646	25.4
26	0.0608	0.0434	5.0	61	0.2946	0.2102	11.8	134	1.2633	0.9016	26.0
27	0.0652	0.0465	5.2	62	0.3036	0.2167	12.0	137	1.3161	0.9393	26.6
28	0.0698	0.0498	5.4	63	0.3127	0.2232	12.2	140	1.3699	0.9777	27.1
29	0.0744	0.0531	5.6	64	0.3220	0.2298	12.4	143	1.4247	1.0168	27.7
30	0.0793	0.0566	5.8	65	0.3313	0.2365	12.6	146	1.4805	1.0566	28.3
31	0.0842	0.0601	6.0	66	0.3408	0.2432	12.8	149	1.5373	1.0972	28.9
32	0.0893	0.0637	6.2	67	0.3504	0.2501	13.0	152	1.5950	1.1384	29.5
33	0.0945	0.0675	6.4	68	0.3602	0.2570	13.2	155	1.6538	1.1803	30.0
34	0.0999	0.0713	6.6	69	0.3700	0.2641	13.4	158	1.7135	1.2229	30.6
35	0.1054	0.0752	6.8	70	0.3800	0.2712	13.6	161	1.7741	1.2662	31.2
36	0.1111	0.0793	7.0	71	0.3901	0.2784	13.8	164	1.8358	1.3102	31.8
37	0.1168	0.0834	7.2	72	0.4003	0.2857	14.0	167	1.8984	1.3549	32.4
38	0.1227	0.0876	7.4	73	0.4107	0.2931	14.2	170	1.9620	1.4003	33.0
39	0.1288	0.0919	7.6	74	0.4212	0.3006	14.3	173	2.0265	1.4463	33.5
40	0.1349	0.0963	7.8	75	0.4317	0.3081	14.5	176	2.0920	1.4930	34.1
41	0.1413	0.1008	7.9	76	0.4425	0.3158	14.7	179	2.1584	1.5405	34.7

Friction Loss Table For 1½" XL Steel Pipe   I.D. = 1.687

Equiv. Feet of Pipe

	LONG EL	90 ELL	45 ELL	TEE	GV	CV		
C–120	2	4	2	8		9		
C–100	1.43	2.85	1.43	5.70	0.00	6.42	0.00	0.00

Q	P	P	V	Q	P	P	V	Q	P	P	V
	PSI/FT	PSI/FT	FT/SEC		PSI/FT	PSI/FT	FT/SEC		PSI/FT	PSI/FT	FT/SEC
GPM	C–100	C–120		GPM	C–100	C–120		GPM	C–100	C–120	
7	0.0026	0.0018	1.0	76	0.2131	0.1521	10.9	145	0.7041	0.5025	20.8
9	0.0041	0.0029	1.3	78	0.2236	0.1596	11.2	148	0.7313	0.5219	21.3
11	0.0060	0.0043	1.6	80	0.2343	0.1672	11.5	151	0.7589	0.5416	21.7
13	0.0081	0.0058	1.9	82	0.2453	0.1750	11.8	154	0.7870	0.5617	22.1
15	0.0106	0.0076	2.2	84	0.2565	0.1830	12.1	157	0.8156	0.5821	22.5
17	0.0133	0.0095	2.4	86	0.2679	0.1912	12.4	160	0.8447	0.6029	23.0
19	0.0164	0.0117	2.7	88	0.2795	0.1995	12.6	163	0.8742	0.6239	23.4
21	0.0197	0.0141	3.0	90	0.2914	0.2079	12.9	166	0.9042	0.6454	23.8
23	0.0233	0.0167	3.3	92	0.3035	0.2166	13.2	169	0.9347	0.6671	24.3
25	0.0272	0.0194	3.6	94	0.3158	0.2254	13.5	172	0.9656	0.6892	24.7
27	0.0314	0.0224	3.9	96	0.3283	0.2343	13.8	175	0.9970	0.7116	25.1
29	0.0359	0.0256	4.2	98	0.3411	0.2434	14.1	178	1.0289	0.7343	25.6
31	0.0406	0.0289	4.5	100	0.3541	0.2527	14.4	181	1.0612	0.7574	26.0
33	0.0455	0.0325	4.7	102	0.3673	0.2621	14.6	184	1.0940	0.7808	26.4
35	0.0508	0.0362	5.0	104	0.3807	0.2717	14.9	187	1.1272	0.8045	26.9
37	0.0563	0.0402	5.3	106	0.3944	0.2815	15.2	190	1.1609	0.8285	27.3
39	0.0620	0.0443	5.6	108	0.4082	0.2914	15.5	193	1.1950	0.8529	27.7
41	0.0680	0.0486	5.9	110	0.4223	0.3014	15.8	196	1.2296	0.8776	28.1
43	0.0743	0.0530	6.2	112	0.4367	0.3116	16.1	199	1.2646	0.9026	28.6
45	0.0808	0.0577	6.5	114	0.4512	0.3220	16.4	202	1.3001	0.9279	29.0
47	0.0876	0.0625	6.8	116	0.4659	0.3325	16.7	205	1.3361	0.9536	29.4
49	0.0946	0.0675	7.0	118	0.4809	0.3432	16.9	208	1.3725	0.9795	29.9
51	0.1019	0.0727	7.3	120	0.4961	0.3541	17.2	211	1.4093	1.0058	30.3
53	0.1094	0.0781	7.6	122	0.5115	0.3651	17.5	214	1.4466	1.0324	30.7
55	0.1172	0.0836	7.9	124	0.5271	0.3762	17.8	217	1.4844	1.0594	31.2
57	0.1252	0.0893	8.2	126	0.5430	0.3875	18.1	220	1.5225	1.0866	31.6
59	0.1334	0.0952	8.5	128	0.5590	0.3990	18.4	223	1.5612	1.1142	32.0
61	0.1419	0.1013	8.8	130	0.5753	0.4106	18.7	226	1.6003	1.1421	32.5
63	0.1506	0.1075	9.0	132	0.5918	0.4223	19.0	229	1.6398	1.1703	32.9
65	0.1596	0.1139	9.3	134	0.6085	0.4343	19.2	232	1.6797	1.1988	33.3
67	0.1688	0.1205	9.6	136	0.6254	0.4463	19.5	235	1.7201	1.2277	33.8
69	0.1782	0.1272	9.9	138	0.6425	0.4585	19.8	238	1.7610	1.2568	34.2
71	0.1879	0.1341	10.2	140	0.6598	0.4709	20.1	241	1.8023	1.2863	34.6
73	0.1978	0.1412	10.5	142	0.6774	0.4834	20.4	244	1.8440	1.3161	35.0
75	0.2079	0.1484	10.8	144	0.6951	0.4961	20.7	247	1.8862	1.3461	35.5

Friction Loss Table For 2" XL Steel Pipe   I.D. = 2.154

Equiv. Feet of Pipe

	LONG EL	90 ELL	45 ELL	TEE	GV	CV	BV			
C-120	3	5	2	10	1	11	6			
C-100	2.14	3.57	1.43	7.13	0.71	7.84	4.28	0.00		

Q GPM	P PSI/FT C-100	P PSI/FT C-120	V FT/SEC	Q GPM	P PSI/FT C-100	P PSI/FT C-120	V FT/SEC	Q GPM	P PSI/FT C-100	P PSI/FT C-120	V FT/SEC
7	0.0008	0.0006	0.6	110	0.1285	0.0917	9.7	213	0.4362	0.3113	18.8
10	0.0015	0.0011	0.9	113	0.1350	0.0964	10.0	218	0.4554	0.3250	19.2
13	0.0025	0.0018	1.1	116	0.1417	0.1012	10.2	223	0.4749	0.3389	19.6
16	0.0036	0.0026	1.4	119	0.1486	0.1060	10.5	228	0.4948	0.3531	20.1
19	0.0050	0.0036	1.7	122	0.1556	0.1110	10.7	233	0.5150	0.3676	20.5
22	0.0065	0.0047	1.9	125	0.1627	0.1162	11.0	238	0.5357	0.3823	21.0
25	0.0083	0.0059	2.2	128	0.1700	0.1214	11.3	243	0.5567	0.3973	21.4
28	0.0102	0.0073	2.5	131	0.1775	0.1267	11.5	248	0.5781	0.4126	21.8
31	0.0123	0.0088	2.7	134	0.1851	0.1321	11.8	253	0.5998	0.4281	22.3
34	0.0146	0.0104	3.0	137	0.1928	0.1376	12.1	258	0.6219	0.4439	22.7
37	0.0171	0.0122	3.3	140	0.2007	0.1432	12.3	263	0.6444	0.4599	23.2
40	0.0198	0.0141	3.5	143	0.2087	0.1490	12.6	268	0.6672	0.4762	23.6
43	0.0226	0.0161	3.8	146	0.2169	0.1548	12.9	273	0.6905	0.4928	24.0
46	0.0256	0.0183	4.1	149	0.2252	0.1607	13.1	278	0.7140	0.5096	24.5
49	0.0288	0.0205	4.3	152	0.2337	0.1668	13.4	283	0.7380	0.5267	24.9
52	0.0321	0.0229	4.6	155	0.2423	0.1729	13.7	288	0.7623	0.5440	25.4
55	0.0356	0.0254	4.8	158	0.2510	0.1792	13.9	293	0.7869	0.5616	25.8
58	0.0393	0.0281	5.1	161	0.2599	0.1855	14.2	298	0.8120	0.5795	26.3
61	0.0432	0.0308	5.4	164	0.2690	0.1920	14.4	303	0.8373	0.5976	26.7
64	0.0472	0.0337	5.6	167	0.2781	0.1985	14.7	308	0.8631	0.6160	27.1
67	0.0513	0.0366	5.9	170	0.2875	0.2052	15.0	313	0.8892	0.6346	27.6
70	0.0557	0.0397	6.2	173	0.2969	0.2119	15.2	318	0.9156	0.6535	28.0
73	0.0602	0.0429	6.4	176	0.3065	0.2187	15.5	323	0.9424	0.6726	28.5
76	0.0648	0.0463	6.7	179	0.3162	0.2257	15.8	328	0.9696	0.6920	28.9
79	0.0696	0.0497	7.0	182	0.3261	0.2327	16.0	333	0.9971	0.7117	29.3
82	0.0746	0.0532	7.2	185	0.3361	0.2399	16.3	338	1.0250	0.7315	29.8
85	0.0797	0.0569	7.5	188	0.3463	0.2471	16.6	343	1.0532	0.7517	30.2
88	0.0850	0.0607	7.8	191	0.3566	0.2545	16.8	348	1.0818	0.7721	30.7
91	0.0905	0.0646	8.0	194	0.3670	0.2619	17.1	353	1.1107	0.7927	31.1
94	0.0961	0.0686	8.3	197	0.3776	0.2695	17.4	358	1.1400	0.8136	31.5
97	0.1018	0.0727	8.5	200	0.3883	0.2771	17.6	363	1.1697	0.8348	32.0
100	0.1077	0.0769	8.8	203	0.3991	0.2848	17.9	368	1.1996	0.8562	32.4
103	0.1138	0.0812	9.1	206	0.4101	0.2927	18.1	373	1.2300	0.8778	32.9
106	0.1200	0.0856	9.3	209	0.4212	0.3006	18.4	378	1.2606	0.8997	33.3
109	0.1263	0.0902	9.6	212	0.4325	0.3087	18.7	383	1.2917	0.9219	33.7

Friction Loss Table For 2½" XL Steel Pipe I.D. = 2.581

Equiv. Feet of Pipe

	LONG EL	90 ELL	45 ELL	TEE	GV	CV	BV			
C-120	4	6	3	12	1	14	7			
C-100	2.85	4.28	2.14	8.56	0.71	9.98	4.99	0.00		

Q GPM	P PSI/FT C-100	P PSI/FT C-120	V FT/SEC	Q GPM	P PSI/FT C-100	P PSI/FT C-120	V FT/SEC	Q GPM	P PSI/FT C-100	P PSI/FT C-120	V FT/SEC
7	0.0003	0.0002	0.4	145	0.0888	0.0634	8.9	283	0.3059	0.2183	17.4
11	0.0008	0.0005	0.7	149	0.0934	0.0666	9.1	290	0.3200	0.2284	17.8
15	0.0013	0.0010	0.9	153	0.0980	0.0700	9.4	297	0.3344	0.2387	18.2
19	0.0021	0.0015	1.2	157	0.1028	0.0734	9.6	304	0.3492	0.2492	18.7
23	0.0029	0.0021	1.4	161	0.1077	0.0769	9.9	311	0.3642	0.2599	19.1
27	0.0040	0.0028	1.7	165	0.1127	0.0805	10.1	318	0.3795	0.2709	19.5
31	0.0051	0.0036	1.9	169	0.1178	0.0841	10.4	325	0.3951	0.2820	19.9
35	0.0064	0.0046	2.1	173	0.1231	0.0878	10.6	332	0.4110	0.2933	20.4
39	0.0078	0.0056	2.4	177	0.1284	0.0916	10.9	339	0.4272	0.3049	20.8
43	0.0094	0.0067	2.6	181	0.1338	0.0955	11.1	346	0.4436	0.3166	21.2
47	0.0110	0.0079	2.9	185	0.1393	0.0994	11.4	353	0.4604	0.3286	21.7
51	0.0128	0.0092	3.1	189	0.1449	0.1034	11.6	360	0.4774	0.3407	22.1
55	0.0148	0.0105	3.4	193	0.1507	0.1075	11.8	367	0.4947	0.3531	22.5
59	0.0168	0.0120	3.6	197	0.1565	0.1117	12.1	374	0.5123	0.3656	22.9
63	0.0190	0.0136	3.9	201	0.1624	0.1159	12.3	381	0.5302	0.3784	23.4
67	0.0213	0.0152	4.1	205	0.1685	0.1202	12.6	388	0.5484	0.3914	23.8
71	0.0237	0.0169	4.4	209	0.1746	0.1246	12.8	395	0.5668	0.4045	24.2
75	0.0262	0.0187	4.6	213	0.1808	0.1290	13.1	402	0.5855	0.4179	24.7
79	0.0289	0.0206	4.8	217	0.1871	0.1336	13.3	409	0.6045	0.4315	25.1
83	0.0316	0.0226	5.1	221	0.1936	0.1382	13.6	416	0.6238	0.4452	25.5
87	0.0345	0.0246	5.3	225	0.2001	0.1428	13.8	423	0.6434	0.4592	26.0
91	0.0375	0.0268	5.6	229	0.2067	0.1476	14.1	430	0.6632	0.4733	26.4
95	0.0406	0.0290	5.8	233	0.2135	0.1524	14.3	437	0.6833	0.4877	26.8
99	0.0438	0.0313	6.1	237	0.2203	0.1572	14.5	444	0.7037	0.5022	27.2
103	0.0471	0.0337	6.3	241	0.2272	0.1622	14.8	451	0.7244	0.5170	27.7
107	0.0506	0.0361	6.6	245	0.2343	0.1672	15.0	458	0.7453	0.5319	28.1
111	0.0541	0.0386	6.8	249	0.2414	0.1723	15.3	465	0.7665	0.5471	28.5
115	0.0578	0.0413	7.1	253	0.2486	0.1774	15.5	472	0.7880	0.5624	29.0
119	0.0616	0.0440	7.3	257	0.2559	0.1827	15.8	479	0.8098	0.5779	29.4
123	0.0655	0.0467	7.5	261	0.2633	0.1879	16.0	486	0.8318	0.5936	29.8
127	0.0695	0.0496	7.8	265	0.2709	0.1933	16.3	493	0.8541	0.6096	30.2
131	0.0736	0.0525	8.0	269	0.2785	0.1987	16.5	500	0.8766	0.6257	30.7
135	0.0778	0.0555	8.3	273	0.2862	0.2042	16.8	507	0.8995	0.6420	31.1
139	0.0821	0.0586	8.5	277	0.2940	0.2098	17.0	514	0.9226	0.6585	31.5
143	0.0865	0.0617	8.8	281	0.3019	0.2155	17.2	521	0.9460	0.6751	32.0

Friction Loss Table For 3" XL Steel Pipe   I.D. = 3.20

Equiv. Feet of Pipe

	LONG EL	90 ELL	45 ELL	TEE	GV	CV	BV	DRY VA
C−120	5	7	3	15	1	16	10	
C−100	3.57	4.99	2.14	10.70	0.71	11.41	7.13	0.00

Q	P	P	V	Q	P	P	V	Q	P	P	V
	PSI/FT	PSI/FT	FT/SEC		PSI/FT	PSI/FT	FT/SEC		PSI/FT	PSI/FT	FT/SEC
GPM	C−100	C−120		GPM	C−100	C−120		GPM	C−100	C−120	
20	0.0008	0.0006	0.8	261	0.0924	0.0660	10.4	502	0.3100	0.2212	20.0
27	0.0014	0.0010	1.1	268	0.0971	0.0693	10.7	511	0.3204	0.2286	20.4
34	0.0021	0.0015	1.4	275	0.1018	0.0727	11.0	520	0.3309	0.2361	20.8
41	0.0030	0.0021	1.6	282	0.1067	0.0761	11.3	529	0.3415	0.2438	21.1
48	0.0040	0.0029	1.9	289	0.1116	0.0797	11.5	538	0.3524	0.2515	21.5
55	0.0052	0.0037	2.2	296	0.1167	0.0833	11.8	547	0.3634	0.2593	21.8
62	0.0065	0.0046	2.5	303	0.1218	0.0869	12.1	556	0.3745	0.2673	22.2
69	0.0079	0.0056	2.8	310	0.1271	0.0907	12.4	565	0.3858	0.2753	22.6
76	0.0094	0.0067	3.0	317	0.1324	0.0945	12.7	574	0.3972	0.2835	22.9
83	0.0111	0.0079	3.3	324	0.1379	0.0984	12.9	583	0.4088	0.2918	23.3
90	0.0129	0.0092	3.6	331	0.1435	0.1024	13.2	592	0.4206	0.3002	23.6
97	0.0148	0.0106	3.9	338	0.1491	0.1064	13.5	601	0.4325	0.3087	24.0
104	0.0168	0.0120	4.2	345	0.1549	0.1105	13.8	610	0.4445	0.3173	24.3
111	0.0190	0.0136	4.4	352	0.1608	0.1147	14.1	619	0.4568	0.3260	24.7
118	0.0213	0.0152	4.7	359	0.1667	0.1190	14.3	628	0.4691	0.3348	25.1
125	0.0237	0.0169	5.0	366	0.1728	0.1233	14.6	637	0.4816	0.3437	25.4
132	0.0262	0.0187	5.3	373	0.1789	0.1277	14.9	646	0.4943	0.3528	25.8
139	0.0288	0.0206	5.5	380	0.1852	0.1322	15.2	655	0.5071	0.3619	26.1
146	0.0316	0.0225	5.8	387	0.1916	0.1367	15.4	664	0.5201	0.3712	26.5
153	0.0344	0.0246	6.1	394	0.1980	0.1413	15.7	673	0.5332	0.3805	26.9
160	0.0374	0.0267	6.4	401	0.2046	0.1460	16.0	682	0.5465	0.3900	27.2
167	0.0405	0.0289	6.7	408	0.2112	0.1508	16.3	691	0.5599	0.3996	27.6
174	0.0437	0.0312	6.9	415	0.2180	0.1556	16.6	700	0.5734	0.4093	27.9
181	0.0470	0.0335	7.2	422	0.2248	0.1605	16.8	709	0.5872	0.4191	28.3
188	0.0504	0.0360	7.5	429	0.2318	0.1654	17.1	718	0.6010	0.4289	28.7
195	0.0539	0.0385	7.8	436	0.2388	0.1705	17.4	727	0.6150	0.4389	29.0
202	0.0575	0.0411	8.1	443	0.2460	0.1756	17.7	736	0.6292	0.4491	29.4
209	0.0613	0.0437	8.3	450	0.2532	0.1807	18.0	745	0.6435	0.4593	29.7
216	0.0651	0.0465	8.6	457	0.2606	0.1860	18.2	754	0.6580	0.4696	30.1
223	0.0691	0.0493	8.9	464	0.2680	0.1913	18.5	763	0.6726	0.4800	30.5
230	0.0732	0.0522	9.2	471	0.2755	0.1966	18.8	772	0.6873	0.4905	30.8
237	0.0773	0.0552	9.5	478	0.2831	0.2021	19.1	781	0.7022	0.5012	31.2
244	0.0816	0.0582	9.7	485	0.2909	0.2076	19.4	790	0.7172	0.5119	31.5
251	0.0860	0.0614	10.0	492	0.2987	0.2132	19.6	799	0.7324	0.5227	31.9
258	0.0905	0.0646	10.3	499	0.3066	0.2188	19.9	808	0.7478	0.5337	32.3

Friction Loss Table For 1" Sch. 30 Dyna-Thread I.D. = 1.087

Equiv. Feet of Pipe

	LONG EL	90 ELL	45 ELL	TEE	GV	CV				
C–120	2	2	1	5		5				
C–100	1.43	1.43	0.71	3.57	0.00	3.57	0.00	0.00		

Q	P	P	V	Q	P	P	V	Q	P	P	V
	PSI/FT	PSI/FT	FT/SEC		PSI/FT	PSI/FT	FT/SEC		PSI/FT	PSI/FT	FT/SEC
GPM	C–100	C–120		GPM	C–100	C–120		GPM	C–100	C–120	
7	0.0220	0.0157	2.4	42	0.6049	0.4317	14.5	77	1.8566	1.3250	26.6
8	0.0281	0.0201	2.8	43	0.6319	0.4510	14.9	78	1.9014	1.3570	27.0
9	0.0350	0.0250	3.1	44	0.6593	0.4705	15.2	79	1.9468	1.3894	27.3
10	0.0425	0.0304	3.5	45	0.6873	0.4905	15.6	80	1.9926	1.4221	27.7
11	0.0507	0.0362	3.8	46	0.7158	0.5109	15.9	81	2.0389	1.4552	28.0
12	0.0596	0.0425	4.2	47	0.7449	0.5316	16.3	82	2.0857	1.4886	28.4
13	0.0691	0.0493	4.5	48	0.7745	0.5527	16.6	83	2.1330	1.5223	28.7
14	0.0793	0.0566	4.8	49	0.8046	0.5742	17.0	84	2.1808	1.5564	29.1
15	0.0900	0.0643	5.2	50	0.8352	0.5961	17.3	85	2.2291	1.5909	29.4
16	0.1015	0.0724	5.5	51	0.8664	0.6183	17.6	86	2.2778	1.6257	29.7
17	0.1135	0.0810	5.9	52	0.8981	0.6409	18.0	87	2.3271	1.6608	30.1
18	0.1262	0.0900	6.2	53	0.9303	0.6639	18.3	88	2.3768	1.6963	30.4
19	0.1394	0.0995	6.6	54	0.9630	0.6873	18.7	89	2.4270	1.7322	30.8
20	0.1533	0.1094	6.9	55	0.9963	0.7110	19.0	90	2.4777	1.7683	31.1
21	0.1678	0.1198	7.3	56	1.0300	0.7351	19.4	91	2.5289	1.8049	31.5
22	0.1829	0.1305	7.6	57	1.0643	0.7596	19.7	92	2.5805	1.8417	31.8
23	0.1986	0.1417	8.0	58	1.0991	0.7844	20.1	93	2.6327	1.8789	32.2
24	0.2148	0.1533	8.3	59	1.1344	0.8096	20.4	94	2.6853	1.9165	32.5
25	0.2317	0.1654	8.6	60	1.1703	0.8352	20.8	95	2.7384	1.9544	32.9
26	0.2491	0.1778	9.0	61	1.2066	0.8611	21.1	96	2.7919	1.9926	33.2
27	0.2671	0.1907	9.3	62	1.2434	0.8874	21.4	97	2.8460	2.0312	33.6
28	0.2857	0.2039	9.7	63	1.2808	0.9141	21.8	98	2.9005	2.0701	33.9
29	0.3049	0.2176	10.0	64	1.3187	0.9411	22.1	99	2.9555	2.1093	34.2
30	0.3246	0.2317	10.4	65	1.3570	0.9685	22.5	100	3.0109	2.1489	34.6
31	0.3449	0.2462	10.7	66	1.3959	0.9963	22.8	101	3.0669	2.1888	34.9
32	0.3658	0.2611	11.1	67	1.4353	1.0244	23.2	102	3.1233	2.2291	35.3
33	0.3872	0.2764	11.4	68	1.4752	1.0528	23.5	103	3.1802	2.2697	35.6
34	0.4092	0.2920	11.8	69	1.5156	1.0816	23.9	104	3.2375	2.3106	36.0
35	0.4317	0.3081	12.1	70	1.5564	1.1108	24.2	105	3.2953	2.3519	36.3
36	0.4548	0.3246	12.5	71	1.5978	1.1404	24.6	106	3.3536	2.3935	36.7
37	0.4785	0.3415	12.8	72	1.6397	1.1703	24.9	107	3.4124	2.4354	37.0
38	0.5027	0.3588	13.1	73	1.6821	1.2005	25.3	108	3.4716	2.4777	37.4
39	0.5274	0.3764	13.5	74	1.7250	1.2311	25.6	109	3.5313	2.5203	37.7
40	0.5527	0.3945	13.8	75	1.7683	1.2621	25.9	110	3.5915	2.5633	38.1
41	0.5786	0.4129	14.2	76	1.8122	1.2934	26.3	111	3.6521	2.6065	38.4

Friction Loss Table For 1¼" Sch. 30 Dyna-Thread   I.D. = 1.416

Equiv. Feet of Pipe

	LONG EL	90 ELL	45 ELL	TEE	GV	CV		
C-120	2	3	1	6		7		
C-100	1.43	2.14	0.71	4.28	0.00	4.99	0.00	0.00

Q GPM	P PSI/FT C-100	P PSI/FT C-120	V FT/SEC	Q GPM	P PSI/FT C-100	P PSI/FT C-120	V FT/SEC	Q GPM	P PSI/FT C-100	P PSI/FT C-120	V FT/SEC
7	0.0061	0.0043	1.4	42	0.1669	0.1191	8.6	77	0.5122	0.3656	15.7
8	0.0078	0.0055	1.6	43	0.1743	0.1244	8.8	78	0.5246	0.3744	15.9
9	0.0097	0.0069	1.8	44	0.1819	0.1298	9.0	79	0.5371	0.3833	16.1
10	0.0117	0.0084	2.0	45	0.1896	0.1353	9.2	80	0.5498	0.3924	16.3
11	0.0140	0.0100	2.2	46	0.1975	0.1410	9.4	81	0.5625	0.4015	16.5
12	0.0164	0.0117	2.4	47	0.2055	0.1467	9.6	82	0.5755	0.4107	16.7
13	0.0191	0.0136	2.7	48	0.2137	0.1525	9.8	83	0.5885	0.4200	16.9
14	0.0219	0.0156	2.9	49	0.2220	0.1584	10.0	84	0.6017	0.4294	17.1
15	0.0248	0.0177	3.1	50	0.2304	0.1645	10.2	87	0.6421	0.4582	17.7
16	0.0280	0.0200	3.3	51	0.2390	0.1706	10.4	90	0.6836	0.4879	18.3
17	0.0313	0.0224	3.5	52	0.2478	0.1768	10.6	93	0.7264	0.5184	19.0
18	0.0348	0.0248	3.7	53	0.2567	0.1832	10.8	96	0.7703	0.5498	19.6
19	0.0385	0.0275	3.9	54	0.2657	0.1896	11.0	99	0.8154	0.5820	20.2
20	0.0423	0.0302	4.1	55	0.2749	0.1962	11.2	102	0.8617	0.6150	20.8
21	0.0463	0.0330	4.3	56	0.2842	0.2028	11.4	105	0.9092	0.6489	21.4
22	0.0505	0.0360	4.5	57	0.2936	0.2096	11.6	108	0.9578	0.6836	22.0
23	0.0548	0.0391	4.7	58	0.3033	0.2164	11.8	111	1.0076	0.7192	22.6
24	0.0593	0.0423	4.9	59	0.3130	0.2234	12.0	114	1.0586	0.7555	23.2
25	0.0639	0.0456	5.1	60	0.3229	0.2304	12.2	117	1.1107	0.7927	23.9
26	0.0687	0.0491	5.3	61	0.3329	0.2376	12.4	120	1.1640	0.8307	24.5
27	0.0737	0.0526	5.5	62	0.3431	0.2448	12.6	123	1.2184	0.8696	25.1
28	0.0788	0.0563	5.7	63	0.3534	0.2522	12.8	126	1.2739	0.9092	25.7
29	0.0841	0.0600	5.9	64	0.3638	0.2597	13.0	129	1.3306	0.9497	26.3
30	0.0896	0.0639	6.1	65	0.3744	0.2672	13.3	132	1.3884	0.9909	26.9
31	0.0952	0.0679	6.3	66	0.3851	0.2749	13.5	135	1.4474	1.0330	27.5
32	0.1009	0.0720	6.5	67	0.3960	0.2826	13.7	138	1.5074	1.0758	28.1
33	0.1068	0.0762	6.7	68	0.4070	0.2905	13.9	141	1.5686	1.1195	28.7
34	0.1129	0.0806	6.9	69	0.4181	0.2984	14.1	144	1.6309	1.1640	29.4
35	0.1191	0.0850	7.1	70	0.4294	0.3065	14.3	147	1.6943	1.2092	30.0
36	0.1255	0.0896	7.3	71	0.4408	0.3146	14.5	150	1.7589	1.2553	30.6
37	0.1320	0.0942	7.5	72	0.4524	0.3229	14.7	153	1.8245	1.3021	31.2
38	0.1387	0.0990	7.7	73	0.4641	0.3312	14.9	156	1.8912	1.3498	31.8
39	0.1455	0.1039	8.0	74	0.4759	0.3397	15.1	159	1.9590	1.3982	32.4
40	0.1525	0.1088	8.2	75	0.4879	0.3482	15.3	162	2.0280	1.4474	33.0
41	0.1596	0.1139	8.4	76	0.5000	0.3568	15.5	165	2.0980	1.4973	33.6

Friction Loss Table For 1½" Sch. 30 Dyna-Thread   I.D. = 1.650

Equiv. Feet of Pipe

	LONG EL	90 ELL	45 ELL	TEE	GV	CV		
C-120	2	4	2	8		9		
C-100	1.43	2.85	1.43	5.70	0.00	6.42	0.00	0.00

Q	P	P	V	Q	P	P	V	Q	P	P	V
	PSI/FT	PSI/FT	FT/SEC		PSI/FT	PSI/FT	FT/SEC		PSI/FT	PSI/FT	FT/SEC
GPM	C-100	C-120		GPM	C-100	C-120		GPM	C-100	C-120	
7	0.0029	0.0021	1.1	42	0.0793	0.0566	6.3	111	0.4785	0.3415	16.7
8	0.0037	0.0026	1.2	44	0.0864	0.0616	6.6	113	0.4945	0.3529	17.0
9	0.0046	0.0033	1.4	46	0.0938	0.0669	6.9	115	0.5108	0.3646	17.3
10	0.0056	0.0040	1.5	48	0.1015	0.0724	7.2	117	0.5274	0.3764	17.6
11	0.0066	0.0047	1.7	50	0.1094	0.0781	7.5	119	0.5442	0.3884	17.9
12	0.0078	0.0056	1.8	52	0.1177	0.0840	7.8	121	0.5612	0.4006	18.2
13	0.0091	0.0065	2.0	54	0.1262	0.0900	8.1	123	0.5785	0.4129	18.5
14	0.0104	0.0074	2.1	56	0.1349	0.0963	8.4	125	0.5960	0.4254	18.8
15	0.0118	0.0084	2.3	58	0.1440	0.1028	8.7	129	0.6318	0.4509	19.4
16	0.0133	0.0095	2.4	60	0.1533	0.1094	9.0	133	0.6685	0.4771	20.0
17	0.0149	0.0106	2.6	62	0.1629	0.1163	9.3	137	0.7062	0.5040	20.6
18	0.0165	0.0118	2.7	64	0.1728	0.1233	9.6	141	0.7448	0.5316	21.2
19	0.0183	0.0130	2.9	66	0.1829	0.1305	9.9	145	0.7844	0.5598	21.8
20	0.0201	0.0143	3.0	68	0.1933	0.1379	10.2	149	0.8249	0.5887	22.4
21	0.0220	0.0157	3.2	70	0.2039	0.1455	10.5	153	0.8663	0.6183	23.0
22	0.0240	0.0171	3.3	72	0.2148	0.1533	10.8	157	0.9087	0.6485	23.6
23	0.0260	0.0186	3.5	74	0.2260	0.1613	11.1	161	0.9520	0.6794	24.2
24	0.0281	0.0201	3.6	76	0.2374	0.1694	11.4	165	0.9962	0.7110	24.8
25	0.0304	0.0217	3.8	78	0.2491	0.1778	11.7	169	1.0413	0.7432	25.4
26	0.0326	0.0233	3.9	80	0.2610	0.1863	12.0	173	1.0874	0.7761	26.0
27	0.0350	0.0250	4.1	82	0.2732	0.1950	12.3	177	1.1343	0.8096	26.6
28	0.0374	0.0267	4.2	84	0.2857	0.2039	12.6	181	1.1822	0.8437	27.2
29	0.0399	0.0285	4.4	86	0.2984	0.2130	12.9	185	1.2310	0.8786	27.8
30	0.0425	0.0304	4.5	88	0.3114	0.2222	13.2	189	1.2807	0.9140	28.4
31	0.0452	0.0322	4.7	90	0.3246	0.2317	13.5	193	1.3313	0.9501	29.0
32	0.0479	0.0342	4.8	92	0.3381	0.2413	13.8	197	1.3828	0.9869	29.6
33	0.0507	0.0362	5.0	94	0.3518	0.2511	14.1	201	1.4352	1.0243	30.2
34	0.0536	0.0383	5.1	96	0.3658	0.2610	14.4	205	1.4885	1.0623	30.8
35	0.0566	0.0404	5.3	98	0.3800	0.2712	14.7	209	1.5426	1.1010	31.4
36	0.0596	0.0425	5.4	100	0.3944	0.2815	15.0	213	1.5977	1.1403	32.0
37	0.0627	0.0447	5.6	102	0.4092	0.2920	15.3	217	1.6536	1.1802	32.6
38	0.0659	0.0470	5.7	104	0.4241	0.3027	15.6	221	1.7105	1.2208	33.2
39	0.0691	0.0493	5.9	106	0.4393	0.3136	15.9	225	1.7682	1.2620	33.8
40	0.0724	0.0517	6.0	108	0.4548	0.3246	16.2	229	1.8268	1.3038	34.4
41	0.0758	0.0541	6.2	110	0.4705	0.3358	16.5	233	1.8863	1.3462	35.0

Friction Loss Table For 2" Sch. 30 Dyna-Thread I.D. = 2.115

Equiv. Feet of Pipe

	LONG EL	90 ELL	45 ELL	TEE	GV	CV	BV	DRY VA
C-120	3	5	2	10	1	11	6	
C-100	2.14	3.57	1.43	7.13	0.71	7.84	4.28	0.00

Q	P	P	V	Q	P	P	V	Q	P	P	V
	PSI/FT	PSI/FT	FT/SEC		PSI/FT	PSI/FT	FT/SEC		PSI/FT	PSI/FT	FT/SEC
GPM	C-100	C-120		GPM	C-100	C-120		GPM	C-100	C-120	
7	0.0009	0.0006	0.6	111	0.1428	0.1019	10.1	215	0.4852	0.3463	19.6
10	0.0017	0.0012	0.9	114	0.1500	0.1071	10.4	219	0.5020	0.3583	20.0
13	0.0027	0.0019	1.2	117	0.1574	0.1123	10.7	223	0.5191	0.3705	20.4
16	0.0040	0.0028	1.5	120	0.1650	0.1177	11.0	227	0.5364	0.3829	20.7
19	0.0055	0.0039	1.7	123	0.1727	0.1232	11.2	231	0.5541	0.3954	21.1
22	0.0072	0.0051	2.0	126	0.1805	0.1288	11.5	235	0.5719	0.4082	21.5
25	0.0091	0.0065	2.3	129	0.1886	0.1346	11.8	239	0.5901	0.4211	21.8
28	0.0112	0.0080	2.6	132	0.1968	0.1404	12.1	243	0.6085	0.4343	22.2
31	0.0135	0.0096	2.8	135	0.2051	0.1464	12.3	247	0.6271	0.4476	22.6
34	0.0160	0.0114	3.1	138	0.2136	0.1525	12.6	251	0.6461	0.4611	22.9
37	0.0187	0.0134	3.4	141	0.2223	0.1587	12.9	255	0.6652	0.4748	23.3
40	0.0216	0.0154	3.7	144	0.2311	0.1650	13.2	259	0.6847	0.4886	23.7
43	0.0247	0.0176	3.9	147	0.2401	0.1714	13.4	263	0.7044	0.5027	24.0
46	0.0280	0.0200	4.2	150	0.2493	0.1779	13.7	267	0.7243	0.5169	24.4
49	0.0315	0.0225	4.5	153	0.2586	0.1845	14.0	271	0.7445	0.5314	24.8
52	0.0351	0.0251	4.8	156	0.2680	0.1913	14.3	275	0.7650	0.5460	25.1
55	0.0390	0.0278	5.0	159	0.2776	0.1981	14.5	279	0.7857	0.5607	25.5
58	0.0430	0.0307	5.3	162	0.2874	0.2051	14.8	283	0.8066	0.5757	25.9
61	0.0472	0.0337	5.6	165	0.2973	0.2122	15.1	287	0.8279	0.5908	26.2
64	0.0516	0.0368	5.8	168	0.3074	0.2194	15.4	291	0.8493	0.6062	26.6
67	0.0561	0.0401	6.1	171	0.3176	0.2267	15.6	295	0.8711	0.6217	27.0
70	0.0609	0.0434	6.4	174	0.3280	0.2341	15.9	299	0.8930	0.6374	27.3
73	0.0658	0.0469	6.7	177	0.3386	0.2416	16.2	303	0.9153	0.6532	27.7
76	0.0709	0.0506	6.9	180	0.3492	0.2493	16.4	307	0.9377	0.6693	28.1
79	0.0761	0.0543	7.2	183	0.3601	0.2570	16.7	311	0.9605	0.6855	28.4
82	0.0816	0.0582	7.5	186	0.3711	0.2648	17.0	315	0.9834	0.7019	28.8
85	0.0872	0.0622	7.8	189	0.3822	0.2728	17.3	319	1.0067	0.7185	29.1
88	0.0929	0.0663	8.0	192	0.3935	0.2809	17.5	323	1.0302	0.7352	29.5
91	0.0989	0.0706	8.3	195	0.4050	0.2890	17.8	327	1.0539	0.7522	29.9
94	0.1050	0.0749	8.6	198	0.4166	0.2973	18.1	331	1.0779	0.7693	30.2
97	0.1113	0.0794	8.9	201	0.4283	0.3057	18.4	335	1.1021	0.7865	30.6
100	0.1177	0.0840	9.1	204	0.4402	0.3142	18.6	339	1.1265	0.8040	31.0
103	0.1243	0.0887	9.4	207	0.4523	0.3228	18.9	343	1.1513	0.8216	31.3
106	0.1311	0.0936	9.7	210	0.4645	0.3315	19.2	347	1.1762	0.8395	31.7
109	0.1381	0.0985	10.0	213	0.4768	0.3403	19.5	364	1.2850	0.9171	33.3

Friction Loss Table For 1" 30 Dyna-Flow   I.D. = 1.197

Equiv. Feet of Pipe

	LONG ELL	90 ELL	45 ELL	TEE	GV	CV
C-120	2	2	1	5		
C-100	1.43	1.43	0.71	3.57		

Q	P	P	V	Q	P	P	V	Q	P	P	V
	PSI/FT	PSI/FT	FT/SEC		PSI/FT	PSI/FT	FT/SEC		PSI/FT	PSI/FT	FT/SEC
GPM	C-100	C-120		GPM	C-100	C-120		GPM	C-100	C-120	
7	0.0137	0.0098	2.0	42	0.3783	0.2700	12.0	77	1.1610	0.8286	22.0
8	0.0176	0.0126	2.3	43	0.3951	0.2820	12.3	78	1.1890	0.8486	22.3
9	0.0219	0.0156	2.6	44	0.4123	0.2943	12.6	79	1.2174	0.8688	22.5
10	0.0266	0.0190	2.9	45	0.4298	0.3067	12.8	80	1.2461	0.8893	22.8
11	0.0317	0.0226	3.1	46	0.4476	0.3195	13.1	81	1.2750	0.9100	23.1
12	0.0373	0.0266	3.4	47	0.4658	0.3324	13.4	82	1.3043	0.9309	23.4
13	0.0432	0.0308	3.7	48	0.4843	0.3456	13.7	83	1.3339	0.9520	23.7
14	0.0496	0.0354	4.0	49	0.5031	0.3591	14.0	84	1.3638	0.9733	24.0
15	0.0563	0.0402	4.3	50	0.5223	0.3728	14.3	85	1.3939	0.9949	24.2
16	0.0635	0.0453	4.6	51	0.5418	0.3867	14.5	86	1.4244	1.0166	24.5
17	0.0710	0.0507	4.8	52	0.5616	0.4008	14.8	87	1.4552	1.0386	24.8
18	0.0789	0.0563	5.1	53	0.5817	0.4152	15.1	88	1.4863	1.0608	25.1
19	0.0872	0.0622	5.4	54	0.6022	0.4298	15.4	89	1.5177	1.0832	25.4
20	0.0959	0.0684	5.7	55	0.6230	0.4446	15.7	90	1.5494	1.1058	25.7
21	0.1049	0.0749	6.0	56	0.6441	0.4597	16.0	91	1.5814	1.1287	26.0
22	0.1144	0.0816	6.3	57	0.6656	0.4750	16.3	92	1.6137	1.1517	26.2
23	0.1242	0.0886	6.6	58	0.6873	0.4905	16.5	93	1.6463	1.1750	26.5
24	0.1343	0.0959	6.8	59	0.7094	0.5063	16.8	94	1.6792	1.1985	26.8
25	0.1449	0.1034	7.1	60	0.7318	0.5223	17.1	95	1.7124	1.2221	27.1
26	0.1558	0.1112	7.4	61	0.7545	0.5385	17.4	96	1.7459	1.2461	27.4
27	0.1670	0.1192	7.7	62	0.7776	0.5550	17.7	97	1.7797	1.2702	27.7
28	0.1787	0.1275	8.0	63	0.8009	0.5716	18.0	98	1.8138	1.2945	28.0
29	0.1907	0.1361	8.3	64	0.8246	0.5885	18.3	99	1.8482	1.3190	28.2
30	0.2030	0.1449	8.6	65	0.8486	0.6057	18.5	100	1.8829	1.3438	28.5
31	0.2157	0.1539	8.8	66	0.8729	0.6230	18.8	101	1.9178	1.3688	28.8
32	0.2287	0.1633	9.1	67	0.8975	0.6406	19.1	102	1.9531	1.3939	29.1
33	0.2421	0.1728	9.4	68	0.9225	0.6584	19.4	103	1.9887	1.4193	29.4
34	0.2559	0.1826	9.7	69	0.9477	0.6764	19.7	104	2.0246	1.4449	29.7
35	0.2700	0.1927	10.0	70	0.9733	0.6946	20.0	105	2.0607	1.4707	30.0
36	0.2844	0.2030	10.3	71	0.9992	0.7131	20.3	106	2.0972	1.4968	30.2
37	0.2992	0.2136	10.6	72	1.0254	0.7318	20.5	107	2.1339	1.5230	30.5
38	0.3144	0.2244	10.8	73	1.0519	0.7507	20.8	108	2.1710	1.5494	30.8
39	0.3298	0.2354	11.1	74	1.0787	0.7699	21.1	109	2.2083	1.5761	31.1
40	0.3456	0.2467	11.4	75	1.1058	0.7892	21.4	110	2.2459	1.6029	31.4
41	0.3618	0.2582	11.7	76	1.1332	0.8088	21.7	111	2.2838	1.6300	31.7

Friction Loss Table For 1¼" Dyna-Flow   I.D. = 1.542

Equiv. Feet of Pipe

	LONG EL	90 ELL	45 ELL	TEE	GV	CV
C-120	2	3	1	6		
C-100	1.43	2.14	0.71	4.28		

Q GPM	P PSI/FT C-100	P PSI/FT C-120	V FT/SEC	Q GPM	P PSI/FT C-100	P PSI/FT C-120	V FT/SEC	Q GPM	P PSI/FT C-100	P PSI/FT C-120	V FT/SEC
7	0.0040	0.0029	1.2	42	0.1102	0.0786	7.2	111	0.6653	0.4748	19.1
8	0.0051	0.0037	1.4	44	0.1201	0.0857	7.6	113	0.6876	0.4908	19.4
9	0.0064	0.0046	1.5	46	0.1304	0.0931	7.9	115	0.7103	0.5070	19.8
10	0.0077	0.0055	1.7	48	0.1411	0.1007	8.3	117	0.7334	0.5234	20.1
11	0.0092	0.0066	1.9	50	0.1521	0.1086	8.6	119	0.7567	0.5401	20.5
12	0.0109	0.0077	2.1	52	0.1636	0.1168	8.9	121	0.7804	0.5570	20.8
13	0.0126	0.0090	2.2	54	0.1754	0.1252	9.3	123	0.8044	0.5741	21.1
14	0.0144	0.0103	2.4	56	0.1876	0.1339	9.6	125	0.8288	0.5915	21.5
15	0.0164	0.0117	2.6	58	0.2002	0.1429	10.0	128	0.8660	0.6180	22.0
16	0.0185	0.0132	2.8	60	0.2132	0.1521	10.3	131	0.9039	0.6451	22.5
17	0.0207	0.0148	2.9	62	0.2265	0.1617	10.7	134	0.9426	0.6727	23.0
18	0.0230	0.0164	3.1	64	0.2402	0.1714	11.0	137	0.9820	0.7008	23.6
19	0.0254	0.0181	3.3	66	0.2543	0.1815	11.3	140	1.0221	0.7295	24.1
20	0.0279	0.0199	3.4	68	0.2687	0.1918	11.7	143	1.0630	0.7587	24.6
21	0.0306	0.0218	3.6	70	0.2835	0.2024	12.0	146	1.1046	0.7884	25.1
22	0.0333	0.0238	3.8	72	0.2987	0.2132	12.4	149	1.1470	0.8186	25.6
23	0.0362	0.0258	4.0	74	0.3142	0.2243	12.7	152	1.1901	0.8494	26.1
24	0.0391	0.0279	4.1	76	0.3301	0.2356	13.1	155	1.2339	0.8806	26.6
25	0.0422	0.0301	4.3	78	0.3464	0.2472	13.4	158	1.2784	0.9124	27.2
26	0.0454	0.0324	4.5	80	0.3630	0.2591	13.8	161	1.3237	0.9447	27.7
27	0.0487	0.0347	4.6	82	0.3799	0.2712	14.1	164	1.3697	0.9776	28.2
28	0.0520	0.0371	4.8	84	0.3973	0.2835	14.4	167	1.4164	1.0109	28.7
29	0.0555	0.0396	5.0	86	0.4149	0.2961	14.8	170	1.4639	1.0448	29.2
30	0.0591	0.0422	5.2	88	0.4330	0.3090	15.1	173	1.5120	1.0791	29.7
31	0.0628	0.0448	5.3	90	0.4514	0.3221	15.5	176	1.5609	1.1140	30.3
32	0.0666	0.0476	5.5	92	0.4701	0.3355	15.8	179	1.6104	1.1494	30.8
33	0.0705	0.0503	5.7	94	0.4892	0.3491	16.2	182	1.6607	1.1853	31.3
34	0.0745	0.0532	5.8	96	0.5086	0.3630	16.5	185	1.7117	1.2217	31.8
35	0.0786	0.0561	6.0	98	0.5284	0.3771	16.8	188	1.7634	1.2586	32.3
36	0.0829	0.0591	6.2	100	0.5485	0.3915	17.2	191	1.8158	1.2960	32.8
37	0.0872	0.0622	6.4	102	0.5690	0.4061	17.5	194	1.8690	1.3339	33.3
38	0.0916	0.0654	6.5	104	0.5898	0.4209	17.9	197	1.9228	1.3723	33.9
39	0.0961	0.0686	6.7	106	0.6109	0.4360	18.2	200	1.9773	1.4112	34.4
40	0.1007	0.0719	6.9	108	0.6324	0.4514	18.6	203	2.0325	1.4506	34.9
41	0.1054	0.0752	7.0	110	0.6543	0.4669	18.9	206	2.0884	1.4905	35.4

Friction Loss Table For 1½" Dyna-Flow   I.D. = 1.752

Equiv. Feet of Pipe

	LONG EL	90 ELL	45 ELL	TEE	GV	CV
C-120	2	4	2	8		
C-100	1.43	2.85	1.43	5.70		

Q GPM	P PSI/FT C-100	P PSI/FT C-120	V FT/SEC	Q GPM	P PSI/FT C-100	P PSI/FT C-120	V FT/SEC	Q GPM	P PSI/FT C-100	P PSI/FT C-120	V FT/SEC
7	0.0022	0.0015	0.9	77	0.1816	0.1296	10.3	147	0.6007	0.4287	19.6
9	0.0034	0.0024	1.2	79	0.1904	0.1359	10.5	149	0.6159	0.4396	19.8
11	0.0050	0.0035	1.5	81	0.1994	0.1423	10.8	151	0.6313	0.4506	20.1
13	0.0068	0.0048	1.7	83	0.2087	0.1489	11.1	153	0.6469	0.4617	20.4
15	0.0088	0.0063	2.0	85	0.2180	0.1556	11.3	155	0.6626	0.4729	20.6
17	0.0111	0.0079	2.3	87	0.2276	0.1625	11.6	157	0.6785	0.4842	20.9
19	0.0136	0.0097	2.5	89	0.2374	0.1694	11.9	159	0.6946	0.4957	21.2
21	0.0164	0.0117	2.8	91	0.2474	0.1766	12.1	161	0.7108	0.5073	21.4
23	0.0194	0.0139	3.1	93	0.2575	0.1838	12.4	165	0.7438	0.5309	22.0
25	0.0227	0.0162	3.3	95	0.2679	0.1912	12.7	169	0.7775	0.5549	22.5
27	0.0261	0.0186	3.6	97	0.2784	0.1987	12.9	173	0.8119	0.5795	23.0
29	0.0298	0.0213	3.9	99	0.2891	0.2063	13.2	177	0.8470	0.6045	23.6
31	0.0337	0.0241	4.1	101	0.3000	0.2141	13.4	181	0.8827	0.6300	24.1
33	0.0379	0.0270	4.4	103	0.3111	0.2220	13.7	185	0.9192	0.6560	24.6
35	0.0422	0.0301	4.7	105	0.3223	0.2301	14.0	189	0.9563	0.6825	25.2
37	0.0468	0.0334	4.9	107	0.3338	0.2382	14.2	193	0.9941	0.7095	25.7
39	0.0516	0.0368	5.2	109	0.3454	0.2465	14.5	197	1.0325	0.7369	26.2
41	0.0566	0.0404	5.5	111	0.3573	0.2550	14.8	201	1.0716	0.7648	26.8
43	0.0618	0.0441	5.7	113	0.3693	0.2635	15.0	205	1.1114	0.7932	27.3
45	0.0672	0.0480	6.0	115	0.3814	0.2722	15.3	209	1.1519	0.8221	27.8
47	0.0729	0.0520	6.3	117	0.3938	0.2811	15.6	213	1.1930	0.8514	28.4
49	0.0787	0.0562	6.5	119	0.4063	0.2900	15.8	217	1.2347	0.8812	28.9
51	0.0847	0.0605	6.8	121	0.4191	0.2991	16.1	221	1.2772	0.9115	29.4
53	0.0910	0.0649	7.1	123	0.4320	0.3083	16.4	225	1.3203	0.9423	30.0
55	0.0975	0.0696	7.3	125	0.4451	0.3176	16.6	229	1.3640	0.9735	30.5
57	0.1041	0.0743	7.6	127	0.4583	0.3271	16.9	233	1.4084	1.0052	31.0
59	0.1110	0.0792	7.9	129	0.4718	0.3367	17.2	237	1.4535	1.0374	31.6
61	0.1180	0.0842	8.1	131	0.4854	0.3464	17.4	241	1.4992	1.0700	32.1
63	0.1253	0.0894	8.4	133	0.4992	0.3563	17.7	245	1.5456	1.1031	32.6
65	0.1327	0.0947	8.7	135	0.5132	0.3662	18.0	249	1.5926	1.1366	33.2
67	0.1404	0.1002	8.9	137	0.5273	0.3763	18.2	253	1.6402	1.1706	33.7
69	0.1483	0.1058	9.2	139	0.5416	0.3866	18.5	257	1.6885	1.2051	34.2
71	0.1563	0.1116	9.5	141	0.5561	0.3969	18.8	261	1.7374	1.2400	34.8
73	0.1645	0.1174	9.7	143	0.5708	0.4074	19.0	265	1.7870	1.2754	35.3
75	0.1730	0.1235	10.0	145	0.5857	0.4180	19.3	269	1.8373	1.3112	35.8

Friction Loss Table For 2" Dyna-Flow I.D. = 2.227

Equiv. Feet of Pipe

	LONG ELL	90 ELL	45 ELL	TEE	GV	CV
C-120	2	5	2	10	1	11
C-100	1.43	3.57	1.43	7.13	0.71	7.84

Q	P	P	V	Q	P	P	V	Q	P	P	V
	PSI/FT	PSI/FT	FT/SEC		PSI/FT	PSI/FT	FT/SEC		PSI/FT	PSI/FT	FT/SEC
GPM	C-100	C-120		GPM	C-100	C-120		GPM	C-100	C-120	
7	0.0007	0.0005	0.6	112	0.1129	0.0806	9.2	251	0.5025	0.3586	20.7
10	0.0013	0.0009	0.8	116	0.1205	0.0860	9.6	255	0.5174	0.3693	21.0
13	0.0021	0.0015	1.1	120	0.1283	0.0916	9.9	259	0.5325	0.3801	21.3
16	0.0031	0.0022	1.3	124	0.1363	0.0973	10.2	263	0.5478	0.3910	21.7
19	0.0042	0.0030	1.6	128	0.1446	0.1032	10.5	267	0.5634	0.4021	22.0
22	0.0056	0.0040	1.8	132	0.1530	0.1092	10.9	271	0.5791	0.4133	22.3
25	0.0070	0.0050	2.1	136	0.1617	0.1154	11.2	275	0.5950	0.4246	22.7
28	0.0087	0.0062	2.3	140	0.1706	0.1218	11.5	279	0.6111	0.4361	23.0
31	0.0105	0.0075	2.6	144	0.1798	0.1283	11.9	284	0.6315	0.4507	23.4
34	0.0124	0.0089	2.8	148	0.1891	0.1350	12.2	289	0.6522	0.4655	23.8
37	0.0146	0.0104	3.0	152	0.1987	0.1418	12.5	294	0.6733	0.4805	24.2
40	0.0168	0.0120	3.3	156	0.2085	0.1488	12.9	299	0.6946	0.4957	24.6
43	0.0192	0.0137	3.5	160	0.2185	0.1559	13.2	304	0.7162	0.5112	25.1
46	0.0218	0.0155	3.8	164	0.2287	0.1632	13.5	309	0.7382	0.5268	25.5
49	0.0245	0.0175	4.0	168	0.2391	0.1706	13.8	314	0.7604	0.5427	25.9
52	0.0273	0.0195	4.3	172	0.2497	0.1782	14.2	319	0.7830	0.5588	26.3
55	0.0303	0.0216	4.5	176	0.2606	0.1860	14.5	324	0.8058	0.5751	26.7
58	0.0334	0.0239	4.8	180	0.2716	0.1939	14.8	329	0.8290	0.5917	27.1
61	0.0367	0.0262	5.0	184	0.2829	0.2019	15.2	334	0.8525	0.6084	27.5
64	0.0401	0.0286	5.3	188	0.2944	0.2101	15.5	339	0.8762	0.6254	27.9
67	0.0436	0.0312	5.5	192	0.3061	0.2185	15.8	344	0.9003	0.6425	28.4
70	0.0473	0.0338	5.8	196	0.3180	0.2269	16.2	349	0.9246	0.6599	28.8
73	0.0512	0.0365	6.0	200	0.3301	0.2356	16.5	354	0.9493	0.6775	29.2
76	0.0551	0.0393	6.3	204	0.3424	0.2444	16.8	359	0.9742	0.6953	29.6
79	0.0592	0.0423	6.5	208	0.3549	0.2533	17.1	364	0.9995	0.7133	30.0
82	0.0634	0.0453	6.8	212	0.3677	0.2624	17.5	369	1.0250	0.7316	30.4
85	0.0678	0.0484	7.0	216	0.3806	0.2716	17.8	374	1.0509	0.7500	30.8
88	0.0723	0.0516	7.3	220	0.3938	0.2810	18.1	379	1.0770	0.7687	31.2
91	0.0769	0.0549	7.5	224	0.4071	0.2905	18.5	384	1.1035	0.7875	31.6
94	0.0817	0.0583	7.7	228	0.4206	0.3002	18.8	389	1.1302	0.8066	32.1
97	0.0866	0.0618	8.0	232	0.4344	0.3100	19.1	394	1.1572	0.8259	32.5
100	0.0916	0.0654	8.2	236	0.4484	0.3200	19.4	399	1.1845	0.8454	32.9
103	0.0967	0.0690	8.5	240	0.4625	0.3301	19.8	404	1.2121	0.8651	33.3
106	0.1020	0.0728	8.7	244	0.4769	0.3404	20.1	409	1.2400	0.8850	33.7
109	0.1074	0.0766	9.0	248	0.4914	0.3507	20.4	414	1.2682	0.9051	34.1

Friction Loss Table For 1" Dyna-Light Sch. 5 I.D. = 1.185

Equiv. Feet of Pipe

	LONG EL	90 ELL	45 ELL	TEE	GV	CV				
C−120	2	2	1	5						
C−100	1.43	1.43	0.71	3.57						

Q	P	P	V	Q	P	P	V	Q	P	P	V
	PSI/FT	PSI/FT	FT/SEC		PSI/FT	PSI/FT	FT/SEC		PSI/FT	PSI/FT	FT/SEC
GPM	C−100	C−120		GPM	C−100	C−120		GPM	C−100	C−120	
7	0.0144	0.0103	2.0	42	0.3973	0.2836	12.2	77	1.2194	0.8703	22.4
8	0.0185	0.0132	2.3	43	0.4150	0.2962	12.5	78	1.2488	0.8913	22.7
9	0.0230	0.0164	2.6	44	0.4330	0.3091	12.8	79	1.2786	0.9125	23.0
10	0.0279	0.0199	2.9	45	0.4514	0.3222	13.1	80	1.3087	0.9340	23.3
11	0.0333	0.0238	3.2	46	0.4701	0.3355	13.4	81	1.3391	0.9557	23.6
12	0.0391	0.0279	3.5	47	0.4892	0.3492	13.7	82	1.3699	0.9777	23.9
13	0.0454	0.0324	3.8	48	0.5087	0.3630	14.0	83	1.4010	0.9999	24.2
14	0.0521	0.0372	4.1	49	0.5284	0.3771	14.3	84	1.4323	1.0223	24.5
15	0.0591	0.0422	4.4	50	0.5486	0.3915	14.6	85	1.4640	1.0449	24.7
16	0.0666	0.0476	4.7	51	0.5690	0.4061	14.8	86	1.4961	1.0677	25.0
17	0.0746	0.0532	4.9	52	0.5898	0.4210	15.1	87	1.5284	1.0908	25.3
18	0.0829	0.0591	5.2	53	0.6110	0.4361	15.4	88	1.5611	1.1141	25.6
19	0.0916	0.0654	5.5	54	0.6325	0.4514	15.7	89	1.5940	1.1377	25.9
20	0.1007	0.0719	5.8	55	0.6543	0.4670	16.0	90	1.6273	1.1614	26.2
21	0.1102	0.0787	6.1	56	0.6765	0.4828	16.3	91	1.6610	1.1854	26.5
22	0.1201	0.0857	6.4	57	0.6990	0.4989	16.6	92	1.6949	1.2096	26.8
23	0.1304	0.0931	6.7	58	0.7219	0.5152	16.9	93	1.7291	1.2341	27.1
24	0.1411	0.1007	7.0	59	0.7451	0.5318	17.2	94	1.7637	1.2587	27.4
25	0.1522	0.1086	7.3	60	0.7686	0.5486	17.5	95	1.7985	1.2836	27.7
26	0.1636	0.1168	7.6	61	0.7925	0.5656	17.8	96	1.8337	1.3087	27.9
27	0.1754	0.1252	7.9	62	0.8167	0.5829	18.0	97	1.8692	1.3341	28.2
28	0.1877	0.1339	8.2	63	0.8412	0.6004	18.3	98	1.9050	1.3596	28.5
29	0.2002	0.1429	8.4	64	0.8661	0.6181	18.6	99	1.9411	1.3854	28.8
30	0.2132	0.1522	8.7	65	0.8913	0.6361	18.9	100	1.9776	1.4114	29.1
31	0.2265	0.1617	9.0	66	0.9168	0.6543	19.2	101	2.0143	1.4376	29.4
32	0.2402	0.1715	9.3	67	0.9427	0.6728	19.5	102	2.0514	1.4640	29.7
33	0.2543	0.1815	9.6	68	0.9689	0.6915	19.8	103	2.0887	1.4907	30.0
34	0.2688	0.1918	9.9	69	0.9954	0.7104	20.1	104	2.1264	1.5176	30.3
35	0.2836	0.2024	10.2	70	1.0223	0.7296	20.4	105	2.1644	1.5447	30.6
36	0.2987	0.2132	10.5	71	1.0494	0.7490	20.7	106	2.2027	1.5720	30.9
37	0.3143	0.2243	10.8	72	1.0769	0.7686	21.0	107	2.2412	1.5996	31.1
38	0.3302	0.2356	11.1	73	1.1048	0.7885	21.2	108	2.2802	1.6273	31.4
39	0.3464	0.2472	11.4	74	1.1329	0.8086	21.5	109	2.3194	1.6553	31.7
40	0.3630	0.2591	11.6	75	1.1614	0.8289	21.8	110	2.3589	1.6835	32.0
41	0.3800	0.2712	11.9	76	1.1902	0.8495	22.1	111	2.3987	1.7120	32.3

Appendix D Friction Loss Tables

Friction Loss Table For 1¼" Dyna-Light Sch. 5   I.D. = 1.530
Equiv. Feet of Pipe

	LONG EL	90 ELL	45 ELL	TEE	GV	CV
C–120	2	3	1	6		
C–100	1.43	2.14	0.71	4.28		

Q	P	P	V	Q	P	P	V	Q	P	P	V
	PSI/FT	PSI/FT	FT/SEC		PSI/FT	PSI/FT	FT/SEC		PSI/FT	PSI/FT	FT/SEC
GPM	C–100	C–120		GPM	C–100	C–120		GPM	C–100	C–120	
7	0.0042	0.0030	1.2	42	0.1145	0.0817	7.3	111	0.6911	0.4932	19.4
8	0.0053	0.0038	1.4	44	0.1248	0.0890	7.7	113	0.7143	0.5098	19.7
9	0.0066	0.0047	1.6	46	0.1355	0.0967	8.0	115	0.7379	0.5266	20.1
10	0.0080	0.0057	1.7	48	0.1466	0.1046	8.4	117	0.7618	0.5437	20.4
11	0.0096	0.0069	1.9	50	0.1580	0.1128	8.7	119	0.7861	0.5610	20.8
12	0.0113	0.0080	2.1	52	0.1699	0.1213	9.1	121	0.8107	0.5786	21.1
13	0.0131	0.0093	2.3	54	0.1822	0.1301	9.4	123	0.8356	0.5964	21.5
14	0.0150	0.0107	2.4	56	0.1949	0.1391	9.8	125	0.8609	0.6145	21.8
15	0.0170	0.0122	2.6	58	0.2080	0.1484	10.1	128	0.8996	0.6420	22.3
16	0.0192	0.0137	2.8	60	0.2214	0.1580	10.5	131	0.9390	0.6701	22.9
17	0.0215	0.0153	3.0	62	0.2353	0.1679	10.8	134	0.9791	0.6988	23.4
18	0.0239	0.0170	3.1	64	0.2495	0.1781	11.2	137	1.0201	0.7280	23.9
19	0.0264	0.0188	3.3	66	0.2641	0.1885	11.5	140	1.0618	0.7578	24.4
20	0.0290	0.0207	3.5	68	0.2791	0.1992	11.9	143	1.1042	0.7881	25.0
21	0.0318	0.0227	3.7	70	0.2945	0.2102	12.2	146	1.1475	0.8190	25.5
22	0.0346	0.0247	3.8	72	0.3103	0.2214	12.6	149	1.1915	0.8504	26.0
23	0.0376	0.0268	4.0	74	0.3264	0.2330	12.9	152	1.2362	0.8823	26.5
24	0.0407	0.0290	4.2	76	0.3429	0.2447	13.3	155	1.2818	0.9148	27.1
25	0.0438	0.0313	4.4	78	0.3598	0.2568	13.6	158	1.3280	0.9478	27.6
26	0.0471	0.0336	4.5	80	0.3771	0.2691	14.0	161	1.3751	0.9814	28.1
27	0.0505	0.0361	4.7	82	0.3947	0.2817	14.3	164	1.4228	1.0155	28.6
28	0.0541	0.0386	4.9	84	0.4127	0.2945	14.7	167	1.4714	1.0501	29.2
29	0.0577	0.0412	5.1	86	0.4310	0.3076	15.0	170	1.5206	1.0853	29.7
30	0.0614	0.0438	5.2	88	0.4498	0.3210	15.4	173	1.5706	1.1210	30.2
31	0.0653	0.0466	5.4	90	0.4689	0.3346	15.7	176	1.6214	1.1572	30.7
32	0.0692	0.0494	5.6	92	0.4883	0.3485	16.1	179	1.6729	1.1939	31.3
33	0.0733	0.0523	5.8	94	0.5081	0.3627	16.4	182	1.7251	1.2312	31.8
34	0.0774	0.0553	5.9	96	0.5283	0.3771	16.8	185	1.7781	1.2690	32.3
35	0.0817	0.0583	6.1	98	0.5489	0.3917	17.1	188	1.8318	1.3074	32.8
36	0.0861	0.0614	6.3	100	0.5698	0.4066	17.5	191	1.8863	1.3462	33.3
37	0.0905	0.0646	6.5	102	0.5910	0.4218	17.8	194	1.9414	1.3856	33.9
38	0.0951	0.0679	6.6	104	0.6126	0.4372	18.2	197	1.9973	1.4255	34.4
39	0.0998	0.0712	6.8	106	0.6346	0.4529	18.5	200	2.0540	1.4659	34.9
40	0.1046	0.0746	7.0	108	0.6569	0.4689	18.9	203	2.1113	1.5069	35.4
41	0.1095	0.0781	7.2	110	0.6796	0.4850	19.2	206	2.1694	1.5483	36.0

Friction Loss Table For 1½" Dyna-Light Sch. 5   I.D. = 1.770

Equiv. Feet of Pipe

	LONG ELL	90 ELL	45 ELL	TEE	GV	CV				
C−120	2	4	2	8						
C−100	1.43	2.85	1.43	5.70						

Q GPM	P PSI/FT C−100	P PSI/FT C−120	V FT/SEC	Q GPM	P PSI/FT C−100	P PSI/FT C−120	V FT/SEC	Q GPM	P PSI/FT C−100	P PSI/FT C−120	V FT/SEC
7	0.0020	0.0015	0.9	77	0.1728	0.1233	10.0	147	0.5715	0.4079	19.2
9	0.0033	0.0023	1.2	79	0.1812	0.1293	10.3	149	0.5860	0.4182	19.4
11	0.0047	0.0034	1.4	81	0.1898	0.1354	10.6	151	0.6006	0.4287	19.7
13	0.0064	0.0046	1.7	83	0.1985	0.1417	10.8	153	0.6154	0.4392	20.0
15	0.0084	0.0060	2.0	85	0.2075	0.1481	11.1	155	0.6304	0.4499	20.2
17	0.0106	0.0075	2.2	87	0.2166	0.1546	11.4	157	0.6455	0.4607	20.5
19	0.0130	0.0093	2.5	89	0.2259	0.1612	11.6	159	0.6608	0.4716	20.7
21	0.0156	0.0111	2.7	91	0.2354	0.1680	11.9	161	0.6763	0.4827	21.0
23	0.0185	0.0132	3.0	93	0.2450	0.1749	12.1	165	0.7077	0.5051	21.5
25	0.0216	0.0154	3.3	95	0.2549	0.1819	12.4	169	0.7398	0.5280	22.0
27	0.0249	0.0177	3.5	97	0.2649	0.1890	12.7	173	0.7725	0.5513	22.6
29	0.0284	0.0203	3.8	99	0.2751	0.1963	12.9	177	0.8059	0.5751	23.1
31	0.0321	0.0229	4.0	101	0.2854	0.2037	13.2	181	0.8399	0.5994	23.6
33	0.0360	0.0257	4.3	103	0.2960	0.2112	13.4	185	0.8745	0.6242	24.1
35	0.0402	0.0287	4.6	105	0.3067	0.2189	13.7	189	0.9098	0.6493	24.7
37	0.0445	0.0318	4.8	107	0.3176	0.2267	14.0	193	0.9458	0.6750	25.2
39	0.0491	0.0350	5.1	109	0.3287	0.2346	14.2	197	0.9824	0.7011	25.7
41	0.0538	0.0384	5.3	111	0.3399	0.2426	14.5	201	1.0196	0.7277	26.2
43	0.0588	0.0420	5.6	113	0.3513	0.2507	14.7	205	1.0574	0.7547	26.7
45	0.0640	0.0457	5.9	115	0.3629	0.2590	15.0	209	1.0959	0.7822	27.3
47	0.0693	0.0495	6.1	117	0.3747	0.2674	15.3	213	1.1350	0.8101	27.8
49	0.0749	0.0534	6.4	119	0.3866	0.2759	15.5	217	1.1748	0.8384	28.3
51	0.0806	0.0575	6.7	121	0.3987	0.2846	15.8	221	1.2152	0.8673	28.8
53	0.0866	0.0618	6.9	123	0.4110	0.2933	16.0	225	1.2562	0.8965	29.4
55	0.0927	0.0662	7.2	125	0.4234	0.3022	16.3	229	1.2978	0.9262	29.9
57	0.0991	0.0707	7.4	127	0.4361	0.3112	16.6	233	1.3400	0.9564	30.4
59	0.1056	0.0754	7.7	129	0.4488	0.3203	16.8	237	1.3829	0.9870	30.9
61	0.1123	0.0801	8.0	131	0.4618	0.3296	17.1	241	1.4264	1.0180	31.4
63	0.1192	0.0851	8.2	133	0.4749	0.3390	17.4	245	1.4705	1.0495	32.0
65	0.1263	0.0901	8.5	135	0.4882	0.3485	17.6	249	1.5152	1.0814	32.5
67	0.1336	0.0953	8.7	137	0.5017	0.3581	17.9	253	1.5606	1.1138	33.0
69	0.1411	0.1007	9.0	139	0.5153	0.3678	18.1	257	1.6065	1.1466	33.5
71	0.1487	0.1061	9.3	141	0.5291	0.3776	18.4	261	1.6531	1.1798	34.1
73	0.1566	0.1117	9.5	143	0.5431	0.3876	18.7	265	1.7003	1.2135	34.6
75	0.1646	0.1175	9.8	145	0.5572	0.3977	18.9	269	1.7480	1.2476	35.1

Appendix D Friction Loss Tables

Friction Loss Table For   2" Dyna-Light Sch. 5   I.D. = 2.245

Equiv. Feet of Pipe

	LONG EL	90 ELL	45 ELL	TEE	GV	CV
C-120	2	5	2	10	1	11
C-100	1.43	3.57	1.43	7.13	0.71	7.84

Q	P	P	V	Q	P	P	V	Q	P	P	V
	PSI/FT	PSI/FT	FT/SEC		PSI/FT	PSI/FT	FT/SEC		PSI/FT	PSI/FT	FT/SEC
GPM	C-100	C-120		GPM	C-100	C-120		GPM	C-100	C-120	
7	0.0006	0.0005	0.6	112	0.1086	0.0775	9.1	251	0.4832	0.3448	20.4
10	0.0012	0.0009	0.8	116	0.1159	0.0827	9.4	255	0.4975	0.3551	20.7
13	0.0020	0.0014	1.1	120	0.1234	0.0880	9.7	259	0.5121	0.3655	21.0
16	0.0030	0.0021	1.3	124	0.1311	0.0936	10.1	263	0.5268	0.3760	21.3
19	0.0041	0.0029	1.5	128	0.1390	0.0992	10.4	267	0.5417	0.3866	21.7
22	0.0053	0.0038	1.8	132	0.1472	0.1050	10.7	271	0.5568	0.3974	22.0
25	0.0068	0.0048	2.0	136	0.1555	0.1110	11.0	275	0.5721	0.4083	22.3
28	0.0084	0.0060	2.3	140	0.1641	0.1171	11.4	279	0.5876	0.4194	22.6
31	0.0101	0.0072	2.5	144	0.1729	0.1234	11.7	284	0.6072	0.4334	23.0
34	0.0120	0.0085	2.8	148	0.1818	0.1298	12.0	289	0.6272	0.4476	23.4
37	0.0140	0.0100	3.0	152	0.1910	0.1363	12.3	294	0.6474	0.4620	23.8
40	0.0162	0.0115	3.2	156	0.2004	0.1431	12.7	299	0.6679	0.4767	24.2
43	0.0185	0.0132	3.5	160	0.2101	0.1499	13.0	304	0.6887	0.4915	24.7
46	0.0209	0.0149	3.7	164	0.2199	0.1569	13.3	309	0.7098	0.5066	25.1
49	0.0235	0.0168	4.0	168	0.2299	0.1641	13.6	314	0.7312	0.5219	25.5
52	0.0263	0.0187	4.2	172	0.2401	0.1714	13.9	319	0.7529	0.5373	25.9
55	0.0291	0.0208	4.5	176	0.2506	0.1788	14.3	324	0.7749	0.5530	26.3
58	0.0321	0.0229	4.7	180	0.2612	0.1864	14.6	329	0.7971	0.5689	26.7
61	0.0353	0.0252	4.9	184	0.2720	0.1942	14.9	334	0.8197	0.5850	27.1
64	0.0386	0.0275	5.2	188	0.2831	0.2020	15.2	339	0.8425	0.6013	27.5
67	0.0420	0.0300	5.4	192	0.2943	0.2101	15.6	344	0.8657	0.6178	27.9
70	0.0455	0.0325	5.7	196	0.3058	0.2182	15.9	349	0.8891	0.6345	28.3
73	0.0492	0.0351	5.9	200	0.3174	0.2265	16.2	354	0.9128	0.6515	28.7
76	0.0530	0.0378	6.2	204	0.3293	0.2350	16.5	359	0.9368	0.6686	29.1
79	0.0569	0.0406	6.4	208	0.3413	0.2436	16.9	364	0.9611	0.6859	29.5
82	0.0610	0.0435	6.6	212	0.3535	0.2523	17.2	369	0.9856	0.7034	29.9
85	0.0652	0.0465	6.9	216	0.3660	0.2612	17.5	374	1.0105	0.7212	30.3
88	0.0695	0.0496	7.1	220	0.3786	0.2702	17.8	379	1.0356	0.7391	30.7
91	0.0740	0.0528	7.4	224	0.3914	0.2794	18.2	384	1.0610	0.7573	31.1
94	0.0785	0.0560	7.6	228	0.4045	0.2887	18.5	389	1.0867	0.7756	31.5
97	0.0832	0.0594	7.9	232	0.4177	0.2981	18.8	394	1.1127	0.7941	32.0
100	0.0880	0.0628	8.1	236	0.4311	0.3077	19.1	399	1.1390	0.8129	32.4
103	0.0930	0.0664	8.4	240	0.4447	0.3174	19.5	404	1.1655	0.8318	32.8
106	0.0981	0.0700	8.6	244	0.4585	0.3273	19.8	409	1.1923	0.8510	33.2
109	0.1033	0.0737	8.8	248	0.4726	0.3373	20.1	414	1.2195	0.8703	33.6

Friction Loss Table For 3/4" CPVC Plastic I.D. = .884

Equiv. Feet of Pipe

	TEE RUN	90 ELL	45 ELL	TEE	CPLG
C-150	1.00	7.00	1.00	3.00	1.00

Q	P	P	V	Q	P	P	V	Q	P	P	V
	PSI/FT	PSI/FT	FT/SEC		PSI/FT	PSI/FT	FT/SEC		PSI/FT	PSI/FT	FT/SEC
GPM	C-150			GPM	C-150			GPM	C-150		
7	0.0284		3.7	42	0.7819		22.0				
8	0.0364		4.2	43	0.8167		22.5				
9	0.0452		4.7	44	0.8522		23.0				
10	0.0550		5.2	45	0.8884		23.5				
11	0.0656		5.8	46	0.9252		24.1				
12	0.0770		6.3	47	0.9628		24.6				
13	0.0893		6.8	48	1.0010		25.1				
14	0.1024		7.3	49	1.0399		25.6				
15	0.1164		7.8	50	1.0795		26.2				
16	0.1312		8.4	51	1.1198		26.7				
17	0.1467		8.9	52	1.1608		27.2				
18	0.1631		9.4	53	1.2024		27.7				
19	0.1802		9.9	54	1.2447		28.2				
20	0.1982		10.5	55	1.2877		28.8				
21	0.2169		11.0	56	1.3314		29.3				
22	0.2364		11.5	57	1.3757		29.8				
23	0.2567		12.0	58	1.4207		30.3				
24	0.2777		12.6	59	1.4663		30.9				
25	0.2995		13.1	60	1.5126		31.4				
26	0.3220		13.6	61	1.5596		31.9				
27	0.3453		14.1	62	1.6072		32.4				
28	0.3693		14.6	63	1.6555		33.0				
29	0.3941		15.2	64	1.7044		33.5				
30	0.4196		15.7								
31	0.4458		16.2								
32	0.4728		16.7								
33	0.5005		17.3								
34	0.5289		17.8								
35	0.5580		18.3								
36	0.5879		18.8								
37	0.6185		19.4								
38	0.6498		19.9								
39	0.6817		20.4								
40	0.7144		20.9								
41	0.7478		21.4								

Friction Loss Table For 1" CPVC Plastic   I.D. = 1.109

Equiv. Feet of Pipe

	TEE RUN	90 ELL	45 ELL	TEE	CPLG				
C-150	1.00	7.00	1.00	5.00	1.00				

Q	P	P	V	Q	P	P	V	Q	P	P	V
	PSI/FT	PSI/FT	FT/SEC		PSI/FT	PSI/FT	FT/SEC		PSI/FT	PSI/FT	FT/SEC
GPM	C-150			GPM	C-150			GPM	C-150		
7	0.0094		2.3	42	0.2592		14.0	77	0.7954		25.6
8	0.0121		2.7	43	0.2707		14.3	78	0.8146		25.9
9	0.0150		3.0	44	0.2824		14.6	79	0.8340		26.3
10	0.0182		3.3	45	0.2944		15.0	80	0.8536		26.6
11	0.0217		3.7	46	0.3067		15.3	81	0.8735		26.9
12	0.0255		4.0	47	0.3191		15.6	82	0.8935		27.3
13	0.0296		4.3	48	0.3318		16.0	83	0.9138		27.6
14	0.0340		4.7	49	0.3447		16.3	84	0.9343		27.9
15	0.0386		5.0	50	0.3578		16.6	85	0.9549		28.2
16	0.0435		5.3	51	0.3712		16.9	86	0.9758		28.6
17	0.0486		5.6	52	0.3847		17.3	87	0.9969		28.9
18	0.0541		6.0	53	0.3985		17.6	88	1.0182		29.2
19	0.0597		6.3	54	0.4126		17.9	89	1.0397		29.6
20	0.0657		6.6	55	0.4268		18.3	90	1.0615		29.9
21	0.0719		7.0	56	0.4413		18.6	91	1.0834		30.2
22	0.0783		7.3	57	0.4560		18.9	92	1.1055		30.6
23	0.0851		7.6	58	0.4709		19.3	93	1.1278		30.9
24	0.0920		8.0	59	0.4860		19.6	94	1.1504		31.2
25	0.0993		8.3	60	0.5013		19.9	95	1.1731		31.6
26	0.1067		8.6	61	0.5169		20.3	96	1.1961		31.9
27	0.1144		9.0	62	0.5327		20.6	97	1.2192		32.2
28	0.1224		9.3	63	0.5487		20.9	98	1.2426		32.6
29	0.1306		9.6	64	0.5649		21.3	99	1.2661		32.9
30	0.1391		10.0	65	0.5814		21.6	100	1.2899		33.2
31	0.1478		10.3	66	0.5980		21.9	101	1.3139		33.6
32	0.1567		10.6	67	0.6149		22.3	102	1.3380		33.9
33	0.1659		11.0	68	0.6320		22.6	103	1.3624		34.2
34	0.1753		11.3	69	0.6493		22.9	104	1.3870		34.6
35	0.1850		11.6	70	0.6668		23.3				
36	0.1949		12.0	71	0.6845		23.6				
37	0.2050		12.3	72	0.7025		23.9				
38	0.2154		12.6	73	0.7206		24.3				
39	0.2260		13.0	74	0.7390		24.6				
40	0.2368		13.3	75	0.7576		24.9				
41	0.2479		13.6	76	0.7764		25.3				

Friction Loss Table For 1¼" PCVC Plastic   I.D. = 1.400

Equiv. Feet of Pipe

	TEE RUN	90 ELL	45 ELL	TEE	CPLG
C-150	1.00	8.00	2.00	6.00	1.00

Q	P	P	V	Q	P	P	V	Q	P	P	V
	PSI/FT	PSI/FT	FT/SEC		PSI/FT	PSI/FT	FT/SEC		PSI/FT	PSI/FT	FT/SEC
GPM	C-150			GPM	C-150			GPM	C-150		
7	0.0030		1.5	43	0.0870		9.0	113	0.5199		23.6
8	0.0039		1.7	45	0.0947		9.4	115	0.5371		24.0
9	0.0048		1.9	47	0.1026		9.8	117	0.5545		24.4
10	0.0059		2.1	49	0.1108		10.2	119	0.5721		24.8
11	0.0070		2.3	51	0.1193		10.6	121	0.5900		25.2
12	0.0082		2.5	53	0.1281		11.1	123	0.6082		25.7
13	0.0095		2.7	55	0.1372		11.5	125	0.6266		26.1
14	0.0109		2.9	57	0.1466		11.9	127	0.6453		26.5
15	0.0124		3.1	59	0.1562		12.3	129	0.6642		26.9
16	0.0140		3.3	61	0.1662		12.7	131	0.6834		27.3
17	0.0156		3.5	63	0.1764		13.1	133	0.7028		27.7
18	0.0174		3.8	65	0.1869		13.6	135	0.7225		28.2
19	0.0192		4.0	67	0.1977		14.0	137	0.7424		28.6
20	0.0211		4.2	69	0.2087		14.4	139	0.7626		29.0
21	0.0231		4.4	71	0.2201		14.8	141	0.7830		29.4
22	0.0252		4.6	73	0.2317		15.2	143	0.8037		29.8
23	0.0273		4.8	75	0.2436		15.6	145	0.8246		30.2
24	0.0296		5.0	77	0.2557		16.1	147	0.8458		30.7
25	0.0319		5.2	79	0.2681		16.5	149	0.8672		31.1
26	0.0343		5.4	81	0.2808		16.9	151	0.8889		31.5
27	0.0368		5.6	83	0.2938		17.3	153	0.9108		31.9
28	0.0394		5.8	85	0.3070		17.7	155	0.9329		32.3
29	0.0420		6.0	87	0.3205		18.1	157	0.9553		32.7
30	0.0447		6.3	89	0.3343		18.6	159	0.9779		33.2
31	0.0475		6.5	91	0.3483		19.0	161	1.0008		33.6
32	0.0504		6.7	93	0.3626		19.4	163	1.0239		34.0
33	0.0533		6.9	95	0.3772		19.8	165	1.0473		34.4
34	0.0564		7.1	97	0.3920		20.2	167	1.0709		34.8
35	0.0595		7.3	99	0.4071		20.6	169	1.0948		35.2
36	0.0626		7.5	101	0.4224		21.1	171	1.1188		35.7
37	0.0659		7.7	103	0.4380		21.5	173	1.1432		36.1
38	0.0692		7.9	105	0.4539		21.9	175	1.1677		36.5
39	0.0726		8.1	107	0.4700		22.3	177	1.1926		36.9
40	0.0761		8.3	109	0.4864		22.7	179	1.2176		37.3
41	0.0797		8.6	111	0.5030		23.1	181	1.2429		37.7

Friction Loss Table For 1½" CPVC Plastic   I.D. = 1.602

Equiv. Feet of Pipe

	TEE RUN	90 ELL	45 ELL	TEE	CPLG
C-150	1.00	9.00	2.00	8.00	1.00

Q GPM	P PSI/FT C-150	P PSI/FT	V FT/SEC	Q GPM	P PSI/FT C-150	P PSI/FT	V FT/SEC	Q GPM	P PSI/FT C-150	P PSI/FT	V FT/SEC
7	0.0016		1.1	43	0.0451		6.8	113	0.2697		18.0
8	0.0020		1.3	45	0.0491		7.2	116	0.2831		18.5
9	0.0025		1.4	47	0.0532		7.5	119	0.2968		19.0
10	0.0030		1.6	49	0.0575		7.8	122	0.3108		19.4
11	0.0036		1.8	51	0.0619		8.1	125	0.3250		19.9
12	0.0043		1.9	53	0.0665		8.4	128	0.3396		20.4
13	0.0049		2.1	55	0.0712		8.8	131	0.3545		20.9
14	0.0057		2.2	57	0.0760		9.1	134	0.3697		21.3
15	0.0064		2.4	59	0.0810		9.4	137	0.3851		21.8
16	0.0072		2.5	61	0.0862		9.7	140	0.4009		22.3
17	0.0081		2.7	63	0.0915		10.0	143	0.4169		22.8
18	0.0090		2.9	65	0.0970		10.4	146	0.4332		23.3
19	0.0100		3.0	67	0.1025		10.7	149	0.4498		23.7
20	0.0110		3.2	69	0.1083		11.0	152	0.4667		24.2
21	0.0120		3.3	71	0.1142		11.3	155	0.4839		24.7
22	0.0131		3.5	73	0.1202		11.6	158	0.5014		25.2
23	0.0142		3.7	75	0.1263		11.9	161	0.5191		25.6
24	0.0153		3.8	77	0.1326		12.3	164	0.5372		26.1
25	0.0166		4.0	79	0.1391		12.6	167	0.5555		26.6
26	0.0178		4.1	81	0.1457		12.9	170	0.5741		27.1
27	0.0191		4.3	83	0.1524		13.2	173	0.5930		27.6
28	0.0204		4.5	85	0.1593		13.5	176	0.6122		28.0
29	0.0218		4.6	87	0.1663		13.9	179	0.6316		28.5
30	0.0232		4.8	89	0.1734		14.2	182	0.6513		29.0
31	0.0246		4.9	91	0.1807		14.5	185	0.6713		29.5
32	0.0261		5.1	93	0.1881		14.8	188	0.6916		29.9
33	0.0277		5.3	95	0.1956		15.1	191	0.7122		30.4
34	0.0292		5.4	97	0.2033		15.4	194	0.7330		30.9
35	0.0308		5.6	99	0.2111		15.8	197	0.7541		31.4
36	0.0325		5.7	101	0.2191		16.1	200	0.7755		31.9
37	0.0342		5.9	103	0.2272		16.4	203	0.7971		32.3
38	0.0359		6.1	105	0.2354		16.7	206	0.8191		32.8
39	0.0377		6.2	107	0.2438		17.0	209	0.8413		33.3
40	0.0395		6.4	109	0.2523		17.4	212	0.8637		33.8
41	0.0413		6.5	111	0.2609		17.7	215	0.8865		34.2

Friction Loss Table For 2" CPVC Plastic   I.D. = 2.003

Equiv. Feet of Pipe

	TEE RUN	90 ELL	45 ELL	TEE	CPLG
C−150	1.00	11.00	2.00	10.00	1.00

Q	P	P	V	Q	P	P	V	Q	P	P	V
	PSI/FT	PSI/FT	FT/SEC		PSI/FT	PSI/FT	FT/SEC		PSI/FT	PSI/FT	FT/SEC
GPM	C−150			GPM	C−150			GPM	C−150		
15	0.0022		1.5	52	0.0216		5.3	157	0.1670		16.0
16	0.0024		1.6	55	0.0240		5.6	162	0.1769		16.5
17	0.0027		1.7	58	0.0265		5.9	167	0.1872		17.0
18	0.0030		1.8	61	0.0290		6.2	172	0.1977		17.5
19	0.0034		1.9	64	0.0317		6.5	177	0.2084		18.0
20	0.0037		2.0	67	0.0345		6.8	182	0.2194		18.5
21	0.0040		2.1	70	0.0375		7.1	187	0.2307		19.1
22	0.0044		2.2	73	0.0405		7.4	192	0.2423		19.6
23	0.0048		2.3	76	0.0436		7.7	197	0.2541		20.1
24	0.0052		2.4	79	0.0469		8.0	202	0.2661		20.6
25	0.0056		2.5	82	0.0502		8.4	207	0.2784		21.1
26	0.0060		2.6	85	0.0537		8.7	212	0.2910		21.6
27	0.0064		2.8	88	0.0572		9.0	217	0.3038		22.1
28	0.0069		2.9	91	0.0609		9.3	222	0.3169		22.6
29	0.0073		3.0	94	0.0646		9.6	227	0.3302		23.1
30	0.0078		3.1	97	0.0685		9.9	232	0.3438		23.6
31	0.0083		3.2	100	0.0725		10.2	237	0.3577		24.1
32	0.0088		3.3	103	0.0765		10.5	242	0.3717		24.7
33	0.0093		3.4	106	0.0807		10.8	247	0.3861		25.2
34	0.0098		3.5	109	0.0850		11.1	252	0.4007		25.7
35	0.0104		3.6	112	0.0894		11.4	257	0.4155		26.2
36	0.0109		3.7	115	0.0939		11.7	262	0.4306		26.7
37	0.0115		3.8	118	0.0984		12.0	267	0.4459		27.2
38	0.0121		3.9	121	0.1031		12.3	272	0.4615		27.7
39	0.0127		4.0	124	0.1079		12.6	277	0.4773		28.2
40	0.0133		4.1	127	0.1128		12.9	282	0.4933		28.7
41	0.0139		4.2	130	0.1178		13.2	287	0.5096		29.2
42	0.0146		4.3	133	0.1228		13.5	292	0.5262		29.7
43	0.0152		4.4	136	0.1280		13.9	297	0.5430		30.3
44	0.0159		4.5	139	0.1333		14.2	302	0.5600		30.8
45	0.0165		4.6	142	0.1386		14.5	307	0.5773		31.3
46	0.0172		4.7	145	0.1441		14.8	312	0.5948		31.8
47	0.0179		4.8	148	0.1497		15.1	317	0.6126		32.3
48	0.0186		4.9	151	0.1553		15.4	322	0.6305		32.8
49	0.0194		5.0	154	0.1611		15.7	327	0.6488		33.3

Friction Loss Table For 2½" CPVC Plastic   I.D. = 2.423

Equiv. Feet of Pipe

	TEE RUN	90 ELL	45 ELL	TEE	CPLG
C–150	2.00	12.00	3.00	12.00	2.00

Q	P	P	V	Q	P	P	V	Q	P	P	V
	PSI/FT	PSI/FT	FT/SEC		PSI/FT	PSI/FT	FT/SEC		PSI/FT	PSI/FT	FT/SEC
GPM	C–150			GPM	C–150			GPM	C–150		
15	0.0009		1.0	122	0.0414		8.5	297	0.2149		20.7
18	0.0012		1.3	127	0.0446		8.8	302	0.2216		21.0
21	0.0016		1.5	132	0.0479		9.2	307	0.2284		21.4
24	0.0020		1.7	137	0.0513		9.5	312	0.2354		21.7
27	0.0025		1.9	142	0.0549		9.9	317	0.2424		22.1
30	0.0031		2.1	147	0.0585		10.2	322	0.2495		22.4
33	0.0037		2.3	152	0.0622		10.6	327	0.2567		22.8
36	0.0043		2.5	157	0.0661		10.9	332	0.2640		23.1
39	0.0050		2.7	162	0.0700		11.3	337	0.2714		23.5
42	0.0058		2.9	167	0.0741		11.6	342	0.2789		23.8
45	0.0065		3.1	172	0.0782		12.0	347	0.2865		24.2
48	0.0074		3.3	177	0.0825		12.3	352	0.2942		24.5
51	0.0083		3.6	182	0.0868		12.7	357	0.3020		24.9
54	0.0092		3.8	187	0.0913		13.0	362	0.3099		25.2
57	0.0101		4.0	192	0.0959		13.4	367	0.3178		25.6
60	0.0111		4.2	197	0.1005		13.7	372	0.3259		25.9
63	0.0122		4.4	202	0.1053		14.1	377	0.3340		26.2
66	0.0133		4.6	207	0.1102		14.4	382	0.3423		26.6
69	0.0144		4.8	212	0.1152		14.8	387	0.3506		26.9
72	0.0156		5.0	217	0.1202		15.1	392	0.3590		27.3
75	0.0168		5.2	222	0.1254		15.5	397	0.3676		27.6
78	0.0181		5.4	227	0.1307		15.8	402	0.3762		28.0
81	0.0194		5.6	232	0.1361		16.2	407	0.3849		28.3
84	0.0208		5.8	237	0.1415		16.5	412	0.3937		28.7
87	0.0222		6.1	242	0.1471		16.8	417	0.4026		29.0
90	0.0236		6.3	247	0.1528		17.2	422	0.4115		29.4
93	0.0251		6.5	252	0.1585		17.5	427	0.4206		29.7
96	0.0266		6.7	257	0.1644		17.9	432	0.4297		30.1
99	0.0282		6.9	262	0.1704		18.2	437	0.4390		30.4
102	0.0297		7.1	267	0.1764		18.6	442	0.4483		30.8
105	0.0314		7.3	272	0.1826		18.9	447	0.4578		31.1
108	0.0331		7.5	277	0.1889		19.3	452	0.4673		31.5
111	0.0348		7.7	282	0.1952		19.6	457	0.4769		31.8
114	0.0365		7.9	287	0.2017		20.0	462	0.4866		32.2
117	0.0383		8.1	292	0.2082		20.3	467	0.4964		32.5

Friction Loss Table For 3" CPVC Plastic   I.D. = 2.951

Equiv. Feet of Pipe

	TEE RUN	90 ELL	45 ELL	TEE	CPLG
C-150	2.00	13.00	4.00	15.00	2.00

Q	P	P	V	Q	P	P	V	Q	P	P	V
	PSI/FT	PSI/FT	FT/SEC		PSI/FT	PSI/FT	FT/SEC		PSI/FT	PSI/FT	FT/SEC
GPM	C-150			GPM	C-150			GPM	C-150		
15	0.0003		0.7	190	0.0360		8.9	365	0.1205		17.1
20	0.0006		0.9	195	0.0378		9.2	375	0.1266		17.6
25	0.0008		1.2	200	0.0396		9.4	385	0.1330		18.1
30	0.0012		1.4	205	0.0414		9.6	395	0.1394		18.5
35	0.0016		1.6	210	0.0433		9.9	405	0.1460		19.0
40	0.0020		1.9	215	0.0453		10.1	415	0.1528		19.5
45	0.0025		2.1	220	0.0472		10.3	425	0.1596		19.9
50	0.0030		2.3	225	0.0492		10.6	435	0.1667		20.4
55	0.0036		2.6	230	0.0513		10.8	445	0.1738		20.9
60	0.0043		2.8	235	0.0533		11.0	455	0.1811		21.4
65	0.0049		3.1	240	0.0555		11.3	465	0.1885		21.8
70	0.0057		3.3	245	0.0576		11.5	475	0.1961		22.3
75	0.0064		3.5	250	0.0598		11.7	485	0.2038		22.8
80	0.0073		3.8	255	0.0620		12.0	495	0.2117		23.2
85	0.0081		4.0	260	0.0643		12.2	505	0.2196		23.7
90	0.0090		4.2	265	0.0666		12.4	515	0.2278		24.2
95	0.0100		4.5	270	0.0690		12.7	525	0.2360		24.6
100	0.0110		4.7	275	0.0713		12.9	535	0.2444		25.1
105	0.0120		4.9	280	0.0738		13.1	545	0.2529		25.6
110	0.0131		5.2	285	0.0762		13.4	555	0.2616		26.0
115	0.0142		5.4	290	0.0787		13.6	565	0.2703		26.5
120	0.0154		5.6	295	0.0812		13.8	575	0.2793		27.0
125	0.0166		5.9	300	0.0838		14.1	585	0.2883		27.5
130	0.0178		6.1	305	0.0864		14.3	595	0.2975		27.9
135	0.0191		6.3	310	0.0891		14.6	605	0.3068		28.4
140	0.0205		6.6	315	0.0917		14.8	615	0.3163		28.9
145	0.0218		6.8	320	0.0944		15.0	625	0.3258		29.3
150	0.0232		7.0	325	0.0972		15.3	635	0.3355		29.8
155	0.0247		7.3	330	0.1000		15.5	645	0.3454		30.3
160	0.0262		7.5	335	0.1028		15.7	655	0.3554		30.7
165	0.0277		7.7	340	0.1056		16.0	665	0.3655		31.2
170	0.0293		8.0	345	0.1085		16.2	675	0.3757		31.7
175	0.0309		8.2	350	0.1115		16.4	685	0.3861		32.2
180	0.0326		8.4	355	0.1144		16.7	695	0.3965		32.6
185	0.0343		8.7	360	0.1174		16.9	705	0.4072		33.1

Appendix D Friction Loss Tables

Friction Loss Table For   1" POZ-LOK   I.D. = 1.0
Equiv. Feet of Pipe

	LONG EL	90 ELL	45 ELL	TEE	GV	CV
C-120	N/A	2	N/A	5	N/A	N/A
C-100	N/A	1.43	N/A	3.57	N/A	N/A

Q	P	P	V	Q	P	P	V	Q	P	P	V
	PSI/FT	PSI/FT	FT/SEC		PSI/FT	PSI/FT	FT/SEC		PSI/FT	PSI/FT	FT/SEC
GPM	C-100	C-120		GPM	C-100	C-120		GPM	C-100	C-120	
7	0.0330	0.0236	2.9	42	0.9081	0.6481	17.2	77	2.7871	1.9891	31.5
8	0.0423	0.0302	3.3	43	0.9485	0.6770	17.6	78	2.8544	2.0372	31.9
9	0.0525	0.0375	3.7	44	0.9898	0.7064	18.0	79	2.9225	2.0858	32.3
10	0.0638	0.0456	4.1	45	1.0318	0.7364	18.4	80	2.9913	2.1349	32.7
11	0.0762	0.0544	4.5	46	1.0746	0.7669	18.8	81	3.0608	2.1845	33.1
12	0.0895	0.0638	4.9	47	1.1182	0.7981	19.2	82	3.1311	2.2346	33.5
13	0.1037	0.0740	5.3	48	1.1626	0.8298	19.6	83	3.2021	2.2853	33.9
14	0.1190	0.0849	5.7	49	1.2078	0.8620	20.0	84	3.2738	2.3365	34.3
15	0.1352	0.0965	6.1	50	1.2538	0.8948	20.4	85	3.3463	2.3882	34.7
16	0.1523	0.1087	6.5	51	1.3006	0.9282	20.8	86	3.4195	2.4405	35.2
17	0.1704	0.1216	6.9	52	1.3482	0.9622	21.3	87	3.4934	2.4932	35.6
18	0.1894	0.1352	7.4	53	1.3965	0.9967	21.7	88	3.5681	2.5465	36.0
19	0.2093	0.1494	7.8	54	1.4457	1.0318	22.1	89	3.6434	2.6003	36.4
20	0.2302	0.1643	8.2	55	1.4956	1.0674	22.5				
21	0.2519	0.1798	8.6	56	1.5463	1.1036	22.9				
22	0.2745	0.1959	9.0	57	1.5977	1.1403	23.3				
23	0.2981	0.2127	9.4	58	1.6500	1.1776	23.7				
24	0.3225	0.2302	9.8	59	1.7030	1.2154	24.1				
25	0.3478	0.2482	10.2	60	1.7568	1.2538	24.5				
26	0.3740	0.2669	10.6	61	1.8113	1.2927	24.9				
27	0.4010	0.2862	11.0	62	1.8667	1.3322	25.3				
28	0.4289	0.3061	11.4	63	1.9227	1.3722	25.8				
29	0.4577	0.3267	11.9	64	1.9796	1.4128	26.2				
30	0.4873	0.3478	12.3	65	2.0372	1.4539	26.6				
31	0.5178	0.3695	12.7	66	2.0955	1.4956	27.0				
32	0.5491	0.3919	13.1	67	2.1547	1.5378	27.4				
33	0.5813	0.4149	13.5	68	2.2145	1.5805	27.8				
34	0.6143	0.4384	13.9	69	2.2751	1.6238	28.2				
35	0.6481	0.4626	14.3	70	2.3365	1.6676	28.6				
36	0.6828	0.4873	14.7	71	2.3986	1.7119	29.0				
37	0.7183	0.5127	15.1	72	2.4615	1.7568	29.4				
38	0.7546	0.5386	15.5	73	2.5251	1.8022	29.8				
39	0.7918	0.5651	15.9	74	2.5895	1.8481	30.2				
40	0.8298	0.5922	16.3	75	2.6546	1.8946	30.7				
41	0.8685	0.6199	16.8	76	2.7205	1.9416	31.1				

Friction Loss Table For 1¼" POZ-LOK I.D. = 1.23

Equiv. Feet of Pipe

	LONG EL	90 ELL	45 ELL	TEE	GV	CV				
C–120	N/A	3	N/A	7	N/A	N/A				
C–100	N/A	2.14	N/A	4.99	N/A	N/A				

Q GPM	P PSI/FT C–100	P PSI/FT C–120	V FT/SEC	Q GPM	P PSI/FT C–100	P PSI/FT C–120	V FT/SEC	Q GPM	P PSI/FT C–100	P PSI/FT C–120	V FT/SEC
7	0.0120	0.0086	1.9	42	0.3314	0.2365	11.3	77	1.0170	0.7258	20.8
8	0.0154	0.0110	2.2	43	0.3461	0.2470	11.6	78	1.0415	0.7433	21.1
9	0.0192	0.0137	2.4	44	0.3612	0.2578	11.9	79	1.0664	0.7611	21.3
10	0.0233	0.0166	2.7	45	0.3765	0.2687	12.2	80	1.0915	0.7790	21.6
11	0.0278	0.0198	3.0	46	0.3921	0.2798	12.4	81	1.1169	0.7971	21.9
12	0.0326	0.0233	3.2	47	0.4080	0.2912	12.7	82	1.1425	0.8154	22.2
13	0.0379	0.0270	3.5	48	0.4242	0.3028	13.0	83	1.1684	0.8339	22.4
14	0.0434	0.0310	3.8	49	0.4407	0.3145	13.2	84	1.1946	0.8526	22.7
15	0.0493	0.0352	4.1	50	0.4575	0.3265	13.5	85	1.2210	0.8714	23.0
16	0.0556	0.0397	4.3	51	0.4746	0.3387	13.8	86	1.2477	0.8905	23.2
17	0.0622	0.0444	4.6	52	0.4919	0.3511	14.0	87	1.2747	0.9098	23.5
18	0.0691	0.0493	4.9	53	0.5096	0.3637	14.3	88	1.3019	0.9292	23.8
19	0.0764	0.0545	5.1	54	0.5275	0.3765	14.6	89	1.3295	0.9488	24.0
20	0.0840	0.0599	5.4	55	0.5457	0.3895	14.9	90	1.3572	0.9686	24.3
21	0.0919	0.0656	5.7	56	0.5642	0.4027	15.1	91	1.3852	0.9886	24.6
22	0.1002	0.0715	5.9	57	0.5830	0.4161	15.4	92	1.4135	1.0088	24.9
23	0.1088	0.0776	6.2	58	0.6021	0.4297	15.7	93	1.4421	1.0292	25.1
24	0.1177	0.0840	6.5	59	0.6214	0.4435	15.9	94	1.4709	1.0498	25.4
25	0.1269	0.0906	6.8	60	0.6410	0.4575	16.2	95	1.5000	1.0705	25.7
26	0.1365	0.0974	7.0	61	0.6609	0.4717	16.5	96	1.5293	1.0915	25.9
27	0.1463	0.1044	7.3	62	0.6811	0.4861	16.8	97	1.5589	1.1126	26.2
28	0.1565	0.1117	7.6	63	0.7016	0.5007	17.0	98	1.5888	1.1339	26.5
29	0.1670	0.1192	7.8	64	0.7223	0.5155	17.3	99	1.6189	1.1554	26.7
30	0.1778	0.1269	8.1	65	0.7433	0.5305	17.6	100	1.6493	1.1771	27.0
31	0.1889	0.1348	8.4	66	0.7646	0.5457	17.8	105	1.8051	1.2883	28.4
32	0.2004	0.1430	8.6	67	0.7862	0.5611	18.1	107	1.8692	1.3341	28.9
33	0.2121	0.1514	8.9	68	0.8081	0.5767	18.4	109	1.9344	1.3806	29.4
34	0.2241	0.1600	9.2	69	0.8302	0.5925	18.6	111	2.0005	1.4278	30.0
35	0.2365	0.1688	9.5	70	0.8526	0.6085	18.9	113	2.0677	1.4757	30.5
36	0.2491	0.1778	9.7	71	0.8752	0.6247	19.2	115	2.1360	1.5244	31.1
37	0.2621	0.1871	10.0	72	0.8982	0.6410	19.5	120	2.3109	1.6493	32.4
38	0.2754	0.1965	10.3	73	0.9214	0.6576	19.7	130	2.6798	1.9125	35.1
39	0.2889	0.2062	10.5	74	0.9449	0.6744	20.0	140	3.0735	2.1936	37.8
40	0.3028	0.2161	10.8	75	0.9686	0.6913	20.3	150	3.4920	2.4922	40.5
41	0.3169	0.2262	11.1	76	0.9927	0.7085	20.5	160	3.9348	2.8083	43.2

Friction Loss Table For 1½" POZ LOK I.D. = 1.56

Equiv. Feet of Pipe

	LONG EL	90 ELL	45 ELL	TEE	GV	CV
C-120	N/A	4	N/A	8	N/A	N/A
C-100	N/A	2.85	N/A	5.70	N/A	N/A

Q GPM	P PSI/FT C-100	P PSI/FT C-120	V FT/SEC	Q GPM	P PSI/FT C-100	P PSI/FT C-120	V FT/SEC	Q GPM	P PSI/FT C-100	P PSI/FT C-120	V FT/SEC
7	0.0038	0.0027	1.2	76	0.3120	0.2227	12.8	145	1.0307	0.7356	24.4
9	0.0060	0.0043	1.5	78	0.3273	0.2336	13.1	148	1.0705	0.7640	24.9
11	0.0087	0.0062	1.8	80	0.3430	0.2448	13.4	151	1.1110	0.7929	25.4
13	0.0119	0.0085	2.2	82	0.3591	0.2563	13.8	154	1.1522	0.8223	25.9
15	0.0155	0.0111	2.5	84	0.3754	0.2679	14.1	157	1.1941	0.8522	26.4
17	0.0195	0.0139	2.9	86	0.3921	0.2799	14.4	160	1.2366	0.8826	26.9
19	0.0240	0.0171	3.2	88	0.4092	0.2920	14.8	163	1.2799	0.9134	27.4
21	0.0289	0.0206	3.5	90	0.4266	0.3044	15.1	166	1.3238	0.9448	27.9
23	0.0342	0.0244	3.9	92	0.4443	0.3171	15.5	169	1.3684	0.9766	28.4
25	0.0399	0.0285	4.2	94	0.4623	0.3299	15.8	172	1.4137	1.0089	28.9
27	0.0460	0.0328	4.5	96	0.4806	0.3430	16.1	175	1.4596	1.0417	29.4
29	0.0525	0.0375	4.9	98	0.4993	0.3564	16.5	178	1.5063	1.0750	29.9
31	0.0594	0.0424	5.2	100	0.5183	0.3699	16.8	181	1.5536	1.1088	30.4
33	0.0667	0.0476	5.5	102	0.5377	0.3837	17.1	184	1.6015	1.1430	30.9
35	0.0743	0.0530	5.9	104	0.5574	0.3978	17.5	187	1.6502	1.1777	31.4
37	0.0824	0.0588	6.2	106	0.5773	0.4121	17.8	190	1.6995	1.2129	31.9
39	0.0908	0.0648	6.6	108	0.5977	0.4266	18.1	193	1.7495	1.2486	32.4
41	0.0996	0.0711	6.9	110	0.6183	0.4413	18.5	196	1.8001	1.2847	32.9
43	0.1088	0.0776	7.2	112	0.6393	0.4562	18.8	199	1.8514	1.3213	33.4
45	0.1183	0.0844	7.6	114	0.6605	0.4714	19.1	202	1.9034	1.3584	33.9
47	0.1282	0.0915	7.9	116	0.6821	0.4868	19.5	205	1.9560	1.3960	34.4
49	0.1385	0.0989	8.2	118	0.7040	0.5025	19.8	208	2.0093	1.4340	34.9
51	0.1492	0.1064	8.6	120	0.7263	0.5183	20.2	211	2.0632	1.4725	35.4
53	0.1602	0.1143	8.9	122	0.7488	0.5344	20.5	214	2.1178	1.5115	35.9
55	0.1715	0.1224	9.2	124	0.7717	0.5508	20.8	217	2.1731	1.5509	36.4
57	0.1832	0.1308	9.6	126	0.7949	0.5673	21.2	220	2.2290	1.5908	36.9
59	0.1953	0.1394	9.9	128	0.8184	0.5841	21.5	223	2.2855	1.6312	37.5
61	0.2077	0.1483	10.2	130	0.8422	0.6011	21.8	226	2.3427	1.6720	38.0
63	0.2205	0.1574	10.6	132	0.8663	0.6183	22.2	229	2.4006	1.7133	38.5
65	0.2336	0.1667	10.9	134	0.8908	0.6357	22.5	232	2.4591	1.7550	39.0
67	0.2471	0.1763	11.3	136	0.9155	0.6534	22.8	235	2.5182	1.7973	39.5
69	0.2609	0.1862	11.6	138	0.9406	0.6713	23.2	238	2.5780	1.8399	40.0
71	0.2751	0.1963	11.9	140	0.9660	0.6894	23.5	241	2.6385	1.8831	40.5
73	0.2896	0.2067	12.3	142	0.9916	0.7077	23.8	244	2.6996	1.9267	41.0
75	0.3044	0.2173	12.6	144	1.0176	0.7263	24.2	247	2.7613	1.9707	41.5

Friction Loss Table For 4" Cement Lined C.I.   I.D. = 3.85

Equiv. Feet of Pipe

	LONG EL	90 ELL	45 ELL	TEE	GV	CV				
C–120	6	10	4	20	2	22				
C–140	7.98	13.30	5.32	26.60	2.66	29.26	0.00	0.00		

Q	P	P	V	Q	P	P	V	Q	P	P	V
	PSI/FT	PSI/FT	FT/SEC		PSI/FT	PSI/FT	FT/SEC		PSI/FT	PSI/FT	FT/SEC
GPM	C–140			GPM	C–140			GPM	C–140		
100	0.0034		2.8	450	0.0552		12.4	800	0.1601		22.1
110	0.0041		3.0	460	0.0575		12.7	810	0.1638		22.3
120	0.0048		3.3	470	0.0598		13.0	820	0.1676		22.6
130	0.0056		3.6	480	0.0622		13.2	830	0.1714		22.9
140	0.0064		3.9	490	0.0646		13.5	840	0.1752		23.2
150	0.0072		4.1	500	0.0671		13.8	850	0.1791		23.4
160	0.0082		4.4	510	0.0696		14.1	860	0.1830		23.7
170	0.0091		4.7	520	0.0721		14.3	870	0.1869		24.0
180	0.0101		5.0	530	0.0747		14.6	885	0.1930		24.4
190	0.0112		5.2	540	0.0774		14.9	900	0.1990		24.8
200	0.0123		5.5	550	0.0800		15.2	915	0.2052		25.2
210	0.0135		5.8	560	0.0827		15.4	930	0.2115		25.6
220	0.0147		6.1	570	0.0855		15.7	945	0.2178		26.1
230	0.0160		6.3	580	0.0883		16.0	960	0.2243		26.5
240	0.0173		6.6	590	0.0911		16.3	975	0.2308		26.9
250	0.0186		6.9	600	0.0940		16.5	990	0.2374		27.3
260	0.0200		7.2	610	0.0969		16.8	1005	0.2441		27.7
270	0.0215		7.4	620	0.0999		17.1	1020	0.2509		28.1
280	0.0230		7.7	630	0.1029		17.4	1035	0.2578		28.5
290	0.0245		8.0	640	0.1059		17.6	1050	0.2647		29.0
300	0.0261		8.3	650	0.1090		17.9	1065	0.2718		29.4
310	0.0277		8.5	660	0.1121		18.2	1080	0.2789		29.8
320	0.0294		8.8	670	0.1153		18.5	1095	0.2861		30.2
330	0.0311		9.1	680	0.1185		18.8	1110	0.2934		30.6
340	0.0329		9.4	690	0.1218		19.0	1125	0.3008		31.0
350	0.0347		9.7	700	0.1250		19.3	1140	0.3082		31.4
360	0.0365		9.9	710	0.1284		19.6	1155	0.3158		31.8
370	0.0384		10.2	720	0.1317		19.9	1170	0.3234		32.3
380	0.0404		10.5	730	0.1351		20.1	1185	0.3311		32.7
390	0.0424		10.8	740	0.1386		20.4	1200	0.3389		33.1
400	0.0444		11.0	750	0.1421		20.7	1215	0.3468		33.5
410	0.0465		11.3	760	0.1456		21.0	1230	0.3548		33.9
420	0.0486		11.6	770	0.1491		21.2	1245	0.3628		34.3
430	0.0508		11.9	780	0.1528		21.5	1260	0.3709		34.7
440	0.0530		12.1	790	0.1564		21.8	1275	0.3791		35.2

Friction Loss Table For 6" Cement Lined C.I. I.D. = 5.89

Equiv. Feet of Pipe

	LONG EL	90 ELL	45 ELL	TEE	GV	CV		
C-120	9	14	7	30	3	32		
C-140	11.97	18.62	9.31	39.90	3.99	42.56	0.00	0.00

Q	P	P	V	Q	P	P	V	Q	P	P	V
	PSI/FT	PSI/FT	FT/SEC		PSI/FT	PSI/FT	FT/SEC		PSI/FT	PSI/FT	FT/SEC
GPM	C-140			GPM	C-140			GPM	C-140		
300	0.0033		3.5	1000	0.0305		11.8	1700	0.0814		20.0
320	0.0037		3.8	1020	0.0316		12.0	1720	0.0832		20.3
340	0.0041		4.0	1040	0.0328		12.3	1740	0.0850		20.5
360	0.0046		4.2	1060	0.0340		12.5	1760	0.0868		20.7
380	0.0051		4.5	1080	0.0352		12.7	1780	0.0886		21.0
400	0.0056		4.7	1100	0.0364		13.0	1800	0.0905		21.2
420	0.0061		4.9	1120	0.0376		13.2	1820	0.0924		21.4
440	0.0067		5.2	1140	0.0389		13.4	1840	0.0942		21.7
460	0.0073		5.4	1160	0.0401		13.7	1870	0.0971		22.0
480	0.0078		5.7	1180	0.0414		13.9	1900	0.1000		22.4
500	0.0085		5.9	1200	0.0427		14.1	1930	0.1029		22.7
520	0.0091		6.1	1220	0.0441		14.4	1960	0.1059		23.1
540	0.0098		6.4	1240	0.0454		14.6	1990	0.1089		23.4
560	0.0104		6.6	1260	0.0468		14.8	2020	0.1120		23.8
580	0.0111		6.8	1280	0.0482		15.1	2050	0.1151		24.2
600	0.0119		7.1	1300	0.0496		15.3	2080	0.1182		24.5
620	0.0126		7.3	1320	0.0510		15.6	2110	0.1214		24.9
640	0.0134		7.5	1340	0.0524		15.8	2140	0.1246		25.2
660	0.0141		7.8	1360	0.0539		16.0	2170	0.1279		25.6
680	0.0149		8.0	1380	0.0553		16.3	2200	0.1312		25.9
700	0.0158		8.2	1400	0.0568		16.5	2230	0.1345		26.3
720	0.0166		8.5	1420	0.0584		16.7	2260	0.1379		26.6
740	0.0175		8.7	1440	0.0599		17.0	2290	0.1413		27.0
760	0.0184		9.0	1460	0.0614		17.2	2320	0.1447		27.3
780	0.0193		9.2	1480	0.0630		17.4	2350	0.1482		27.7
800	0.0202		9.4	1500	0.0646		17.7	2380	0.1517		28.0
820	0.0211		9.7	1520	0.0662		17.9	2410	0.1553		28.4
840	0.0221		9.9	1540	0.0678		18.1	2440	0.1589		28.7
860	0.0231		10.1	1560	0.0694		18.4	2470	0.1625		29.1
880	0.0241		10.4	1580	0.0711		18.6	2500	0.1662		29.5
900	0.0251		10.6	1600	0.0728		18.9	2530	0.1699		29.8
920	0.0261		10.8	1620	0.0745		19.1	2560	0.1736		30.2
940	0.0272		11.1	1640	0.0762		19.3	2590	0.1774		30.5
960	0.0283		11.3	1660	0.0779		19.6	2620	0.1812		30.9
980	0.0294		11.5	1680	0.0796		19.8	2650	0.1851		31.2

Friction Loss Table For 8" Cement Lined C.I.   I.D. = 7.98

Equiv. Feet of Pipe

	LONG EL	90 ELL	45 ELL	TEE	GV	CV		
C-120	13	18	9	35	4	45		
C-140	17.29	23.94	11.97	46.55	5.32	59.85	0.00	0.00

Q	P	P	V	Q	P	P	V	Q	P	P	V
	PSI/FT	PSI/FT	FT/SEC		PSI/FT	PSI/FT	FT/SEC		PSI/FT	PSI/FT	FT/SEC
GPM	C-140			GPM	C-140			GPM	C-140		
400	0.0010		2.3	1450	0.0108		8.4	2500	0.0297		14.5
430	0.0011		2.5	1480	0.0112		8.6	2530	0.0303		14.7
460	0.0013		2.7	1510	0.0117		8.8	2560	0.0310		14.9
490	0.0015		2.8	1540	0.0121		8.9	2590	0.0317		15.0
520	0.0016		3.0	1570	0.0125		9.1	2620	0.0324		15.2
550	0.0018		3.2	1600	0.0130		9.3	2650	0.0330		15.4
580	0.0020		3.4	1630	0.0134		9.5	2680	0.0337		15.6
610	0.0022		3.5	1660	0.0139		9.6	2710	0.0344		15.7
640	0.0024		3.7	1690	0.0144		9.8	2750	0.0354		16.0
670	0.0026		3.9	1720	0.0149		10.0	2790	0.0363		16.2
700	0.0028		4.1	1750	0.0153		10.2	2830	0.0373		16.4
730	0.0030		4.2	1780	0.0158		10.3	2870	0.0383		16.7
760	0.0033		4.4	1810	0.0163		10.5	2910	0.0393		16.9
790	0.0035		4.6	1840	0.0168		10.7	2950	0.0403		17.1
820	0.0038		4.8	1870	0.0173		10.9	2990	0.0413		17.4
850	0.0040		4.9	1900	0.0179		11.0	3030	0.0423		17.6
880	0.0043		5.1	1930	0.0184		11.2	3070	0.0434		17.8
910	0.0046		5.3	1960	0.0189		11.4	3110	0.0444		18.1
940	0.0049		5.5	1990	0.0195		11.6	3150	0.0455		18.3
970	0.0051		5.6	2020	0.0200		11.7	3190	0.0466		18.5
1000	0.0054		5.8	2050	0.0206		11.9	3230	0.0477		18.8
1030	0.0058		6.0	2080	0.0211		12.1	3270	0.0488		19.0
1060	0.0061		6.2	2110	0.0217		12.3	3310	0.0499		19.2
1090	0.0064		6.3	2140	0.0223		12.4	3350	0.0510		19.5
1120	0.0067		6.5	2170	0.0228		12.6	3390	0.0521		19.7
1150	0.0071		6.7	2200	0.0234		12.8	3430	0.0533		19.9
1180	0.0074		6.9	2230	0.0240		12.9	3470	0.0544		20.1
1210	0.0077		7.0	2260	0.0246		13.1	3510	0.0556		20.4
1240	0.0081		7.2	2290	0.0252		13.3	3550	0.0568		20.6
1270	0.0085		7.4	2320	0.0258		13.5	3590	0.0579		20.8
1300	0.0088		7.5	2350	0.0265		13.6	3630	0.0591		21.1
1330	0.0092		7.7	2380	0.0271		13.8	3670	0.0604		21.3
1360	0.0096		7.9	2410	0.0277		14.0	3710	0.0616		21.5
1390	0.0100		8.1	2440	0.0284		14.2	3750	0.0628		21.8
1420	0.0104		8.2	2470	0.0290		14.3	3790	0.0641		22.0

Appendix D Friction Loss Tables

Friction Loss Table For 10" Cement Lined C.I.   I.D. = 9.97

Equiv. Feet of Pipe

	LONG EL	90 ELL	45 ELL	TEE	GV	CV				
C-120	16	22	11	50	5	55				
C-140	21.28	29.26	14.63	66.50	6.65	73.15	0.00	0.00		

Q	P	P	V	Q	P	P	V	Q	P	P	V
	PSI/FT	PSI/FT	FT/SEC		PSI/FT	PSI/FT	FT/SEC		PSI/FT	PSI/FT	FT/SEC
GPM	C-140			GPM	C-140			GPM	C-140		
500	0.0007		2.1	2250	0.0105		9.3	4000	0.0305		16.4
550	0.0008		2.3	2300	0.0110		9.5	4050	0.0313		16.7
600	0.0009		2.5	2350	0.0114		9.7	4100	0.0320		16.9
650	0.0011		2.7	2400	0.0119		9.9	4150	0.0327		17.1
700	0.0012		2.9	2450	0.0123		10.1	4200	0.0334		17.3
750	0.0014		3.1	2500	0.0128		10.3	4250	0.0342		17.5
800	0.0016		3.3	2550	0.0133		10.5	4300	0.0349		17.7
850	0.0017		3.5	2600	0.0138		10.7	4350	0.0357		17.9
900	0.0019		3.7	2650	0.0143		10.9	4450	0.0372		18.3
950	0.0021		3.9	2700	0.0148		11.1	4550	0.0388		18.7
1000	0.0024		4.1	2750	0.0153		11.3	4650	0.0404		19.1
1050	0.0026		4.3	2800	0.0158		11.5	4750	0.0420		19.5
1100	0.0028		4.5	2850	0.0163		11.7	4850	0.0436		19.9
1150	0.0030		4.7	2900	0.0168		11.9	4950	0.0453		20.4
1200	0.0033		4.9	2950	0.0174		12.1	5050	0.0470		20.8
1250	0.0036		5.1	3000	0.0179		12.3	5150	0.0488		21.2
1300	0.0038		5.3	3050	0.0185		12.5	5250	0.0505		21.6
1350	0.0041		5.6	3100	0.0191		12.7	5350	0.0523		22.0
1400	0.0044		5.8	3150	0.0196		13.0	5450	0.0541		22.4
1450	0.0047		6.0	3200	0.0202		13.2	5550	0.0560		22.8
1500	0.0050		6.2	3250	0.0208		13.4	5650	0.0579		23.2
1550	0.0053		6.4	3300	0.0214		13.6	5750	0.0598		23.6
1600	0.0056		6.6	3350	0.0220		13.8	5850	0.0617		24.1
1650	0.0059		6.8	3400	0.0226		14.0	5950	0.0637		24.5
1700	0.0063		7.0	3450	0.0232		14.2	6050	0.0657		24.9
1750	0.0066		7.2	3500	0.0239		14.4	6150	0.0677		25.3
1800	0.0070		7.4	3550	0.0245		14.6	6250	0.0697		25.7
1850	0.0073		7.6	3600	0.0251		14.8	6350	0.0718		26.1
1900	0.0077		7.8	3650	0.0258		15.0	6450	0.0739		26.5
1950	0.0081		8.0	3700	0.0264		15.2	6550	0.0761		26.9
2000	0.0085		8.2	3750	0.0271		15.4	6650	0.0782		27.3
2050	0.0089		8.4	3800	0.0278		15.6	6750	0.0804		27.8
2100	0.0093		8.6	3850	0.0285		15.8	6850	0.0826		28.2
2150	0.0097		8.8	3900	0.0291		16.0	6950	0.0849		28.6
2200	0.0101		9.0	3950	0.0298		16.2	7050	0.0872		29.0

Friction Loss Table For 12" Cement Lined C.I. I.D. = 11.99

Equiv. Feet of Pipe

	LONG ELL	90 ELL	45 ELL	TEE	GV	CV		
C-120	18	27	13	60	6	65		
C-140	23.94	35.91	17.29	79.80	7.98	86.45	0.00	0.00

Q	P	P	V	Q	P	P	V	Q	P	P	V
	PSI/FT	PSI/FT	FT/SEC		PSI/FT	PSI/FT	FT/SEC		PSI/FT	PSI/FT	FT/SEC
GPM	C-140			GPM	C-140			GPM	C-140		
500	0.0003		1.4	3950	0.0122		11.2	7400	0.0388		21.0
600	0.0004		1.7	4050	0.0127		11.5	7500	0.0398		21.3
700	0.0005		2.0	4150	0.0133		11.8	7600	0.0408		21.6
800	0.0006		2.3	4250	0.0139		12.1	7700	0.0418		21.9
900	0.0008		2.6	4350	0.0145		12.4	7800	0.0428		22.2
1000	0.0010		2.8	4450	0.0152		12.7	7900	0.0438		22.5
1100	0.0011		3.1	4550	0.0158		12.9	8000	0.0448		22.7
1200	0.0013		3.4	4650	0.0164		13.2	8100	0.0459		23.0
1300	0.0016		3.7	4750	0.0171		13.5	8200	0.0469		23.3
1400	0.0018		4.0	4850	0.0178		13.8	8300	0.0480		23.6
1500	0.0020		4.3	4950	0.0184		14.1	8400	0.0491		23.9
1600	0.0023		4.5	5050	0.0191		14.4	8500	0.0502		24.2
1700	0.0026		4.8	5150	0.0199		14.6	8600	0.0513		24.5
1800	0.0028		5.1	5250	0.0206		14.9	8700	0.0524		24.7
1900	0.0031		5.4	5350	0.0213		15.2	8800	0.0535		25.0
2000	0.0035		5.7	5450	0.0220		15.5	8900	0.0546		25.3
2100	0.0038		6.0	5550	0.0228		15.8	9000	0.0558		25.6
2200	0.0041		6.3	5650	0.0236		16.1	9100	0.0569		25.9
2300	0.0045		6.5	5750	0.0243		16.3	9200	0.0581		26.2
2400	0.0048		6.8	5850	0.0251		16.6	9300	0.0592		26.4
2500	0.0052		7.1	5950	0.0259		16.9	9400	0.0604		26.7
2600	0.0056		7.4	6050	0.0267		17.2	9500	0.0616		27.0
2700	0.0060		7.7	6150	0.0276		17.5	9600	0.0628		27.3
2800	0.0064		8.0	6250	0.0284		17.8	9700	0.0640		27.6
2900	0.0069		8.2	6350	0.0292		18.1	9800	0.0653		27.9
3000	0.0073		8.5	6450	0.0301		18.3	9900	0.0665		28.1
3100	0.0078		8.8	6550	0.0310		18.6	10000	0.0678		28.4
3200	0.0082		9.1	6650	0.0319		18.9	10100	0.0690		28.7
3300	0.0087		9.4	6750	0.0327		19.2	10200	0.0703		29.0
3400	0.0092		9.7	6850	0.0336		19.5	10300	0.0716		29.3
3500	0.0097		10.0	6950	0.0346		19.8	10400	0.0729		29.6
3600	0.0102		10.2	7050	0.0355		20.0	10500	0.0742		29.9
3700	0.0108		10.5	7150	0.0364		20.3	10600	0.0755		30.1
3800	0.0113		10.8	7250	0.0374		20.6	10700	0.0768		30.4
3900	0.0119		11.1	7350	0.0383		20.9	10800	0.0781		30.7

Appendix D Friction Loss Tables

Friction Loss Table For   6" D.I. Class 50   I.D. = 6.28
Equiv. Feet of Pipe

	LONG EL	90 ELL	45 ELL	TEE	GV	CV				
C–120	9	14	7	30	3	32				
C–140	11.97	18.62	9.31	39.90	3.99	42.56	0.00	0.00		

Q	P	P	V	Q	P	P	V	Q	P	P	V
	PSI/FT	PSI/FT	FT/SEC		PSI/FT	PSI/FT	FT/SEC		PSI/FT	PSI/FT	FT/SEC
GPM	C–140			GPM	C–140			GPM	C–140		
300	0.0024		3.1	1000	0.0223		10.4	1700	0.0596		17.6
320	0.0027		3.3	1020	0.0232		10.6	1720	0.0609		17.8
340	0.0030		3.5	1040	0.0240		10.8	1740	0.0622		18.0
360	0.0034		3.7	1060	0.0249		11.0	1760	0.0635		18.2
380	0.0037		3.9	1080	0.0257		11.2	1780	0.0649		18.4
400	0.0041		4.1	1100	0.0266		11.4	1800	0.0662		18.7
420	0.0045		4.4	1120	0.0275		11.6	1820	0.0676		18.9
440	0.0049		4.6	1140	0.0284		11.8	1840	0.0690		19.1
460	0.0053		4.8	1160	0.0294		12.0	1870	0.0711		19.4
480	0.0057		5.0	1180	0.0303		12.2	1900	0.0732		19.7
500	0.0062		5.2	1200	0.0313		12.4	1930	0.0753		20.0
520	0.0067		5.4	1220	0.0322		12.6	1960	0.0775		20.3
540	0.0071		5.6	1240	0.0332		12.9	1990	0.0797		20.6
560	0.0076		5.8	1260	0.0342		13.1	2020	0.0820		20.9
580	0.0081		6.0	1280	0.0352		13.3	2050	0.0842		21.2
600	0.0087		6.2	1300	0.0363		13.5	2080	0.0865		21.6
620	0.0092		6.4	1320	0.0373		13.7	2110	0.0889		21.9
640	0.0098		6.6	1340	0.0384		13.9	2140	0.0912		22.2
660	0.0103		6.8	1360	0.0394		14.1	2170	0.0936		22.5
680	0.0109		7.0	1380	0.0405		14.3	2200	0.0960		22.8
700	0.0115		7.3	1400	0.0416		14.5	2230	0.0984		23.1
720	0.0122		7.5	1420	0.0427		14.7	2260	0.1009		23.4
740	0.0128		7.7	1440	0.0438		14.9	2290	0.1034		23.7
760	0.0134		7.9	1460	0.0450		15.1	2320	0.1059		24.0
780	0.0141		8.1	1480	0.0461		15.3	2350	0.1084		24.4
800	0.0148		8.3	1500	0.0473		15.5	2380	0.1110		24.7
820	0.0155		8.5	1520	0.0484		15.8	2410	0.1136		25.0
840	0.0162		8.7	1540	0.0496		16.0	2440	0.1163		25.3
860	0.0169		8.9	1560	0.0508		16.2	2470	0.1189		25.6
880	0.0176		9.1	1580	0.0520		16.4	2500	0.1216		25.9
900	0.0184		9.3	1600	0.0533		16.6	2530	0.1243		26.2
920	0.0191		9.5	1620	0.0545		16.8	2560	0.1271		26.5
940	0.0199		9.7	1640	0.0557		17.0	2590	0.1298		26.8
960	0.0207		9.9	1660	0.0570		17.2	2620	0.1326		27.2
980	0.0215		10.2	1680	0.0583		17.4	2650	0.1354		27.5

Friction Loss Table For 8" D.I. Class 50   I.D. = 8.51

Equiv. Feet of Pipe

	LONG EL	90 ELL	45 ELL	TEE	GV	CV		
C-120	13	18	9	35	4	45		
C-140	17.29	23.94	11.97	46.55	5.32	59.85	0.00	0.00

Q GPM	P PSI/FT C-140	P PSI/FT	V FT/SEC	Q GPM	P PSI/FT C-140	P PSI/FT	V FT/SEC	Q GPM	P PSI/FT C-140	P PSI/FT	V FT/SEC
400	0.0009		2.3	1450	0.0101		8.2	2500	0.0277		14.1
430	0.0011		2.4	1480	0.0105		8.4	2530	0.0283		14.3
460	0.0012		2.6	1510	0.0109		8.5	2560	0.0289		14.4
490	0.0014		2.8	1540	0.0113		8.7	2590	0.0296		14.6
520	0.0015		2.9	1570	0.0117		8.9	2620	0.0302		14.8
550	0.0017		3.1	1600	0.0121		9.0	2650	0.0308		15.0
580	0.0019		3.3	1630	0.0125		9.2	2680	0.0315		15.1
610	0.0020		3.4	1660	0.0130		9.4	2710	0.0321		15.3
640	0.0022		3.6	1690	0.0134		9.5	2750	0.0330		15.5
670	0.0024		3.8	1720	0.0139		9.7	2790	0.0339		15.7
700	0.0026		4.0	1750	0.0143		9.9	2830	0.0348		16.0
730	0.0028		4.1	1780	0.0148		10.0	2870	0.0357		16.2
760	0.0031		4.3	1810	0.0152		10.2	2910	0.0367		16.4
790	0.0033		4.5	1840	0.0157		10.4	2950	0.0376		16.6
820	0.0035		4.6	1870	0.0162		10.6	2990	0.0386		16.9
850	0.0038		4.8	1900	0.0167		10.7	3030	0.0395		17.1
880	0.0040		5.0	1930	0.0172		10.9	3070	0.0405		17.3
910	0.0043		5.1	1960	0.0176		11.1	3110	0.0415		17.6
940	0.0045		5.3	1990	0.0182		11.2	3150	0.0425		17.8
970	0.0048		5.5	2020	0.0187		11.4	3190	0.0435		18.0
1000	0.0051		5.6	2050	0.0192		11.6	3230	0.0445		18.2
1030	0.0054		5.8	2080	0.0197		11.7	3270	0.0455		18.5
1060	0.0057		6.0	2110	0.0202		11.9	3310	0.0465		18.7
1090	0.0060		6.2	2140	0.0208		12.1	3350	0.0476		18.9
1120	0.0063		6.3	2170	0.0213		12.2	3390	0.0486		19.1
1150	0.0066		6.5	2200	0.0219		12.4	3430	0.0497		19.4
1180	0.0069		6.7	2230	0.0224		12.6	3470	0.0508		19.6
1210	0.0072		6.8	2260	0.0230		12.8	3510	0.0519		19.8
1240	0.0076		7.0	2290	0.0235		12.9	3550	0.0530		20.0
1270	0.0079		7.2	2320	0.0241		13.1	3590	0.0541		20.3
1300	0.0083		7.3	2350	0.0247		13.3	3630	0.0552		20.5
1330	0.0086		7.5	2380	0.0253		13.4	3670	0.0563		20.7
1360	0.0090		7.7	2410	0.0259		13.6	3710	0.0575		20.9
1390	0.0093		7.8	2440	0.0265		13.8	3750	0.0586		21.2
1420	0.0097		8.0	2470	0.0271		13.9	3790	0.0598		21.4

Appendix D Friction Loss Tables

Friction Loss Table For 10" D.I. Class 50 I.D. = 10.52

Equiv. Feet of Pipe

	LONG EL	90 ELL	45 ELL	TEE	GV	CV		
C-120	16	22	11	50	5	55		
C-140	21.28	29.26	14.63	66.50	6.65	73.15	0.00	0.00

Q GPM	P PSI/FT C-140	P PSI/FT	V FT/SEC	Q GPM	P PSI/FT C-140	P PSI/FT	V FT/SEC	Q GPM	P PSI/FT C-140	P PSI/FT	V FT/SEC
500	0.0005		1.8	2250	0.0081		8.3	4000	0.0235		14.8
550	0.0006		2.0	2300	0.0084		8.5	4050	0.0241		15.0
600	0.0007		2.2	2350	0.0088		8.7	4100	0.0246		15.1
650	0.0008		2.4	2400	0.0091		8.9	4150	0.0252		15.3
700	0.0009		2.6	2450	0.0095		9.0	4200	0.0257		15.5
750	0.0011		2.8	2500	0.0099		9.2	4250	0.0263		15.7
800	0.0012		3.0	2550	0.0102		9.4	4300	0.0269		15.9
850	0.0013		3.1	2600	0.0106		9.6	4350	0.0275		16.1
900	0.0015		3.3	2650	0.0110		9.8	4450	0.0286		16.4
950	0.0016		3.5	2700	0.0114		10.0	4550	0.0298		16.8
1000	0.0018		3.7	2750	0.0118		10.2	4650	0.0311		17.2
1050	0.0020		3.9	2800	0.0122		10.3	4750	0.0323		17.5
1100	0.0022		4.1	2850	0.0126		10.5	4850	0.0336		17.9
1150	0.0023		4.2	2900	0.0130		10.7	4950	0.0349		18.3
1200	0.0025		4.4	2950	0.0134		10.9	5050	0.0362		18.7
1250	0.0027		4.6	3000	0.0138		11.1	5150	0.0375		19.0
1300	0.0029		4.8	3050	0.0142		11.3	5250	0.0389		19.4
1350	0.0032		5.0	3100	0.0147		11.4	5350	0.0403		19.8
1400	0.0034		5.2	3150	0.0151		11.6	5450	0.0417		20.1
1450	0.0036		5.4	3200	0.0156		11.8	5550	0.0431		20.5
1500	0.0038		5.5	3250	0.0160		12.0	5650	0.0446		20.9
1550	0.0041		5.7	3300	0.0165		12.2	5750	0.0460		21.2
1600	0.0043		5.9	3350	0.0169		12.4	5850	0.0475		21.6
1650	0.0046		6.1	3400	0.0174		12.6	5950	0.0490		22.0
1700	0.0048		6.3	3450	0.0179		12.7	6050	0.0506		22.3
1750	0.0051		6.5	3500	0.0184		12.9	6150	0.0521		22.7
1800	0.0054		6.6	3550	0.0189		13.1	6250	0.0537		23.1
1850	0.0056		6.8	3600	0.0194		13.3	6350	0.0553		23.5
1900	0.0059		7.0	3650	0.0199		13.5	6450	0.0569		23.8
1950	0.0062		7.2	3700	0.0204		13.7	6550	0.0586		24.2
2000	0.0065		7.4	3750	0.0209		13.8	6650	0.0602		24.6
2050	0.0068		7.6	3800	0.0214		14.0	6750	0.0619		24.9
2100	0.0071		7.8	3850	0.0219		14.2	6850	0.0636		25.3
2150	0.0075		7.9	3900	0.0224		14.4	6950	0.0654		25.7
2200	0.0078		8.1	3950	0.0230		14.6	7050	0.0671		26.0

Friction Loss Table For 12" D.I. Class 50 I.D. = 12.58

Equiv. Feet of Pipe

	LONG EL	90 ELL	45 ELL	TEE	GV	CV		
C-120	18	27	13	60	6	65		
C-140	23.94	35.91	17.29	79.80	7.98	86.45	0.00	0.00

Q GPM	P PSI/FT C-140	P PSI/FT	V FT/SEC	Q GPM	P PSI/FT C-140	P PSI/FT	V FT/SEC	Q GPM	P PSI/FT C-140	P PSI/FT	V FT/SEC
500	0.0002		1.3	3950	0.0096		10.2	7400	0.0307		19.1
600	0.0003		1.5	4050	0.0101		10.5	7500	0.0315		19.4
700	0.0004		1.8	4150	0.0105		10.7	7600	0.0323		19.6
800	0.0005		2.1	4250	0.0110		11.0	7700	0.0331		19.9
900	0.0006		2.3	4350	0.0115		11.2	7800	0.0339		20.1
1000	0.0008		2.6	4450	0.0120		11.5	7900	0.0347		20.4
1100	0.0009		2.8	4550	0.0125		11.8	8000	0.0355		20.7
1200	0.0011		3.1	4650	0.0130		12.0	8100	0.0363		20.9
1300	0.0012		3.4	4750	0.0135		12.3	8200	0.0371		21.2
1400	0.0014		3.6	4850	0.0141		12.5	8300	0.0380		21.4
1500	0.0016		3.9	4950	0.0146		12.8	8400	0.0388		21.7
1600	0.0018		4.1	5050	0.0152		13.0	8500	0.0397		22.0
1700	0.0020		4.4	5150	0.0157		13.3	8600	0.0406		22.2
1800	0.0022		4.6	5250	0.0163		13.6	8700	0.0414		22.5
1900	0.0025		4.9	5350	0.0169		13.8	8800	0.0423		22.7
2000	0.0027		5.2	5450	0.0174		14.1	8900	0.0432		23.0
2100	0.0030		5.4	5550	0.0180		14.3	9000	0.0441		23.2
2200	0.0033		5.7	5650	0.0186		14.6	9100	0.0450		23.5
2300	0.0035		5.9	5750	0.0193		14.9	9200	0.0460		23.8
2400	0.0038		6.2	5850	0.0199		15.1	9300	0.0469		24.0
2500	0.0041		6.5	5950	0.0205		15.4	9400	0.0478		24.3
2600	0.0044		6.7	6050	0.0212		15.6	9500	0.0488		24.5
2700	0.0048		7.0	6150	0.0218		15.9	9600	0.0497		24.8
2800	0.0051		7.2	6250	0.0225		16.1	9700	0.0507		25.1
2900	0.0054		7.5	6350	0.0231		16.4	9800	0.0517		25.3
3000	0.0058		7.7	6450	0.0238		16.7	9900	0.0526		25.6
3100	0.0061		8.0	6550	0.0245		16.9	10000	0.0536		25.8
3200	0.0065		8.3	6650	0.0252		17.2	10100	0.0546		26.1
3300	0.0069		8.5	6750	0.0259		17.4	10200	0.0556		26.3
3400	0.0073		8.8	6850	0.0266		17.7	10300	0.0566		26.6
3500	0.0077		9.0	6950	0.0274		17.9	10400	0.0577		26.9
3600	0.0081		9.3	7050	0.0281		18.2	10500	0.0587		27.1
3700	0.0085		9.6	7150	0.0288		18.5	10600	0.0597		27.4
3800	0.0090		9.8	7250	0.0296		18.7	10700	0.0608		27.6
3900	0.0094		10.1	7350	0.0303		19.0	10800	0.0618		27.9

Appendix D Friction Loss Tables

Friction Loss Table For   4" D.I. Class 52   I.D. = 4.22

Equiv. Feet of Pipe

	LONG EL	90 ELL	45 ELL	TEE	GV	CV		
C-120	6	10	4	20	2	22		
C-140	7.98	13.30	5.32	26.60	2.66	29.26	0.00	0.00

Q GPM	P PSI/FT C-140	P PSI/FT	V FT/SEC	Q GPM	P PSI/FT C-140	P PSI/FT	V FT/SEC	Q GPM	P PSI/FT C-140	P PSI/FT	V FT/SEC
100	0.0022		2.3	450	0.0353		10.3	800	0.1024		18.4
110	0.0026		2.5	460	0.0368		10.6	810	0.1048		18.6
120	0.0031		2.8	470	0.0383		10.8	820	0.1072		18.8
130	0.0036		3.0	480	0.0398		11.0	830	0.1096		19.0
140	0.0041		3.2	490	0.0413		11.2	840	0.1121		19.3
150	0.0046		3.4	500	0.0429		11.5	850	0.1145		19.5
160	0.0052		3.7	510	0.0445		11.7	860	0.1170		19.7
170	0.0058		3.9	520	0.0461		11.9	870	0.1196		20.0
180	0.0065		4.1	530	0.0478		12.2	885	0.1234		20.3
190	0.0072		4.4	540	0.0495		12.4	900	0.1273		20.7
200	0.0079		4.6	550	0.0512		12.6	915	0.1313		21.0
210	0.0086		4.8	560	0.0529		12.9	930	0.1353		21.3
220	0.0094		5.0	570	0.0547		13.1	945	0.1393		21.7
230	0.0102		5.3	580	0.0565		13.3	960	0.1435		22.0
240	0.0110		5.5	590	0.0583		13.5	975	0.1476		22.4
250	0.0119		5.7	600	0.0601		13.8	990	0.1519		22.7
260	0.0128		6.0	610	0.0620		14.0	1005	0.1561		23.1
270	0.0137		6.2	620	0.0639		14.2	1020	0.1605		23.4
280	0.0147		6.4	630	0.0658		14.5	1035	0.1649		23.8
290	0.0157		6.7	640	0.0678		14.7	1050	0.1693		24.1
300	0.0167		6.9	650	0.0697		14.9	1065	0.1738		24.4
310	0.0177		7.1	660	0.0717		15.1	1080	0.1784		24.8
320	0.0188		7.3	670	0.0738		15.4	1095	0.1830		25.1
330	0.0199		7.6	680	0.0758		15.6	1110	0.1877		25.5
340	0.0210		7.8	690	0.0779		15.8	1125	0.1924		25.8
350	0.0222		8.0	700	0.0800		16.1	1140	0.1972		26.2
360	0.0234		8.3	710	0.0821		16.3	1155	0.2020		26.5
370	0.0246		8.5	720	0.0843		16.5	1170	0.2069		26.9
380	0.0258		8.7	730	0.0864		16.8	1185	0.2118		27.2
390	0.0271		9.0	740	0.0886		17.0	1200	0.2168		27.5
400	0.0284		9.2	750	0.0909		17.2	1215	0.2218		27.9
410	0.0297		9.4	760	0.0931		17.4	1230	0.2269		28.2
420	0.0311		9.6	770	0.0954		17.7	1245	0.2321		28.6
430	0.0325		9.9	780	0.0977		17.9	1260	0.2373		28.9
440	0.0339		10.1	790	0.1000		18.1	1275	0.2425		29.3

Friction Loss Table For 6" D.I. Class 52 I.D. = 6.16

Equiv. Feet of Pipe

	LONG EL	90 ELL	45 ELL	TEE	GV	CV				
C-120	9	14	7	30	3	32				
C-140	11.97	18.62	9.31	39.90	3.99	42.56	0.00	0.00		

Q	P	P	V	Q	P	P	V	Q	P	P	V
	PSI/FT	PSI/FT	FT/SEC		PSI/FT	PSI/FT	FT/SEC		PSI/FT	PSI/FT	FT/SEC
GPM	C-140			GPM	C-140			GPM	C-140		
300	0.0026		3.2	1000	0.0245		10.8	1700	0.0654		18.3
320	0.0030		3.4	1020	0.0254		11.0	1720	0.0669		18.5
340	0.0033		3.7	1040	0.0264		11.2	1740	0.0683		18.7
360	0.0037		3.9	1060	0.0273		11.4	1760	0.0698		19.0
380	0.0041		4.1	1080	0.0283		11.6	1780	0.0713		19.2
400	0.0045		4.3	1100	0.0292		11.8	1800	0.0727		19.4
420	0.0049		4.5	1120	0.0302		12.1	1820	0.0742		19.6
440	0.0054		4.7	1140	0.0312		12.3	1840	0.0758		19.8
460	0.0058		5.0	1160	0.0323		12.5	1870	0.0781		20.1
480	0.0063		5.2	1180	0.0333		12.7	1900	0.0804		20.5
500	0.0068		5.4	1200	0.0344		12.9	1930	0.0828		20.8
520	0.0073		5.6	1220	0.0354		13.1	1960	0.0852		21.1
540	0.0078		5.8	1240	0.0365		13.4	1990	0.0876		21.4
560	0.0084		6.0	1260	0.0376		13.6	2020	0.0900		21.8
580	0.0090		6.2	1280	0.0387		13.8	2050	0.0925		22.1
600	0.0095		6.5	1300	0.0398		14.0	2080	0.0951		22.4
620	0.0101		6.7	1320	0.0410		14.2	2110	0.0976		22.7
640	0.0107		6.9	1340	0.0421		14.4	2140	0.1002		23.1
660	0.0114		7.1	1360	0.0433		14.6	2170	0.1028		23.4
680	0.0120		7.3	1380	0.0445		14.9	2200	0.1054		23.7
700	0.0127		7.5	1400	0.0457		15.1	2230	0.1081		24.0
720	0.0134		7.8	1420	0.0469		15.3	2260	0.1108		24.3
740	0.0140		8.0	1440	0.0481		15.5	2290	0.1136		24.7
760	0.0148		8.2	1460	0.0494		15.7	2320	0.1163		25.0
780	0.0155		8.4	1480	0.0506		15.9	2350	0.1191		25.3
800	0.0162		8.6	1500	0.0519		16.2	2380	0.1220		25.6
820	0.0170		8.8	1520	0.0532		16.4	2410	0.1248		26.0
840	0.0178		9.0	1540	0.0545		16.6	2440	0.1277		26.3
860	0.0186		9.3	1560	0.0558		16.8	2470	0.1306		26.6
880	0.0194		9.5	1580	0.0572		17.0	2500	0.1336		26.9
900	0.0202		9.7	1600	0.0585		17.2	2530	0.1366		27.3
920	0.0210		9.9	1620	0.0599		17.4	2560	0.1396		27.6
940	0.0219		10.1	1640	0.0612		17.7	2590	0.1426		27.9
960	0.0227		10.3	1660	0.0626		17.9	2620	0.1457		28.2
980	0.0236		10.6	1680	0.0640		18.1	2650	0.1488		28.5

Friction Loss Table For   8" D.I. Class 52   I.D. = 8.39

Equiv. Feet of Pipe

	LONG EL	90 ELL	45 ELL	TEE	GV	CV				
C-120	13	18	9	35	4	45				
C-140	17.29	23.94	11.97	46.55	5.32	59.85	0.00	0.00		

Q	P	P	V	Q	P	P	V	Q	P	P	V
	PSI/FT	PSI/FT	FT/SEC		PSI/FT	PSI/FT	FT/SEC		PSI/FT	PSI/FT	FT/SEC
GPM	C-140			GPM	C-140			GPM	C-140		
400	0.0010		2.3	1450	0.0108		8.4	2500	0.0297		14.5
430	0.0011		2.5	1480	0.0112		8.6	2530	0.0303		14.7
460	0.0013		2.7	1510	0.0117		8.8	2560	0.0310		14.9
490	0.0015		2.8	1540	0.0121		8.9	2590	0.0317		15.0
520	0.0016		3.0	1570	0.0125		9.1	2620	0.0324		15.2
550	0.0018		3.2	1600	0.0130		9.3	2650	0.0330		15.4
580	0.0020		3.4	1630	0.0134		9.5	2680	0.0337		15.6
610	0.0022		3.5	1660	0.0139		9.6	2710	0.0344		15.7
640	0.0024		3.7	1690	0.0144		9.8	2750	0.0354		16.0
670	0.0026		3.9	1720	0.0149		10.0	2790	0.0363		16.2
700	0.0028		4.1	1750	0.0153		10.2	2830	0.0373		16.4
730	0.0030		4.2	1780	0.0158		10.3	2870	0.0383		16.7
760	0.0033		4.4	1810	0.0163		10.5	2910	0.0393		16.9
790	0.0035		4.6	1840	0.0168		10.7	2950	0.0403		17.1
820	0.0038		4.8	1870	0.0173		10.9	2990	0.0413		17.4
850	0.0040		4.9	1900	0.0179		11.0	3030	0.0423		17.6
880	0.0043		5.1	1930	0.0184		11.2	3070	0.0434		17.8
910	0.0046		5.3	1960	0.0189		11.4	3110	0.0444		18.1
940	0.0049		5.5	1990	0.0195		11.6	3150	0.0455		18.3
970	0.0051		5.6	2020	0.0200		11.7	3190	0.0466		18.5
1000	0.0054		5.8	2050	0.0206		11.9	3230	0.0477		18.8
1030	0.0058		6.0	2080	0.0211		12.1	3270	0.0488		19.0
1060	0.0061		6.2	2110	0.0217		12.3	3310	0.0499		19.2
1090	0.0064		6.3	2140	0.0223		12.4	3350	0.0510		19.5
1120	0.0067		6.5	2170	0.0228		12.6	3390	0.0521		19.7
1150	0.0071		6.7	2200	0.0234		12.8	3430	0.0533		19.9
1180	0.0074		6.9	2230	0.0240		12.9	3470	0.0544		20.1
1210	0.0077		7.0	2260	0.0246		13.1	3510	0.0556		20.4
1240	0.0081		7.2	2290	0.0252		13.3	3550	0.0568		20.6
1270	0.0085		7.4	2320	0.0258		13.5	3590	0.0579		20.8
1300	0.0088		7.5	2350	0.0265		13.6	3630	0.0591		21.1
1330	0.0092		7.7	2380	0.0271		13.8	3670	0.0604		21.3
1360	0.0096		7.9	2410	0.0277		14.0	3710	0.0616		21.5
1390	0.0100		8.1	2440	0.0284		14.2	3750	0.0628		21.8
1420	0.0104		8.2	2470	0.0290		14.3	3790	0.0641		22.0

Friction Loss Table For 10" D.I. Class 52 I.D. = 10.40

Equiv. Feet of Pipe

	LONG EL	90 ELL	45 ELL	TEE	GV	CV				
C–120	16	22	11	50	5	55				
C–140	21.28	29.26	14.63	66.50	6.65	73.15	0.00	0.00		

Q	P	P	V	Q	P	P	V	Q	P	P	V
	PSI/FT	PSI/FT	FT/SEC		PSI/FT	PSI/FT	FT/SEC		PSI/FT	PSI/FT	FT/SEC
GPM	C–140			GPM	C–140			GPM	C–140		
500	0.0005		1.9	2250	0.0086		8.5	4000	0.0249		15.1
550	0.0006		2.1	2300	0.0089		8.7	4050	0.0254		15.3
600	0.0007		2.3	2350	0.0093		8.9	4100	0.0260		15.5
650	0.0009		2.5	2400	0.0097		9.1	4150	0.0266		15.7
700	0.0010		2.6	2450	0.0100		9.3	4200	0.0272		15.9
750	0.0011		2.8	2500	0.0104		9.4	4250	0.0278		16.1
800	0.0013		3.0	2550	0.0108		9.6	4300	0.0284		16.2
850	0.0014		3.2	2600	0.0112		9.8	4350	0.0290		16.4
900	0.0016		3.4	2650	0.0116		10.0	4450	0.0303		16.8
950	0.0017		3.6	2700	0.0120		10.2	4550	0.0316		17.2
1000	0.0019		3.8	2750	0.0124		10.4	4650	0.0329		17.6
1050	0.0021		4.0	2800	0.0129		10.6	4750	0.0342		18.0
1100	0.0023		4.2	2850	0.0133		10.8	4850	0.0355		18.3
1150	0.0025		4.3	2900	0.0137		11.0	4950	0.0369		18.7
1200	0.0027		4.5	2950	0.0142		11.1	5050	0.0383		19.1
1250	0.0029		4.7	3000	0.0146		11.3	5150	0.0397		19.5
1300	0.0031		4.9	3050	0.0151		11.5	5250	0.0411		19.8
1350	0.0033		5.1	3100	0.0155		11.7	5350	0.0426		20.2
1400	0.0036		5.3	3150	0.0160		11.9	5450	0.0441		20.6
1450	0.0038		5.5	3200	0.0165		12.1	5550	0.0456		21.0
1500	0.0041		5.7	3250	0.0169		12.3	5650	0.0471		21.4
1550	0.0043		5.9	3300	0.0174		12.5	5750	0.0487		21.7
1600	0.0046		6.0	3350	0.0179		12.7	5850	0.0502		22.1
1650	0.0048		6.2	3400	0.0184		12.8	5950	0.0518		22.5
1700	0.0051		6.4	3450	0.0189		13.0	6050	0.0535		22.9
1750	0.0054		6.6	3500	0.0194		13.2	6150	0.0551		23.2
1800	0.0057		6.8	3550	0.0199		13.4	6250	0.0568		23.6
1850	0.0060		7.0	3600	0.0205		13.6	6350	0.0585		24.0
1900	0.0063		7.2	3650	0.0210		13.8	6450	0.0602		24.4
1950	0.0066		7.4	3700	0.0215		14.0	6550	0.0619		24.8
2000	0.0069		7.6	3750	0.0221		14.2	6650	0.0637		25.1
2050	0.0072		7.7	3800	0.0226		14.4	6750	0.0655		25.5
2100	0.0076		7.9	3850	0.0232		14.5	6850	0.0673		25.9
2150	0.0079		8.1	3900	0.0237		14.7	6950	0.0691		26.3
2200	0.0082		8.3	3950	0.0243		14.9	7050	0.0710		26.6

Appendix D Friction Loss Tables

Friction Loss Table For 12" D.I. Class 52   I.D. = 12.46

Equiv. Feet of Pipe

	LONG EL	90 ELL	45 ELL	TEE	GV	CV				
C-120	18	27	13	60	6	65				
C-140	23.94	35.91	17.29	79.80	7.98	86.45	0.00	0.00		

Q	P	P	V	Q	P	P	V	Q	P	P	V
	PSI/FT	PSI/FT	FT/SEC		PSI/FT	PSI/FT	FT/SEC		PSI/FT	PSI/FT	FT/SEC
GPM	C-140			GPM	C-140			GPM	C-140		
500	0.0002		1.3	3950	0.0101		10.4	7400	0.0322		19.5
600	0.0003		1.6	4050	0.0106		10.7	7500	0.0330		19.7
700	0.0004		1.8	4150	0.0110		10.9	7600	0.0338		20.0
800	0.0005		2.1	4250	0.0115		11.2	7700	0.0346		20.3
900	0.0007		2.4	4350	0.0120		11.5	7800	0.0355		20.5
1000	0.0008		2.6	4450	0.0126		11.7	7900	0.0363		20.8
1100	0.0009		2.9	4550	0.0131		12.0	8000	0.0372		21.1
1200	0.0011		3.2	4650	0.0136		12.2	8100	0.0380		21.3
1300	0.0013		3.4	4750	0.0142		12.5	8200	0.0389		21.6
1400	0.0015		3.7	4850	0.0147		12.8	8300	0.0398		21.9
1500	0.0017		3.9	4950	0.0153		13.0	8400	0.0407		22.1
1600	0.0019		4.2	5050	0.0159		13.3	8500	0.0416		22.4
1700	0.0021		4.5	5150	0.0165		13.6	8600	0.0425		22.6
1800	0.0024		4.7	5250	0.0171		13.8	8700	0.0434		22.9
1900	0.0026		5.0	5350	0.0177		14.1	8800	0.0444		23.2
2000	0.0029		5.3	5450	0.0183		14.3	8900	0.0453		23.4
2100	0.0031		5.5	5550	0.0189		14.6	9000	0.0462		23.7
2200	0.0034		5.8	5650	0.0195		14.9	9100	0.0472		24.0
2300	0.0037		6.1	5750	0.0202		15.1	9200	0.0482		24.2
2400	0.0040		6.3	5850	0.0208		15.4	9300	0.0491		24.5
2500	0.0043		6.6	5950	0.0215		15.7	9400	0.0501		24.7
2600	0.0046		6.8	6050	0.0222		15.9	9500	0.0511		25.0
2700	0.0050		7.1	6150	0.0229		16.2	9600	0.0521		25.3
2800	0.0053		7.4	6250	0.0236		16.5	9700	0.0531		25.5
2900	0.0057		7.6	6350	0.0243		16.7	9800	0.0541		25.8
3000	0.0061		7.9	6450	0.0250		17.0	9900	0.0552		26.1
3100	0.0064		8.2	6550	0.0257		17.2	10000	0.0562		26.3
3200	0.0068		8.4	6650	0.0264		17.5	10100	0.0572		26.6
3300	0.0072		8.7	6750	0.0272		17.8	10200	0.0583		26.9
3400	0.0076		9.0	6850	0.0279		18.0	10300	0.0593		27.1
3500	0.0081		9.2	6950	0.0287		18.3	10400	0.0604		27.4
3600	0.0085		9.5	7050	0.0294		18.6	10500	0.0615		27.6
3700	0.0089		9.7	7150	0.0302		18.8	10600	0.0626		27.9
3800	0.0094		10.0	7250	0.0310		19.1	10700	0.0637		28.2
3900	0.0098		10.3	7350	0.0318		19.4	10800	0.0648		28.4

Friction Loss Table For 4" PVC Class 150   I.D. = 4.24

Equiv. Feet of Pipe

	LONG EL	90 ELL	45 ELL	TEE	GV	CV		
C-120	6	10	4	20	2	22		
C-150	9.06	15.10	6.04	30.20	3.02	33.22	0.00	0.00

Q	P	P	V	Q	P	P	V	Q	P	P	V
	PSI/FT	PSI/FT	FT/SEC		PSI/FT	PSI/FT	FT/SEC		PSI/FT	PSI/FT	FT/SEC
GPM	C-150			GPM	C-150			GPM	C-150		
100	0.0019		2.3	445	0.0298		10.1	790	0.0860		18.0
110	0.0022		2.5	455	0.0310		10.3	800	0.0881		18.2
120	0.0026		2.7	465	0.0323		10.6	810	0.0901		18.4
130	0.0031		3.0	475	0.0336		10.8	820	0.0922		18.6
140	0.0035		3.2	485	0.0349		11.0	830	0.0943		18.9
150	0.0040		3.4	495	0.0362		11.3	840	0.0964		19.1
160	0.0045		3.6	505	0.0376		11.5	850	0.0985		19.3
170	0.0050		3.9	515	0.0390		11.7	860	0.1007		19.6
180	0.0056		4.1	525	0.0404		11.9	875	0.1039		19.9
190	0.0062		4.3	535	0.0418		12.2	890	0.1073		20.2
200	0.0068		4.5	545	0.0433		12.4	905	0.1106		20.6
210	0.0074		4.8	555	0.0448		12.6	920	0.1141		20.9
220	0.0081		5.0	565	0.0463		12.8	935	0.1175		21.3
230	0.0088		5.2	575	0.0478		13.1	950	0.1210		21.6
240	0.0095		5.5	585	0.0494		13.3	965	0.1246		21.9
250	0.0102		5.7	595	0.0509		13.5	980	0.1282		22.3
260	0.0110		5.9	605	0.0525		13.8	995	0.1318		22.6
270	0.0118		6.1	615	0.0541		14.0	1010	0.1355		23.0
280	0.0126		6.4	625	0.0558		14.2	1025	0.1393		23.3
290	0.0135		6.6	635	0.0574		14.4	1040	0.1431		23.6
300	0.0143		6.8	645	0.0591		14.7	1055	0.1469		24.0
310	0.0152		7.0	655	0.0608		14.9	1070	0.1508		24.3
320	0.0162		7.3	665	0.0626		15.1	1085	0.1548		24.7
330	0.0171		7.5	675	0.0643		15.3	1100	0.1587		25.0
340	0.0181		7.7	685	0.0661		15.6	1115	0.1628		25.4
350	0.0191		8.0	695	0.0679		15.8	1130	0.1668		25.7
360	0.0201		8.2	705	0.0697		16.0	1145	0.1710		26.0
370	0.0211		8.4	715	0.0715		16.3	1160	0.1751		26.4
380	0.0222		8.6	725	0.0734		16.5	1175	0.1793		26.7
390	0.0233		8.9	735	0.0753		16.7	1190	0.1836		27.1
400	0.0244		9.1	745	0.0772		16.9	1205	0.1879		27.4
410	0.0256		9.3	755	0.0791		17.2	1220	0.1922		27.7
420	0.0267		9.5	765	0.0811		17.4	1235	0.1966		28.1
430	0.0279		9.8	775	0.0830		17.6	1250	0.2011		28.4
440	0.0291		10.0	785	0.0850		17.8	1265	0.2056		28.8

Friction Loss Table For 6" PVC Class 150 I.D. = 6.08

Equiv. Feet of Pipe

	LONG EL	90 ELL	45 ELL	TEE	GV	CV				
C–120	9	14	7	30	3	32				
C–150	13.59	21.14	10.57	45.30	4.53	48.32	0.00	0.00		

Q	P	P	V	Q	P	P	V	Q	P	P	V
	PSI/FT	PSI/FT	FT/SEC		PSI/FT	PSI/FT	FT/SEC		PSI/FT	PSI/FT	FT/SEC
GPM	C–150			GPM	C–150			GPM	C–150		
300	0.0025		3.3	990	0.0226		10.9	1680	0.0601		18.6
320	0.0028		3.5	1010	0.0234		11.2	1700	0.0614		18.8
340	0.0031		3.8	1030	0.0243		11.4	1720	0.0627		19.0
360	0.0035		4.0	1050	0.0252		11.6	1740	0.0641		19.2
380	0.0038		4.2	1070	0.0261		11.8	1760	0.0655		19.5
400	0.0042		4.4	1090	0.0270		12.1	1780	0.0668		19.7
420	0.0046		4.6	1110	0.0279		12.3	1800	0.0682		19.9
440	0.0050		4.9	1130	0.0288		12.5	1820	0.0696		20.1
460	0.0055		5.1	1150	0.0298		12.7	1850	0.0718		20.5
480	0.0059		5.3	1170	0.0308		12.9	1880	0.0740		20.8
500	0.0064		5.5	1190	0.0317		13.2	1910	0.0762		21.1
520	0.0069		5.7	1210	0.0327		13.4	1940	0.0784		21.5
540	0.0074		6.0	1230	0.0337		13.6	1970	0.0806		21.8
560	0.0079		6.2	1250	0.0348		13.8	2000	0.0829		22.1
580	0.0084		6.4	1270	0.0358		14.0	2030	0.0852		22.4
600	0.0089		6.6	1290	0.0368		14.3	2060	0.0876		22.8
620	0.0095		6.9	1310	0.0379		14.5	2090	0.0900		23.1
640	0.0101		7.1	1330	0.0390		14.7	2120	0.0924		23.4
660	0.0107		7.3	1350	0.0401		14.9	2150	0.0948		23.8
680	0.0113		7.5	1370	0.0412		15.1	2180	0.0973		24.1
700	0.0119		7.7	1390	0.0423		15.4	2210	0.0997		24.4
720	0.0125		8.0	1410	0.0434		15.6	2240	0.1023		24.8
740	0.0132		8.2	1430	0.0446		15.8	2270	0.1048		25.1
760	0.0138		8.4	1450	0.0457		16.0	2300	0.1074		25.4
780	0.0145		8.6	1470	0.0469		16.3	2330	0.1100		25.8
800	0.0152		8.8	1490	0.0481		16.5	2360	0.1126		26.1
820	0.0159		9.1	1510	0.0493		16.7	2390	0.1153		26.4
840	0.0167		9.3	1530	0.0505		16.9	2420	0.1180		26.8
860	0.0174		9.5	1550	0.0517		17.1	2450	0.1207		27.1
880	0.0182		9.7	1570	0.0530		17.4	2480	0.1235		27.4
900	0.0189		10.0	1590	0.0542		17.6	2510	0.1262		27.8
920	0.0197		10.2	1610	0.0555		17.8	2540	0.1290		28.1
940	0.0205		10.4	1630	0.0568		18.0	2570	0.1319		28.4
960	0.0213		10.6	1650	0.0581		18.2	2600	0.1347		28.7
980	0.0222		10.8	1670	0.0594		18.5	2630	0.1376		29.1

Friction Loss Table For 8" PVC Class 150   I.D. = 7.98

Equiv. Feet of Pipe

	LONG EL	90 ELL	45 ELL	TEE	GV	CV		
C-120	13	18	9	30	4	45		
C-150	19.63	27.18	13.59	45.30	6.04	67.95	0.00	0.00

Q GPM	P PSI/FT C-150	P PSI/FT	V FT/SEC	Q GPM	P PSI/FT C-150	P PSI/FT	V FT/SEC	Q GPM	P PSI/FT C-150	P PSI/FT	V FT/SEC
400	0.0011		2.6	1440	0.0120		9.2	2480	0.0328		15.9
430	0.0013		2.8	1470	0.0125		9.4	2510	0.0336		16.1
460	0.0015		3.0	1500	0.0130		9.6	2540	0.0343		16.3
490	0.0016		3.1	1530	0.0134		9.8	2570	0.0351		16.5
520	0.0018		3.3	1560	0.0139		10.0	2600	0.0358		16.7
550	0.0020		3.5	1590	0.0144		10.2	2630	0.0366		16.9
580	0.0022		3.7	1620	0.0149		10.4	2660	0.0374		17.1
610	0.0025		3.9	1650	0.0155		10.6	2690	0.0382		17.3
640	0.0027		4.1	1680	0.0160		10.8	2730	0.0392		17.5
670	0.0029		4.3	1710	0.0165		11.0	2770	0.0403		17.8
700	0.0032		4.5	1740	0.0170		11.2	2810	0.0414		18.0
730	0.0034		4.7	1770	0.0176		11.4	2850	0.0425		18.3
760	0.0037		4.9	1800	0.0181		11.6	2890	0.0436		18.5
790	0.0040		5.1	1830	0.0187		11.7	2930	0.0447		18.8
820	0.0042		5.3	1860	0.0193		11.9	2970	0.0458		19.1
850	0.0045		5.5	1890	0.0199		12.1	3010	0.0470		19.3
880	0.0048		5.6	1920	0.0205		12.3	3050	0.0481		19.6
910	0.0051		5.8	1950	0.0210		12.5	3090	0.0493		19.8
940	0.0055		6.0	1980	0.0216		12.7	3130	0.0505		20.1
970	0.0058		6.2	2010	0.0223		12.9	3170	0.0517		20.3
1000	0.0061		6.4	2040	0.0229		13.1	3210	0.0529		20.6
1030	0.0065		6.6	2070	0.0235		13.3	3250	0.0542		20.9
1060	0.0068		6.8	2100	0.0241		13.5	3290	0.0554		21.1
1090	0.0072		7.0	2130	0.0248		13.7	3330	0.0566		21.4
1120	0.0075		7.2	2160	0.0254		13.9	3370	0.0579		21.6
1150	0.0079		7.4	2190	0.0261		14.1	3410	0.0592		21.9
1180	0.0083		7.6	2220	0.0268		14.2	3450	0.0605		22.1
1210	0.0087		7.8	2250	0.0274		14.4	3490	0.0618		22.4
1240	0.0091		8.0	2280	0.0281		14.6	3530	0.0631		22.7
1270	0.0095		8.2	2310	0.0288		14.8	3570	0.0644		22.9
1300	0.0099		8.3	2340	0.0295		15.0	3610	0.0658		23.2
1330	0.0104		8.5	2370	0.0302		15.2	3650	0.0671		23.4
1360	0.0108		8.7	2400	0.0309		15.4	3690	0.0685		23.7
1390	0.0113		8.9	2430	0.0316		15.6	3730	0.0699		23.9
1420	0.0117		9.1	2460	0.0323		15.8	3770	0.0713		24.2

Appendix D Friction Loss Tables

Friction Loss Table For 10" PVC Class 150   I.D. = 9.79

Equiv. Feet of Pipe

	LONG EL	90 ELL	45 ELL	TEE	GV	CV		
C-120	16	22	11	50	5	55		
C-150	24.16	33.22	16.61	75.50	7.55	83.05	0.00	0.00

Q GPM	P PSI/FT C-150	P PSI/FT	V FT/SEC	Q GPM	P PSI/FT C-150	P PSI/FT	V FT/SEC	Q GPM	P PSI/FT C-150	P PSI/FT	V FT/SEC
500	0.0006		2.1	2250	0.0101		9.6	4000	0.0294		17.1
550	0.0007		2.3	2300	0.0106		9.8	4050	0.0301		17.3
600	0.0009		2.6	2350	0.0110		10.0	4100	0.0308		17.5
650	0.0010		2.8	2400	0.0114		10.2	4150	0.0315		17.7
700	0.0012		3.0	2450	0.0119		10.4	4200	0.0322		17.9
750	0.0013		3.2	2500	0.0123		10.7	4250	0.0329		18.1
800	0.0015		3.4	2550	0.0128		10.9	4300	0.0336		18.3
850	0.0017		3.6	2600	0.0132		11.1	4350	0.0343		18.6
900	0.0019		3.8	2650	0.0137		11.3	4450	0.0358		19.0
950	0.0021		4.1	2700	0.0142		11.5	4550	0.0373		19.4
1000	0.0023		4.3	2750	0.0147		11.7	4650	0.0388		19.8
1050	0.0025		4.5	2800	0.0152		11.9	4750	0.0404		20.3
1100	0.0027		4.7	2850	0.0157		12.2	4850	0.0420		20.7
1150	0.0029		4.9	2900	0.0162		12.4	4950	0.0436		21.1
1200	0.0032		5.1	2950	0.0167		12.6	5050	0.0452		21.5
1250	0.0034		5.3	3000	0.0173		12.8	5150	0.0469		22.0
1300	0.0037		5.5	3050	0.0178		13.0	5250	0.0486		22.4
1350	0.0039		5.8	3100	0.0183		13.2	5350	0.0503		22.8
1400	0.0042		6.0	3150	0.0189		13.4	5450	0.0521		23.2
1450	0.0045		6.2	3200	0.0194		13.6	5550	0.0539		23.7
1500	0.0048		6.4	3250	0.0200		13.9	5650	0.0557		24.1
1550	0.0051		6.6	3300	0.0206		14.1	5750	0.0575		24.5
1600	0.0054		6.8	3350	0.0212		14.3	5850	0.0594		24.9
1650	0.0057		7.0	3400	0.0218		14.5	5950	0.0613		25.4
1700	0.0060		7.2	3450	0.0223		14.7	6050	0.0632		25.8
1750	0.0064		7.5	3500	0.0230		14.9	6150	0.0651		26.2
1800	0.0067		7.7	3550	0.0236		15.1	6250	0.0671		26.7
1850	0.0071		7.9	3600	0.0242		15.4	6350	0.0691		27.1
1900	0.0074		8.1	3650	0.0248		15.6	6450	0.0711		27.5
1950	0.0078		8.3	3700	0.0254		15.8	6550	0.0732		27.9
2000	0.0082		8.5	3750	0.0261		16.0	6650	0.0752		28.4
2050	0.0085		8.7	3800	0.0267		16.2	6750	0.0774		28.8
2100	0.0089		9.0	3850	0.0274		16.4	6850	0.0795		29.2
2150	0.0093		9.2	3900	0.0280		16.6	6950	0.0816		29.6
2200	0.0097		9.4	3950	0.0287		16.8	7050	0.0838		30.1

Friction Loss Table For 1/2" PVC Class 150   I.D. = 11.65

Equiv. Feet of Pipe

	LONG EL	90 ELL	45 ELL	TEE	GV	CV		
C-120	18	27	13	60	6	65		
C-150	27.18	40.77	19.63	90.60	9.06	98.15	0.00	0.00

Q GPM	P PSI/FT C-150	P PSI/FT	V FT/SEC	Q GPM	P PSI/FT C-150	P PSI/FT	V FT/SEC	Q GPM	P PSI/FT C-150	P PSI/FT	V FT/SEC
500	0.0003		1.5	3950	0.0123		11.9	7400	0.0393		22.3
600	0.0004		1.8	4050	0.0129		12.2	7500	0.0403		22.6
700	0.0005		2.1	4150	0.0135		12.5	7600	0.0413		22.9
800	0.0006		2.4	4250	0.0141		12.8	7700	0.0423		23.2
900	0.0008		2.7	4350	0.0147		13.1	7800	0.0433		23.5
1000	0.0010		3.0	4450	0.0153		13.4	7900	0.0444		23.8
1100	0.0012		3.3	4550	0.0160		13.7	8000	0.0454		24.1
1200	0.0014		3.6	4650	0.0166		14.0	8100	0.0465		24.4
1300	0.0016		3.9	4750	0.0173		14.3	8200	0.0475		24.7
1400	0.0018		4.2	4850	0.0180		14.6	8300	0.0486		25.0
1500	0.0021		4.5	4950	0.0187		14.9	8400	0.0497		25.3
1600	0.0023		4.8	5050	0.0194		15.2	8500	0.0508		25.6
1700	0.0026		5.1	5150	0.0201		15.5	8600	0.0519		25.9
1800	0.0029		5.4	5250	0.0208		15.8	8700	0.0530		26.2
1900	0.0032		5.7	5350	0.0216		16.1	8800	0.0542		26.5
2000	0.0035		6.0	5450	0.0223		16.4	8900	0.0553		26.8
2100	0.0038		6.3	5550	0.0231		16.7	9000	0.0565		27.1
2200	0.0042		6.6	5650	0.0239		17.0	9100	0.0576		27.4
2300	0.0045		6.9	5750	0.0246		17.3	9200	0.0588		27.7
2400	0.0049		7.2	5850	0.0254		17.6	9300	0.0600		28.0
2500	0.0053		7.5	5950	0.0263		17.9	9400	0.0612		28.3
2600	0.0057		7.8	6050	0.0271		18.2	9500	0.0624		28.6
2700	0.0061		8.1	6150	0.0279		18.5	9600	0.0636		28.9
2800	0.0065		8.4	6250	0.0288		18.8	9700	0.0648		29.2
2900	0.0069		8.7	6350	0.0296		19.1	9800	0.0661		29.5
3000	0.0074		9.0	6450	0.0305		19.4	9900	0.0673		29.8
3100	0.0079		9.3	6550	0.0314		19.7	10000	0.0686		30.1
3200	0.0083		9.6	6650	0.0323		20.0	10100	0.0699		30.4
3300	0.0088		9.9	6750	0.0332		20.3	10200	0.0712		30.7
3400	0.0093		10.2	6850	0.0341		20.6	10300	0.0725		31.0
3500	0.0098		10.5	6950	0.0350		20.9	10400	0.0738		31.3
3600	0.0104		10.8	7050	0.0359		21.2	10500	0.0751		31.6
3700	0.0109		11.1	7150	0.0369		21.5	10600	0.0764		31.9
3800	0.0115		11.4	7250	0.0378		21.8	10700	0.0778		32.2
3900	0.0120		11.7	7350	0.0388		22.1	10800	0.0791		32.5

Computerized Hydraulic Program Documentation*

"THE" Sprinkler Demo Program, Version 1.4

SYSTEM REQUIREMENTS

"THE" Sprinkler Demo Program requires the computer disk included in the back cover of this book and:
- An IBM PC, XT, 286, 386, 486, Pentium, or PS-2 computer
- An MCA, CGA, EGA, VGA or Hercules Graphics Adapter and Monitor
- MS-DOS version 3.0 or higher
- 512K random access memory (RAM)
- A hard disk drive (If you don't have a hard disk drive, call FPE Software, 410-586-2128)
- A 3½" floppy disk drive
- A printer
- A math co-processor is recommended, (but not required), since it will reduce calculation time by 200 to 500%

---

*Adapted from information supplied by Kevin Kimmel, P.E. and Mark Lentocha, P. E. of FPE Software, Inc.

FEATURES OF THE PROGRAM

The program is limited to 35 pipes and nodes, sufficient to solve the hydraulics problems assigned in Chapter 6 this book. An expanded program with a maximum of 600 pipes and nodes is available from FPE Software.

The following is a list of keyboard features:

- The bottom of the screen displays information applicable to the current window for function key designations, arrow key usage, and current filename in use.
- ↑, ↓, →, and ← keys control the movement of the cursor within the highlight bar.
- The **Enter** (↵) key selects items or acts as an acknowledgement to accept input data.
- The **PgUp** and **PgDn** keys display the previous and next screen of data respectively.
- The **Ctrl PgUp** and **Ctrl PgDn** key combinations apply to the **Pipe Input Data Window** and **Node Input Data Window**.
- The **Home** and **End** keys move the cursor to the beginning and end of data in the highlight bar, respectively.
- If the user wishes to exit a window, press **Esc** and control will be returned to where the user left off prior to selecting that option.
- When the **Ins** key is pressed, the cursor enlarges and any text existing at the present cursor position shifts to the right as new text is entered. If the highlight bar width is full, the excess data moving to the right is lost. Pressing the **Ins** key once the cursor is enlarged, restores the cursor to normal size and any text existing at the present cursor location is overwritten when new text is input.
- When the **Del** key is pressed, the character at the cursor location will be erased and the text to the right of the cursor moves to the left one space.
- Pressing the **BackSpace** key erases the character at the cursor location and moves the cursor to the left one space.

Figure E-1 shows a keyboard quick reference chart, which covers all function key operations within the Pipe and Node Input Data Windows.

Appendix E Computerized Hydraulic Program Documentation

Pipe Input Data Window	Node Input Data Window

Pipe Input Data Window

- **Ctrl F1** - On-Line Help
- **F1** - Cursor Notes
- **F2** - Edit Menu
    - **F1** - Search
    - **F2** - Block Replace
    - **F3** - Block Insert Row(s)
    - **F4** - Block Delete Row(s)
    - **F5** - Block Erase
    - **F6** - Block Move
    - **F7** - Block Copy
- **F3** - Get End Node
- **F4** - Pipe Status
    - **F1** - Normal: -
    - **F2** - Closed: <>
    - **F3** - Check Valve = ->
    - **F4** - Check Valve = <-
    - **F5** - Fire Pump: PMP (Pump)
    - **F6** - F.D.P.L.D. (Flow Dependent Pressure Loss Device)
- **F5** - Fitting Window
- **F6** - Plot Pump Curve
- **F7** - Plot F.D.P.L.D. Curve
- **F8** - Not Used
- **F9** - Results CRT
- **F10** - Calc. Data
- **Alt F1** - Alt Function Menu
- **Alt C** - Cover Sheet
- **Alt F** - Fitting Data
- **Alt G** - Grid Data Generator
- **Alt I** - Input Data to PRT/File
- **Alt L** - Load Data
- **Alt N** - Node Input Data Window
- **Alt O** - Output Data to PRT/File
- **Alt P** - Pipe Diameter Info
- **Alt R** - Reset ALL Data
- **Alt S** - Save Data
- **Alt T** - Tree Generator Program

Node Input Data Window

- **Ctrl F1** - On-Line Help
- **F1** - Cursor Notes
- **F2** - Edit Menu
    - **F1** - Search
    - **F2** - Block Replace
    - **F3** - Block Insert Row(s)
    - **F4** - Block Delete Row(s)
    - **F5** - Block Erase
    - **F6** - Block Move
    - **F7** - Block Copy
- **F3** - Pipe Nodes
- **F4** - Source Window
- **F5** - Peak Data
- **F6** - Not Used
- **F7** - Not Used
- **F8** - Not Used
- **F9** - Results CRT
- **F10** - Calc. Data
- **Alt F1** - Alt Function Menu
- **Alt C** - Cover Sheet
- **Alt F** - Not Used
- **Alt G** - Grid Data Generator
- **Alt I** - Input Data to PRT/File
- **Alt L** - Load Data
- **Alt N** - Node Input Data Window
- **Alt O** - Output Data to PRT/File
- **Alt P** - Pipe Input Data Window
- **Alt R** - Reset ALL Data
- **Alt S** - Save Data
- **Alt T** - Tree Generator Program

Figure E-1 *Keyboard quick reference. When two keys are listed, such as* **ALT T**, *pressing both keys simultaneously will select the window listed. (Courtesy of FPE Software, Inc.)*

MINIMUM DATA REQUIRED

The program requires a minimum amount of sprinkler system information before it performs a calculation and automatically checks to make sure the user has provided all information before beginning a calculation. If additional information is required, the program notifies the user and places the input cursor at the location of the incomplete data. Minimum data required for a sprinkler system calculation includes:

Node Data
- Hydraulic reference node ID's
- Hydraulic reference node ID elevation
- K-factor/hydrant flow associated with each flowing node ID defined

Pipe Data
- Begin/end node ID's of pipe segment
- Length of pipe segment
- Diameter character ID of pipe segment
- Pipe type character ID of pipe segment
- Fitting description of pipe segment
- Hazen-Williams coefficient of pipe segment

Supply Data
- Supply data defined at a hydraulic reference node ID

Other Data Necessary for Specific Calculation Types
- Fire pump curve data defined at a hydraulic pipe segment
- Sprinkler area coverage at hydraulic reference node ID
- Closed pipe defined at a hydraulic pipe segment
- Check valves defined at a hydraulic pipe segment
- Flow dependent pressure loss device at a hydraulic pipe segment
- Peaking data for gridded system calculations

DEMO SET-UP

- Turn on your computer.
- Insert the disk in drive A or B. From that drive, type: **INSTALL** and press **Enter** (↵).
- An installation program will display several information screens and automatically extract the necessary files.
- When the DOS prompt appears you are ready to begin.

STARTING THE PROGRAM

Three start-up options are available to configure the program units to the calculation performed and the equipment used. The following start-up options can be controlled at the DOS prompt when first starting the program:

To Start Program Type:	Units
THE14	English (gpm-psi)
THE14 (space bar) M1	Metric (lpm-bar)
THE14 (space bar) M2	Metric (lpm-Kpa)

The following example is presented in English units. Sprinkler systems may be calculated in metric by restarting the program from the DOS prompt using the parameters specified above. In order to change an option after start-up, exit the program and restart the program.

INPUT AND CALCULATION PROCEDURE

This appendix provides step-by-step instruction for the input and computer calculation of Figure 6-17, solved in Chapter 6 using the manual hydraulic calculation method. The computer calculation proceeds as follows:

Step 1. Start-up and Input Pipe Information
Step 2. Input Node Information
Step 3. Input Source Information
Step 4. Calculate System
Step 5. Review Results

Pipe information is entered into the program first to permit the reuse of data for node information, eliminating several manual steps. Node ID's are entered for Pipe Begin Node/End Node data and can be automatically retrieved within the Node Data window. Elevation, K-factor and area of coverage data is added manually, with editing shortcuts provided for ease of data entry.

As with the manual hydraulic calculation method used in Chapter 6, the program uses the Hazen-Williams formula to calculate friction loss. Elevation head gain or loss is calculated within the piping network, making it necessary to know the elevation of each hydraulic node from a common elevation datum.

Step 1: Start-Up and Input Pipe Information

- Once started, the program displays the *Main Menu*, shown in Figure E-2.
- Each heading has subchoices from which you choose.
- At the start of the program, the position of the highlight bar is preset to *Files* with a subchoice of *Load*.

		"THE" Sprinkler Program by FPE Software, Inc.		
Files	Input/Edit	Output	Set-up	Quit
Load				
Save				
Rename				
Copy				
Erase				
DOS				

Figure E-2 *"THE" Sprinkler Program Main Menu Display, files submenu. (Courtesy of FPE Software, Inc.)*

- To move to another *Main Menu* heading use the LEFT ARROW (←) or RIGHT ARROW (→) keys on your keyboard.
- As the *Main Menu* heading selection is changed, its corresponding submenu choices are displayed.
- To highlight one of the submenu choices, use the UP ARROW (↑) or DOWN ARROW (↓) keys on your keyboard.
- To select a submenu choice, position the highlight bar over the submenu item and press **Enter**.

The sprinkler system for which we will enter data is depicted in Figure 6-17.

- Press the (→) key once, moving to the Input/Edit *Main Menu* option, highlight the *Pipes* submenu option, and press **Enter** (↵). You are now in the **Pipe Input Data Window**, and your screen should look like Figure E-3.

Enter data using the following keystrokes, beginning in the first row, and entering keystrokes as presented below from left to right. The **F3** function keystroke is used to duplicate the previous end node entry in the current begin node piping row.

Begin Node	End Node	Length	Diam	Type	Fitting Description	C-Value
1 (↵)	2 (↵)	12 (↵)	1 (↵)	(↵)	(↵)	(↵)
(F3)	3 (↵)	12 (↵)	1Q (↵)	(↵)	(↵)	(↵)
(F3)	4 (↵)	12 (↵)	1Q (↵)	(↵)	(↵)	(↵)
(F3)	A T (↵)	18 (↵)	1H (↵)	(↵)	T (↵)	(↵)
(F3)	A B (↵)	1.5 (↵)	2 (↵)	(↵)	T (↵)	(↵)
5 (↵)	6 (↵)	12 (↵)	1 (↵)	(↵)	(↵)	(↵)
(F3)	7 (↵)	12 (↵)	1Q (↵)	(↵)	(↵)	(↵)
(F3)	8 (↵)	12 (↵)	1Q (↵)	(↵)	(↵)	(↵)
(F3)	BT (↵)	18 (↵)	1H (↵)	(↵)	T (↵)	(↵)
(F3)	BB (↵)	1.5 (↵)	2 (↵)	(↵)	T (↵)	(↵)
9 (↵)	10 (↵)	12 (↵)	1 (↵)	(↵)	(↵)	(↵)
(F3)	11 (↵)	12 (↵)	1Q (↵)	(↵)	(↵)	(↵)
(F3)	12 (↵)	12 (↵)	1Q (↵)	(↵)	(↵)	(↵)
(F3)	CT (↵)	18 (↵)	1H (↵)	(↵)	T (↵)	(↵)
(F3)	CB (↵)	1.5 (↵)	2 (↵)	(↵)	T (↵)	(↵)
13 (↵)	DT (↵)	18 (↵)	1H (↵)	(↵)	T (↵)	(↵)
(F3)	DB (↵)	1.5 (↵)	2 (↵)	(↵)	T (↵)	(↵)
AB (↵)	BB (↵)	10 (↵)	3 (↵)	1 (↵)	(↵)	(↵)
(F3)	CB (↵)	10 (↵)	3 (↵)	1 (↵)	(↵)	(↵)
(F3)	DB (↵)	10 (↵)	3 (↵)	1 (↵)	(↵)	(↵)
(F3)	E (↵)	10+10+10+10+5+ (↵)	3 (↵)	1 (↵)	(↵)	(↵)
(F3)	G (↵)	58.5+20+ (↵)	4 (↵)	1 (↵)	GE20 (↵)	(↵)
(F3)	H (↵)	50+5+ (↵)	6 (↵)	DI (↵)	TE (↵)	14 (↵)
(F3)	I (↵)	20 (↵)	10 (↵)	DI (↵)	T (↵)	14 (↵)
(F3)	J (↵)	10 (↵)	6 (↵)	DI (↵)	EG (↵)	14 (↵)

Data Entry for Figure 6-17. The symbol(↵) indicates the **Enter** *Key.*

"THE" Sprinkler Program

Pipe Input Data

Begin Node	End Node	+ " = Length	Diam	Type	Fitting Description	C-Value
	–			40		120
	–			40		120
	–			40		120
	–			40		120
	–			40		120
	–			40		120
	–			40		120
	–			40		120
	–			40		120
	–			40		120
	–			40		120
	–			40		120
	–			40		120
	–			40		120
	–			40		120
	–			40		120
	–			40		120
	–			40		120

Figure E-3 *Pipe Input Data Window (EMPTY). (Courtesy of FPE Software, Inc.)*

- Now is a good time to save your data file. Press **Alt** and **S** keys simultaneously. The "*Save File Name*" window will appear with a path and partial (*.the) filename.
- The file name is chosen by the user and is limited to eight characters, a period, and a "THE" extension. The "THE" extension quickly identifies sprinkler data files for loading. An example file name is: *example.the*.
- Press **Enter** (↵) to accept the file name and save the input data to disk.

Step 2: Input Node Information

- Press the **Alt N** key combination to display the **Node Input Data Window** as shown on Figure E-4.

Appendix E Computerized Hydraulic Program Documentation

"THE" Sprinkler Program				
Node Input Data				
Node	Elevation	K-Factor	Area

Figure E-4 *Node Input Data Window. (Courtesy FPE Software, Inc.)*

- Press **F3** for Node Data retrieval from the data entered within the Pipe Data Window.
- The question "All Current Node Data will be lost, proceed (Y/N)?" will appear. Currently none exists so press **Y** for yes.
- The question "Are nodes to be sorted (Y/N)?" will appear. Press **N** for no. The Node ID's input within the Pipe Data window will be placed within the Node column of the Node Data window, in their order of entry, excluding all duplications.
- Press **Enter** (↵) to move the input window to row 1 of the Elevation column. As an entry shortcut you may enter symbols representing specific data that can be replaced for the exact input value required. An example is sprinkler head elevations. They are equal to 21.5 feet above the warehouse floor in accordance with Figure 6-18. Placing a 1 in the elevation column for each sprinkler node is easier to input than 21.5, saving three keystrokes per sprinkler. The "REPLACE" editing feature can then be used to replace all symbols "1" in the elevation column with 21.5 making the computer do the work and easing the entry keystrokes required. For elevations, the "1" input will represent the sprinkler elevation of 21.5 feet above the finished floor. The "2" input will represent the crossmain elevation of 20 feet above finished floor. Follow the keystroke guidance provided below.

Node Elevations

- Press **1**, then press the ↓ key. Repeat as follows: **1,↓,1,↓,1,↓,1,↓,2,↓**. The input window should now be on row 7. The current row counter is at the bottom left corner of the screen.
- Press **1,↓,1,↓,1,↓,1,↓,1,↓,2,↓**. The input window should now be on row 13.
- Press **1,↓,1,↓,1,↓,1,↓,1,↓,2,↓**. The input window should now be on row 19.
- Press **1,↓,1,↓,2,↓,2,↓**. The input window should now be on row 23.
- Press **0,↓,-,5,↓,-,5,↓,0**. The input window should now be on row 26.
- Press **F2** to display the Edit Window Menu.
- Press **F2** again to select the Block Replace edit option. At the top of the screen, instructions in black lettering on a white background ask you to "Move Cell to Start Location, Press ↵".
- Press **Enter** (↵), which will anchor the edit highlight window to row 26. The top of the screen instructions will now ask you to "Highlight Range to Replace, Press ↵".
- Press **PgUp**,↑,↑,↑,↑,↑,↑ in order to highlight all of the Elevation column up to row 1.
- Press **Enter** (↵) to accept the highlighted edit area. The question "Conditional or All, C/A" will appear.
- Press **C** for conditional replace. Now the question box "Enter Current String" will appear, with the number 1 inside of it. This is the first character or string you wish to replace.
- Press **Enter** (↵) to accept the current string for replacement. Immediately below the Current String Window will appear the question box "Enter New String". This is the sprinkler head elevation in feet discussed earlier.
- Press **21.5** and then **Enter** (↵).

  Each number "1" within the highlighted edit box has now been replaced with 21.5. The edit box is now located at row 1 of the elevation column. Repeat the exact replacement procedure above for the "2", which represents the crossmain elevation of 20 feet above finished floor.

- Press **F2** to display the Edit Window Menu.
- Press **F2** again to select the Block Replace edit option.
- Press **Enter** (↵), which will anchor the edit highlight window to row 1.
- Press **PgDn**,↓,↓, in order to highlight the Elevation column down to row 22.
- Press **Enter** (↵) to accept the highlighted edit area. The question "Conditional or All, C/A" will appear.
- Press **C** for conditional replace. Now the question box "Enter Current String" will appear, with the number "2" inside of it. This is the second and last elevation character or string you wish to replace.

Appendix E Computerized Hydraulic Program Documentation

- Press **Enter** (↵) to accept the current string for replacement. Immediately below the Current String Window will appear the question box "Enter New String". This is the crossmain elevation in feet.
- Press **20** and then **Enter** (↵). Each number "2" within the highlighted edit box has now been replaced with 20. The edit box is now located on row 22 of the elevation column. All elevations have now been input.
- Now is another good time to save your data file to disk. Press **Alt** and **S** keys simultaneously. The "Save File Name" window will appear with the path and filename previously input.
- Press **Enter** (↵) to accept this path and file name. You will be prompted "File already exists. OK to overwrite (Y/N)?".
- Press **Y** for yes, and save the most up-to-date input data to file.

K-Factor Input

Move to the K-factor column for input and follow a similar code input entry procedure as follows:

- Press **PgUp**, ↑, ↑ and **Enter** (↵) to move the input box to row 1 of the K-factor column. The desired K-factor is 5.6 for all sprinklers within the remote area, thus an entry symbol or code of "1" can be used to quickly represent the K-factor value.
- Press **1**,↓,**1**,↓,**1**,↓,**1**,↓,↓,↓. The input window should now be on row 7.
- Press **1**,↓,**1**,↓,**1**,↓,**1**,↓,↓,↓. The input window should now be on row 13.
- Press **1**,↓,**1**,↓,**1**,↓,**1**,↓,↓,↓. The input window should now be on row 19.
- Press **1**. The input window should still be on row 19.
- Press **F2** to display the Edit Window Menu.
- Press **F2** again to select the Block Replace edit option. At the top of the screen, instructions in black lettering on a white background ask you to "Move Cell to Start Location, Press ↵".
- Press **Enter** (↵), which will anchor the edit highlight window to row 19. The top of the screen instructions will now ask you to "Highlight Range to Replace, Press ↵".
- Press ↑ arrow key as many times necessary in order to highlight all of the K-factor column up to row 1.
- Press **Enter** (↵) to accept the highlighted edit area. The question "Conditional or All, C/A" will appear.
- Press **C** for conditional replace. Now the question box "Enter Current String" will appear, with the number 1 inside of it.
- Press **Enter** (↵) to accept the current string for replacement. Immediately below the Current String Window will appear the question box "Enter New String". This is the sprinkler head K-factor.

- Press **5.6** and then **Enter** (↵).

  Each number "1" within the highlighted edit box has now been replaced with 5.6. The input window is now located at row 1 of the K-factor column.

- Now is another good time to save your data file to disk. It is a recommended practice to save input data frequently since temporary loss of power is beyond the control of any program. Press **Alt** and **S** keys simultaneously. The "Save File Name" window will appear with the path and filename previously input.

- Press **Enter** (↵) to accept this path and file name. You will be prompted "File already exists. OK to overwrite (Y/N)?".

- Press **Y** for yes, and save the most updated input data to file.

Area of Coverage

- Press **Enter** (↵) to move the input box to row 1 of the Area of Coverage column. The area of coverage for all sprinklers within the remote area is equal to 10 times 12 or 120 sq. ft. An entry symbol or code of "1" can be used to represent this common area of coverage value.

- Press **1**,↓,**1**,↓,**1**,↓,**1**,↓,↓,↓. The input window should now be on row 7.

- Press **1**,↓,**1**,↓,**1**,↓,**1**,↓,↓,↓. The input window should now be on row 13.

- Press **1**,↓,**1**,↓,**1**,↓,**1**,↓,↓,↓. The input window should now be on row 19.

- Press **1**. The input window should still be on row 19.

- Press **F2** to display the Edit Window Menu.

- Press **F2** again to select the Block Replace edit option. At the top of the screen, instructions in black lettering on a white background ask you to "Move Cell to Start Location, Press ↵".

- Press **Enter** (↵), which will anchor the edit highlight window to row 19. The top of the screen, instructions will now ask you to "Highlight Range to Replace, Press ↵".

- Press ↑ arrow key as many times necessary in order to highlight all of the K-factor column up to row 1.

- Press **Enter** (↵) to accept the highlighted edit area. The question "Conditional or All, C/A" will appear.

- Press **C** for conditional replace. Now the question box "Enter Current String" will appear, with the number 1 inside of it.

- Press **Enter** (↵) to accept the current string for replacement. Immediately below the Current String Window will appear the question box "Enter New String". This is the sprinkler head area of coverage.

- Press **120** and then **Enter** (↵). Each number "1" within the highlighted edit box has now been replaced with 120. The input window is now located at row 1 of the area of coverage column.

Appendix E Computerized Hydraulic Program Documentation

Step 3: Input Source Information

To input flow test information:

- Press the **PgDn** key once and the ↓ key until the input cursor is on row 26 representing node ID "J". This is the location where the water supply data provided in Figure 6-17 will be input. The static pressure is 60 psi and the residual pressure is 35 psi with 700 gpm flowing.
- Press the **F4** key to display the Source Window (Figure E-5). The input cell will be adjacent to the Static pressure label.
- Input **60** and press **Enter** (↵). The input cell will be adjacent to the Residual pressure label.
- Input **35** and press **Enter** (↵). The input cell will be adjacent to the Residual flow label.
- Input **700** and press **Enter** (↵). The input cell will be adjacent to the Hose Allowance label.
- Input **250** and press **Enter** (↵).
- Press **F10** to accept the source data input and return to the **Node Input Data Window**. Pressing the **Alt P** key combination will redisplay the **Pipe Input Data Window**.
- Now is another good time to save your data file to disk. Press **Alt** and **S** keys simultaneously. The "Save File Name" window will appear with the path and filename previously input.
- Press **Enter** (↵) to accept this path and file name. You will be prompted "File already exists. OK to overwrite (Y/N)?".
- Press **Y** for yes, and save the most updated input data.

Step 4: Calculate System

We are now ready to begin our hydraulic calculation of the Figure 6-17 sprinkler system.

Source Window		
Node: J		
Static:	psi	
Residual:	psi @	gpm
Hose Allowance:	gpm	
F10 - Accept/Exit	ESC - Abort	

Figure E-5 *Source Window. (Courtesy of FPE Software, Inc.)*

Demand (Minimum Density) or Supply (Maximum Density) Calc. (D/S)?
ESC - Abort

Figure E-6 *Calculation Query Window.*

- Press the **F10** key. The program will ask you to select a sprinkler DEMAND calculation or sprinkler SUPPLY calculation as shown in Figure E-6. We will perform a DEMAND calculation to determine the minimum required flow and pressure to adequately feed the sprinkler system created.
- Press **D**, to perform a DEMAND sprinkler calculation on the system. The status of all Pipe and Node data will be checked. If a problem exists, an error message will be displayed and the cursor will be positioned at the row and column with incorrect data. The system data input should calculate without problems. If an error exists, read the error screen provided and check your data against the information provided. Once the problem has been corrected, press the **F10** key, followed by **D** for DEMAND calculation to return to the current point of calculation.
- Once a DEMAND calculation has been chosen, a **Demand Window** (Figure E-7) will appear. Questions must be answered relative to the DEMAND calculation to be performed. The first is the minimum design density (gpm/sq. ft.), specified as 0.15 gpm/sq. ft. in Figure 6-17.
- Press **0.15** and **Enter** (↵), to confirm the design density. The input cell is now directly adjacent to "Remote Friction Node:". Space is provided for the user to specify a specific Node ID of the hydraulically most remote sprinkler. If left blank, the program will automatically find the most remote flowing node and perform its calculations accordingly. For this example, this field will be left blank and the most remote flowing node ID will be determined.
- Press **Enter** (↵). The input cell will query the user for the minimum operating pressure (psi) or flow (gpm) at the remote flowing node ID. We will

Demand Window	
Design Density:	gpm/sq.ft.
Remote Friction Node:	
Minimum Flow or Pressure (F/P)?	
F10 - Calc ESC - Abort	

Figure E-7 *Demand Window. (Courtesy of FPE Software, Inc.)*

input a minimum flow of 18 gallons per minute (120 sq. ft. sprinkler coverage times 0.15 gpm/sq. ft. design density).
- Press the **F** key, and the **Minimum Flow Window** will appear.
- Input **18**, and press **Enter** (⏎). This represents 18.0 gpm minimum gallonage required at the most remote sprinkler.
- Pressing the **F10** key will allow the calculation to proceed, while the **Esc** key cancels the calculation procedure and returns to the **Pipe Input Data Window** or **Node Input Data Window**.
- Press **F10** to begin the calculation. Once the calculation routine has begun, a **Trial/Continuity Window** will be posted in the middle of the screen. Demand calculation continuity imbalance is set at a maximum of 0.01 gpm and the calculation will proceed until the maximum continuity imbalance shown within the window is below 0.01 gpm. The system will take approximately 34 trials until meeting the 0.01 gpm maximum continuity imbalance criteria. These 34 trials, on a 33Mhz 486DX with math co-processor will take about half a second. A computer with no co-processor will take about 10 to 15 seconds depending on the speed of the computer's hard disk.

Step 5: Review the Results

Figure E-8 displays the **Source Summary Window**, the first output window to be displayed.

	"THE" Sprinkler Program			
	Source Summary			
NODE	REQUIRED (gpm)	VS. (psi)	AVAILABLE (psi)	SAFETY FACTOR (psi)
J	515.21	37.92	45.82	7.90
12	Maximum velocity is to CT	12.61 ft/sec in Row 14, pipe segment and having a diameter of 1H in.		
	Continue with results (Y/N)?			

Figure E-8 *Source Summary Window. Courtesy of FPE Software, Inc.*

- The **Source Summary Window** provides the requirements at the source of system supply. If the system requires revision, press **N** to the question "Continue with results (Y/N)?", and you will be returned to the **Pipe Input Data Window** for system editing and recalculation.
- Press **Y** to continue displaying calculation results on the screen. Figure E-9 displays the **Node Output Data Window** results for the Figure 6-17 sprinkler system calculation.
- From this window, output can be reviewed but not changed. The **PgUp**, **PgDn**, ↑, ↓, and **Ctrl PgDn** keys can be used to move through the output information easily.
- Press the **Alt P** key combination to review the **Pipe Output Data Window** (Figure E-10).

"THE" Sprinkler Program

Node Output Data

Node	Elevation	K Factor	Pressure	Actual Flow	Min. Flow	Excess
1	21.5	5.6	10.33	18.00	18.00	0.00
2	21.5	5.6	11.62	19.09	18.00	1.09
3	21.5	5.6	12.90	20.11	18.00	2.11
4	21.5	5.6	15.77	22.24	18.00	4.24
AT	21.5		21.16			
AB	20		22.52			
5	21.5	5.6	10.36	18.03	18.00	0.03
6	21.5	5.6	11.65	19.12	18.00	1.12
7	21.5	5.6	12.94	20.15	18.00	2.15
8	21.5	5.6	15.82	22.27	18.00	4.27
BT	21.5		21.23			
BB	20		22.58			
9	21.5	5.6	10.48	18.13	18.00	0.13
10	21.5	5.6	11.79	19.22	18.00	1.22
11	21.5	5.6	13.09	20.26	18.00	2.26
12	21.5	5.6	16.00	22.40	18.00	4.40
CT	21.5		21.46			
CB	20		22.82			

Figure E-9 *Node Output Data Window. (Courtesy of FPE Software, Inc.)*

"THE" Sprinkler Program

Pipe Output Data

Begin Node		End Node	Flow	Friction	Actual Diam.	Total Length	Velocity
1	<	2	18.00	1.28	1.049	12.00	6.68
2	<	3	37.08	1.29	1.380	12.00	7.95
3	<	4	57.20	2.87	1.380	12.00	12.27
4	<	AT	79.44	5.39	1.610	26.00	12.52
AT	<	AB	79.44	0.71	2.067	11.50	7.60
5	<	6	18.03	1.29	1.049	12.00	6.69
6	<	7	37.14	1.29	1.380	12.00	7.97
7	<	8	57.29	2.88	1.380	12.00	12.29
8	<	BT	79.56	5.40	1.610	26.00	12.54
BT	<	BB	79.56	0.71	2.067	11.50	7.61
9	<	10	18.13	1.30	1.049	12.00	6.73
10	<	11	37.36	1.30	1.380	12.00	8.01
11	<	12	57.62	2.91	1.380	12.00	12.36
12	<	CT	80.01	5.46	1.610	26.00	12.61
CT	<	CB	80.01	0.72	2.067	11.50	7.65
13	<	DT	26.20	0.69	1.610	26.00	4.13
DT	<	DB	26.20	0.09	2.067	11.50	2.51
AB	<	BB	79.44	0.07	3.260	10.00	3.05

Figure E-10 *Pipe Output Data Window. Courtesy of FPE Software, Inc.*

The **Alt N** key combination can be used to return to the **Node Output Data Window**.

- Press the (**Esc**) key once to view the **Node Input Data Window**.
- To obtain a hardcopy printout of the calculated results, press (**Alt**) **O**, to initiate the Output Forms Window.
- Press ↓,↓,↓,↓,↓ to move the input cell adjacent to the "NFPA 13" row and the "Ptr" column.
- Press **X**, then (**Ctrl**) **Enter** (↵) simultaneously. Program defaults are set to LPT1, but these can be changed in the Main Menu Setup Option.
- Press **Y** to the "OK to output[Y/N]?" query on the bottom line of the screen. The output will be sent to the printer.

- Press **Esc** to return to the **Node Input Data Window**. All output forms presented conform to the requirements of NFPA 13. The columns "Dsk" and "Scr" send whatever output form is marked to diskette and the video screen, respectively. Please review these output forms and options at your leisure: They are all fully operational within the disk provided with this book.

The results of the computer calculation output is shown in Figure E-11. A Supply vs. Demand curve is now reviewed on-screen, to pictorialize the 7.9 psi safety margin expressed on the hydraulic summary window. Note the similarity in results to the manual hydraulic calculation performed in Chapter 6.

- Press the **Esc** key once, returning to the Main Menu bar.
- Press the → key to move to the Output menu item.
- Press **S** to select the Supply Graph Output option.
- Press **F6** key to select the Send graph to Screen Only option. The Graph Defaults window will now be displayed. Placing a value greater than 0 in the axis maximum defaults, forces the graphing routine to use these values as the axis maximum. The Label Points "Y or N" determines whether the demand and hose requirements are labeled on the displayed graph.
- Press **Enter** (↵), **Enter** (↵), **Enter** (↵) to retain default values and display the graph on screen (Figure E-12).
- Press any key to return to the Output Main Menu bar.
- For submittal purposes a cover sheet is available. Press **C** to activate the Cover Sheet submenu option. Follow the on-screen menu options to fill out the sections of the cover sheet that apply. Print the completed cover sheet.

The demonstration calculation of Figure 6-17 is now complete. Return to the main menu by pressing the **Esc** key. Press the → key to move to the Quit main menu option. Press **Enter** (↵) and the prompt "Data has been modified. Save Changes (Y/N)?" will appear. Press **Y** and **Enter** (↵) if you desire to retain the editing you have done to the Figure 6-17 data file. Pressing **N** will terminate the program without saving the data file and return to the DOS prompt.

Feel free to explore the program disk using the Keyboard Quick Reference chart, Figure E-1 as reference. This disk is intended for educational purposes only and shall not be used for the performance of calculations in a professional capacity.

Appendix E Computerized Hydraulic Program Documentation

"THE" Sprinkler Program by FPE Software, Inc.

File: | | | | | | | | | | Page: 1

Reference	Nozzle Type & Location	Flow in gpm	Pipe Size in.	Fitting & Devices	Pipe Eqiv. Length	Friction Loss psi/ft		Req. psi	Notes		
1	5.6	q	18.00	1.049		lgth	12.00		Pt	10.33
to		Q	18.00	120		ftg	0.00	0.107	Pf	1.28
2						tot	12.00		Pe	0.00
2	5.6	q	19.09	1.380		lgth	12.00		Pt	11.62
to		Q	37.08	120		ftg	0.00	0.107	Pf	1.29
3						tot	12.00		Pe	0.00
3	5.6	q	20.11	1.380		lgth	12.00		Pt	12.90
to		Q	57.20	120		ftg	0.00	0.239	Pf	2.87
4						tot	12.00		Pe	0.00
4	5.6	q	22.24	1.610		lgth	18.00		Pt	15.77
to		Q	79.44	120	T	ftg	8.00	0.207	Pf	5.39
AT						tot	26.00		Pe	0.00
AT		q	0.00	2.067		lgth	1.50		Pt	21.16
to		Q	79.44	120	T	ftg	10.00	0.061	Pf	0.71
AB						tot	11.50		Pe	0.65
5	5.6	q	18.03	1.049		lgth	12.00		Pt	10.36
to		Q	18.03	120		ftg	0.00	0.107	Pf	1.29
6						tot	12.00		Pe	0.00
6	5.6	q	19.12	1.380		lgth	12.00		Pt	11.65
to		Q	37.14	120		ftg	0.00	0.108	Pf	1.29
7						tot	12.00		Pe	0.00
7	5.6	q	20.15	1.380		lgth	12.00		Pt	12.94
to		Q	57.29	120		ftg	0.00	0.240	Pf	2.88
8						tot	12.00		Pe	0.00
8	5.6	q	22.27	1.610		lgth	18.00		Pt	15.82
to		Q	79.56	120	T	ftg	8.00	0.208	Pf	5.40
BT						tot	26.00		Pe	0.00
BT		q	0.00	2.067		lgth	1.50		Pt	21.23
to		Q	79.56	120	T	ftg	10.00	0.062	Pf	0.71
BB						tot	11.50		Pe	0.65
9	5.6	q	18.13	1.049		lgth	12.00		Pt	10.48
to		Q	18.13	120		ftg	0.00	0.109	Pf	1.30
10						tot	12.00		Pe	0.00
10	5.6	q	19.22	1.380		lgth	12.00		Pt	11.79
to		Q	37.36	120		ftg	0.00	0.109	Pf	1.30
11						tot	12.00		Pe	0.00
11	5.6	q	20.26	1.380		lgth	12.00		Pt	13.09
to		Q	57.62	120		ftg	0.00	0.242	Pf	2.91
12						tot	12.00		Pe	0.00

Figure E-11 *Computer calculation file output for Figure 6-17. (Courtesy of FPE Software, Inc.)*

"THE" Sprinkler Program by FPE Software, Inc.

File: Page: 2

Reference	Nozzle Type & Location	Flow in gpm	Pipe Size in.	Fitting & Devices	Pipe Eqiv. Length	Friction Loss psi/ft	Req. psi	Notes		
12 to CT	5.6	q Q	22.40 80.01	1.610 120	T	lgth ftg tot	18.00 8.00 26.00	0.210	Pt Pf Pe	16.00 5.46 0.00
CT to CB		q Q	0.00 80.01	2.067 120	T	lgth ftg tot	1.50 10.00 11.50	0.062	Pt Pf Pe	21.46 0.72 0.65
13 to DT	5.6	q Q	26.20 26.20	1.610 120	T	lgth ftg tot	18.00 8.00 26.00	0.027	Pt Pf Pe	21.90 0.69 0.00
DT to DB		q Q	0.00 26.20	2.067 120	T	lgth ftg tot	1.50 10.00 11.50	0.008	Pt Pf Pe	22.60 0.09 0.65
AB to BB		q Q	0.00 79.44	3.260 120		lgth ftg tot	10.00 0.00 10.00	0.007	Pt Pf Pe	22.52 0.07 0.00
BB to CB		q Q	0.00 159.00	3.260 120		lgth ftg tot	10.00 0.00 10.00	0.024	Pt Pf Pe	22.58 0.24 0.00
CB to DB		q Q	0.00 239.01	3.260 120		lgth ftg tot	10.00 0.00 10.00	0.051	Pt Pf Pe	22.82 0.51 0.00
DB to E		q Q	0.00 265.21	3.260 120	T	lgth ftg tot	45.00 15.00 60.00	0.062	Pt Pf Pe	23.34 3.72 0.00
E to G		q Q	0.00 265.21	4.260 120	GE20	lgth ftg tot	78.50 32.00 110.50	0.017	Pt Pf Pe	27.06 1.86 8.67
G to H		q Q	0.00 265.21	6.275 140	TE	lgth ftg tot	55 69.02 124.02	0.002	Pt Pf Pe	37.59 0.24 2.17
H to I		q Q	0.00 265.21	10.400 140	T	lgth ftg tot	20 79.66 99.66	0.000	Pt Pf Pe	40.00 0.02 0.00
I to J	SOURCE	q Q	0.00 265.21	6.275 140	EG	lgth ftg tot	10 26.67 36.67	0.002	Pt Pf Pe	40.01 0.07 −2.17

Figure E-11 *(Cont.)*

Appendix E Computerized Hydraulic Program Documentation

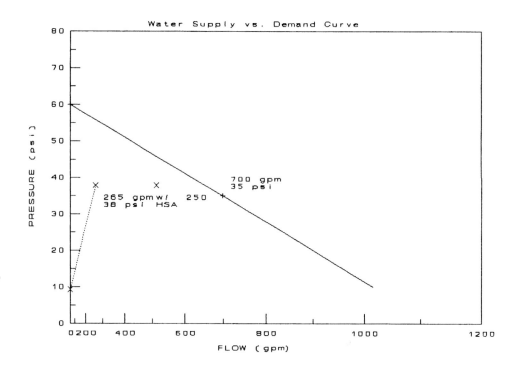

Figure E-12 *Supply graph output, (Courtesy FPE Software, Inc.)*

Appendix F

Selected NFPA Standards

The following standards published by the National Fire Protection Association (Quincy, Massachusetts) contain information pertinent to the design of water-based fire protection systems.

NFPA 13, *Standard for the Installation of Sprinkler Systems.*

NFPA 13D, *Standard on the Installation of Sprinkler Systems in One and Two Family Dwellings and Manufactured Homes.*

NFPA 13R, *Standard on the Installation of Sprinkler Systems in Residential Occupancies up to and Including Four Stories in Height.*

NFPA 14, *Standard for the Installation of Standpipe and Hose Systems.*

NFPA 15, *Standard for Water Spray Fixed Systems for Fire Protection.*

NFPA 20, *Standard for the Installation of Centrifugal Fire Pumps.*

NFPA 22, *Standard for the Installation of Water Tanks for Private Fire Protection.*

NFPA 24, *Standard for Private Service Mains and Their Appurtenances.*

NFPA 61A, *Standard for the Prevention of Fire and Dust Explosions in Facilities Manufacturing and Handling Starch.*

NFPA 61B, *Standard for the Prevention of Fires and Explosions in Grain Elevators and Facilities Handling Bulk Raw Agricultural Commodities.*

NFPA 61C, *Standard for the Prevention of Fire and Dust Explosions in Feed Mills.*

NFPA 61D, *Standard for the Prevention of Fire and Dust Explosions in the Milling of Agricultural Commodities for Human Consumption.*

NFPA 68, *Guide for Venting of Deflagrations.*

NFPA 69, *Standard on Explosions Prevention Systems.*

NFPA 70, *National Electrical Code.*

NFPA 72, *National Fire Alarm Code.*

NFPA 77, *Recommended Practice on Static Electricity.*

NFPA 170, *Standard on Fire Safety Symbols.*

NFPA 214, *Standard on Water Cooling Towers.*

NFPA 231, *Standard on General Storage.*

NFPA 231C, *Standard on Rack Storage of Materials.*

NFPA 231D, *Standard for Storage of Rubber Tires.*

NFPA 231F, *Standard for Storage of Rolled Paper.*

NFPA 321, *Standard on Basic Classification of Flammable and Combustible Liquids.*

NFPA 325M, *Fire Hazard Properties of Flammable Liquids, Gases, and Volatile Solids.*

NFPA 495, *Explosive Materials Code.*

NFPA 750, *Standard for the Installation of Water Mist Fire Protection Systems.*

Acronyms

AFSA	American Fire Sprinkler Association	NFSA	National Fire Sprinkler Association
AHJ	Authority having jurisdiction	NICET	National Institute for Certification in Engineering Technologies
ANSI	American National Standards Institute		
ASTM	American Society for the Testing of Materials	NIST	National Institute of Standards and Technology
BLEVE	Boiling liquid-expanding vapor explosion	NSPE	National Society of Professional Engineers
CAD	Computer-aided design	OS&Y	Outside screw and yoke valves
CPVC	Chlorinated polyvinyl chloride	PVC	Polyvinyl chloride
ESFR	Early suppression fast response	RTI	Response time index
FAA	Federal Aviation Administration	SFPE	Society of Fire Protection Engineers
HVAC	Heating, ventilating, and air conditioning	SSP	Standard spray pendent
		SSU	Standard spray upright
NFPA	National Fire Protection Association	UL	Underwriters Laboratories

Glossary

Antifreeze system A sprinkler system where an additive is mixed with water to prevent freezing.

Appendix Portion of the standard containing explanatory material and generally using the word "should."

Approved Meets the minimum requirements of the authority having jurisdiction.

Architectural drawings Drawings that show dimensions of walls, floors, ceilings, and other building features.

Area protection The application of high-speed water spray over the surface area of a room or over the surface area of an object.

Authority having jurisdiction (AHJ) The individual or agency placed in responsible charge of reviewing and approving drawings and completed installations.

Autoignition Ignition and sustained combustion that could occur in cases where the fuel could ignite without an outside ignition source.

Backfill The dirt or other material used to cover an underground pipe after installation.

Backflow preventer A device designed to prevent sprinkler system water from being drawn backward into a public water supply.

Bar joist A webbed member supported by I-beams.

Bar joist bearing dimension The depth of the top of a joist.

Bay Space between the beams along the column lines in exposed construction.

Bearing area The area of the vertical plane of the thrust block onto which the fitting must rest.

BLEVE A boiling liquid-expanding vapor explosion that may occur when a portion of a flammable liquid storage tank is unwetted by water spray, causing that portion of the tank surface to rupture.

Body of the standard The portion of an NFPA standard containing mandatory requirements for sprinkler system design and installation.

Boring table Compilation of data obtained by drilling cores of earth at several strategic locations.

Branch line K-factor A K-factor that represents the flow of all flowing sprinklers on a branch line in the design area.

Branch lines Sprinkler pipes that have sprinklers installed directly on them.

Btu (British thermal unit) A unit of heat energy needed to raise the temperature of 1 pound of water 1°F at constant pressure of 1 atmosphere.

Building footings Underground structural elements that support building walls and floor slabs.

Building sections Cutaway views through a building.

Butterfly valve Listed indicating control valve that has a dial showing whether the valve is open or closed.

Cathodic protection An electronic method for combating pipe corrosion.

Cavitation Destructive vibrations caused by the churning of a fire pump when it is pumping air instead of water.

Centrifugal fire pump A pump that develops pressure by forcing water to rotate around a shaft, creating centrifugal pressure that boosts the water supply.

c-factor The Hazen-Williams roughness coefficient.

City water supply A network of underground pipes pressurized and maintained by a public water authority, usually supplied by a reservoir or aboveground water storage tank.

Civil drawings Scaled drawings that coordinate underground utilities entering and leaving a building or group of buildings.

Civil section Portion of contract drawings that contains site and underground utility drawings, primarily originated by civil engineers and landscape architects.

Class A fires Fires involving ordinary combustibles.

Class B fires Fires involving flammable liquids.

Class C fires Fires involving electrical hazards.

Client's value system A system directing a client to demand a code-compliant fire protection system for the least possible price.

Coating The condition that occurs when water floats and provides a barrier between the surface of a flammable liquid and the combustible vapor above the surface.

Code of Ethics for Engineers A compilation of rules and obligations published by the National Society of Professional Engineers used by fire protection engineers to resolve ethical conflicts in fire protection system design.

Codes Mandatory requirements suitable for adoption into law.

Commodity The item being stored plus its packaging and the pallet upon which it rests.

Company manual A written set of company regulations that requires your signature.

Company values Values that require a fire protection professional to behave in a way that would meet the goals of the company and bring credit to the company.

Computer-aided design (CAD) A computerized method of preparing a drawing.

Contour lines Lines indicating the elevation of the finished exterior grade.

Contract drawings A set of plans that describe a project in pictorial form.

Coordination meetings Meetings where the exact locations of mechanical equipment are determined before installation.

Cross bracing Supports that provide structural rigidity between bar joists.

Cut-in Procedure involving cutting into the existing underground main, removing a segment of existing pipe, installing a new underground tee, and reconnecting the completed assembly.

Deflagration A fire in which the expanding flame front travels at a velocity less than the speed of sound.

Deflector Metal plate that deflects and distributes water from the sprinkler.

Deflector distance The dimension from the face of a ceiling to the sprinkler deflector.

Deluge sprinkler system A system that uses open sprinklers or nozzles so that all flow water is discharged when the deluge valve actuates.

Deluge water mist systems Water mist systems with open nozzles that discharge water mist simultaneously from all nozzles on the system.

Depth of cover Lineal distance from the top of an underground pipe to the finished grade.

Design area An area whose size is related to the occupancy, wherein all sprinklers in the area are expected to actuate.

Design density The quantity of water per square foot of design area that has been found by experience to be effective in controlling a fire.

Detailed specification In-depth requirements for the design of a fire protection system, such that very little latitude is available for interpretation or alternative design proposals.

Detection releasing system A system that supervises the area being protected and is capable of actuating a fire protection system when a fire is detected.

Detonation A fire in which the expanding flame front travels at a velocity greater than the speed of sound.

Dilution Reducing the concentration of a water-soluble flammable liquid by mixing it with water.

Direct impingement Direct application of water spray to all exposed surfaces.

Domestic plumbing piping Piping that connects water service to fountains, sinks, and toilets.

Domestic water supply The water supply for toilets and sinks.

Draft curtains Barriers suspended from the roof of a structure to limit the spread of smoke.

Drawing Graphical representation of a designer's ideas.

Drop Vertical pipes that drop from the branch lines to sprinklers mounted in the ceiling.

Dry pipe sprinkler system A system where piping is pressurized with air, and with heat-responsive elements at all sprinklers.

Dry pipe water mist systems Water mist systems with air-filled piping in the ambient condition and nozzles with individual heat-responsive actuating devices.

Electric detection system A system employing detectors that send electronic signals to a control panel when a fire is detected.

Empirical evidence The use of field experience or history as the basis for a design standard or criteria.

Encapsulation The wrapping of a commodity in plastic on the sides and top in order for the commodity to remain rigid while being transported by forklift trucks.

Equivalent length A length of pipe of a given diameter whose friction loss is equivalent to the friction loss of a pipe of differing diameter.

Equivalent fitting length A length of straight pipe that has the same friction loss as a fitting where the water changes direction.

ESFR (early suppression fast response) sprinkler A sprinkler with a quick response element and an extra large orifice that allows for larger amounts of water to be applied early in the fire's development.

Ethical dilemma A situation where pressure is placed upon an individual to do something that is not ethically proper, or where it is not immediately obvious which option would be ethically proper.

Explosion The sudden overpressurization of a vessel or enclosure beyond its maximum pressure potential.

Explosion suppression systems Systems designed in accordance with NFPA 69 for enclosed vessels.

Extended coverage sprinkler A sprinkler designed to discharge water over larger areas than standard sprinklers.

Field check A thorough survey of a proposed fire protection system, using a completed fire protection system design as a basis.

Finish schedule A contract drawing that lists all rooms in a building and provides details of several room features.

Fire control Application of water to limit the size of a fire and allow safe occupant egress and structural protection until the fire service arrives.

Fire hydrant A device that provides a water supply to fire department pumpers for use in combating structure fires.

Fire load density A function of the quantity, arrangement, combustibility, and rate of heat release of the combustibles in a space.

Fire protection engineer A person who is qualified by experience and education to perform fire protection system design.

Fire protection system design Design of a fire protection system based on engineering criteria that may not always coincide with criteria found in an accepted national standard.

Fire protection system layout The act of following the requirements of a standard to execute a drawing.

Fire protection technicians Persons with the knowledge, skill, and training to perform fire protection layout, using accepted national standards.

Fire pump A specially designed and listed pump that increases the pressure of the water serving a fire protection system.

Fire service response time The time it takes for the fire department to arrive at the fire scene after notification.

Firestop systems Sealants for fire protection piping penetrations through walls or other building components.

Fire suppression Application of water such that the release of heat from a fire is sharply reduced and the regrowth of the fire is prevented.

Fire tetrahedron A concept that considers the inhibition of chain reactions in addition to the three points of the fire triangle.

Fire triangle A concept identifying the interdependent entities of oxygen, heat, and fuel as they relate to sustaining combustion.

Flash point The temperature at which a flammable liquid ignites vapor from the fuel for a brief period.

Flow hydrant A hydrant selected to measure the water flow available from the water supply.

Flush mounted lights Lights that are mounted flush to and below a ceiling.

Foundation plans Plans that show floor and wall structural details and sectional views.

Framing plans Plans that show beam and joist location, size, and elevation.

Friction loss The frictional resistance of the interior walls of a pipe to the flow of water.

Friction loss tables Compilations of data representing the results of numerous friction loss calculations.

Geographically most remote area The area whose linear distance from the sprinkler system riser is the longest.

Gravity tank An elevated water tank that utilizes the force of gravity to provide pressure to a water supply.

Grid system System of branch lines interconnected by crossmains on both ends of the branch lines.

Guides Informative but nonbinding NFPA documents.

Hazen-Williams formula Standard formula for calculating friction loss in pipes.

Head The pressure of water described in terms of feet of water column instead of psi.

Heat detector A device that sends a signal to a control panel when it detects a predetermined temperature level.

High-pressure water mist systems Systems where the pressures encountered by the system piping are 500 psi or greater.

HVAC drawings Heating, ventilating, and air conditioning drawings.

Hydrant cap with bourdon gauge A device mounted to a fire hydrant to measure the water pressure.

Hydrant coefficient A factor that describes the nature of the opening from which water flows from a fire hydrant.

Hydraulically most demanding area A design area protected by a sprinkler system that demands more water for fire suppression than any other design area on the system.

Hydrokinetics The study of water (hydro) in motion (kinetic).

Hydrostatics The study of water (hydro) at rest (static).

I-beam A solid steel structural member that looks like an "I" from its end.

In-rack sprinklers Sprinklers with water shields to prevent the heat responsive elements from being wetted by the roof sprinklers.

Inspector's test connection Test valve used to simulate the flow of the hydraulically most demanding sprinkler.

Intermediate-pressure water mist systems Systems where the pressures encountered by the system piping are between 175 psi and 500 psi.

Interpolating Selecting a friction loss value between two values that appear on a friction loss table.

Invert elevation An elevation referencing the bottom of an underground pipe with respect to the reference elevation.

Jockey pump A very small pump that boosts the static pressure within the piping system when the fire pump is not running.

K-factor A unique orifice coefficient that is calculated for a specific sprinkler.

Large drop sprinklers Sprinklers with deflectors that create large water drops intended to penetrate a fire plume without evaporating.

Legal/ethical conflicts Conflicts related to the congruence between our system of laws and our system of ethical values.

Limited area dwellings Dwellings that do not exceed 2,000 square feet in total area.

Linear heat detection system Heat-sensitive electric wiring that can supervise and activate a water spray system.

Line-of-sight A visual path above a ceiling created by removing ceiling tiles at regular intervals along a length of dimension of particular interest, such as the location of a sprinkler feedmain, and by using a flashlight and a telescoping elevation pole.

Liquidated damages Monetary fines for failure to perform completion of work within a specific time frame.

Listed Included in a document published by an approved testing laboratory, such as Underwriters Laboratories.

Glossary

Local application Applying water mist to a specific object.

Logarithmic graph paper Lined paper used to show pressure in psi on the Y axis and flow in gpm on the X axis, with spacing between flow values increasing with flow.

Looped system A system with crossmains connected at two or more locations.

Low-pressure water mist systems Systems where the pressures encountered by the system piping are 175 psi or less.

Mechanical general requirements Requirements applying to all mechanical trades.

Metric Conversion Act of 1975 Legislation that created a requirement for conversion to the metric system for all federal projects by 1992.

Millisecond A thousandth of a second.

Mineral oil An oil with a flash point of about 300°F that is used in transformers.

Munitions Hazardous commodities used in ammunition.

National Fire Protection Association (NFPA) The principal source and publisher of fire protection standards in the United States.

NICET The National Institute for Certification in Engineering Technologies.

Node A hydraulic reference point that correlates a point of interest on a plan to a point in a hydraulic calculation.

Nonabsorbing ground area Concrete slabs not equipped with drainage pits onto which a transformer may be mounted.

Occupancy A function of the expected level of severity of a fire in a particular building or room.

Old style sprinkler A sprinkler that sprays 40 to 60% of its water upward, with the remainder sprayed below.

Orifice plate A disk designed to restrict the flow of water in a water spray system.

OS&Y valve A listed indicating control valve that is determined to be in the open or closed position by observing the position of the stem with respect to the valve wheel.

Overspray Water that misses the target surface.

Pallet A wooden, plastic, or metal platform used for transporting a commodity to its storage location.

Performance objective An engineering basis for a predetermined end result of a given design.

Performance specification The minimum information necessary to estimate, design, and install a fire protection system.

Personal values Values obtained from parents, peers, mentors, television, film, the media, the community, and schools.

Piloted ignition Ignition and sustained combustion from an outside source such as a spark or match.

Pilot head detection system A system that consists of closed sprinklers connected to piping supervised by a pressure switch.

Pipe diameter measurement gauge A device used to determine the diameter of an existing pipe into which you are tapping a new sprinkler main.

Pipe schedule A chart in NFPA 13 that specifies pipe size as a function of a specific maximum number of sprinklers that can be supplied by the pipe.

Pitot pressure The pressure reading (in psi) as shown on the bourdon gauge attached to the pitot tube.

Pitot tube with bourdon gauge A device inserted into a water stream to the measure the pressure of the water flowing.

Plan job Design performed using new architectural plans as the basis for design.

Plan reference elevation Sea level, or an elevation specifically chosen for a building or group of buildings.

Plenum space A space above a suspended ceiling that is kept under negative pressure for return air exhaust.

Point of contact A person who meets you at the job site, shows you the area to be surveyed, and remains available if any questions arise.

Point protection The application of concentrated high-speed water spray onto a likely point of ignition.

Post indicating valve A listed indicating valve with a shaft extending through a wall or from an underground fire protection main.

Preaction sprinkler system A system that uses closed sprinklers and pipes filled with pressurized air for the protection of valuable and irreplaceable property.

Preaction water mist systems Water mist systems having nozzles with individual heat-responsive actuating devices, closed heads, and piping filled with pressurized air.

Prefabrication Cutting all pipe, installing fittings, and sometimes cutting hanger rods in advance of installation.

Pressure switches Electrical devices that indicate high or low pressure in a water tank or in a sprinkler system.

Pressure tank An enclosed water storage vessel that contains water pressurized by compressed air.

Profile plan A plan that shows reference elevations of utilities with respect to finished grade.

Propellants Hazardous commodities used in rockets.

Propriety information Unique information or data that are the sole property of the company.

Pump relief valve A valve installed to prevent overpressurization and damage to sprinkler components.

Pyrotechnics Hazardous commodities used in fireworks or flares.

Quality assurance administrator A senior level supervisory person who is an expert on the subject matter contained on a drawing.

Quality assurance procedure An independent review of each fire protection design before it leaves the office for approval and again before construction.

Quick opening device A device installed to rapidly remove air from a system and speed the operation of the dry pipe valve.

Quick response sprinkler A sprinkler with a specially designed response element that allows the rapid transfer of heat to the heat-responsive element.

Radiator A unit on a power transformer in which hot oil is circulated for cooling by fans.

Rain leaders Rainwater drainage piping.

Rate compensated detector A detector that measures the rate of temperature rise, sending a signal when the temperature rises a predetermined rate of temperature change within a predetermined time.

Rationalization An attempt to justify one's own action that is known to be wrong.

Reaction time The time from activation of a detector to water flow at nozzles.

Recessed fixtures Lights whose faces are flat to the ceiling and whose bodies protrude into the ceiling space.

Recommended practices An NFPA document that provides nonmandatory advice.

Rectangular prism An imaginary box that simplifies any discontinuities, making it unnecessary to calculate minor irregularities on a transformer surface.

Redirection Sweeping standing oil from the surfaces of a transformer to a less hazardous area.

Reference grid A system of parallel reference lines aligned either to magnetic north or plant north.

Reflected ceiling plan A plan view of a suspended ceiling.

Residual pressure The pressure reading (in psi) at the test hydrant, read while water is flowing from the flow hydrant.

Return diffuser A ceiling fixture used to draw stale air from a room.

Riser nipple Short vertical pipe that connects the crossmain with each branch line.

Room design method A design area where only the sprinklers in a particular room need to be calculated.

Rundown The accumulation of water at lower elevations of a vessel, applied from nozzles at higher elevations of the vessel.

Rupture A situation in which a tank shell is heated to its failure point, causing the shell to deform and burst.

Sanctioned coordination meeting A meeting required by the contract documents.

Scope Part of a specification that describes what is and what is not covered by the contract documents.

Shall Identifies NFPA wording that constitutes a mandatory requirement.

Should Identifies an NFPA recommendation or suggestion, but not a mandatory requirement.

Glossary

Sidewall sprinkler A sprinkler with a specially designed deflector that allows the sprinkler to discharge water from a wall-mounted position.

Significant figures Digits to the right of the decimal point.

Simple loops Loops with one inflow, one outflow, and two paths for water flow.

Societal values Values consisting of the public expectation that all fire protection professionals competently discharge their responsibilities to assure the safety of all who enter a building protected by a fire protection system.

Solenoid-operated ultra high speed water spray system A system consisting of a preprimed discharge system with solenoid-actuated nozzles, a wet pilot system supervised by solenoid valves, a flame detection system, and a central control panel.

Solenoids Switches used to open and close electrically actuated fire protection valves.

Specification divisions Broad categories of building component groupings.

Specifications A written description of project requirements.

Specification sections Detailed requirements for each division.

Spray sprinkler A sprinkler designed to discharge all of its water downward, with a deflector that creates smaller water droplets than old style sprinklers.

Sprigs Vertical pipes rising from a branch line to sprinklers extended above the branch line.

Squib-actuated system A preprimed ultra high speed water spray system with a squib-operated deluge valve; a system of piping preprimed with water, nozzles with rupture discs, and a flame detection system.

Standards Mandatory NFPA requirements that may be used by an approving authority to approve a fire protection system.

Standard spray pendant (SSP) A sprinkler mounted below the branch line, usually at or below the surface of a suspended ceiling.

Standard spray upright (SSU) A sprinkler mounted upright above a branch line, usually in a room with exposed structural elements.

Standpipe A vertical piping riser that serves to supply water to hose connections for fire department or occupant use.

Static pressure The pressure available from a water supply with no water flowing.

Steam smothering The process of a fire turning droplets of water into steam, cooling the flame.

Structural drawings Drawings that provide details related to the floors, roof, and structural elements of a building.

Supply diffuser A ceiling fixture used to distribute fresh air to a room.

Surface cooling The cooling of a flammable liquid that has a flash point greater than 125°F.

Survey A thorough investigation of a building and its components for the purpose of taking detailed measurements of the building to serve as reference for a fire protection drawing.

Survey job A project involving an existing building for which plans cannot be obtained.

Survey reference elevation The lowest, flattest, most reliable elevation that can be found.

Survey strategy The approach by which you begin your survey with the layout of structural elements, such as building columns and structural beams, then reference all measurements to these elements.

Suspended fixtures Lights suspended from a ceiling by rods or chains.

Swing joints Piping joints that allow a nozzle to be adjusted in all directions.

Tamper switches Electronic devices that monitor sprinkler valve position.

Tank farm A group of closely spaced flammable liquid tanks.

Tapping sleeve A device that allows piping to be easily reconnected once a cut-in is complete.

Telescoping elevation measurement pole A device used to measure elevations in the 30- to 50-foot range.

Temperature control Reducing room temperature during combustion to allow safe egress and reduced damage.

Test hydrant A fire hydrant selected for observing and recording water pressures available to the water supply.

Thrust block A poured block of concrete that restrains an underground pipe at points where it changes direction, such as at elbows, tees, fire hydrants, and caps.

Total flooding Flooding an entire compartment with water mist.

Transformer case The transformer tank that contains oil to cool the electric coils.

Transformer loop Water spray piping designed for rigidity and enhanced hydraulic characteristics.

Transformer outline drawing A dimensional shop drawing of a transformer.

Trapeze hanger A piece of angle iron or pipe that spans two structural members to facilitate the hanging of crossmains and feedmains.

Tree system System of dead-end branch lines centrally fed by a crossmain.

Underwriters Laboratories (UL) The laboratory that tests components to assure that their function corresponds to their performance objective.

Unit heaters Modular units that project heat.

Ultra high speed water spray system A system designed to discharge water within 100 milliseconds of detection for the protection of unconfined areas where overpressurization is not the primary concern.

Values of the fire protection profession Values summarized by the *Code of Responsibility for NICET-Certified Engineering Technicians and Technologists* and the *Code of Ethics for Engineers*.

Vapor exhaust A ventilation system to remove vapors to a point 25% below the lower flammable limit of the vapor.

Water flow switches Electronic devices that indicate water movement in a sprinkler pipe.

Water flow test Procedure to determine that a water supply of sufficient flow and pressure is available for a water-based fire protection system.

Water hammer Excessive pressure caused by rapid valve or hydrant closure.

Water level switches Devices that indicate the level of water in a fire protection water storage tank.

Water mist A fine water spray having droplets less than 1000 microns at a distance of 3.3 feet from the discharge nozzle.

Water mist system An automatic water-based fire protection system with nozzles capable of distributing water mist to a variety of hazards.

Water spray system A fixed pipe system permanently connected to a water supply providing directional water spray protection over a protected area.

Water wastage Water that flows from a water spray nozzle but does not directly impinge upon the target area.

Wet pipe sprinkler system A system that has water in the pipes in the ambient or normal condition, and with heat-responsive elements at all sprinklers.

Wet pipe water mist systems Water mist systems with nozzles having individual heat-responsive actuating devices and with piping filled with water in the ambient or normal condition.

Whistle-blowing An external action where an individual exposes a situation perceived to be unethical.

Wind wastage Water droplets that have been swept away by wind.

Zoned application systems Systems where a volume is protected by several distinct zones, each with its own detection system.

Additional Readings

Alpert, R. L., "Incentive for Use of Misting Sprays as a Fire Suppression Flooding Agent," in *Proceedings of the Water Mist Fire Suppression Workshop*, National Institute of Standards and Technology, Gaithersburg, Maryland, March 1–2, 1993.

Bachalo, W. D., "Advances in Spray Drop Size and Velocity Measurement Capabilities for the Characterization of Fire Protection Systems," in *Proceedings of the Water Mist Fire Suppression Workshop*, National Institute of Standards and Technology, Gaithersburg, Maryland, March 1–2, 1993.

Brannigan, Francis L., "Building Construction for the Fire Service," 3rd edition, National Fire Protection Association, Quincy, Massachusetts, 1992.

Brock, Pat D., *Fire Protection Hydraulics and Water Supply Analysis*. Fire Protection Publications, Oklahoma State University, Stillwater, Oklahoma, 1990.

Bryan, John L., *Automatic Sprinkler and Standpipe Systems*. National Fire Protection Association, Quincy, Massachusetts, 1990.

Bryan, John L., *Fire Suppression and Detection Systems*. Macmillan, 1993.

Bugbee, Percy, "Principles of Fire Protection," pp. 60–69. National Fire Protection Association, Quincy, Massachusetts, May 1982.

Butz, J. R., "Application of Fine Mists to Hydrogen Deflagrations," in *Proceedings of the Halon Alternatives Technical Working Conference*, co-sponsored by the University of New Mexico, New Mexico Engineering Research Institute, Center for Global Environmental Technologies, National Association of Fire Equipment Distributors, Inc., Halon Alternative Research Corporation, Fire Suppression Systems Association, and Hughes Associates, Inc., pp. 364–373, Albuquerque, New Mexico, May 11–13, 1993.

Chow, W. K., "On the Evaporation Effect of a Sprinkler Water Spray," *Fire Technology*, vol. 25, no. 4, pp. 364–373, November 1989.

Daugherty, Robert L., Franzini, Joseph B., and Finnemore, John E., *Fluid Mechanics with Engineering Applications*, 8th ed. McGraw-Hill, New York, New York, 1984.

DeGeorge, R. T., *Business Ethics*. Macmillan, New York, 1982.

Department of the Army, *AMC Safety Manual*, C-1, AMC-R-385-100, March 16, 1990.

Department of Defense, *Military Handbook—Fire Protection for Facilities—Engineering, Design, and Construction*, MIL-HDBK-1008B, January 15, 1994.

Department of Defense, *Ammunition and Explosives Safety Standards*, Document DOD 6055.9-STD, July, 1984.

Detector Electronics Corporation, *Munitions Manufacturing*, Form 91-1005, 6-87.

Detector Electronics Corporation, Munitions Capabilities Brochure, Form 92-1004, 1986.

Dundas, P. H., *The Scaling of Sprinkler Discharge Prediction of Drop Size*. Factory Mutual Research Corp., Norwood, Massachusetts, 1974.

Evans, D., and Pfenning, D., "Water Sprays Suppress Gas-Well Blowout Fires," *Oil and Gas Journal*, pp. 80–86, April 29, 1985.

Fadorsen, Gary A., "Ultra High Speed Fire Suppression for Explosives Facilities," in *American Chemical Society Symposium for Design Considerations for Toxic Chemical and Explosives Facilities, Ralph A. Scott, and L. J. Doemeny, eds., American Chemical Society, Washington, DC, 1987.

Fleming, R. P., "New Interest in Water Mist," *NFSA Sprinkler Quarterly*, no. 84, p. 22, Fall 1993.

Gagnon, Robert M., "Performing a Successful Building Survey," *Fire Protection Contractor*, vol. 16, no. 12, pp. 8–10, December 1993.

Gagnon, Robert M., "Ultra High Speed Deluge Fire Suppression Systems," *Fire Protection Contractor*, June 1993.

Gagnon, Robert M., "Ultra High Speed Suppression Systems for Explosive Hazards," Chapter 6-15 in *Fire Protection Handbook*, 18th ed. National Fire Protection Association, Quincy, Massachusetts, 1997.

Gagnon, Robert M., "Water Mist Fire Suppression Systems—Theory and Applications," *Fire Protection Contractor Magazine*, May 1994, pp. 36–42, June 1994, pp. 42–47.

Gagnon, Robert M., "Design of Special Hazard and Fire Alarm Systems," Delmar Publishers, Albany, New York, 1997.

Gamiero, V. M., "Fine Water Spray Fire Suppression Alternative to Halon 1301 in Gas Turbine Enclosures," in *Proceedings of the Halon Alternatives Technical Working Conference*, co-sponsored by the University of New Mexico, New Mexico Engineering Research Institute, Center for Global Environmental Technologies, National Association of Fire Equipment Distributors, Inc., Halon Alternative Research Corporation, Fire Suppression Systems Association, and Hughes Associates, Inc., pp. 364–373, Albuquerque, New Mexico, May 11–13, 1993.

Gieck, Kurt, *Engineering Formulas*, 5th ed., McGraw-Hill, New York, New York, 1986.

Goedeke, A. D., and Fadorsen, G. A., *Evaluation of State-of-the-Art High Speed Deluge Systems Presently in Service at Various U.S. Ammunitions Plants*, Document WL-TR-93-3510, Wright Laboratory, Wright-Patterson AFB, Ohio, September 1993.

Haberman, William L., and John, James A., *Engineering Thermodynamics with Heat Transfer*, 2nd ed. Allyn and Bacon, Boston, Massachusetts.

Hayes, W. D., Jr., *Literature Survey on Drop Size Data, Measuring Equipment, and a Discussion of the Significance of Drop Size in Fire Extinguishment*. Report Number NBSIR 85-3100-1. National Institute of Standards and Technology, Gaithersburg, Maryland, July 1985.

Heilbroner, Robert L., *In the Name of Profit*. Doubleday, Garden City, New York, 1972.

Hickey, Harry E., *Hydraulics for Fire Protection*. National Fire Protection Association, Quincy, Massachusetts, 1980.

Hill, R. G., Marker, T. M., and Sarkos, C. P., "Evaluation and Optimization of On-Board Water Spray Fire Suppression Systems in Aircraft," in *Proceedings of the Water Mist Fire Suppression Workshop*, National Institute of Standards and Technology, Gaithersburg, Maryland, March 1–2, 1993.

Hills, A. T., Simpson, T., and Smith, D. P., "Water Mist Fire Protection Systems for Telecommunications Switchgear and Other Electronic Facilities," in *Proceedings of the Water Mist Fire Suppression Workshop*, National Institute of Standards and Technology, Gaithersburg, Maryland, March 1–2, 1993.

Jackman, L. A., Glockling, J. L. D., and Nolan, P. F., "Water Sprays: Characteristics and Effectiveness," in *Proceedings of the Halon Alternatives Technical Working Conference*, co-sponsored by the University of New Mexico, New Mexico Engineering Research Institute, Center for Global Environmental Technologies, National Association of Fire Equipment Distributors, Inc., Halon Alternative Research Corporation, Fire Suppression Systems Association, and Hughes Associates, Inc., pp. 364–373, Albuquerque, New Mexico, May 11–13, 1993.

Kruse, D. H., "Typical Configurations of High Speed Deluge Systems for Various LAP Equipment," in *Proceedings of the Annual Meeting of the Load, Assemble and Packaging Section, American Defense Preparedness Association*, Colorado Springs, Colorado, March 8–10, 1988.

Lawson, J. R., Walton, W. D., and Evans, D. D., *Measurement of Droplet Size in Sprinkler Sprays*. National Institute of Standards and Technology, Gaithersburg, Maryland, 1988.

Lee, S., and Sichel, M., "Evaporation of Liquid Droplets in a Confined Medium," in *Chemical and Physical Processes in Combustion*, 20th Fall Technical Meeting, combined technical meetings of the Combustion Institute/Eastern States Section and NBS/CFR Annual Conference on Fire Research, Gaithersburg, Maryland, 1987.

Lev, Y., "Cooling Sprays for Hot Surfaces," *Fire Prevention*, no. 222, pp. 42–47, 1989.

Levine, R. S., *Navy Safety Center Data on the Effects of Fire Protection Systems on Electrical Equipment*. National Institute of Standards and Technology, Gaithersburg, Maryland, 1991.

Lindeburg, Michael R., *Engineering Unit Conversions*. Professional Publications, Belmont, California, 1988.

Lindorff, Dave, "Engineers' Duty to Speak Out," *The Nation*, June 28, 1986.

Liu, S. T., *Analytical and Experimental Study of Evaporative Cooling and Room Fire Suppression by Corridor Sprinkler System*. National Institute for Standards and Technology, Gaithersburg, Maryland, 1977.

Loyd, R. A. (ed.), *Minutes of the Rapid Action Fire Protection System Seminar*, U.S. Army Armament, Munitions, and Chemical Command, Rock Island, Illinois, October 23–24, 1984.

Loyd, R. A., "Ultra High Speed Deluge Systems," in *Proceedings of the 22nd Department of Defense Explosives Safety Seminar*, U.S. Army Armament, Munitions, and Chemical Command, Rock Island, Illinois, August 26–28, 1986.

Loyd, R. A., "Ultra High Speed Deluge Systems for Ordnance Operations," in *Proceedings of the Federal Fire Forum*, U.S. Army Armament, Munitions, and Chemical Command, Rock Island, Illinois, November 3, 1989.

Loyd, R. A., "Design and Installation of Ultra High-Speed Deluge Systems," in *Proceedings of the 24th Department of Defense Explosives Safety Seminar*, U.S. Army Armament, Munitions, and Chemical Command, Rock Island, Illinois, August 27–30, 1990.

Lugar, J. R., *Water Mist Fire Protection*. David W. Taylor Naval Ship Research and Development Center, Bethesda, Maryland, 1979.

Marioff Hi-Fog Oy, *Hi-Fog for Machinery Spaces*, video, 1993, Marioff, P.O. Box 25, 01511 Vantaa, Finland.

Marioff Hi-Fog Oy, *Hi-Fog Fire Test*, video, 1993, Marioff, P.O. Box 25, 01511 Vantaa, Finland.

Marlatt, F. P., "Maryland Fire and Rescue Institute Hosts High-Pressure Sprinkler Testing Program," *MFRI Bulletin*, vol. 24, no. 5, November 1993.

Martin, M., and Schinzinger, R., *Ethics in Engineering*. McGraw-Hill, New York, 1983.

Marttila, P., "Water Mist in Total Flooding Applications," in *Proceedings of the Halon Alternatives Technical Working Conference*, co-sponsored by the University of New Mexico, New Mexico Engineering Research Institute, Center for Global Environmental Technologies, National Association of Fire Equipment Distributors, Inc., Halon Alternative Research Corporation, Fire Suppression Systems Association, and Hughes Associates, Inc., pp. 364–373, Albuquerque, New Mexico, May 11–13, 1993.

Mawhinney, J. R., *Fine Water Spray Suppression Project*. National Fire Laboratory, Institute for Research in Construction, National Research Council of Canada, May 5, 1992.

Mawhinney, J. R., "Engineering Criteria for Water Mist Fire Suppression Systems," in *Proceedings of the Water Mist Fire Suppression Workshop*, National Institute of Standards and Technology, Gaithersburg, Maryland, March 1–2, 1993.

Mawhinney, J. R., "Characteristics of Water Mists for Fire Suppression in Enclosures," in *Proceedings of the Halon Alternatives Technical Working Conference*, co-sponsored by the University of New Mexico, New Mexico Engineering Research Institute, Center for Global Environmental Technologies, National Association of Fire Equipment Distributors, Inc., Halon Alternative Research Corporation, Fire Suppression Systems Association, and Hughes Associates, Inc., pp. 364–373, Albuquerque, New Mexico, May 11–13, 1993.

Mawhinney, J. R., "Water Mist Suppression Systems May Solve an Array of Fire Protection Problems," *NFPA Journal*, vol. 88, no. 3, May/June 1994.

McCuen, Richard H., "The Ethical Dimension of Professionalism," *Engineering Issues*, vol. 105 (E11), pp. 89–105, April 1979.

McCuen, Richard H., *Hydrologic Analysis and Design*, 2nd ed., Chapter 16. Prentice-Hall, Englewood Cliffs, New Jersey, 1996.

The Montreal Protocol on Substances That Deplete the Ozone Layer, Final Act, United Nations Environment Program, HMSO, CM977, September 1987.

National Fire Protection Association, *Fire Protection Handbook*, 17th edition. Quincy, Massachusetts, 1991.

National Institute for Standards and Technology, *FPEtool (includes DETACT)*. Available through the One Stop Data Stop, National Fire Protection Association, Quincy, Massachusetts.

NFSA, *The Hydraulics Handbook*. National Fire Sprinkler Association, Patterson, New York, 1996.

O'Hern, T. J., and Rader, D. J., "Practical Application of In Situ Aerosol Measurement," in *Proceedings of the Halon Alternatives Technical Working Conference*, co-sponsored by the University of New Mexico, New Mexico Engineering Research Institute, Center for Global Environmental Technologies, National Association of Fire Equipment Distributors, Inc., Halon Alternative Research Corporation, Fire Suppression Systems Association, and Hughes Associates, Inc., pp. 364–373, Albuquerque, New Mexico, May 11–13, 1993.

Papavergos, P. G., *Fine Water Sprays for Fire Protection—A Halon Replacement Option*. British Petroleum Ventures, BP Research, Sunbury Research Center, 1990.

Perry, Tekla S., "Knowing How to Blow the Whistle," *IEEE Spectrum*, September 1981.

Pfenning, D., and Evans, D., "Suppression of Gas Well Blowout Fires Using Water Sprays—Large and Small Scale Studies," in *Proceedings of the Production Session*, American Petroleum Institute, pp. 1–12, San Antonio, Texas, September 11–13, 1984.

Reischl, U., "Water Fog Stream Heat Radiation," *Fire Technology*, vol. 15, pp. 262–270, November 1979.

Rosander, M., and Giselsson, K., "Making the Best Use of Water for Fire Extinguishing Purposes," *Fire*, pp. 43–46, October 1984.

Ross, S. D., *Moral Decision: An Introduction to Ethics*. Freeman, San Francisco, California, 1972.

SFPE, *The SFPE Handbook of Fire Protection Engineering*, second edition. Society of Fire Protection Engineers and the National Fire Protection Association, Quincy, Massachusetts, 1995.

Spring, D. J., Simpson, T., Smith, D. P., and Ball, D. N., "New Applications of Aqueous Agents for Fire Suppression," in *Proceedings of the Halon Alternatives Technical Working Conference*, co-sponsored by the University of New Mexico, New Mexico Engineering Research Institute, Center for Global Environmental Technologies, National Association of Fire Equipment Distributors, Inc., Halon Alternative Research Corporation, Fire Suppression Systems Association, and Hughes Associates, Inc., pp. 364–373, Albuquerque, New Mexico, May 11–13, 1993.

Turner, A. R. F., "Water Mist in Marine Applications," in *Proceedings of the Water Mist Fire Suppression Workshop*, National Institute of Standards and Technology, Gaithersburg, Maryland, March 1–2, 1993.

Underwriters Laboratories, *Fire Protection Equipment Directory*. Underwriters Laboratories, Inc., Northbrook, Illinois, 1996.

Wass, Harold S., Jr., *Sprinkler Hydraulics*. Improved Risk Mutual Insurance, White Plains, New York, 1983.

Watkins, B. O., and Meador, R., *Technology and Human Values*. Ann Arbor Science, Ann Arbor, Michigan, 1977.

Wilson, Rexford, and Ledoux, Edward F., "High-Speed Protection for Personnel in Oxygen-Enriched Atmospheres," *Fire Journal*, pp. 23–24, March 1968.

Technical Reports on Ultra High Speed Water Spray Systems

These reports can be ordered from the Defense Technical Information Center, Cameron Station, Alexandria, Va. 22314, (703)-274-6733. Documents that are approved for public release are noted (APR). Documents that contain confidential or proprietary information and whose distribution is limited to government agencies only are noted (DL).

Analysis of Mixer Bay Designs for Pyrotechnic Operations, No. AD-E401-602, November, 1986. (APR)

Deluge Systems in Army Ammunition Plants, prepared by Science Applications, Inc., for the U.S. Army Munitions Production Base Modernization Agency, June 30, 1981.

Design of a Deluge System to Extinguish Lead Azide Fires, No. AD-E400-204, August 1978. (APR)

Design of a Water Deluge System to Extinguish M-1 Propellant Fires in Closed Conveyors, No. AD-E400-216, September, 1978. (APR)

Development of a Water Deluge System to Extinguish M-1 Propellant Fires, No. AD-E400-217, September, 1978. (APR)

Dynamic Model of a Water Deluge System for Propellant Fires, No. AD-E400-315, May, 1979. (APR)

Engineering Guide for Fire Protection and Detection Systems at Army Ammunition Plants, Vol. I (Selection and Design), No. AD-E400-531, December, 1980. (APR)

Engineering Guide for Fire Protection and Detection Systems at Army Ammunition Plants, Vol. II (Testing and Inspection), No. AD-E400-874, December, 1982. (DL)

Evaluation of an Improved Fire Suppression System for Pyrotechnic Mixes, No. AD-E401-569, September, 1986. (DL)

Evaluation of Pyrotechnic Fire Suppression Systems for Six Pyrotechnic Compositions, No. AD-E401-306, March 1985. (APR)

Feasibility Study to Develop a Water Deluge System for Conveyor Lines Transporting High Explosives, No. 4889, August, 1975. (APR)

Fire Suppression System Safety Evaluation, No. AD-E401-083, December, 1983. (APR)

Minutes of the Rapid Action Fire Protection Systems Seminar, U.S. Army Armament, Munitions, and Chemical Command, October 23–24, 1984.

On-Site Survey and Analysis of Pyrotechnic Mixer Bays, No. AD-E401-141, February, 1984. (APR)

Technical Report on the Testing of Ultraviolet Actuated Deluge Systems Utilizing Solid State Controllers and Detonator-Actuated Valves to Extinguish Black Powder Fires, Day and Zimmerman, Lone Star Division, November, 1986.

Water Deluge Fire Protection System for Conveyor Lines Transporting High Explosives, No. AD-E400-034, December, 1977. (APR)

Index

Acronyms, 439
Actual distance between sprinklers, determining, 102
AHJ (Authority having jurisdiction), 2, 3, 58, 73, 74, 110, 141, 280
American National Standards Institute (ANSI), 82
American Society for the Testing of Materials (ASTM), 82
Antifreeze system, 174
Appendix, of the standard, 74
Approved sprinkler system, 73
Architectural plans, 8–12
   building sections, 10
   finish schedule, 12
   reflected ceiling plans, 10–12
Area protected by each sprinkler (A), 96, 98
Area protection, 250, 251
Asbestos, 39
Authority having jurisdiction. *See* AHJ
Autoignition, 225
Automatic sprinkler systems, 70–113
   developing sprinkler plan for approval, 87–110
   myth vs. fact, 71–72
   sprinkler system components, 81–87
   sprinkler system definitions, 73–74
      "approved" vs. "listed," 73
      "fire control" vs. "fire suppression," 74
      "shall" vs. "should," 73–74
   sprinkler system types, 78–81
   sprinkler types, 74–78
   standardized design, 72–73

Backfill, 51
Backflow preventers, 174
Bar joist, 12–13
Bar joist bearing dimension, 12–13
Bay, 98
Bearing area, 51, 52
BLEVE (boiling liquid-expanding explosion), 226
Body of the standard, 73

Boiling contents, prevention of, 227–229
Boring table, 6–7
Branch line K-factor, 154
Branch line logic, 91–92
Branch lines, 13
   determining distance between, 100
   determining number of, 98–99
Btu (British Thermal Unit), 115
Building elevations, 39
Building footings, 45–46
Building outline, tracing or scanning, 89–90
Building sections, 10
Butterfly valve, 86, 87

CAD. *See* Computer-aided design
Carbon dioxide systems, 272
Cathodic protection, 26
Cavitation, 183
Ceiling measurements, 39–40
Centrifugal fire pump, 182
c-factor, 126, 324–325
City water supply, 44, 65
Civil drawings, 3, 5
Civil section, of contract drawings, 45
Class A fires, 272
Class B fires, 272
Class C fires, 272
Client's value system, 292
Coating, of transformer, 200
Code of Ethics for Engineers, xxvii, 295–296, 297–298
Code of Responsibility for NICET-Certified Engineering Technicians and Technologists, xxvii, 293–294, 297
Codes, NFPA, xxv–xxvi
Combination ultra high speed system, 250, 252
Commodities, classification of, 176
Communication, xxviii
Company manual, 291
Company values, 290–291

Computer-aided design (CAD), xxxiv–xxxvi
Computerized hydraulic calculation programs, 160
Computerized hydraulic program documentation, 415–435
Computer specification database, 24
Contour lines, 5
Contract drawings, 1–22
   architectural plans, 8–12
   cover sheet, 2–3, 4
   electrical drawings, 19–20
   fire protection drawings, 18–19
   HVAC drawings, 15–17
   plumbing drawings, 17–18
   site drawings, 3, 5–8
   structural drawings, 12–15
Contract specifications, 2, 23–34
   sample specification, 29–32
   specification format, 25–29
      divisions, 25
      sections, 25
   specification types, 24–25
      computer specification database, 24
      detailed specifications, 24–25
      performance specifications, 24, 25
Coordination meetings, 26, 283
Coordination of fire protection plan, 281–286
   coordination drawings, 285
   coordination failures, 285
   coordination meeting, 283
   coordination negotiations, 284
   prefabrication of fire protection systems, 282–283
   putting agreements into action, 284–285
   sanctioning a coordination meeting, 283–284
   scheduling coordination process, 282
Cover sheet, 2–3, 4
Cross bracing, 13
Crossmain, 84, 92
   location of, 103–104
Cut-in, 47, 48

Index

Deflagration, 226, 249
Deflector, 75, 76
Deflector distance, 104
Deluge system, 79, 81
Deluge water mist systems, 267
Depth of cover, 6, 48, 49, 50
Design area, 140
Design density, 140
DETACT computer program, 78
Detailed specifications, 24–25
Detectors for ultra high speed water spray systems, 250
Detonation, 249
Dilution, 199
Dimensioning sprinklers and piping on drawing, 102
Direct impingement, 206
Discharge system, 255
Domestic plumbing piping, 18
Domestic water supply, 166
Draft curtains, 90
Drafting equipment, xxx
Drawing, xxviii
Drawing building detail, 90
Drawing fire protection systems, xxviii–xxvi
   blank page, xxix
   computer-aided design (CAD), xxxiv–xxxvi
   drafting equipment, xxx
   drawing arrangement, xxxiii–xxxiv
   drawing media, xxxii–xxxiii
   drawing parallel lines, xxxi
   drawing surface, xxx–xxxi
   scales, xxxi–xxxii
Drawing index, 3, 4
Drawings
   civil, 3, 5
   contract, 1–22
   coordination, 285
   electrical, 19–20
   fire protection, 18–19
   HVAC, 15–17
   outline, 200–202
   site, 3, 5–8
   structural, 12–15
Droplet size, 263–266
   and the fire service, 265
   measuring, 266
   performance objectives, 265
   water mist pressure and, 266
Drops, 104, 105
Dry pipe systems, 78–79
Dry pipe water mist systems, 268

Early suppression fast response (ESFR) sprinkler system, 74, 78, 179

Electrical drawings, 19–20
Electric detection systems, 217
Empirical evidence, 140
Encapsulation, 177
Equivalent fitting length, 154
Equivalent length, 133
ESFR (early suppression fast response) sprinkler, 74, 78
   for storage applications, 179
Ethical dilemma, 289
   resolving, 299–301
      rationalization, 299–300
      whistle-blowing, 300–301
Ethics, 288–301
Explosion, 226
Explosion suppression systems, 247–248
Exposure protection, 226, 263
Extended coverage sprinkler, 75

Field check, 40–41, 110
Finish schedule, 12
Fire, classification of, 272
Fire control, 74, 263
Fire extinguishment, 263
Fire hose streams, 156
Fire hydrants, 52–53, 54
   flow hydrant, 57
   test (pressure) hydrant, 57
Fire load density, 95, 97
Fire protection as a profession, xxvi–xxvii
Fire protection contract drawings, 18–19
Fire protection engineer, xxvii
Fire Protection Equipment Directory (UL), 73
Fire protection survey procedure, 35–42
   obtain basic information, 36
   determine general building layout, 36–37
   bring proper survey equipment, 37–38
   survey building details, 38–39
   develop system design strategy, 39–40
      building elevations, 39
      ceiling measurements, 39–40
      water supply information, 40
      draw the building and lay out the system, 40
   field check the drawing, 40–41
   inspect system after installation, 41
Fire protection system design, xxvii
Fire protection system layout, xxvii
Fire protection technicians, xxvii
Fire pumps, 44, 66, 67, 182–189
   calculating pressure differentials, 184–186

   combining a pump and city supply, 187–188
   components, 183–184
   graphing fire pump supply, 186–187
   sizing a fire pump, 188–189
   when to add, 186
Fire service response time, 74
Firestop systems, 27
Fire suppression, 74, 263
Fire tetrahedron, 115
Fire triangle, 115, 116
Fittings, 82–84
Flash point, 225
Flow hydrant, 57, 59, 61
Flush mounted lights, 19
Foundation plans, 12
Framing plans, 12–13
   used to dimension a sprinkler system, 13–15
Freeze protection for residential sprinkler systems, 173–176
Friction loss, 126–137
   analysis of loop calculation results, 136–137
   reducing, 129
   in simple loops—equivalent length, 132–136
   in simple pipe loops, 129–132
   tables, 128, 328–414

Glossary, 441–448
Gravity tanks, 44, 65, 189–191
   calculation, 189–190
   graphing a gravity tank supply, 190–191
Grid system, 93, 94–95
Guides, NFPA, xxv, xxvi

Halon systems, 272
Hangers, 84, 92
   determining spacing of, 106, 108
   selecting, 105–106
   types of, 107
Hazard class of occupancy, 95–96, 97
Hazen-Williams formula, 126–127, 128, 214
Head, as term for water pressure, 121
Heat detectors, 218
High-piled storage occupancies, sprinkler protection for, 176–179
   ESFR sprinklers, 179
High-pressure water mist systems, 266
HVAC drawings, 15–17
   conceptual, 15
   shop, 15
Hydrant cap with bourdon gauge, 56
Hydrant coefficient (C_d), 58, 60
Hydraulic calculation, 110, 114–162

calculation method for sprinkler systems, 137–160
  choosing the hydraulically most demanding area, 138
  configuration of sprinklers in design area, 145
  determining friction loss, 147–148
  determining length of design area, 142–143
  determining minimum pressure at the hydraulically most demanding sprinkler, 146–147
  example calculation, 148–156
  extended coverage sprinklers and the design area, 159–160
  fire hose streams, 156
  graphing results, 156–157
  hand calculations vs. computer calculations, 137–138
  minimum flow at the hydraulically most demanding sprinkler, 146
  number of sprinklers flowing along length of design area, 143–145
  quick response sprinklers and the design area, 158–159
  selection of hydraulic density, 139–142
  selection of occupancy, 138–139
  submitting calculations for approval, 157
  velocity pressure calculations, 157–158
computerized hydraulic calculation programs, 160
fundamentals of hydraulics, 115–137
Hydraulic calculation applications
  fire pumps, 182–189
  gravity tanks, 189–191
  high-piled storage occupancies, 176–179
  residential sprinkler systems, 165–176
  standpipe and hose systems, 179–182
  transformers, 213–217
Hydraulics, fundamentals of, 115–137
  advantages of water, 115–116
  friction loss, 126–137
  hydrokinetics, 123–126
  hydrostatics, 120–123
  properties of water, 116–120
Hydrokinetics, 123–126
Hydrostatics, 120–123

I-beam, 12, 13
In-rank sprinklers, 177–178
Inspector's test connection, 108
Intermediate-pressure water mist systems, 266
Interpolating, 154
Invert elevation, 6

Jockey pump, 183

Large drop sprinklers, 78
Legal/ethical conflicts, 299
Limited area dwellings, 170
Linear heat detection system, 221
Line-of-sight, 39–40
Liquidated damages, 26
Listed sprinkler components, 73
Local application of water mist, 267
Location map, 2, 3
Logarithmic graph paper, 63
Loop calculation results, analysis of, 136–137
Looped system, 93, 95, 96
Low-pressure water mist systems, 266

Maximum allowable distance between sprinklers, 100–101
Mechanical general requirements, 28
Metric Conversion Act of 1975, xxiv
Metric conversions, xxiv, Appendix A
Metric units, xxiv–xxv
Mineral oil, 197
Minimum number of sprinklers on each branch line, 101–102
Munitions, 247

National Fire Protection Association (NFPA), xxiii
  publications, xxv–xxvi
National Institute for Certification in Engineering Technologies (NICET), xxvii, 293
National Society of Professional Engineers, xxvii
NFPA (National Fire Protection Association), xxiii
NFPA standards, selected listing, 436–437
  13. Installation of Sprinkler Systems, 71, 165, 176
  13D. Installation of Sprinkler Systems in One and Two Family Dwellings and Manufactured Homes, 165, 274
  13R. Installation of Sprinkler Systems in Residential Occupancies up to and Including Four Stories in Height, 165
  14. Installation of Standpipe and Hose Systems, 179
  15. Water Spray Fixed System for Fire Protection, 194, 225, 247
  20. Installation of Centrifugal Fire Pumps, 66, 182
  22. Water Tanks for Private Fire Protection, 65, 66, 189
  24. Private Service Mains and Their Appurtenances, 44
  69. Explosion Suppression Systems, 247
  170. Fire Safety Symbols, 3, 46
  214. Water Cooling Towers, 196
  231. General Storage, 176
  231C. Rack Storage of Materials, 176
  231D. Storage of Rubber Tires, 176
  231F. Storage of Rolled Paper, 176
  750. Installation of Water Mist Fire Protection Systems, 263
NFPA 13 design checklist, 109–110
NICET (National Institute for Certification in Engineering Technologies), xxvii, 293
Node, 148
Nonabsorbing ground area, 209
  and protection of tanks and vessels, 229–230

Occupancy, 95–96
  classifications, 97
Old style sprinklers, 74–75, 76, 77
Orifice plates, 214
OS&Y (outside screw and yoke) valves, 85, 86
Outline drawings, 200–202
Overspray, 232

Pallet, 176
Pendent sprinkler, 75, 76, 77
Performance objective, 2
Performance specifications, 24, 25
Personal valves, 290
Piloted ignition, 225
Pilot head detection system, 217–218
Pilot system, 256, 258, 259
Pipe diameter measurement gauge, 38
Pipe diameters, 326–327
Pipe schedule designs, 164–165
Piping, 82
  assigning elevations to, 104–105
Pitot pressure, 61, 62
Pitot tube with bourdon gauge, 56, 61, 62
Plan job, 8
Plenum space, 17
Plumbing drawings, 17–18
Point of contact, 36
Point protection, 250, 251
Pool fires, 225
  containment, 225
Post indicating valve, 86, 87

Index

Power-generating plant, water spray protection for, 194–196
Preaction systems, 79, 80
Preaction water mist systems, 268
Prefabrication of system, 282–283
Preprimed piping system, 252–253
Pressure differentials, calculating, 184–186
Pressure hydrant, 57
Pressure spray fires, 225–226
Pressure switches, 19
Pressure tank, 44, 66
Professional development, 301–303
Professional standards of conduct, 293–299
    Code of Ethics for Engineers, 295–296, 297–298
    Code of Responsibility for NICET-Certified Engineering Technicians and Technologists, 293–294, 297
Profession of engineering, 298–299
Profile plan, 6, 8
Project schedule, 88–89
Propellants, 247
Proprietary information, 290
Pump relief valve, 183
Pyrotechnic materials, 247

Quality assurance procedures, 280–281
    quality assurance administrator, 280
    quality assurance review, 280–281
Quick opening device, 79
Quick response sprinkler, 75, 76

Radiator, of power transformer, 197
Rain leaders, 18
Rankine cycle, 196–197
Rate compensated detectors, 218, 221
Rationalization of unethical behavior, 299–300
Reaction time, 249
Recessed fixtures, 19
Recommended practices, NFPA, xxv, xxvi
Rectangular prism, 206
Redirection, 200
Reference elevation, 5, 39
Reference grid, 6, 7
Reference tables, 307–323
Reflected ceiling plans, 10–12
Renovation architectural drawings, 10
Residential sprinkler systems, 165–176
    freeze protection for, 173–176
    one- and two-family dwellings and manufactured homes, 165–173
    residential occupancies up to and including four stories, 173

residential occupancies greater than four stories, 173
Residual pressure, 61
Response time
    of solenoid-operated ultra high speed water spray system, 260
    of squib-actuated ultra high speed water spray system, 254–255
Response time index (RTI), 78
Responsibilities of fire protection professional, 290–293, 301
    company values, 290–291
    personal values, 290
    societal values, 291–292
    values of the client, 292–293
    values of the fire protection profession, 293
Return diffuser, 17
Riser nipple, 104
Risers, locating, 105, 106
Room design method, 141
Rounding, xxv
Rundown, 227
Rupture, of tank shell, 226

Sanctioned coordination meeting, 283
Scale, xxxi–xxxii, 89
Scope, 26
Shall, as indication of mandatory requirement, 73
Should, as indication of recommendation or suggestion, 73
Sidewall sprinkler, 75, 77
Significant figures, xxv
Simple loops, 129
Site drawings, 3, 5–8
Societal values, 291–292
Soffit, 175
Solenoid-operated ultra high speed water spray system, 255–260
    discharge system, 255
    factors influencing speed of pilot-actuated system, 260
    pilot system, 256, 258, 259
    response time, 260
    sequence of operation for pilot-actuated system, 258, 260
    system arrangement, 255, 256, 257, 258
Solenoids, 19
Specifications. See Contract specifications
Split flow formula, 131
Spray sprinkler, 74–75, 76
Sprigs, 104, 105
Sprinkler coverage, 75
    verifying, 102–103

Sprinkler response, 75, 78
Sprinkler system components, 81–87
    fittings, 82–84
    hangers, 84
    piping and tubing, 82
    sprinklers, 81
    valves, 84–87
Sprinkler systems. See Automatic sprinkler systems; specific types of systems
Sprinkler system types, 78–81
    deluge systems, 79, 81
    dry pipe systems, 78–79
    preaction systems, 79, 80
    wet pipe systems, 78
Sprinkler types, 74–78
    old style and spray sprinklers, 74–75
    sprinkler coverage, 75
    sprinkler response, 75, 78
    sprinklers for storage applications, 78
    upright, pendent, and sidewall sprinklers, 75
Squib-actuated ultra high speed water spray system, 252–255
    definition, 252
    design, 255
    preprimed piping system, 252–253
    squib-actuated valve, 253–254
    system response time, 254–255
Standards, NFPA, xxv, xxvi. See also NFPA standards
Standard spray pendent (SSP) sprinkler, 75, 76, 77
Standard spray upright (SSU) sprinkler, 75, 76, 77
Standpipe and hose systems, 179–182
Static pressure, of water supply, 58
Steam smothering, 198–199
Structural drawings, 12–15
    branch line orientation, 13
    foundation plans, 12
    framing plans, 12–13
    using framing plans to dimension a sprinkler system, 13–15
Students, responsibilities of, 301
Supply diffuser, 17
Surface cooling, 198, 227
Survey job, 10
Survey procedure. See Fire protection survey procedure
Survey strategy, 38
Suspended fixtures, 19
Swing joints, 212
Symbol and abbreviation index, 3
System configurations, 93–95
    grid system, 93, 94–95
    looped system, 93, 95, 96
    tree system, 93–94

System response time, 254–255

Tamper switches, 19
Tank farm, 226
Tanks, water spray protection of. *See* Water spray system design for flammable liquid storage tanks and vessels
Tapping sleeve, 47, 48
Telescoping elevation measurement pole, 38
Temperature control, 263
Temperature ratings, 90, 91
Test hydrant, 57, 59
Thrust block, 47, 48
Total flooding, 267
Transformer case, 197
Transformer loop, 204
Transformers. *See* Water spray system design for transformers
Trapeze hanger, 105–106, 108
Tree system, 93–94
Tubing, 82

Ultra high speed water spray systems, 246–261
  definition, 247
  deflagrations and detonations, 249
  design approaches, 250
  detectors for, 250
  and explosion suppression systems, 247–248
  limitations of human senses, 249–250
  reaction time, 249
  types of, 250
    solenoid-operated system, 255–260
    squib-actuated system, 252–255
  uses of, 249
  water, and protection of explosive hazards, 248–249
Underground piping design procedure, 43–69
  add invert piping elevations and plan dimensions, 47–50
  detail locations of fire hydrants, 52–53
  detail thrust blocks and piping restraints, 51–52, 53
  detail water supply information on plans, 65–67
  determine pipe routing, 46
  draw cut-in detail, 47, 48
  obtain reference material and trace or scan background, 45–46
  perform water flow test, 53–64
  specify piping and fittings, 50–51
Underwriters Laboratories (UL), 73
Unit heaters, 90

Upright sprinkler, 75, 76, 77

Values of the fire protection profession, 293
Valves, 84–87
Vapor exhaust, 200
Velocity pressure calculations, 157–158
Vessels, water spray protection of. *See* Water spray system design for flammable liquid storage tanks and vessels
Vicinity map, 2, 3

Wall post indicator valve, 86, 87
Water
  advantages of, 115–116
  and explosive hazards, 248–249
  properties of, 116–120
Water flow switches, 19
Water flow test, 53–64
  procedure, 57–64
  protocol, 56–57
  tools, 56
Water hammer, 53
Water level switches, 19
Water mist system design, 262–278
  applications, 268–276
    aircraft, 271–272
    commercial and residential, 273–276
    electrical equipment, 272–273
    submarine, 269–270
    surface ship, 270–271
  calculation criteria, 268
  defined, 263
  performance objectives, 263
  water mist droplets, 263–266
    droplet size, 263–264
    droplet size and the fire service, 265
    droplet size performance objectives, 265
    measuring droplet size, 266
    water mist pressure and droplet size, 266
  water mist system types, 267–268
Water spray system design for flammable liquid storage tanks and vessels, 224–245
  hazards associated with tanks and vessels, 225–226
    deflagration and rupture, 226
    piloted and autoignition, 225
    pool fire containment, 225
    pool fires, 225
    pressure spray fires, 225–226
  vessel protection mechanisms, 226–230

    exposure protection, 226
    prevention of boiling contents, 227–229
    rundown, 227
    supplementary protection, 229–230
    surface cooling, 227
  vessel water spray design procedure, 230–243
    estimating minimum water supply requirement, 231
    estimation of water supply adequacy, 232–233
    loop and nozzle spacing procedure, 232–243
    pressure estimation, 233
    surface area calculation, 230–231
    total water requirement, 232
Water spray system design for transformers, 193–223
  transformer design procedure, 200–221
    calculation of transformer surface area, 206–208
    commence layout, 202–204
    compile reference data, 200–202
    detection systems, 217–218, 221
    hydraulic calculation procedure for transformers, 213–217
    recommendations for water spray system layout, 209–213
    transformer loop design, 204–206
    water spray for nonabsorbing ground surfaces, 208–209
  water spray protection for transformers, 197–200
    extinguishment/control mechanism, 198–200
    oil recovery and reclaim, 198
    personnel and other hazards, 198
    power generation, 198
    transformer hazards, 197
  water spray systems, 194–197
    Rankine cycle, 196–197
    water spray protection for power-generating plant, 194–196
Water supply information, 40
Water supply methods, 44
  city water supply, 44, 65
  fire pump, 44, 66, 67
  gravity tank, 44, 65
  pressure tank, 44, 66
Water wastage, 232
Wet pipe systems, 78, 79
Wet pipe water mist systems, 267
Whistle-blowing, 300–301
Wind wastage, 232

Zoned application system, 267, 268

Responder Magazine

THE MANAGEMENT LEADERSHIP SOURCE

For All Of The Emergency Services

The *Management Leadership Source* for All The Emergency Services delivers high impact issues each month. Don't Miss Out! Join Responder Magazine in changing the way things have always been done. Dynamic leadership topics for each issue means a fresh perspective on emergency service management.

Subscribe Today!
CALL
1-800-200-1901
Fax (813) 288-8452

Responder
Management Leadership Source

Responder Publications, Inc.
Rocky Point Centre - Suite 530
3030 North Rocky Point Dr. West
Tampa, Florida 33607

License Agreement for Delmar Publishers
an International Thomson Publishing company

Educational Software/Data

You the customer, and Delmar incur certain benefits, rights, and obligations to each other when you open this package and use the software/data it contains. BE SURE YOU READ THE LICENSE AGREEMENT CAREFULLY, SINCE BY USING THE SOFTWARE/DATA YOU INDICATE YOU HAVE READ, UNDERSTOOD, AND ACCEPTED THE TERMS OF THIS AGREEMENT.

Your rights:

1. You enjoy a non-exclusive license to use the enclosed software/data on a single microcomputer that is not part of a network or multi-machine system in consideration for payment of the required license fee, (which may be included in the purchase price of an accompanying print component), or receipt of this software/data, and your acceptance of the terms and conditions of this agreement.

2. You own the media on which the software/data is recorded, but you acknowledge that you do not own the software/data recorded on them. You also acknowledge that the software/data is furnished "as is," and contains copyrighted and/or proprietary and confidential information of Delmar Publishers or its licensors.

3. If you do not accept the terms of this license agreement you may return the media within 30 days. However, you may not use the software during this period.

There are limitations on your rights:

1. You may not copy or print the software/data for any reason whatsoever, except to install it on a hard drive on a single microcomputer and to make one archival copy, unless copying or printing is expressly permitted in writing or statements recorded on the diskette(s).

2. You may not revise, translate, convert, disassemble or otherwise reverse engineer the software data except that you may add to or rearrange any data recorded on the media as part of the normal use of the software/data.

3. You may not sell, license, lease, rent, loan, or otherwise distribute or network the software/data except that you may give the software/data to a student or and instructor for use at school or, temporarily at home.

Should you fail to abide by the Copyright Law of the United States as it applies to this software/data your license to use it will become invalid. You agree to erase or otherwise destroy the software/data immediately after receiving note of Delmar Publishers' termination of this agreement for violation of its provisions.

Delmar gives you a LIMITED WARRANTY covering the enclosed software/data. The LIMITED WARRANTY can be found in this package and/or the instructor's manual that accompanies it.

This license is the entire agreement between you and Delmar Publishers interpreted and enforced under New York law.

Limited Warranty

Delmar Publishers warrants to the original licensee/purchaser of this copy of microcomputer software/data and the media on which it is recorded that the media will be free from defects in material and workmanship for ninety (90) days from the date of original purchase. All implied warranties are limited in duration to this ninety (90) day period. THEREAFTER, ANY IMPLIED WARRANTIES, INCLUDING IMPLIED WARRANTIES OF MERCHANTABILITY AND FITNESS FOR A PARTICULAR PURPOSE ARE EXCLUDED. THIS WARRANTY IS IN LIEU OF ALL OTHER WARRANTIES, WHETHER ORAL OR WRITTEN, EXPRESSED OR IMPLIED.

If you believe the media is defective, please return it during the ninety day period to the address shown below. A defective diskette will be replaced without charge provided that it has not been subjected to misuse or damage.

This warranty does not extend to the software or information recorded on the media. The software and information are provided "AS IS." Any statements made about the utility of the software or information are not to be considered as express or implied warranties. Delmar will not be liable for incidental or consequential damages of any kind incurred by you, the consumer, or any other user.

Some states do not allow the exclusion or limitation of incidental or consequential damages, or limitations on the duration of implied warranties, so the above limitation or exclusion may not apply to you. This warranty gives you specific legal rights, and you may also have other rights which vary from state to state. Address all correspondence to:

Delmar Publishers
3 Columbia Circle
P.O. Box 15015
Albany, NY 12212-5015